SPSS 16.0 Guide to Data Analysis

Marija J. Norušis

Prentice Hall
A division of Pearson Education
1 Lake Street
Upper Saddle River, NJ 07458

For more information about SPSS® software products, please visit our Web site at *http://www.spss.com* or contact

SPSS Inc.
233 South Wacker Drive, 11th Floor
Chicago, IL 60606-6412
Tel: (312) 651-3000
Fax: (312) 651-3668

SPSS 16.0 Guide to Data Analysis
Published by Prentice Hall Inc., 2008
Upper Saddle River, NJ 07458

ISBN-13: 978-0-13-606136-6
ISBN-10: 0-13-606136-2

Preface

The goal of this book is to provide an unintimidating introduction to data analysis and to SPSS. This edition focuses on topics that interest today's students—including the role of the Internet in society. The book can be used either as a supplementary text or as a primary text in an introductory course in data analysis. It is designed for use with SPSS 16.0, including the Student Version.

The *SPSS Statistical Procedures Companion* and the *SPSS Advanced Statistical Procedures Companion* cover many of the more advanced statistical procedures in SPSS 16.0 that are not discussed in this book.

Sample chapters and detailed descriptions of the contents of all three books can be found at *www.norusis.com*.

Data Files

Since the best way to learn about data analysis is to actually do it, this book uses real data to solve a variety of problems. Data from the General Social Survey are used to determine who uses the Internet, and for how long, each week. Data from the Impact of the Internet on Library Use study allows students to examine the role of the Internet in traditional institutions, such as libraries. Data from the Stanford Institute for the Quantitative Study of Society allows students to investigate the effect of the Internet on socialization. But of course there's more to life than the Internet. Completion times from the Chicago marathon, opinions of the criminal justice system, and an ABC survey of manners are also analyzed. The variety of data files used should make this book appealing to a broad range of students and teachers. The data files used in the chapters and in the exercises are included with this book.

Using This Book

This book is divided into four parts: "Getting Started with SPSS," "Describing Data," "Testing Hypotheses," and "Examining Relationships." Each chapter presents a problem and introduces statistical techniques useful for solving it. Detailed descriptions of the SPSS procedures used are also provided. Each chapter also contains extensive exercises that use the data files described above, as well as others. Instructors can choose exercises that best suit the interests and sophistication of their students. Solutions to even-numbered statistical concepts exercises are in Appendix F.

Send Me Comments

I am always eager to receive feedback and suggestions from readers of my books. Please send your suggestions and comments about the book (not the software) to *marija@norusis.com*.

Acknowledgments

I wish to thank numerous users of the previous editions of the *SPSS Guide to Data Analysis* for their comments and suggestions. Tom Pasquarello of the State University of New York at Cortland has offered many excellent suggestion and corrections. The book has benefited greatly from them. I am most grateful to the contributors of the data files used in this edition. In particular, I thank George D'Elia of the State University of New York at Buffalo for data from his study of the impact of the Internet on public library use, funded by the Institute of Museum and Library Services, and Norman Nie and D. Sunshine Hillygus for data from the Stanford Institute for the Quantitative Study of Society. I also thank Elizabeth Stapleton for the Chicago Marathon data, Richard Shekelle for the Western Electric data, the late Harry Roberts for the bank salary data, Howard Corwin for the renal data, Roger Johnson for the body fat data, and the Inter-university Consortium for Political and Social Research for permission to distribute the ABC Manners poll data. I remain grateful to the General Social Survey for providing such a wealth of data.

I also wish to thank Yvonne Smith, JoAnn Ziebarth, Dave O'Neil, and Rhonda Smith of SPSS for editorial and production support and David Nichols for many helpful suggestions and corrections. Finally, I wish to thank my husband, Bruce Stephenson, for his varied contributions to the preparation of this book.

Marija J. Norušis

Contents

Part 1 Getting Started with SPSS

Part 3 Testing Hypotheses

Part 4 Examining Relationships

Appendices

A Obtaining Charts in SPSS 579

B Transforming and Selecting Data 597

Introduction

"What does it all mean?" is a question that used to be relegated to philosophy classes. But now you're probably more likely to hear it mumbled by people peering at computer screens. Gathering information has become a national pastime. Personal, corporate, and research data are comfortably lounging on CD-ROMs, while their anxious keepers try to figure out how to transform them into useful conclusions and recommendations. It's unlikely that you'll be able to avoid this activity. Whether you're a corporate CEO, a zoo manager, or a research scientist, dealing with data is part of the job.

The goal of this book is to acquaint you with data analysis—the art of examining, summarizing, and drawing conclusions from data. You probably think that the word *art* is a misprint. Creativity isn't usually associated with classes in statistics or data analysis—but it should be. Data analysis is much more than knowing what some exotic terms and statistical formulas mean. Good data analysis involves a mixture of common sense, technical expertise, and curiosity. It's knowing what questions to ask and how best to answer them. Analyzing data isn't a rote activity—every dataset you encounter is in some way unique.

The best way to learn about data analysis is to actually do it. In this book, you'll analyze a variety of datasets. You'll see how Internet users differ from those who don't use the Internet. You'll determine whether Internet users visit public libraries less often than non-users. You'll see whether running a marathon alters certain chemicals found in the blood and whether having a college degree makes people more likely to work longer hours or open doors for strangers. You'll try to predict life expectancy for people in different countries, based on various characteristics of the countries.

Statistical software is essential for analyzing data. It lets you channel your energy into thinking about a problem instead of being preoccupied with computational details. But that doesn't mean that you need to know less about the concepts underlying data analysis. It means that you must know more. Simple achievements, such as being able to calculate a variety of statistical tests on a pocket calculator or mimic arcane computer commands, count for little these days. Instead, you must learn how best to harness the powerful statistical techniques at your fingertips to solve problems intelligently.

About This Book

This book is divided into four parts: "Getting Started with SPSS," "Describing Data," "Testing Hypotheses," and "Examining Relationships." In each of these parts, you're introduced to tools that you can apply to a broad range of problems. Most data analysis tasks require you to select and combine approaches from the different parts of the book. For example, examining your data values is always the first step to solving any problem, no matter how complicated or simple.

Getting Started with SPSS

The statistical program that you will use to analyze data is called SPSS. SPSS runs on a variety of platforms. The operations and examples in this book have been verified with SPSS 16.0 for Windows but should also apply to other operating systems. Chapter 2 provides a brief overview of how to operate the SPSS system (including the Student Version). Using SPSS, you'll be able to produce graphical displays and statistical analyses easily. But the program can't do what's most important—select the appropriate procedures and interpret their results. That's left up to you.

Describing Data

The first step of any statistical analysis is displaying and summarizing the data values. How many people use the Internet? How old are they? How much money do they earn? You can also look at several items together. How many men and how many women find life exciting? What are their average ages? Are college graduates more satisfied with their jobs than people without college degrees? How quickly did people run the Chicago marathon? You also identify values that appear to be unusual—ages in the 100s, incomes in the billions—and check the original records to make sure that these values are not the result of errors in coding or entering the data. You don't want to waste time analyzing incorrect data.

An important part of describing and summarizing data is making graphical displays. Some of these displays, such as pie charts and bar charts, are useful for presenting your results. Other more specialized displays help you to see more complex features of the distribution of data values. As you'll learn, it can be particularly important to see these features when analyzing data.

Testing Hypotheses

Sometimes all you need to do is describe your data, since you have information available for everyone (or everything) that you're interested in drawing conclusions about. That's not usually the case, though. Instead, you want to draw conclusions about much larger groups of people or things than those for whom you have data. You want to know whether two treatments for a disease are equally effective for all people with the disease, not just for those in your study. You want to know how Internet users differ from non-users, not just in your sample but in the United States population. You don't want to restrict your conclusions to those who participated in your survey. To test hypotheses about a larger group of people or things based on the results observed in your sample, you have to learn about statistical hypothesis testing. That's what this part of this book is about.

Examining Relationships

How are sales related to marketing expenditures? Does diastolic blood pressure increase with body weight? Can you predict life expectancy for a country from its birth rate? What are good predictors of salary: education, years of work experience, job seniority, type of position? There are many different ways to study and model the relationships between variables. You can compute statistics that measure how strongly two variables are related. You can build a mathematical model to predict values of one variable from values of other variables. In the last part of the book, you'll learn about linear regression analysis, a powerful and frequently used tool for modeling relationships between variables.

Let's Get Started

Although it is possible to read through this entire book without actually attempting any of the analyses yourself, the book has been designed with a hands-on approach in mind. Each chapter provides specific instructions so that you can work along in SPSS. To get started using SPSS, turn to Chapter 2. SPSS may look intimidating when you first start it up, but you'll quickly realize how easy it is to get SPSS to do the work of calculating statistical results, leaving you free to focus on the more important question: "What do the numbers mean?"

2 An Introductory Tour of SPSS

This chapter provides a quick, guided tour of SPSS 16.0. It applies to both the full system and the Student Version. (For a more extensive guided tour, run the online tutorials that were installed along with SPSS.) In order to use SPSS, you need to know how to do the following things, all of which are included in this chapter's tour:

- Open a data file.
- Run a graphical or statistical procedure.
- Examine the results.

The chapter concludes with an overview of three of the features that make SPSS 16.0 easy to use: the toolbar, the online tutorial, and the comprehensive Help system.

Starting SPSS

Note: The following instructions for starting SPSS apply to the Windows operating system. If you are running SPSS on a different system, instructions may vary.

The easiest way to run SPSS for Windows is by using the Start button. During the installation of SPSS, the setup procedure adds SPSS to the menu that appears when you click the Start button, as shown in Figure 2.1.

Figure 2.1 SPSS on the Start menu

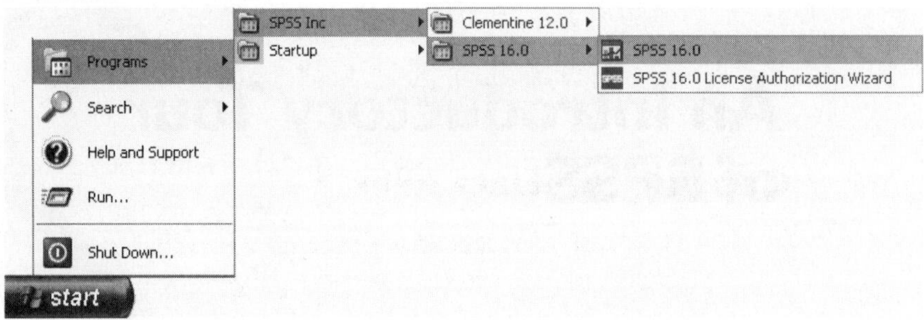

Always use the left mouse button unless the right one is specifically indicated.

▶ To start SPSS, click Start to display the Windows Start menu. Then click [All] Programs > SPSS Inc > SPSS 16.0 > SPSS 16.0.

The SPSS Data Editor window is displayed, as shown in Figure 2.2. You can move it, like any other window, by clicking and dragging its title bar, or resize it by clicking and dragging its sides or corners.

Figure 2.2 SPSS Data Editor window

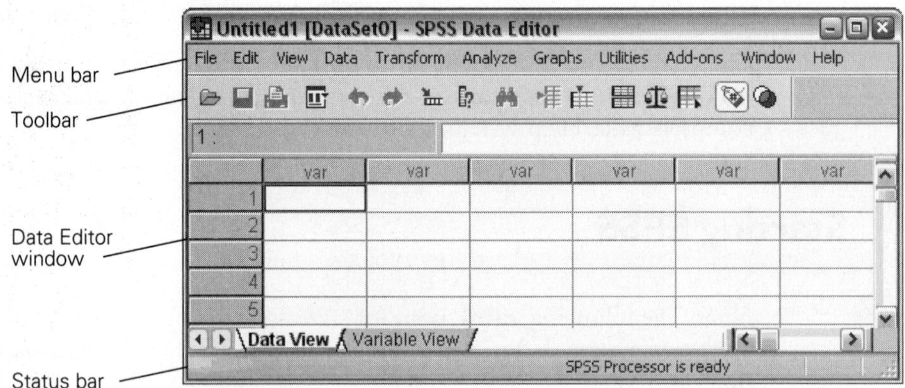

Help Is Always at Hand

This is a book about statistical data analysis. It will not attempt to point out all of the different things that you can do with SPSS. If you're interested in learning more about the software, you can usually find answers to your questions in the SPSS Help system. There is an overview of the Help system at the end of this chapter. For tool buttons on a toolbar, just point the mouse at the button, and a description appears both in a ToolTip and at the left end of the status bar.

See "Contextual Help" on p. 28 for more information.

Copying the Data Files

This book is packaged with the data files that you'll be using. To follow along, you need to copy the data files onto your hard disk. If you don't know how to do this, follow these steps:

▶ Insert the CD-ROM into the appropriate drive.

▶ Double-click the My Computer icon on the Windows desktop to open the My Computer window.

▶ Double-click the icon for the appropriate drive in the My Computer window.

▶ You will see a list of data files in the window that opens. Press Ctrl-A to select them all, and then press Ctrl-C to copy them.

▶ Double-click the icon for your hard disk in the My Computer window. (If you want to tuck the folder of data files away inside some other folder, double-click the folder that you want to put it in.)

▶ Make a new folder on your hard disk to hold the data files. From the menu bar on your hard disk window choose:

File
 New ▶
  Folder

▶ Rename the new folder to something memorable, such as *GDA Data Files.*

▶ Double-click the icon for the new folder to open it into a window, and then press Ctrl-V to paste the files that you copied. Wait for the copy to complete.

Opening a Data File

The SPSS Data Editor window displays your working data file. You don't have one yet—that's why the Data Editor is empty. If you have data of your own that are not in the computer yet, you can type the numbers right into the Data Editor. If the data are already in a spreadsheet or database file, you can probably read that file into SPSS. The data used in this book are already in the form of SPSS data files. To use them for the exercises, or just to follow along in the analysis, simply open the appropriate data file.

To open a data file:

▶ Click the left mouse button on the word File on the SPSS Data Editor menu bar.

▶ Move the cursor down to the Open command. The File Open submenu appears, as shown in Figure 2.3.

Figure 2.3 Opening a data file

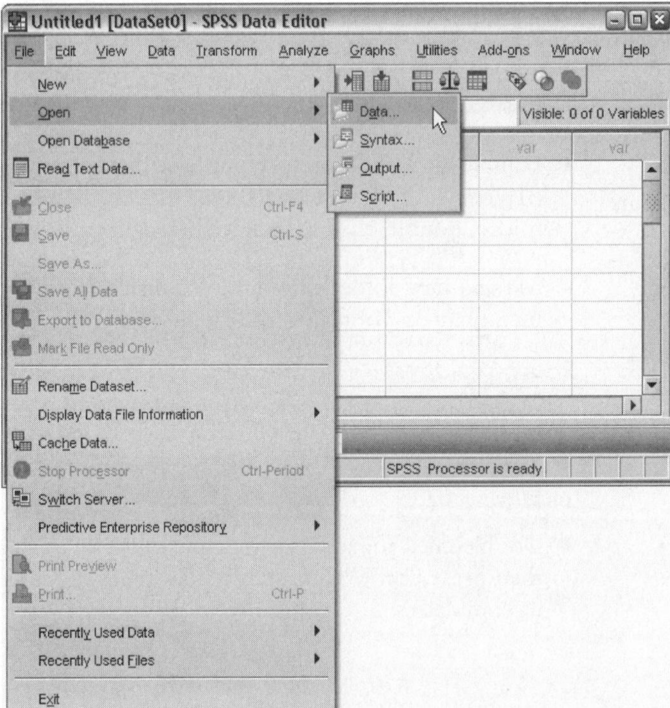

▶ Move the cursor over to **Data** on the File Open submenu, and click the left mouse button. The Open Data dialog box appears, listing the available data files in the current directory, as shown in Figure 2.4.

Figure 2.4 Open Data dialog box

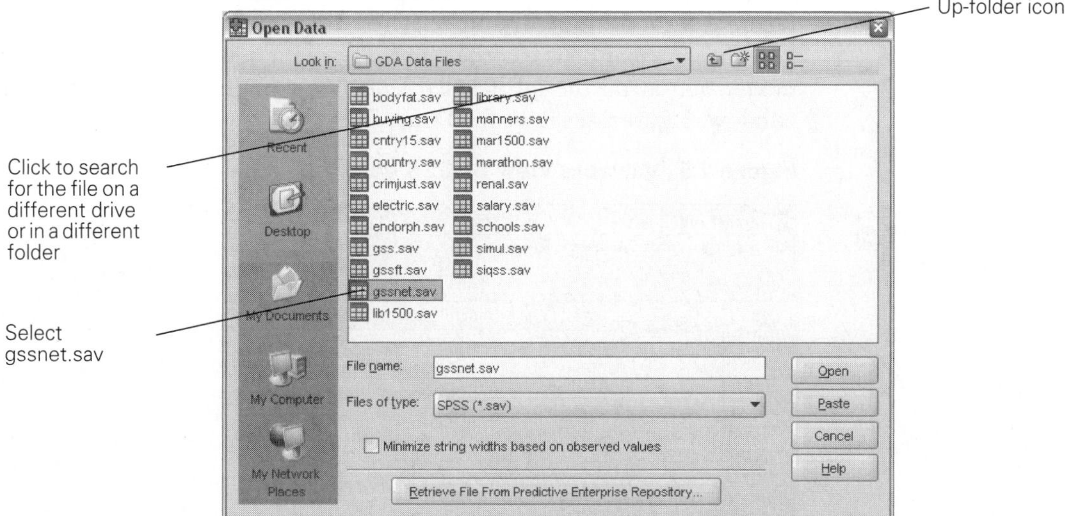

▶ Click the *gssnet.sav* data file where it appears in the list.

▶ Click **Open**.

? *What if the gssnet.sav file doesn't appear?* Only files in the current drive and directory are listed. The file that you want may be either in another directory or saved on a different drive.

To look in a parent folder (one that contains the current folder), click the up-folder icon, as shown in Figure 2.4.

To look in a subfolder (one contained in the current folder), double-click it in the list.

To look on a different drive, click the up-folder icon repeatedly until you reach My Computer, and then double-click the desired drive icon and continue down through the folder hierarchy on that drive. ■■■

When SPSS has finished reading the data file, it displays information about the data in the Data Editor, as shown in Figure 2.5. This particular data file contains selected information for 984 people who were interviewed by the General Social Survey, which periodically asks a broad range of questions to a sample of adults in the United States population. Figure 2.5 shows the Variable View of the data file, which summarizes how the information is recorded in it. To see the actual data, click the left mouse button on the tab labeled **Data View** at the bottom left of the window. Figure 2.6 shows the Data View.

To view the data in the Data Editor, from the menus choose:

Window
gssnet.sav

If your screen displays all numbers rather than value labels, such as Male and Female, in the cells, from the menus choose:

View
Value Labels

Figure 2.5 Variable View of GSS data

*gssnet.sav [DataSet1] - SPSS Data Editor

File Edit View Data Transform Analyze Graphs Utilities Add-ons Window Help

	Name	Type	Width	Dec	Label	Values	Missing
1	age	Numeric	2	0	Age of respondent	{98, DK}...	0, 99
2	sex	Numeric	1	0	Respondent's sex	{1, Male}...	None
3	educ	Numeric	2	0	Highest year of school complete	{97, NAP}...	97, 99, 98
4	degree	Numeric	1	0	Resp's highest degree	{0, Less than h	7, 9, 8
5	usecomp	Numeric	1	0	Use computer?	{0, No}...	9
6	usenet	Numeric	1	0	Use internet ?	{0, No}...	9
7	usewww	Numeric	1	0	Use www other than e-mail?	{0, No}...	9
8	usemail	Numeric	1	0	Use e-mail?	{0, No}...	9
9	netcat	Numeric	2	0	Weekly internet use (hours)	{-9, Time unkn	-1, -9

Data View / Variable View

SPSS Processor is ready

Figure 2.6 Data View of GSS data

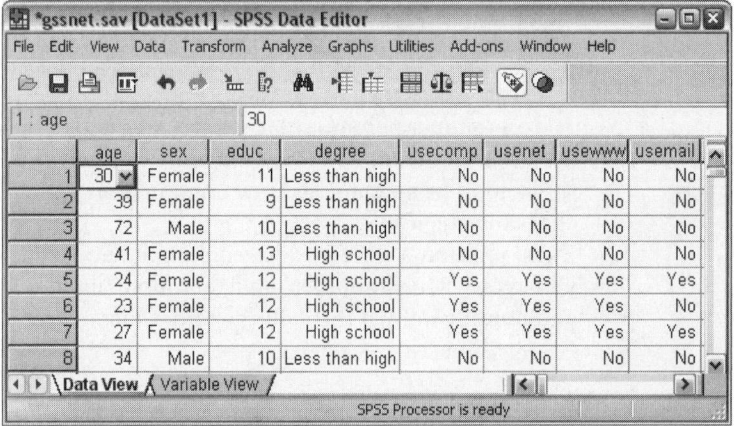

*gssnet.sav [DataSet1] - SPSS Data Editor

File Edit View Data Transform Analyze Graphs Utilities Add-ons Window Help

1 : age 30

	age	sex	educ	degree	usecomp	usenet	usewww	usemail
1	30	Female	11	Less than high	No	No	No	No
2	39	Female	9	Less than high	No	No	No	No
3	72	Male	10	Less than high	No	No	No	No
4	41	Female	13	High school	No	No	No	No
5	24	Female	12	High school	Yes	Yes	Yes	Yes
6	23	Female	12	High school	Yes	Yes	Yes	No
7	27	Female	12	High school	Yes	Yes	Yes	Yes
8	34	Male	10	Less than high	No	No	No	No

Data View / Variable View

SPSS Processor is ready

Statistical Procedures

You can use the Frequencies procedure to summarize and display values for the variables in your data file. (Chapter 4 discusses frequency tables in detail. At this point, you'll just use Frequencies to see how SPSS procedures work.) To open the Frequencies dialog box, click the left mouse button on the word Analyze on the SPSS menu bar, as shown in Figure 2.7.

Figure 2.7 Analyze menu

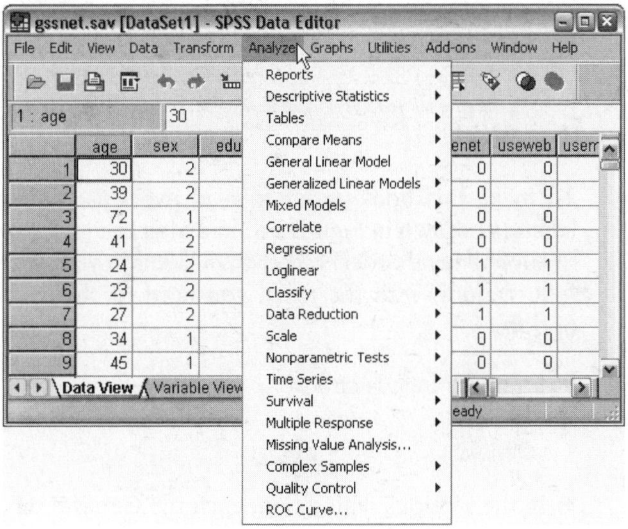

You can find the Analyze menu either in the Data Editor window or the Viewer window. Either one works.

▶ Click **Descriptive Statistics** on the Analyze menu.

▶ Click **Frequencies.**

From now on, we'll use shorthand to indicate menu selections. The following shorthand represents the steps we just took:

Analyze
 Descriptive Statistics ▶
  Frequencies...

The menu selection opens the Frequencies dialog box, as shown in Figure 2.8.

Figure 2.8 Frequencies dialog box with default variable labels

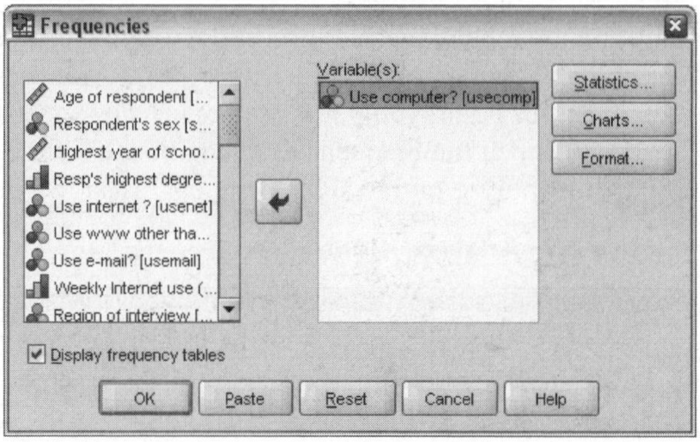

To make this book easier to read, we'll use variable names instead of labels (as shown in Figure 2.8) in dialog boxes. To display variable names (in alphabetical order) rather than labels in your dialog boxes (so you can follow along with the text), you need to change one of SPSS's default options.

▶ From the menus choose:

Edit
  Options...

▶ In the Options dialog box, click the **General** tab.

▶ In the Variable Lists group, select **Display names** and select **Alphabetical**.

▶ Click **OK**.

The effect of the changed option is shown in Figure 2.9.

? *How can I tell what's in a data file?*

From the File menu choose:

Display Data File Information ▶
  Working File

You'll now have information about each of the variables in the data file. Try it. ■■■

Figure 2.9 Frequencies dialog box

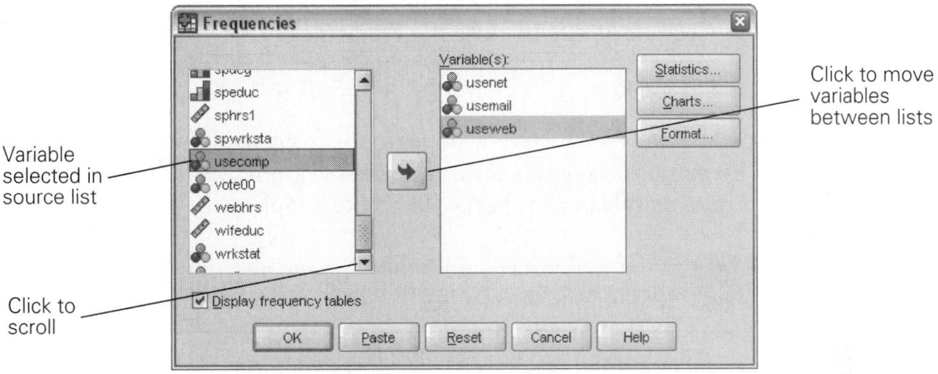

To use this dialog box:

As a shortcut to scroll the source list, click in the list and type the letter u. This scrolls to the first variable beginning with the letter u.

▶ Click *usenet* in the scroll list and then click [▶] , or drag it into the Variable(s) list.

This moves *usenet* into the Variable(s) list.

▶ Scroll down the source list until you see *usemail* and move it into the Variable(s) list.

▶ Click *useweb* and move it into the Variable(s) list as well.

▶ Click Charts.

This opens the Frequencies Charts dialog box, as shown in Figure 2.10. Here you can request charts along with your frequency tables.

Figure 2.10 Frequencies Charts dialog box

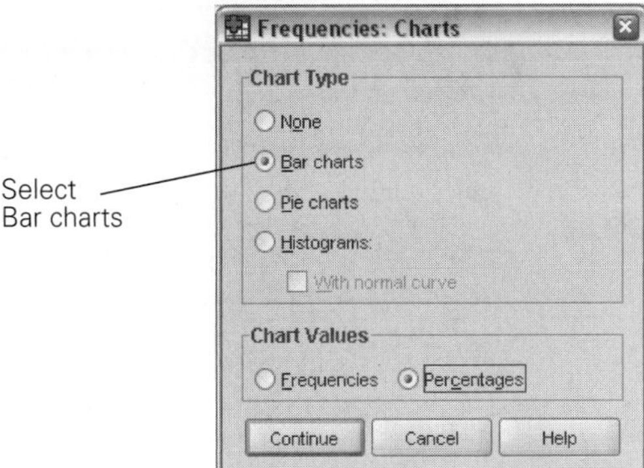

▶ Select Bar charts and Percentages, as shown in Figure 2.10.

▶ Click Continue to close the Frequencies Charts dialog box.

▶ Click OK.

SPSS brings the Viewer window to the front, as shown in Figure 2.11. The Viewer now contains several pieces of output, including three frequency tables and three bar charts. Let's take a look at the output in the Viewer.

? *What are the symbols that appear in front the variable names?* SPSS selects the symbol to use for each variable based on the scale on which the variable is measured. Variables whose values have no meaningful order are identified with three balloons. Variables whose values can be ordered from small to large but not on a scale with equal intervals are labeled with three bars of increasing size. Variables measured on a scale with equal intervals are identified with a ruler. You'll learn more about this in Chapter 5. ■■■

The Viewer Window

The Viewer window is where you see the statistics and graphics—the **output**—from your work in SPSS. As shown in Figure 2.11, the Viewer window is split into two parts, or **panes**. (A piece of a window is often called a pane in computer software, just as it is at your local hardware store.)

Figure 2.11 Viewer window

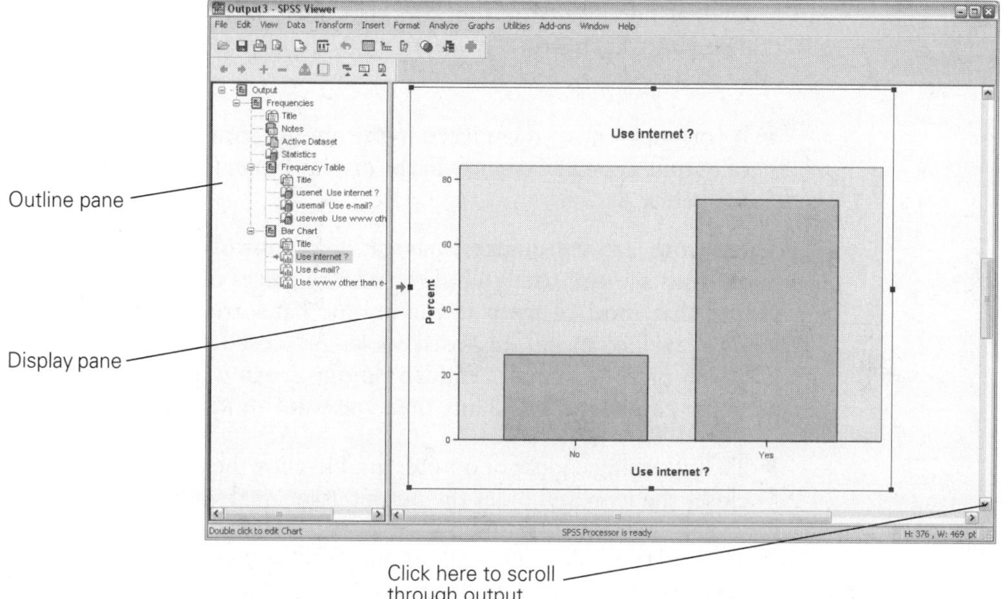

The left side (the **outline pane**) contains an outline view of all of the different pieces of output in the Viewer, whether or not they are currently visible. The right side (the **display pane**) contains the output itself.

▶ To change the sizes of the two panes (for example, to make the display pane wider), just point the mouse at the line that divides them, click the left mouse button, and drag the line to the left or right.

It's possible to ignore the outline pane and simply scroll through the output displayed in the display pane on the right side of the Viewer. The outline view offers some handy tricks, however.

The Outline Pane

Individual portions of output are associated with "book" icons in the outline pane. Each icon represents a particular piece of output, such as a table of statistics or a chart.

▶ If you click one of these icons in the outline pane, the associated piece of output appears instantly in the display pane. (But it may be hidden! See below.)

These icons are the quickest navigational controls in the Viewer. The book icons are also used to hide or display pieces of output temporarily. Notice that most of them in the outline pane are "open book" icons, while a few look more like closed books. A "closed book" icon represents a hidden piece of output. Hidden output doesn't appear in the display pane but can be recovered any time you want to look at it.

▶ To hide a single piece of output, double-click the open book icon. This closes the icon and hides the output associated with it.

▶ To display a hidden piece of output, double-click the closed book icon. This opens the icon and displays the output associated with it.

▶ To hide *all* of the output from a procedure, such as Frequencies, click the little box containing a minus sign to the left of the procedure name. That whole part of the outline collapses, and the minus sign changes to a plus sign to show you that more output is hiding there. Click the plus sign to show it all again.

You will find that you can do a lot of things in fairly obvious ways by playing with the outline pane. Try rearranging the output (click the left mouse button on a book icon, drag it to a different place in the outline, and then release the mouse button) or deleting part of the output (click the icon and press the Delete key). The SPSS Help system can tell you all the details.

The Display Pane

The display pane shows as much of the SPSS output as can fit in it. To see more, you can either scroll the pane or use the outline pane to jump around.

The output in the display pane includes several different kinds of objects: tables of numbers (actually a special kind of table, called a **pivot table**); charts; and bits of text, such as titles. You have complete control over the appearance, and even the content, of most of these objects.

▶ To change something about an object, double-click it in the display pane.

Double-clicking an object opens an editor that is specially designed to modify it. The appearance of the object changes to show that you are editing it. The menu bar may change. If the object is a chart, a special chart-editing window opens to offer you a powerful set of tools for changing the chart's appearance. Let's look at these objects in the Viewer.

Viewer Objects

You can choose whether output is labeled by variable name, variable label, or both. From the menus choose:

Edit
Options...

Click the Output Labels tab.

In the outline pane, the first line is a container for the entire batch of output. It's simply called Output. There might be a line below it called Log, which isn't going to be discussed in this book. The next line, Frequencies, is a heading that contains all of the various kinds of output produced by the Frequencies procedure that you just ran. In order, they are:

- Title. The title of the procedure, which is simply text.
- Notes. Notes are usually hidden, so this probably looks like a closed book in the outline pane.
- Active Dataset. Since you can have more than one data file open in SPSS, this identifies the file used for this procedure.
- Statistics. This is a pivot table that reports the number of cases, or "observations," that were processed by the Frequencies procedure. Most procedures start by producing such a table. The icon is an open book, so if you click it, the display pane will show you what it looks like.
- A frequency table for the variable *usenet*, the first variable analyzed. Frequency tables are discussed in Chapter 4. Notice that the icon in the outline pane is labeled *Use Internet?*, which is a descriptive label that was assigned to the variable *usenet* when the data file was set up.
- A frequency table for the variable *usemail*, whose icon is labeled *Use e-mail?* in the outline pane.
- A frequency table for the variable *useweb*, whose icon is labeled *Use WWW other than e-mail?* in the outline pane.
- A bar chart for *usenet*.
- A bar chart for *usemail*.
- A bar chart for *useweb*.

Let's see what these pivot tables and charts are like.

Pivot Tables

First, we'll look at a pivot table. Most of SPSS's tabular and statistical output appears in the Viewer in the form of pivot tables.

Figure 2.12 Pivot table in the Viewer

usenet Use internet ?

		Frequency	Percent	Valid Percent	Cumulative Percent
Valid	0 No	255	25.9	26.1	26.1
	1 Yes	723	73.5	73.9	100.0
	Total	978	99.4	100.0	
Missing	9 Unknown	6	.6		
Total		984	100.0		

Click here to display the frequency table for usenet

▶ In the outline under Frequency Table, click the icon for the pivot table labeled *Use Internet?*. The table instantly appears in the display pane, with an arrow pointing to it, as shown in Figure 2.12.

▶ Move the mouse to the display pane and double-click on the table itself to indicate that you want to edit it.

Figure 2.13 Activated pivot table

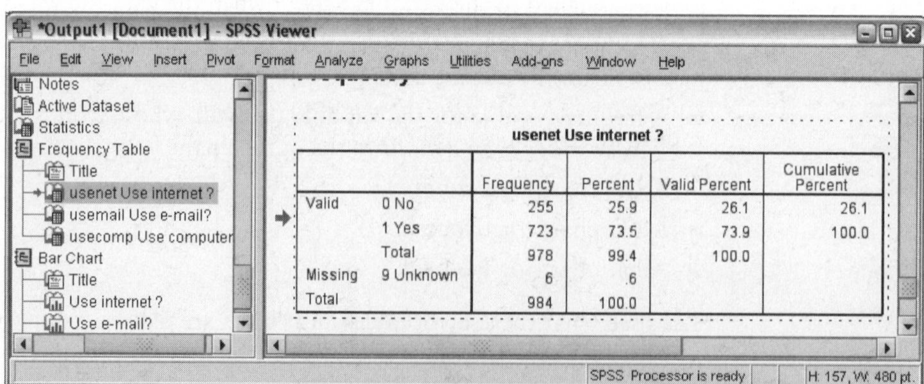

usenet Use internet ?

		Frequency	Percent	Valid Percent	Cumulative Percent
Valid	0 No	255	25.9	26.1	26.1
	1 Yes	723	73.5	73.9	100.0
	Total	978	99.4	100.0	
Missing	9 Unknown	6	.6		
Total		984	100.0		

Not a lot seems to happen in Figure 2.13. The pivot table is now surrounded by a dotted line to indicate that it is active in the Pivot Table Editor. The SPSS toolbar vanishes, and if you watch carefully, the menu bar changes—there is now a Pivot menu.

Double-clicking a pivot table lets you edit it "in place"—that is, right where it sits in the display pane of the Viewer. If you need more room, select the pivot table by clicking once with the left mouse button, and from the menus choose:

Edit
  Edit Content ▸
    In Separate Window

This command opens the pivot table into a window of its own.

When you are editing a pivot table, you can change almost anything about it that you want. If you don't like the label, just double-click it. It reappears as highlighted text. Type in the label the way you want it, perhaps *Number of Internet Users*, and click somewhere else to enter the new label. To change the font of the title or make it bold or italic, click the title and from the menus choose:

Format
  Table Properties ▸
    Cell Formats

Then choose a different font or a bold or italic style.

If you don't like the way numbers are displayed in the pivot table, make sure that the Pivot Table Editor is active (by double-clicking the table in the display pane), and then either make a selection from the Format menu or *right-click* one of the numbers in the table to pop up a context menu for it. Most of the things that you might want to change can be found in either of these menus under Table Properties or Cell Properties. Check out TableLooks, too, to see how you can apply consistent sets of formatting to whole tables.

Changing fonts and styles and even the text of the labels in a table can make a big difference in the way the table looks. An SPSS pivot table lets you do much more than that, however. You can change the basic organization of the data presented in the table. The Pivot menu (which appears only when you have double-clicked a pivot table in the display frame to activate it) gives you access to powerful tools for reorganizing the table. To get a feel for these tools, activate a pivot table, and from the menus choose:

Pivot
  Transpose Rows and Columns

The same information is displayed. The different codes or responses to the question, which were laid out vertically, are now laid out horizontally; the different types of statistical summaries, which were laid out horizontally, are now laid out vertically.

To see the pivot table as it was before, simply transpose the rows and columns again.

This example is a very simple pivot table. Multidimensional tables offer many more structural possibilities. You can explore those in the SPSS online Help system, or, if you like, to see how things work you can build a complex table and start pivoting.

Charts

Now let's take a closer look at a chart. In the Viewer's outline pane, click the icon for the bar chart of *Use Internet?*. This brings the bar chart into view in the display pane.

- ▶ To change the size of the selected chart, point the mouse at one of the "handles" at its corners or sides, click the left mouse button, and drag the handle to resize the chart.

- ▶ To change anything about the bar chart, bring it into the Chart Editor by double-clicking it in the Viewer display pane.

Figure 2.14 SPSS Chart Editor

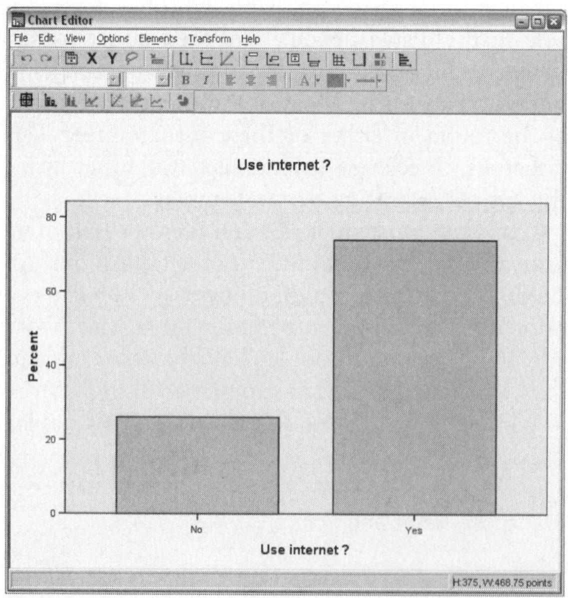

Warning! The choices that are available in the Chart Editor depend on which elements of the chart you selected. If you do not see what you are looking for, select another item by clicking it. To learn more, from the Help menu choose Topics, or choose Tutorial to begin the tutorial "Creating and Editing Charts."

As you see in Figure 2.14, the SPSS Chart Editor displays the chart in a new window.

Most things that you might want to do to a chart can be done directly in the Chart Editor.

- Use one of the Chart Editor menus to add items, such as lines or point labels.
- Use the Properties dialog boxes, which appear when you double-click a chart or one of its elements, to specify options for the chart, such as colors or line styles.

To leave the Chart Editor, close its window. SPSS updates the chart displayed in the display pane of the Viewer to reflect the changes you made.

The Data Editor Window

If your screen shows labels instead of numbers, from the menus choose View and deselect Value Labels.

Let's take a closer look at the Data Editor (see Figure 2.15). You can select it from the Window menu or simply click on it if any part of it is visible on your screen.

If you've ever used a spreadsheet, the Data Editor should look familiar. It's just an array of rows and columns. In the Data Editor, each row is a case, and each column is a variable. Cases and variables are fundamental concepts in data analysis. It's time we stopped to define them.

Figure 2.15 Data Editor window

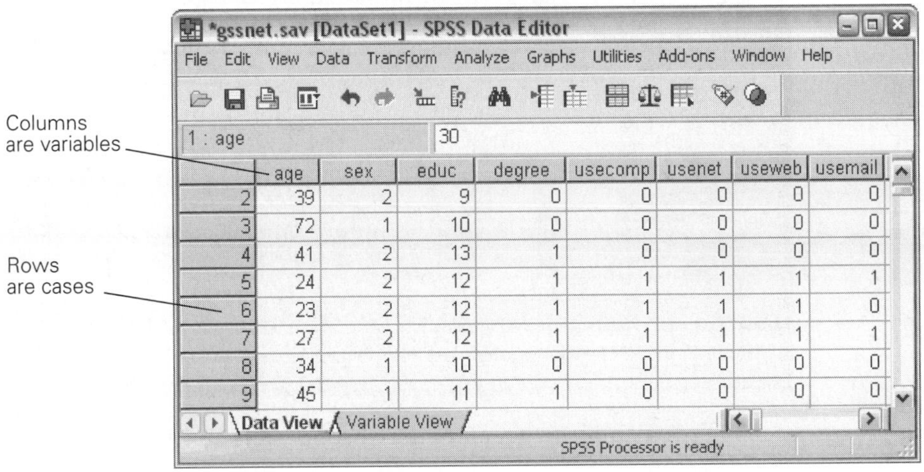

*gssnet.sav [DataSet1] - SPSS Data Editor

File Edit View Data Transform Analyze Graphs Utilities Add-ons Window Help

1 : age 30

Columns are variables

Rows are cases

	age	sex	educ	degree	usecomp	usenet	useweb	usemail
2	39	2	9	0	0	0	0	0
3	72	1	10	0	0	0	0	0
4	41	2	13	1	0	0	0	0
5	24	2	12	1	1	1	1	1
6	23	2	12	1	1	1	1	0
7	27	2	12	1	1	1	1	1
8	34	1	10	0	0	0	0	0
9	45	1	11	1	0	0	0	0

Data View / Variable View

SPSS Processor is ready

Cases (rows) are the people who participate in a survey or experiment. (Another word often used is **observation.**) Actually, a case need not be a person. It can be anything. If you're doing experiments on rats, the case is the individual rat. If you're studying the beef content of hamburgers, each hamburger is a case. Generally speaking, the case is the unit for which you take measurements.

Variables (columns) are the different items of information that you collect for your cases. Think about the way you conduct a survey. You ask each person for the same type of information: date of birth, sex, marital status, education, views on whatever subjects your survey is about. Each item for which you record an answer is known as a variable. The answer a particular person gives is known as the **value** for that variable. Year of birth is a variable; responses such as 1952 or 1899 are values for that variable.

The intersection of the row and the column is called a **cell.** Each cell holds the value of a particular case for a particular variable. You can edit values in the Data Editor as follows:

▶ Click in one of the cells with the mouse.

The **cell editor** displays the value for the selected cell, as shown in Figure 2.16.

Figure 2.16 Data Editor with cell selected

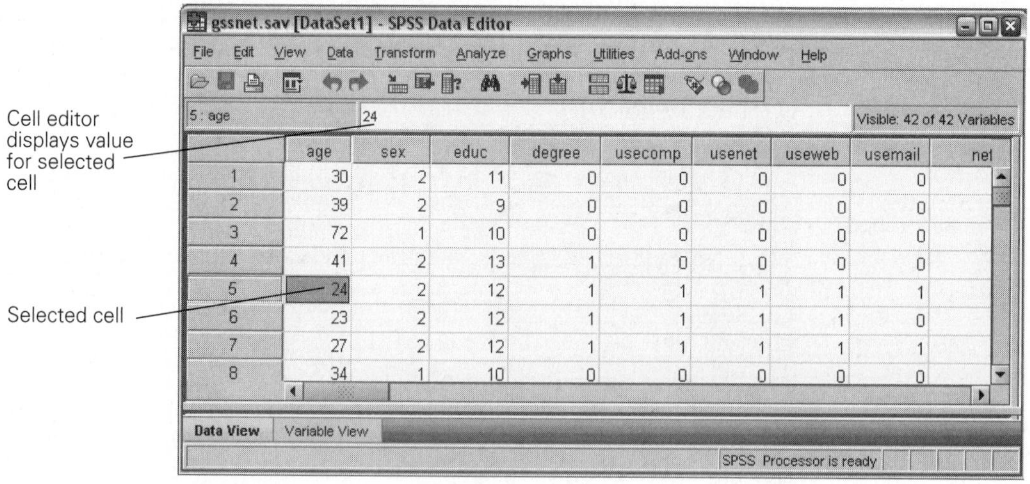

▶ Type a number to replace the existing value and press ↵Enter.

The new value appears in the cell editor as you type it, but the value in the cell is not updated until you press ↵Enter.

▶ Change another value in the cell editor, but instead of pressing ↵Enter, press Esc.

When you press Esc rather than ↵Enter, the original value in the cell remains unchanged.

Entering Non-Numeric Data

If you try to enter anything other than a number into a numeric variable cell, your computer will probably beep at you.

All of the variables in *gssnet.sav* are **numeric**; that is, they are made up of numbers. To enter non-numeric data into a column, you must first specify the variable type in the Variable View of the data file.

For example, suppose that you want to type the name of each respondent in the survey. To do this, you must create a new variable and define it as a string variable. (**String** is a common computer term meaning that the variable contains text—words or characters—rather than numbers.)

▶ Click the Variable View tab at the bottom left of the Data Editor.

▶ Scroll to the bottom of the Variable View, to the last variable in the data file.

▶ In the *Name* column, type **respname** as a new variable name. Press Tab⇥.

In the *Type* column, click the gray box with the word Numeric and then click the button with the three dots. The Variable Type dialog box appears (see Figure 2.17).

Figure 2.17 Variable Type dialog box

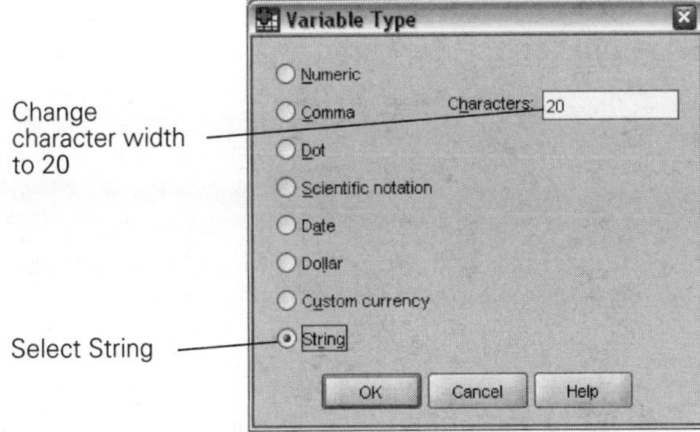

When you select String as the variable type, SPSS changes the measurement alternative to Nominal. In the list of variables, SPSS also uses the symbol "a" to identify string variables.

▶ Select the String option at the bottom of the dialog box. Notice that the default width is 8. Since you might need to enter names longer than 8 characters, change the width to 20. This will be the maximum length of the data values for *respname*.

▶ Click OK to define *respname* as a string variable.

▶ Click the Data View tab at the bottom left to enter the data values.

You have now defined a string variable called *respname*, 20 characters in width. To see all of the data you entered for *respname*, you'll need to make its column wider in the Data View.

▶ Move the mouse cursor over the dividing line to the right of the variable *respname* in the column headers for the Data View.

▶ Click the left mouse button and drag the dividing line to the right to make the column wider.

▶ Release the mouse button when the column is wide enough to show your data values. The width of the column in the Data View affects only the way that the data values are displayed.

Clearing the Data Editor without Saving Changes

You have changed some of the values in the GSS data. You don't want these changes to be permanent because you want to get correct statistical results when you use the data again.

▶ To clear the Data Editor *without* saving your changes, from the menus choose:

File
 New ▶
  Data...

This opens a new (second) Data Editor window. Reactivate (click) the original Data Editor window, and from the menus choose:

File
 Close

SPSS asks if the contents of the Data Editor should be saved.

▶ Click No.

This closes the modified GSS dataset so that you won't forget and save it later on when you exit SPSS.

The SPSS Online Tutorial

SPSS comes with an online tutorial that shows you how to use many of the features of the software. To run the tutorial, choose Tutorial from the SPSS Help menu.

If SPSS is already running, you can start the tutorial by selecting Tutorial from the Help menu.

Figure 2.18 SPSS online tutorial main menu

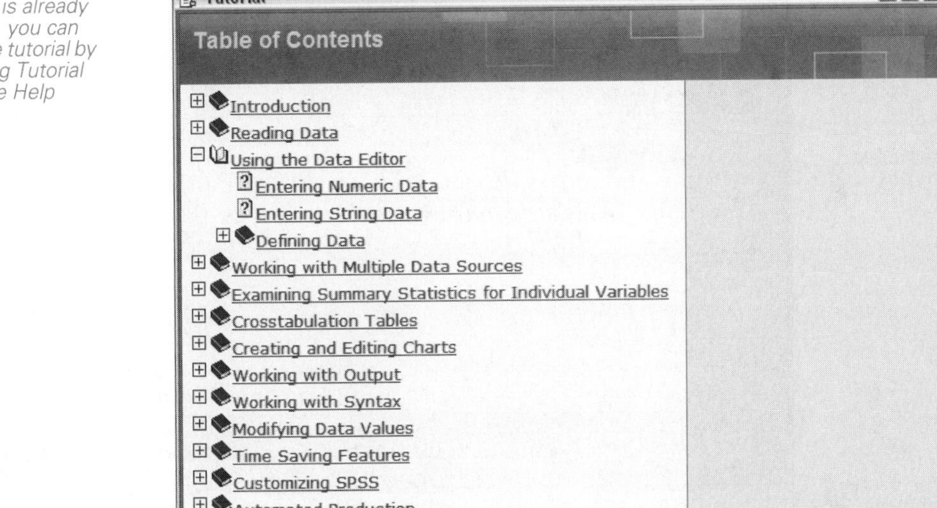

The tutorial is a Help file. It includes an overview and detailed sections on defining data, using the menus and dialog boxes, editing charts, getting help, and other topics (see Figure 2.18).

You can walk through the whole tutorial, or you can choose the specific topics that interest you at the moment. If you need a more complete orientation to SPSS and how it works than this chapter can provide, the online tutorial is the best place to start.

The SPSS Toolbar

The tutorial helps you when you're getting started with SPSS. The toolbar provides shortcuts for commonly used procedures and functions. You will very likely use the toolbar more and more as you become familiar with SPSS.

Figure 2.19 SPSS toolbar

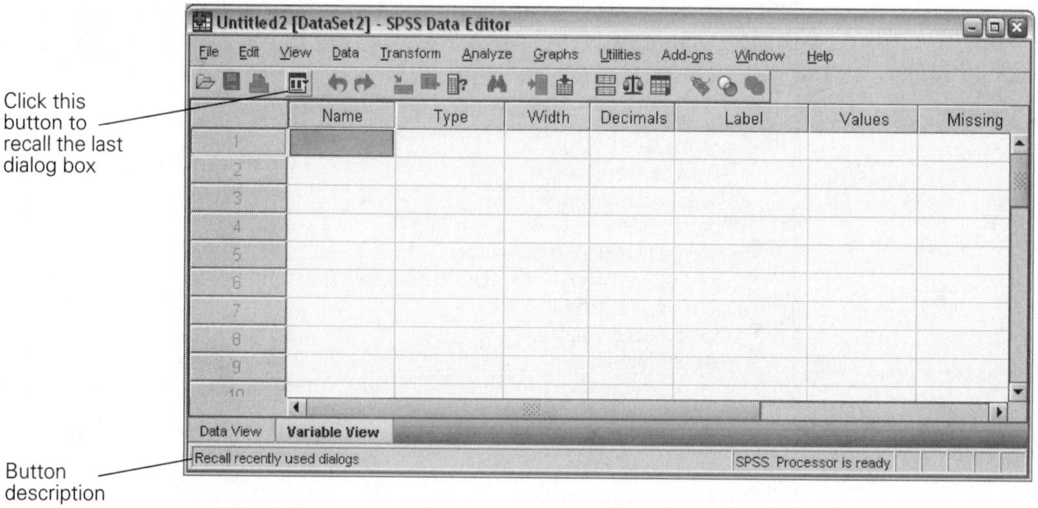

Click this button to recall the last dialog box

Button description

Move the cursor over any tool (don't click) to display in a ToolTip and, in the status bar, a description of what the tool does, as shown in Figure 2.19. The toolbar provides shortcuts. Everything in the toolbar can be done in other ways, too, so you don't have to remember all of the tools—just use the ones that you find helpful.

The toolbar is different in every window. SPSS toolbars are easy to use:

- To activate a tool, just click its button.
- To move the toolbar, point the mouse somewhere on it but not on a button. Click the mouse button and drag the toolbar. You can drag it to other parts of the window; you can even drag it outside the window so that it becomes a detached, floating toolbar.
- To add or remove individual buttons, to hide or show a toolbar, or to create a new toolbar, choose Toolbars from the View menu. You'll find that you can do almost anything to arrange the toolbar to suit the way you like to work.

The SPSS Help System

SPSS comes with an extensive online Help system, held together by a Help window, with its Contents, Index, Search, and Favorites tabs (see Figure 2.20). Click the **Contents** tab for an organized view of what's in the Help system (like a book's table of contents). Click the **Index** tab for an alphabetical list of all of the topics and concepts in the Help system. Click the **Search** tab if you want to search through the SPSS 16.0 Help file. (This is useful as a last resort, when you're sure something must be in there but you haven't been able to find it any other way.) Click the **Favorites** tab to create a list of frequently visited Help topics.

Figure 2.20 SPSS Help system Contents tab

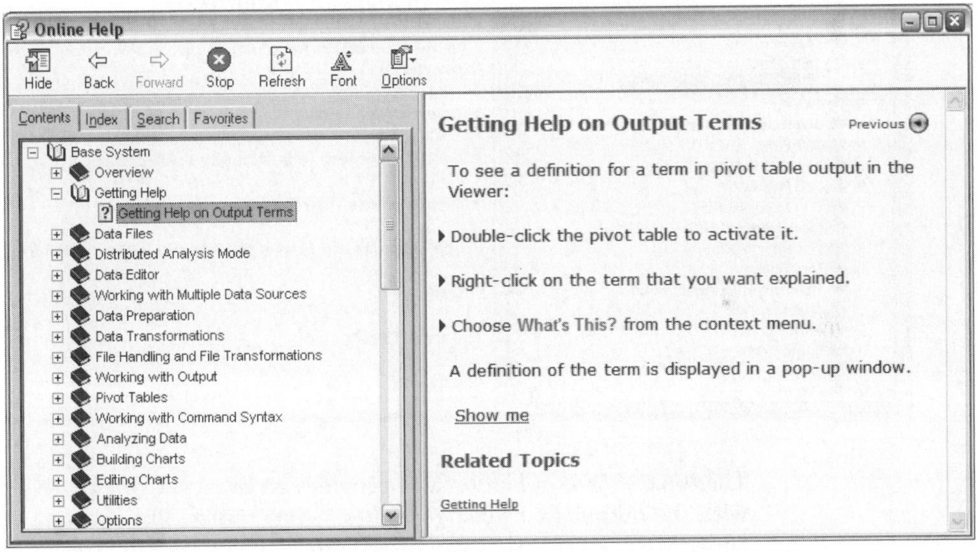

The Contents tab (see Figure 2.20) displays a list of book icons, each of which can be opened to show the Help topics inside. Some books contain other books, in addition to individual Help topics. As you see in the figure, the SPSS Help system includes topics on how to use the Help system itself, a good place to start if you're unfamiliar with SPSS Help. It also includes tutorials that show you how to use SPSS, topics on data and output management in SPSS, statistical and graphical analysis, and a host of other topics.

The SPSS Help system explains the SPSS software program but does not try to teach statistics. If that's what you want, though, you're in the right place—just keep reading this book!

▶ For more information on the Help system, from the SPSS menus choose:

Help
 Topics

Double-click the **Getting Help** book icon to open it, as shown above, and then click one of the topics to display the Help.

Figure 2.21 Help for a topic

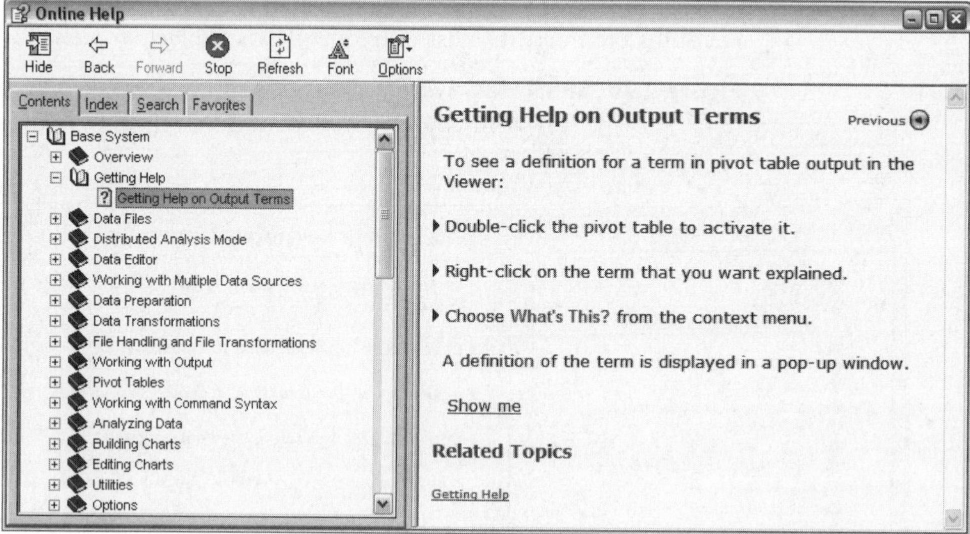

The topic shown in Figure 2.21 provides an overview of how to get help with the output that SPSS generates. Some topics, like this one, have a Show Me link that takes you directly into one of the tutorials, at just the right place!

Contextual Help

Another handy part of the Help system is **contextual help**, also known as "right-mouse-button help." In Windows, clicking the right mouse button on an object pops up the context menu, which lets you carry out common actions instantly. When contextual help is available, the first item on the context menu reads What's This?. Selecting it pops up a short Help message describing the object you clicked on.

Contextual help is particularly useful in *activated* pivot tables, where the names of statistics are attached to contextual topics that briefly tell you what the statistic means. Figure 2.22 shows contextual help for the frequency column.

Figure 2.22 Contextual help for a statistical term

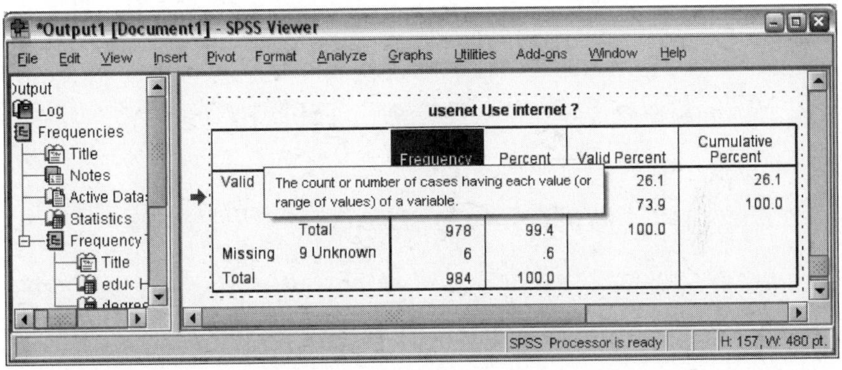

Click either mouse button (anywhere) to clear a pop-up message from contextual help.

What's Next?

Once you get comfortable with the Help system, you'll find that it can tell you just about anything you need to know about how to use SPSS. Chapter 3 examines important considerations in designing a survey or an experiment. To start learning how (and why) to use SPSS as a tool to solve your problems, turn to Chapter 4.

3 Sources of Data

What's important when designing a survey or an experiment?

- How should you ask a question?
- How should you select participants?
- How should you assign people to experimental groups?
- Why is it important to include a comparison group?

Know Your Data

As soon as you get a little experience with SPSS, you'll see how easy it is to feed data into SPSS and generate loads of information. Unfortunately, it's just as easy to generate loads of nonsense. Computer programs can't tell the difference between good data and bad, but you must. If you are analyzing data that have been gathered by someone else, you must learn exactly how the data were obtained and how they were recorded. You must identify the limitations of the data for answering your questions. If you are going to be collecting your own data, you must determine what data you will collect, from whom, and how.

Surveys and experiments are the two most common sources of data. In a survey, you ask questions of people or observe them. In an experiment, you actually do something and then observe the response. In this book, you'll analyze data from several different surveys.

Survey Data

You're eating dinner, the phone rings, and an inquisitive voice on the other end asks what you think about birdseed. An interviewer with a clipboard obstructs your path in the shopping mall, anxious to hear your opinion on soaps. You buy a cheap radio and find it packaged with a registration form that asks more about you than your undergraduate college application. Even your Web surfing is littered with requests for information. (Remember, those cookies are not made for eating.) All of the above are examples of surveys, which are among the most frequently used and abused methods for gathering data.

The concept behind a survey is fairly simple: you record information about people or objects of interest. You ask questions or take measurements. You don't do anything that might affect the responses of the participants. It seems straightforward enough, but it's not. The major difficulties are constructing the questions so that they will yield the desired information, selecting participants who represent the larger pool of people that you want to draw conclusions about, and convincing the chosen participants to actually participate. Let's consider each of these problems.

Asking the Question

All surveys require participants to answer questions or to allow various measurements or observations to be made. Asking and answering questions is something you've done since your toddler days, so you might think it's not a big deal, but it is in survey research. The way in which you ask questions affects the answers that you get.

You don't want to influence the response a person gives. Asking a respondent whether they agree with the statement that the president could do a better job will paint a different picture than asking if they agree with the statement that the president is doing a good job. You want to structure the question in such a way that an answer is not suggested. For example, you might ask a person to rank the president's performance on a scale from 1 to 5, where 1 is poor and 5 is excellent. The interviewer should always appear nonjudgmental, making sure the respondent doesn't try to give answers designed to please the interviewer.

Eliciting even simple information requires thought. For example, "How old are you?" is an unimaginative question that you've been asked many times. You may recall that you haven't always answered it truthfully. When you were 12 years old, you may have promoted yourself to the more glamorous teenage years. When you reach middle age, you may want to turn back the clock. Or someday, you simply may not remember the magic number anymore. Does that mean that you shouldn't ask about age in a survey? Of course not. It means that you have to think about the best, practical way to obtain reliable information about age. Asking to see a birth certificate is one way of obtaining accurate ages, but it's obviously not a realistic tactic for most surveys, while asking people to give their birth date as well as their age is. People are much less likely to manipulate their birth dates than their ages, since a birth date doesn't change annually, and it's work to fudge a birth year to match a desired age.

Although determining age has its problems, they are fairly minor compared to the difficulties encountered in getting information about sensitive topics such as drug use or criminal records. Even obtaining information about routine activities, such as time spent on the Internet, poses challenges. Skilled researchers devote a lot of effort to designing the survey questionnaire. Many surveys involve focus groups that help shape how the desired information is to be elicited. An actual questionnaire is pretested on a random sample of people from the population to which the survey will be administered.

Measuring Time

In this book, you'll be working with three surveys that obtained information about Internet usage in very different ways. In the General Social Survey (GSS), people were asked the following questions:

- Do you personally ever use a computer at home, at work, or at some other location?
- About how many minutes or hours per week do you spend sending and answering electronic mail, or e-mail?
- Other than for e-mail, do you ever use the Internet or World Wide Web?
- Not counting e-mail, about how many minutes or hours per week do you use the Web? (Include time that you spend visiting regular Web sites and using interactive Internet services like chat rooms, Usenet groups, discussion forums, bulletin boards, and the like.)

Although the GSS questions seem to focus on how much people use the Internet, do you really think people's reports of weekly time spent on the Internet are completely accurate? Do you know how much time you spend on the Internet each week? Is it the same for most weeks? If you don't spend much time online, would it bother you to admit that? If you think you spend too much time online when you should be studying, would you be tempted to trim your reported hours?

The Stanford Institute for the Quantitative Study of Society (SIQSS) study measured Internet use in a much more detailed manner. Researchers used a diary approach. In diary studies, people are asked to record detailed information about particular events that take place during a day. For example, in a food-diary study, people are asked to record everything they eat during a specified time period. In a time-diary study, people are asked to record how they spend their time. Diary studies are by nature intrusive and time consuming. You may decide to pass up that chocolate bar if you have to tell a nutritionist about it. Or, you may just decide to eat the chocolate bar and forget to report it.

Norman H. Nie and his colleagues at SIQSS sent electronic time-diary questionnaires to approximately 6000 Americans aged 18 years and older. Respondents were asked about their activities during six randomly selected hours of the previous day—one in each of six time-periods: night, early morning, late morning, afternoon, early evening, and late evening. People received the questionnaire on different days of week, so that all days would be equally represented. Since people were asked to report on only six hours of the day, the questionnaire was more manageable and the researchers were able to get detailed information about activities during the selected six hours. Since it asked about activities the prior day, it didn't intrude on selection of activities during the reported times. Such an approach should result in more reliable information about Internet usage and how it relates to other activities such as spending time with one's family, watching television, and sleeping. Six hours is a large chunk of one's waking time, but is it enough to form an accurate snapshot of a person's day? And of course, there is the question of does one really remember what they were doing in a particular hour of a previous day, for how long, and with whom?

Selecting Participants

When you conduct a survey or perform an experiment, you seldom want to restrict your conclusions to just the participants. Usually, you want to draw conclusions about a much larger group of people or objects. If you're a psychology student examining the effect of stress on eating behavior, you don't want to restrict your conclusions to the 50 sophomores that you actually weighed. You probably want to draw conclusions about all college students or perhaps even all adults. If you're doing market research on anchovy-flavored toothpaste or self-washing coffee cups and query 1000 people about these innovative products, you're not really interested in what these particular 1000 people think, you're interested in whether shoppers in general would find these products appealing. You want the 1000 people in your survey to represent the much larger group of potential buyers. The people or objects included in your study are called the **sample**. The people or objects about whom you want to draw conclusions collectively are called the **population**.

Defining a population may seem straightforward, but often it isn't. Suppose that you are a personnel manager, and you want to study why employees are absent from work. You are probably content to draw conclusions about the employees in your company. All employees at your company are then the population of interest. However, if you're a graduate student writing a thesis about the same topic, you face a much more complicated problem. Do you want to draw conclusions about professionals, laborers, or clerical staff? About men or women? Which part of the world is of interest—a particular city, a country, or the world as a whole? No doubt, you (and your advisor) would be delighted if you could come up with an enlightened explanation for absenteeism that would apply to all types of workers everywhere. You know that's not a realistic expectation. You must narrow your population of interest so that the problem would be tractable.

Even when the population of interest seems to be well defined, you may not be able to study it. If you're evaluating a new method for weight loss, you would ideally like to draw conclusions about how well it works for all overweight people. Unfortunately, people who don't want to lose weight or who have been disheartened by past failures may not agree to participate in your study. Your population may have to be people who are interested in losing weight using your method. This population may be lighter, younger, or healthier than the population of all overweight people. Your conclusions from studying people who want to lose weight may not apply to the population of all overweight people. For example, if your treatment has unpleasant side effects, motivated people might be willing to tolerate such inconvenience to reach their goal. People who are resigned to being overweight might not accept such drawbacks. Your new method may work quite differently for those who are motivated than for those who are not.

Selecting a Sample

There are many different samples that can be selected from the same population. Think of all the ways that you can choose 2200 adults from the population of adults in the United States. You can include the first 2200 names in your local phone book. You can get your college to give you the names of 2200 graduates. You can get a directory of U.S. lawyers and select the required number from the directory. Common sense tells you that there are serious problems with all of these samples. If you use your local phone book, you are excluding people who live outside of your area. College graduates from a particular school may be different from college graduates from other schools and from individuals with less education. A sample composed entirely of lawyers would certainly not be representative of the U.S. adult population as a whole.

So what is a good sample? Obviously you want it to be selected from the entire population of interest. If you're studying U.S. adults, you want your sample to include people of all ages from all areas of the U.S. You want to make sure that you are not excluding any particular types of people and that all people have the same chance of being selected. This does not mean that you should go out and select people that you think are "typical" of the U.S. population. Such samples, called **judgment samples**, are fraught with problems, since the criteria for inclusion depend on someone's notion of who should be selected. There's no way to determine how well the sample really represents the population. Different people will select different samples from the same population based on their preconceived notion of what the population is really like. The same is true for convenience samples. A **convenience sample** selects anyone who's convenient to include. Interviewers who stand in malls, subway stations, and similar places are pursuing convenience samples. Any warm body that is willing to answer questions is fair game. Drawing statistically valid conclusions from a convenience sample is impossible, since the people included are not representative of any population except those willing to be interviewed at a particular location at a particular time.

If judgment doesn't result in a "good" sample, what does? The answer may appear strange at first. The mechanism for selecting a sample should be chance. Not haphazard chance, but carefully contemplated chance. Sounds like a contradiction, but it's not. Contemplated chance means that you have a method to ensure that each person in the population of interest has an equal probability of being selected. Surveys are not alone in relying on chance: lotteries of all kinds depend on chance for outcomes that are fair, and many urgent disputes are resolved by the toss of a coin. When samples are selected by chance, you actually know a lot about them because mathematically you can calculate the likelihood of various types of samples. This is a topic that you'll study in great detail in Part 3 of this book.

? *Wouldn't it be simpler and cheaper to just rely on volunteers?*
Yes. Surveys that rely on volunteers are easy to do. That's why they're everywhere. Unfortunately such surveys are of very limited use. They tell you almost nothing about any population, except that of people who are willing to volunteer to answer questions. It is well known that volunteers can differ in important ways from people who don't volunteer. For example, when a newspaper or radio station asks you to call and register your opinion on some topic, you can be quite certain that the results don't reflect the views of the population of all readers or listeners. Would you bother to make a phone call to register your opinion on the certification of statisticians? Of course not, unless you're terribly bored or a have a strong opinion on the topic. People who volunteer their opinions are often very different from those who don't. That's why when you read opinions of recipes, books, and CDs online, you don't really know how to interpret them. You know that these are not randomly selected individuals voicing their concerns. The opinions belong to a special class of people who may differ in important ways from others. ■■■

Let's look at how the data were collected for the surveys that you'll be analyzing in this book.

General Social Survey

The General Social Survey (GSS) has been conducted regularly since 1972 by NORC, a social science research organization at the University of Chicago. The population of interest is all adults living in the U.S. but not in institutions such as mental hospitals and college dormitories. Members of the military are also excluded. A carefully trained interviewer visits each selected household and questions the chosen person, called the "respondent," about present and past experiences, behaviors, and opinions.

The methods that NORC uses to select people for inclusion in the sample is complex. First, a random sample of cities and counties is selected, then a random sample of neighborhoods, then a random sample of households, and finally a random adult person within each selected household. Each household in the U.S. has the same probability of being included in the survey. (However, since the number of adults in a household varies, the probability that a particular adult is selected is not the same for persons living in different-sized households. To simplify analyses in this book, we'll treat the General Social Survey sample as being equivalent to a random sample of the U.S. population.)

Random-Digit Dialing

Surveys that target the entire U.S. population are very expensive and costly, if, like the GSS, they require interviewers to visit each household. The telephone provides a much cheaper and quicker entry point to U.S. households. You don't need detailed maps of neighborhoods, enumeration of households, cars, or travel time. As you've undoubtedly noticed, telephone surveys are intrusive and annoying. They still require interviewers to ask the questions. However, people are much more likely to answer questions on the phone than return a mail questionnaire.

The most common technique for conducting telephone surveys is random-digit dialing (RDD). With RDD, all residential telephone numbers in the U.S. have an equal probability of being selected. Computers randomly generate the numbers and dial them. (That explains the annoying gap between when you pick up the phone and the interviewer asks about your welfare.) Of course, telephone surveys exclude households without phones, currently estimated to be about 7% of U.S. households, so strictly speaking all U.S. households are not eligible for inclusion. Also households with many phones are more likely to be included than households with only one phone. Since only one person is selected from a household, people who reside in households with many members are less likely to be included than people who reside alone or in smaller households. The increased use of cell phones and caller ID also complicates the problem. Random digit dialing of cell phones is currently not allowed, so people who rely only on their cell phones are automatically excluded, as are people who speak only to prescreened callers. Many telephone surveys are conducted only in English and thus exclude people who can't respond in English.

Even if a randomly selected telephone rings, there's no assurance that it will be answered or that the person answering it will cooperate with the interviewer in either answering questions or asking the selected household member to come to the phone. People who answer all telephone calls are different from people who screen calls or are seldom home. That's why most RDD surveys make repeated calls to numbers that are not answered. It's also not a "random" adult who answers the phone. Women are much more likely to answer phones than are men. That's why many RDD surveys ask to speak to household members in particular age groups or of a particular gender.

Data for the Impacts of the Internet on Public Library Use survey (Rodger et al., 2000) that you'll be analyzing were obtained by RDD of the 48 contiguous states. It was conducted in either English or Spanish. Only adults 18 years or older were included, since participation of minors requires parental consent as well as different types of questions. The Attitudes Toward Crime and Punishment in Vermont survey data (Doble and Green, 1999) were also obtained by RDD. Up to three callbacks were made to numbers that were not answered. Only adults aged 18 years and older were included. Additionally, the survey imposed a constraint that half of the sample were to be men and half women. The ABC Manners Express survey (ABC News, 1999) was also done with RDD.

Internet Surveys

The Internet is the latest technology being explored for survey research. Many sites have pop-up windows asking you to give your opinion on all kinds of subjects from your favorite hobby to the war against terrorism. The major drawback to using the Internet is obvious—only about half of U.S. households have Internet access. The households that have access differ from those that don't. (You'll see how when you analyze the GSS data.)

The Stanford Institute for the Quantitative Study of Society study (Nie et al., 2003) used an innovative combination of random-digit dialing and Web access technology to select their sample and to disseminate questionnaires. First, random-digit dialing was used to select a sample of approximately 6000 Americans aged 18 years and older. All subjects were then provided with Web TV to give them Internet access. Periodic questionnaires, including the time-diary study, were then conducted over the Internet. This approach provided the convenience of Internet questionnaires without restricting the population to people who already had Internet access.

Designing Experiments

Unlike a survey, an experiment involves actually doing something to people, animals, or objects. Instead of asking people whether they think chicken soup is effective for preventing colds, you give them chicken soup and measure how many colds they get. Sometimes you study the same subject before and after an experimental treatment and determine if there has been a change. Or you might take several groups of people, do something different to each of the groups, and then compare the results.

Experimentation on people and animals poses ethical questions that deserve careful thought. Human experimentation requires informed consent from participants. Risks must be carefully and exhaustively detailed. Institutions have committees that regulate experiments involving humans or animals, and the federal government has taken strong measures when institutions do not comply with guidelines for human experimentation.

In experiments as well as surveys, the subjects must come from the population that you're interested in. (This is a lot easier to do for animals than people!) To properly assess the effect of different treatments, you must ensure that the groups receiving the treatments are as similar as possible. The best way to do this is to use chance to assign subjects to the different groups. This doesn't guarantee that the groups will have the same characteristics, but it does minimize the bias that is introduced when treatments are assigned to subjects by the investigator. (A study is termed biased if it favors one result over another. For example, if healthier patients receive the new treatment while sicker patients receive the old treatment, the study is biased in favor of the new treatment.)

Random Assignment

Using chance to assign treatments to individuals does not mean that treatments are assigned haphazardly. As in surveys, chance requires a carefully selected systematic approach. If you want to study the effect of personal computer use on grades, you can't let classroom teachers decide which of their students receive personal computers and which do not. They might assign the computers only to the brighter students to reward them for past performance, or they might assign them only to the low-achieving students with the hope that the computers might help them. Any evaluation of the effect of the personal computers would be tainted by the differences between the students.

? *Wouldn't it be simpler to observe samples of people who have already experienced, and not experienced, the intervention that you're interested in?* It would be simpler but unless you assign people randomly to receive or not receive the intervention, you cannot be sure that the groups don't differ in other important ways. For example, if you want to study the effect of eating red meat on serum cholesterol, you can't simply compare people who eat red meat regularly with people who don't. Meat eaters may differ in many ways from people who don't eat red meat. Any difference you may observe between the groups might be attributable to factors such as income, alcohol consumption, or exercise. By assigning people randomly to treatment groups, you decrease the likelihood that the groups will differ in important ways before the start of the experiment. ■■■

Random numbers are the preferred method for assigning experimental treatments to subjects. You can't make up your own table of numbers that you think are random; you have to use random numbers generated by computers. Computers don't have birthdays, license plates, children, or any other reason to prefer one number over another. Every number has the same chance of being selected. A simple method for assigning people to experimental conditions is to assign everyone a random number and then put people with even numbers in one group and those with odd numbers in the other. Or you can order the numbers from smallest to largest and assign people with numbers in the lower half to one group and numbers in the upper half to the other. You can use all sorts of systems based on random numbers to assign subjects to groups, even in very complicated experimental designs.

? *Why is randomness so important? Does it really matter?*

Absolutely. Unless you use a procedure that assigns your subjects to experimental conditions randomly, the results of your study may be difficult or impossible to interpret. Many assignment schemes that appear random to the inexperienced researcher turn out to have hidden flaws. For example, researchers at a hospital compared two treatments for a particular disease. Patients who were admitted on even-numbered days received one treatment, and those admitted on odd-numbered days received the other. Sounds random, but it failed. The number of patients admitted with the disease on even days gradually became larger than the number admitted on odd days. Why? Physicians figured out the scheme and admitted their patients on days when the procedure that they preferred was being used. This introduced bias, since physicians were making the decisions as to which patients received which treatment. That may have resulted in less sick patients being admitted on even days and sicker patients on odd-numbered days, or vice versa. ■■■

Minimizing Bias

In experiments, just as in surveys, you must be careful not to let your prejudices influence the results. Some events, such as death, are not disputable. Others, such as "improvement," are not as clear. You must make sure that the endpoints that are being measured are well defined and unambiguous. Don't ask, "Are you better today?" Determine what constitutes better, and ask about the components, for example, freedom from pain, ability to sleep without interruption, performing activities of daily living, and so on.

The best way to make sure that measurements are obtained without bias is to make sure that neither the subject nor the evaluator is aware of the experimental procedure that a person has undergone. In medical studies, when a patient doesn't know which treatment he or she is receiving, the study is called **single blind**. If neither the researcher nor the patient knows, the study is termed **double blind**.

People respond favorably to any treatment that they think will help them, even if it's a sugar pill. That is known as the **placebo effect**. If you think you've been given a "magic" pill to help you stay awake in class, you may be more alert than someone who hasn't been given the pill, even if the pill is totally ineffective. That's why it's important to make sure that all people are treated as similarly as possible. If one groups gets a pill, so should the other, even if it contains only sugar.

If you're evaluating a new treatment, you should make sure to include a group that doesn't receive the new treatment, known as a **control group**. For medical studies, the control group might receive the usual treatment for a condition. In studies of new instructional methods, the control group would receive standard instruction. Unless you have a control group that is being observed at the same time and under the same conditions as your experimental group, you will not be able to draw unbiased conclusions about the new method. A "new" treatment may have better survival rates, not because it is better but because patients are being diagnosed earlier today than they were in the past. Similarly, students may perform better in statistics classes today, not because the teachers are better, but because students today are more industrious.

Summary

What's important when designing a survey or an experiment?

- You should carefully formulate the questions that you're going to ask.
- You must determine what the population of interest is and select a random sample of objects or people from that population.
- You must make sure that you don't bias your sample by making it more likely that some members are included than others.
- You must collect information in an objective fashion. The procedures for gathering information must be standardized.
- If several different experimental conditions are being compared, you must randomly assign the subjects to groups.
- You must prevent subjects and researchers from allowing their personal prejudices to influence the outcome of the investigation.

What's Next?

Designing surveys and experiments is a complicated undertaking. The role of this chapter was to not to teach you how to conduct studies, but to make you aware of the complexity of this area. In the next chapter, you'll actually start analyzing data.

Exercises

Statistical Concepts

1. A candidate for political office is interested in determining what percentage of a city's voters support him. He obtains bids from two survey organizations to conduct a poll. Both organizations plan to canvass about 1000 residents. The first, using a register of households, proposes to select a random sample of 400 household and then question all registered voters in the household. The second proposes to randomly select 1000 people from the population. Explain to the candidate which poll will probably be more informative and why.

2. The principal of the high school that you attended is interested in determining why some graduates are successful and others not. He commissions you to develop a plan for studying this question.

   a. What is your population?

   b. Discuss several ways for selecting a sample.

   c. How would you define "success"?

   d. If you do a mailing to graduates and receive questionnaires returned as "undeliverable," discuss the problem with excluding such people from your study.

   e. If you do a mail survey, how will you deal with people who do not return your form? That is, how will you deal with the problem of nonresponse?

   f. Comment on the strategy of distributing questionnaires at a class reunion.

3. If there are 50 children in a classroom, how would you randomly select 10 of them to participate in a study?

4. Which of the following procedures would you expect to result in a random sample of a city's adult population?

   a. Random-digit dialing. A computer places calls to randomly generated phone numbers.

   b. Random selection of 10 places of employment and then random selection of employees within each.

   c. Selection of every fifth person entering a grocery store.

   d. Randomly selecting children in all schools and then including their parents in the survey.

5. In the 1936 presidential race between Roosevelt and Landon, the *Literary Digest*, a magazine that ran the largest polls of that time, predicted a Landon victory on the basis of 2,376,523 mail questionnaires (out of about 10,000,000 mailed.) In fact, Roosevelt won by a margin of 19 percentage points. What possible reasons can you think for their missing the mark so badly?

6. In the 1954 clinical trial of the Salk polio vaccine, many different study designs were considered. Discuss the advantages and disadvantages of each of the following designs:

   a. Select a random sample of children and vaccinate them. Compare their polio rate to children in the U.S.

   b. Vaccinate children whose parents have volunteered them for the study and then compare the polio rate for vaccinated children with the rate for unvaccinated children in the same area.

   c. Vaccinate children in one city and compare their polio rate to children in another city.

7. U.S. employment and unemployment statistics are based on results from the monthly Current Population Survey. The question, "Were you unemployed?" is not included in the survey. Why not?

8. Suppose you want to determine how successful entrepreneurs differ from the general population.

   a. What problem do you see with taking a random sample of the general population?

   b. There are many more complex types of random samples than those in which all members of the population have the same chance of inclusion. (In all of them, however, every member of the population has a known, non-zero chance of inclusion. Analysis of the data from such surveys incorporates this information.) Suggest an alternative sampling strategy to deal with the problems that you mentioned in your response to question 8a.

Counting Responses 4

How can you summarize the various responses people give to a question?

- What is a frequency table, and what can you learn from it?
- How can you tell from a frequency table if there have been errors in coding or entering data?
- What are percentages and cumulative percentages?
- What are pie charts and bar charts, and when do you use them?
- When do you use a histogram?
- What are the mode and the median?
- What do percentiles tell you?

In the next three chapters, you'll analyze some of the Internet questions from the General Social Survey. You'll determine how many people use the Internet and for how long. You'll also see whether people who use the Internet differ from people who don't.

To determine how many people use the Internet, you have to count the responses they gave to the General Social Survey questions. Whenever you ask a number of people to answer the same questions, or when you measure the same characteristics for several people or objects, you want to know how frequently the possible responses or values occur. This can be as simple as just counting up the number of *yes* or *no* responses to a question. Or it can be considerably more complicated—for example, if you've asked people to report their annual income to the nearest penny. In this case, simply counting the number of times each unique income occurs may not be a useful summary of the data. In this chapter, you'll use the Frequencies procedure to summarize and display values for one variable at a time. You'll also learn to select appropriate statistics and charts to summarize different types of data.

▶ The data analyzed in this chapter are in the *gssnet.sav* data file. For instructions on how to obtain the Frequencies output shown in the chapter, see "How to Obtain a Frequency Table" on p. 65.

Describing Variables

To see what's actually involved in examining and summarizing data, you'll use the variables described in Table 4.1. They are based on data obtained from the General Social Survey. Some of the variables in the original General Social Survey have been modified to make them easier to analyze.

All of these variables are defined as numeric in SPSS, but in most cases the numbers are just codes for non-numeric information. Value labels for each variable specify what the codes really mean.

In the SPSS Data Editor, to display (or hide) value labels, from the menus choose:

View
Value Labels

Table 4.1 Variables based on the General Social Survey

Variable Name	Description
age	Age in years
educ	Years of education
usecomp	"Do you personally ever use a computer at home, at work, or at some other location?" (1=*Yes*, 0=*No*)
usenet	"Other than for e-mail, do you ever use the Internet or World Wide Web?" (1=*Yes*, 0=*No*)
usemail	E-mail use (1=*Yes*, 0=*No*)
emailhrs	"About how many minutes or hours per week do you spend sending and answering electronic mail or e-mail?" (The result is in given in hours.)
webhrs	"Not counting e-mail, about how many minutes or hours per week do you use the Web? Include time you spend visiting regular Web sites and time spent using interactive Internet services such as chat rooms, Usenet groups, discussion forums, bulletin boards, and the like." (The result is given in hours.)
nethrs	Sum of *emailhrs* and *webhrs*
netcat	Times on the Internet, grouped
region	Region of the interview

Start by looking at the variable *usenet*, which tells you if a person ever uses the Internet. Since there are only two possible responses, you can easily count how many people gave each of them.

A Simple Frequency Table

In Figure 4.1, you see the frequency table for the Internet use variable.

Figure 4.1 Frequency table of Internet use

To obtain this frequency table, from the menus choose:

Analyze
Descriptive Statistics ▶
Frequencies...

In the Frequencies dialog box, select the variable usenet, as shown in Figure 4.13.

		Frequency	Percent	Valid Percent	Cumulative Percent
Valid	0 No	255	25.9	26.1	26.1
	1 Yes	723	73.5	73.9	100.0
	Total	978	99.4	100.0	
Missing	9 Unknown	6	.6		
Total		984	100.0		

The response "Yes" was chosen by 723 people. This response is coded in the data file as the number 1.

From a frequency table, you can tell how frequently people gave each response. The first row is for the response *No* (coded in the data with the value 0). The second row is for the response *Yes* (coded in the data with the number 1). To determine how many people gave each response, look at the column labeled *Frequency*. You find that 723 people use the Internet and 255 do not. In the row labeled *Total*, you see that 978 people selected one of the two possible valid responses.

The second part of the table, labeled *Missing*, tells you how many people did not select one of the two valid answers. The row labeled *Unknown* is for responses that were illegible, lost, or not recorded by the interviewer. When the data file was defined, code 9, for *Unknown*, was identified as a missing-value code. In the *Frequency* column, you see that *Unknown* was recorded for six people.

In the last row of the frequency table, you see that a total of 984 people participated in the survey. Of these, six had a response identified as *Missing*. The other 978 provided a valid response.

? *Were there only 984 people included in the General Social Survey in 2004?* No. The General Social Survey was administered to 2812 people in 2004. However, not all people were asked the questions about Internet use. The file *gssnet.sav* contains only respondents who were asked the computer questions. For additional datasets on Internet use, visit the WebUse center maintained by the University of Maryland at *www.webuse.umd.edu.* ■■■

To display both values and value labels in a pivot table, from the menus choose:

Edit
  Options...

Click the Output Labels tab, and in the Pivot Table Labeling group, from the Variable Values in Labels Shown As drop-down list, select Values and Labels.

In the frequency table shown, the rows are identified with numeric codes as well as their value labels, which are descriptions of the codes assigned when you define a variable. If you don't assign these descriptions, only the codes are shown. If your codes are not inherently meaningful, you should assign value labels to them so that the output is easier to understand. Assigning a value label once is much easier than repeatedly having to look up the meanings of codes.

Only codes for responses actually selected by the participants are included in the frequency table. If the response *No* had not been selected by anyone, it would not be included in the table. Similarly, if you accidentally enter a code that does not correspond to a valid response—say a code of 4, 6, or 7 for the computer use variable—you will find it as a row in the frequency table. That's why frequency tables are useful for detecting mistakes in the data file. If you find wrong codes in your data values, you must correct the data file before proceeding.

? *Can you have only one code for missing responses?* No. In fact, it's a good idea to use different codes for different types of missing values. For example, if you do not ask all people all questions, you want to use one missing value code to indicate that a question was not asked, a different value code to indicate that the question was asked but the answer is not available, and a third code to indicate that a respondent refused to answer the question. You should always carefully examine the causes of missing data values because missing data can make the results of your survey impossible to interpret. ■■■

Percentages

To change the number of decimal places shown in the output, double-click the pivot table to activate it, select the cell or column of interest, and choose:

Format
  Cell Properties...

Then in the Format value tab change the Decimals

A frequency count alone is not a very good summary of the data. For example, if you want to compare your results to those of another survey, it won't do you much good to know only that 550 people in that survey use the Internet. From the count alone, you can't tell if the other survey's results are similar to yours. To compare the two surveys, you must convert the observed counts to percentages.

From a **percentage,** you can tell what proportion of people in the survey gave each of the responses. Unlike counts, percentages can be compared across surveys with different numbers of cases. You compute a percentage by dividing the number of cases that gave a particular response by the total number of cases. Then you multiply the result by 100.

In Figure 4.1, you find percentages in the column labeled *Percent*. Note that the people who gave the response *Yes* are 73.5% of the 984 people in your survey. The 255 people who gave the response *No* are 25.9% of your sample. The six people whose responses are not available are 0.6% of the total sample. The sum of the percentages over all the possible responses, including *Unknown,* is 100%. If the percentage of cases with missing values is large, including all cases in the denominator of the percentage calculations can result in percentages that are difficult to interpret.

Percentages Based on Valid Responses

Consider Figure 4.2, which shows how much time people spend online in a week. There are three missing value codes. A code of –1 is used for someone who doesn't use the Internet at all. These people were not asked how long they spend online in a week. An code of –3 is used when you don't know if someone uses the Internet. A code of –9 is used for Internet users whose time on the Internet is unknown.

In the column labeled *Percent*, you see that almost 15% of the sample used the Internet for more than 16 hours each week.

In the Frequencies dialog box, select the variable netcat, as shown in Figure 4.13.

Figure 4.2 Frequency table of weekly Internet use

		Frequency	Percent	Valid Percent	Cumulative Percent
Valid	1 4 hours or less	201	20.4	32.6	32.6
	2 4+ to 8 hours	136	13.8	22.0	54.6
	3 8+ to 16 hours	136	13.8	22.0	76.7
	4 16+hours	144	14.6	23.3	100.0
	Total	617	62.7	100.0	
Missing	-9 Time unknown	106	10.8		
	-3 Internet use unknown	6	.6		
	-1 Not internet user	255	25.9		
	Total	367	37.3		
Total		984	100.0		

This percentage is difficult to interpret. It can be small if few people in the sample use the Internet (many cases with missing values) or if most people who use the Internet use it for less than 16 hours. You want to distinguish between these two alternatives. The column labeled *Valid Percent* in Figure 4.2 shows that only 23.3% of people who use the Internet use it for more than 16 hours per week. (Pretty hard to believe, isn't it?!) The percentages shown in the *Valid Percent* column use only cases that have valid (nonmissing) responses in the denominator. These percentages are much easier to interpret than are the percentages based on all cases. They tell you what percentage of the people who gave valid responses gave each of them. Valid percentages sum to 100 over the valid response categories.

? *What if I want to treat Not Internet user as a valid response?* Go to the Data Editor, select Variable View, and find the variable *netcat*. Click in the cell that you use to define missing values. (It is in the column labeled *Missing*.) Click again to open the Missing Values dialog box. Remove the value –1 from the missing values list. At any point, you can change which values you want to consider as missing by adding or removing values from the missing values list. ■■■

Problems with Missing Data

Removing people who aren't asked a question from the calculation of percentages is not troublesome. They don't make interpretation of the results difficult. However, if a lot of people who are asked the question refuse to answer, that can be a problem. Consider the following situation. You conduct an employee satisfaction survey among 100 employees and find that 55 of them rate themselves as satisfied, four rate themselves as unsatisfied, and the remaining 41 decline to answer your question. That means that 55% of the polled employees consider themselves satisfied. However, if you exclude those who refused to answer from the denominator, 93% of the employees who answered the question consider themselves satisfied.

Which is the correct conclusion? Unfortunately, you don't know. It's possible that you have a company full of satisfied employees, many of whom don't like to answer questions. It's also possible that almost half of your employees are unhappy but are wary of voicing their dissatisfaction. When your data have many missing values because of people refusing to answer questions, it may be difficult, if not impossible, to draw correct conclusions. When you report percentages based on cases with nonmissing values, you should also report the percentage of cases that refused to give an answer.

Cumulative Percentages

There's one more percentage of interest in the frequency table. It's called the **cumulative percentage**. For each row of the frequency table, the cumulative percentage tells you the percentage of people who gave that response and any response that precedes it in the frequency table. It is the sum of the valid percentages for that row and all rows before it.

The cumulative percentage for an Internet usage time of eight hours or less is 54.6%. That means that about half of all people who were asked the question and who gave a valid response used the Internet for eight or fewer hours a week. Almost 77% of the people use the Internet for 16 or fewer hours a week. Cumulative percentages are most useful when there is an underlying order to the codes assigned to a variable. If you have a frequency table of religious preference, the codes used to designate religions are arbitrary, and it makes no sense to say that the cumulative percentage for Presbyterians is 40%.

Sorting Frequency Tables

Unless you specify otherwise, SPSS produces a frequency table in which the order of the rows corresponds to the values of the codes you assign to the responses. The first row is for the smallest code found in the data values, and the last is for the largest. Codes that have been declared missing are at the end of the table.

When you have several possible responses and the codes are not arranged in a meaningful order, you may want to rearrange the frequency table so that it's easier to use. You can determine the order of the rows in the table based on the frequency of values in the data. For example, Figure 4.3 shows a frequency table for the region of the country in which the interviews were conducted. The table is sorted in descending order of frequencies. Look at the column labeled *Frequency*. The frequencies go from largest to smallest.

In the Frequencies dialog box, click Format. Then select Descending counts, as shown in Figure 4.14.

Figure 4.3 Frequency table sorted by counts

		Frequency	Percent	Valid Percent	Cumulative Percent
Valid	South Atlantic	211	21.4	21.4	21.4
	E. Nor. Central	180	18.3	18.3	39.7
	Pacific	137	13.9	13.9	53.7
	Middle Atlantic	127	12.9	12.9	66.6
	West South Central	112	11.4	11.4	77.9
	Mountain	67	6.8	6.8	84.8
	W. Nor. Central	66	6.7	6.7	91.5
	E. South Central	57	5.8	5.8	97.3
	New England	27	2.7	2.7	100.0
	Total	984	100.0	100.0	

Table is sorted by the counts in the Frequency column

Sorting a frequency table will usually change the values in the *Cumulative Percent* column, since the cumulative percentages depend on the order of the rows in the table. In the table, you see that the largest number of interviews were conducted in the South Atlantic region, and the smallest number were conducted in New England.

Pie Charts

The information in a frequency table is easier to see if you turn it into a visual display, such as a bar chart or a pie chart. Figure 4.4 shows a pie chart of Internet usage. There is a "slice" for each row of the frequency table. The largest slice is for four hours or less.

If you have many small slices in a pie chart, you can combine them into an *other* category. For example, Figure 4.5 shows a pie chart for the region of the country. All regions with fewer than 10% of the respondents are combined into a single slice.

From the menus choose:

Graphs
Legacy Dialogs ▶
Pie...

Select Summaries for groups of cases. Move netcat into Define Slices by. Click OK. Activate the chart by double-clicking it. From the Elements menu, choose Show Data Labels. Move netcat into Displayed and click Apply.

Figure 4.4 Pie chart of weekly Internet use

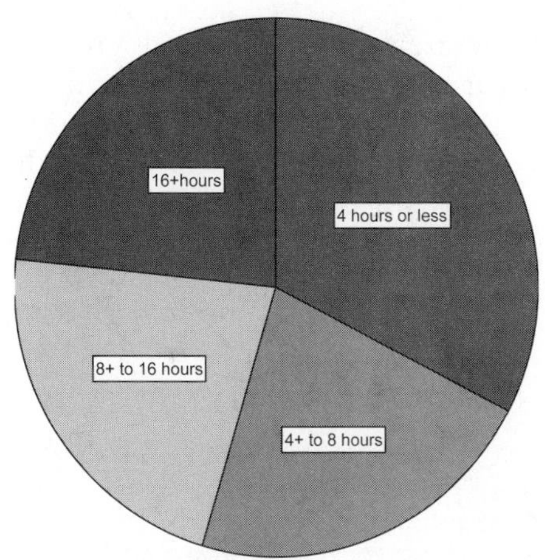

To collapse categories in a pie chart after it has been created, see "Modifying Chart Options" on p. 588 in Appendix A.

Figure 4.5 Pie chart of region of interview with categories collapsed

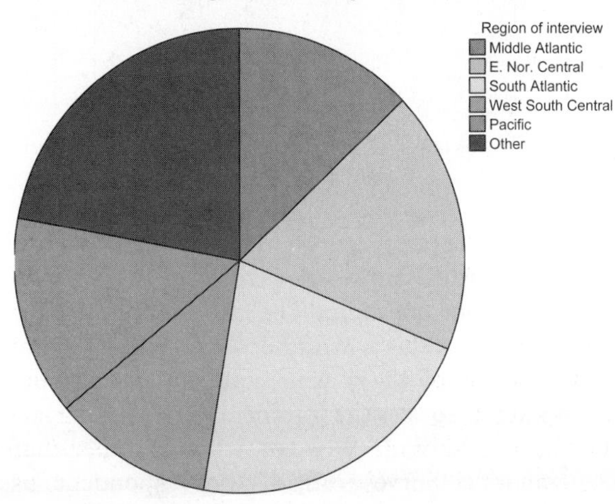

Bar Charts

In a pie chart, the size of the slice depends on the number of cases in the category. In a bar chart, the length of a bar depends on the number of cases in the category. Figure 4.6 shows a bar chart for the Internet use variable. You have as many bars as you did slices in the pie chart. The tallest bar is for people who use the Internet for four hours or less each week.

In the Frequencies Charts dialog box, select Bar charts, as shown in Figure 4.16. For chart values, select Percentages.

Figure 4.6 Bar chart of weekly Internet use

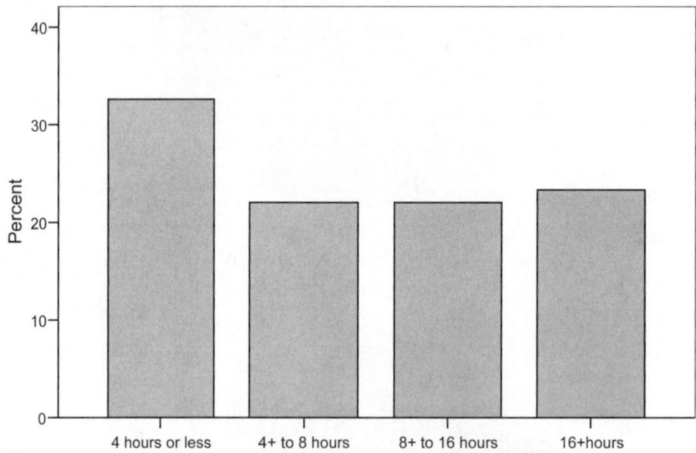

You can make pie charts and bar charts to display values other than just counts. Figure 4.7 shows the percentage of respondents who answered *Yes* to each of four questions. The information is much easier to understand when presented this way rather than in four separate frequency tables.

You see that 74% of your sample use a computer. Everyone who uses a computer also uses the Internet. Sixty-four percent use e-mail, and 70% use the World Wide Web for activities other than e-mail. In the 2000 General Social Survey, only 47% of respondents used the Internet!

Figure 4.7 Bar chart of computer and Internet use

Appendix A contains instructions for making this chart.

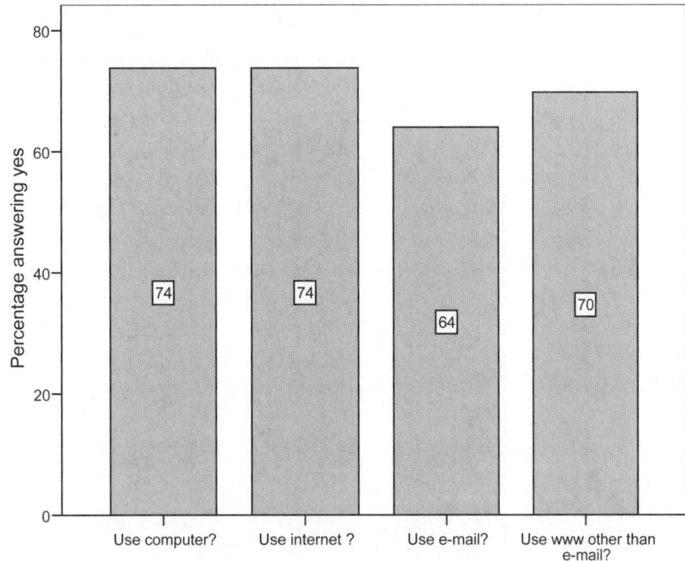

Summarizing Internet Time

Although you can produce frequency tables for any kind of data, a frequency table becomes less useful as the number of possible responses increases. For example, you can construct a frequency table for the actual number of hours people spend online in a week, as shown in Figure 4.8. (The beginning and end of the table are not shown to save space.) There is a row in the frequency table for every value of Internet time that occurs in the data file. Notice that in the frequency table, nice integer values of Internet time occur more frequently than fractional values. Do you think that weekly Internet use occurs primarily in hour units? Of course not. People don't really measure the exact time they spend on the Internet in a week. They provide a "guesstimate" to the nearest hour, even when they are asked for hours and minutes. For many variables, you'll find that larger numbers of cases occur at "nice" numbers if people are doing the measuring or reporting. For example, blood pressure values cluster at last digits of 0 and 5.

In the Frequencies dialog box, select nethrs.

Figure 4.8 Frequency table of weekly Internet use

		Frequency	Percent	Valid Percent	Cumulative Percent
Valid					
	4.75	1	.1	.2	33.7
	5.00	37	3.8	6.0	39.7
	5.12	1	.1	.2	39.9
	5.25	2	.2	.3	40.2
	5.50	4	.4	.6	40.8
	5.67	1	.1	.2	41.0
	6.00	30	3.0	4.9	45.9
	6.17	1	.1	.2	46.0
	6.50	1	.1	.2	46.2
	7.00	25	2.5	4.1	50.2
	7.50	1	.1	.2	50.4
	8.00	26	2.6	4.2	54.6
	Total	617	62.7	100.0	
Missing	-9.00 Time unknown	106	10.8		
	-3.00 Internet use unknown	6	.6		
	-1.00 Not internet user	255	25.9		
	Total	367	37.3		
Total		984	100.0		

Histograms

You won't find pie charts and bar charts of the *nethrs* variable to be useful either. There will be as many slices and bars as there are distinct times. The arrangement of the values in the charts can be troublesome as well. Both bar charts and pie charts arrange bars and slices in ascending order of the values. However, if a particular value doesn't occur, an empty space is not left for it. That means that in a bar chart, the bar for eight hours may be right next to the bar for 10 hours. You won't see a gap to remind you that values of 9 don't occur in your data.

A better display for a variable such as *nethrs*, for which it makes sense to group adjacent values, is a histogram. A **histogram** looks like a bar chart, except that each bar represents a range of values. For example, a single bar may represent all people who use e-mail between one and three hours a week. In a histogram, the bars are plotted on a numerical scale that is determined by the observed range of your data.

In the Frequencies dialog box, click Charts. Then select Histograms, as shown in Figure 4.16.

Edit the axes as described in Appendix A.

Figure 4.9 Histogram of weekly Internet use

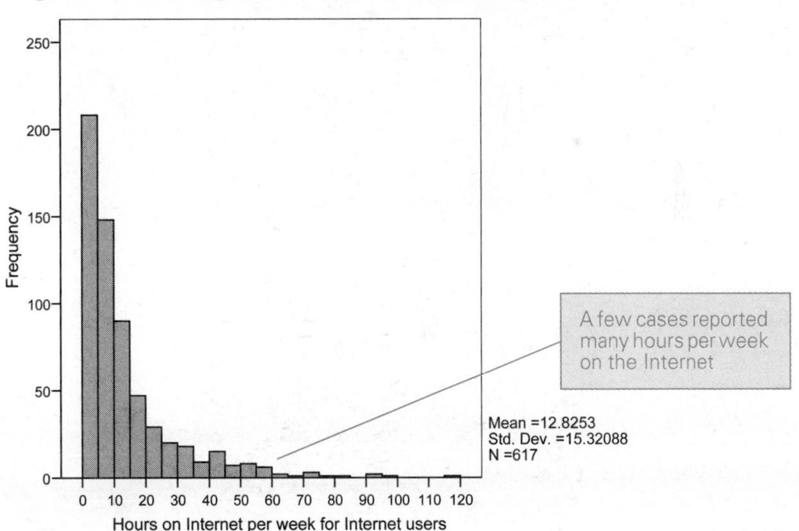

Figure 4.9 shows a histogram for the Internet time variable. Time values are on the horizontal axis, and frequencies are on the vertical axis. The first bar represents cases with values between 0 and 5. The middle value in this interval, the **midpoint**, is 2.5 hours. In the histogram, you see that over 200 cases had values in this interval. Similarly, the second bar represents cases with times between 5 and 10 hours. This bar represents about 150 cases.

? *Why are all of the values bunched together?* When SPSS makes a histogram, it chooses a scale based on the smallest and largest observed data values. If you have data values that are far removed from the rest, most of the distribution will be bunched together in order to fit the outlying data points on the same plot. You can edit the histogram to specify a narrower range for the values to be plotted. That will spread out your histogram. Figure 4.10 shows a histogram for weekly time on the Internet when 40 hours is the largest time plotted. You see more detail now. Note, however, that the statistics displayed on the histogram are for all cases, even if they are excluded from the histogram. ■■■

Figure 4.10 Truncated histogram of weekly Internet use

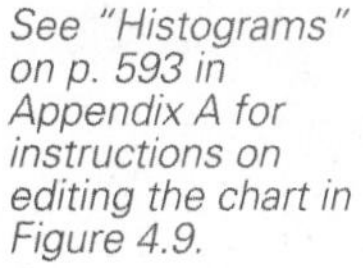
See "Histograms" on p. 593 in Appendix A for instructions on editing the chart in Figure 4.9.

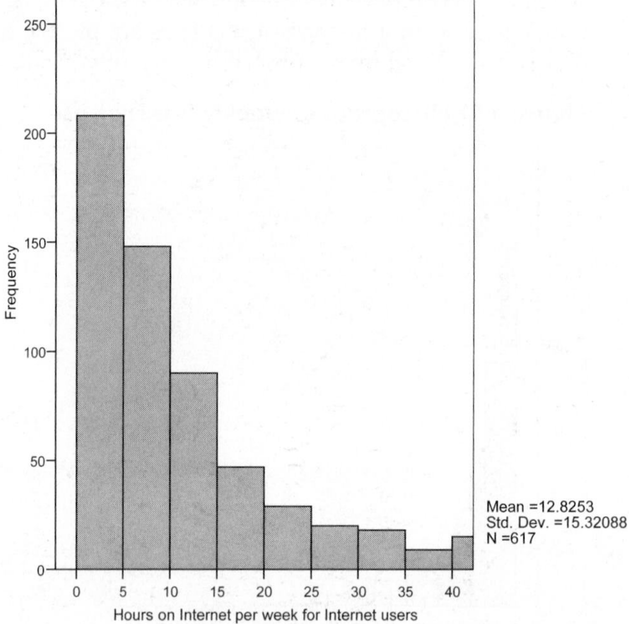

A histogram tells you about the distribution of the data values. That is, it tells you how likely various values are. From it, you can see whether the cases cluster around a central value. You can also see whether large and small values are equally likely and whether there are values far removed from the rest. Make sure to examine any unusual values to determine that they are not the result of errors in data recording or data entry. This is important not only for understanding the data you've collected but also for choosing appropriate statistical techniques for analyzing them. In

Figure 4.9, you see that the Internet time distribution has a peak in the first interval. Additionally, you can see that the distribution of times is not symmetric but has a "tail" extending to the longer times. That's because the smallest value of Internet time possible is restricted to 0, while large values are constrained only by the number of hours in a week.

? *What's a symmetric distribution?* A distribution is **symmetric** if a vertical line going through its center divides it into two halves that are mirror images of each other. Figure 4.11 shows an approximately symmetric distribution of a hypothetical age variable. Note that small and large values of age are almost equally likely. ■■■

Figure 4.11 Approximately symmetric distribution

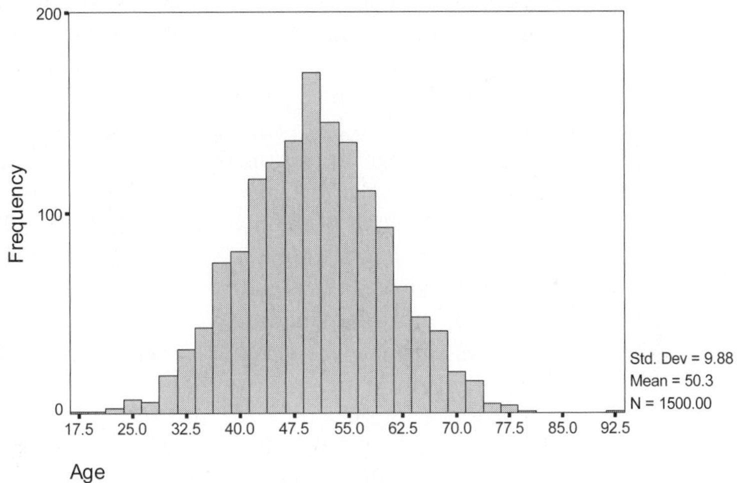

Mode and Median

You can use a variety of statistics to further summarize the information in a frequency table. In Chapter 5, you'll learn about a large number of such summary statistics. In the remainder of this chapter, you'll focus on summary measures that are easily obtained from the frequency table.

The **mode** is defined as the most frequently occurring value in your data. In the frequency table in Figure 4.8, the mode is three hours spent on the Internet, since that's the row that has the largest frequency (38 cases). About 6% of all people who use the Internet said they spend three hours a week online. The mode is not a very good summary measure for a variable that can have many values, since several values can be tied for "largest frequency" and the frequency need not represent a large percentage of the cases. For example, five hours was also selected by nearly 6% of the respondents. There's little reason to prefer three hours to five hours as a summary measure.

? *What is the mode for the region variable?* In Figure 4.3, you see that *South Atlantic* is the mode for the region of the country, since it was mentioned the most frequently. When a frequency table is sorted by descending counts, the first value is always a mode. If succeeding values have the same count, they are also modes. ■■■

Scales on which variables are measured are discussed in Chapter 5.

If you can meaningfully order your data values from smallest to largest, you can compute additional summary measures. These measures are better than the mode, since they make use of the additional information about the order of the data values. For example, the **median** is the value that is greater than half of the data values and less than the other half.

You calculate the median by finding the middle value when values for all cases are ordered from smallest to largest. If you have an odd number of cases, the median is just the middle value. If you have an even number of cases, the median is the value midway between the two middle ones. For example, the median of the five values 12, 34, 57, 92, and 100 is 57. For the six numbers 13, 20, 40, 60, 89, and 123, the median is 50, because 50 is the value midway between 40 and 60, the two middle numbers. (Add the two middle values and divide by 2.)

To display the median and mode along with your frequency table, in the Frequencies dialog box, click Statistics. Then select Median and Mode (see Figure 4.15).

You can calculate the median very easily from a frequency table. Find the first value for which the cumulative percentage exceeds or is equal to 50%. For the Internet time variable in Figure 4.8, the median is seven hours. That means that half of the people in your sample who use the Internet do so for seven hours or less a week. That's much more useful information than knowing that three hours is the mode.

Percentiles

When you calculate the median, you find the number that splits the sample into two equal parts. Half of the cases have values smaller than the median, and the other half have values larger than the median. You can compute values that split the sample in other ways. For example, you can find the value below which 25% of the data values fall. Such values are called **percentiles,** since they tell you the percentage of cases with values below and above them. Twenty-five percent of the cases have values smaller than the 25th percentile, and 75% of the cases have values larger than the 25th percentile. The median is the 50th percentile, since 50% of the cases have values less than the median, and 50% have values greater than the median. Together, the 25th, 50th, and 75th percentiles are known as **quartiles**, since they split the sample into four groups with approximately equal numbers of cases.

In the Frequencies dialog box, click Statistics. Then select Quartiles, as shown in Figure 4.15.

Figure 4.12 Percentiles for Internet variables

		NETHRS Hours on Internet per week for Internet users	EMAILHRS Hours of e-mail per week for Internet users	WEBHRS Hours on the WWW per week for Internet users
N	Valid	617	617	617
	Missing	367	367	367
Percentiles	25	3.0000	1.0000	1.0000
	50	7.0000	2.0000	4.0000
	75	15.5000	7.0000	8.0000

Figure 4.12 shows the 25th, 50th, and 75th percentiles for the Internet variables. They are computed only for people who use the Internet. In Figure 4.12, you see that 25% of people use the Internet for three hours or less. Only 25% of people use it for 15.5 hours or more. The median time for e-mail use is two hours. For World Wide Web use, it is four hours a week.

Although frequency tables are a simple statistical tool, by using them you can learn much about your data. From simple tables, you learned that about three quarters of your sample use computers and the Internet. Slightly less, 64%, use e-mail. From the percentile values, you saw that the reported median weekly time on the Internet is about seven hours. The median weekly time spent on e-mail is two hours.

Summary

How can you summarize the various responses people give to a question?

- A frequency table tells you how many people (cases) selected each of the responses to a question. It contains the number and percentage of the people who gave each response, as well as the number of people for whom responses are not available.
- If you find codes in the frequency table that weren't used in your coding scheme, you know that an error in data coding or data entry has occurred.
- A count can be transformed into a percentage by dividing it by the total number of responses and multiplying by 100.
- A cumulative percentage is the percentage of cases with values less than or equal to a particular value.
- Pie charts and bar charts are graphical displays of counts.
- A histogram is a graphical display of counts for ranges of data values.
- The mode is the data value that occurs most frequently.
- The median is the middle value when data values are arranged from smallest to largest.
- Percentiles are values below which and above which a certain percentage of case values fall.

What's Next?

In this chapter, you learned how to use a frequency table to summarize the values of a variable with a small number of distinct categories. You also learned about pie charts and bar charts, which are visual displays of a frequency table. You used a histogram to summarize a variable whose values can be meaningfully ordered from smallest to largest. You also saw how to use a frequency table to compute several summary statistics. In the next chapter, you'll learn about additional statistics that are useful for describing the values of a variable. You'll also learn about the scales on which variables are measured.

How to Obtain a Frequency Table

This section shows you how to use SPSS to count how frequently each value of a variable occurs in your data. The Frequencies procedure tabulates the different values that occur for a variable and produces statistics and charts based on these tabulations. In addition, the Frequencies command can:

- Calculate percentages of cases having each value of a variable.
- Calculate descriptive statistics for individual variables ("univariate" statistics).
- Produce high-resolution bar charts, pie charts, and histograms showing the distribution of values for individual variables.

▶ To open the Frequencies dialog box (see Figure 4.13), from the menus choose:

Analyze
  Descriptive Statistics ▶
    Frequencies...

Figure 4.13 Frequencies dialog box

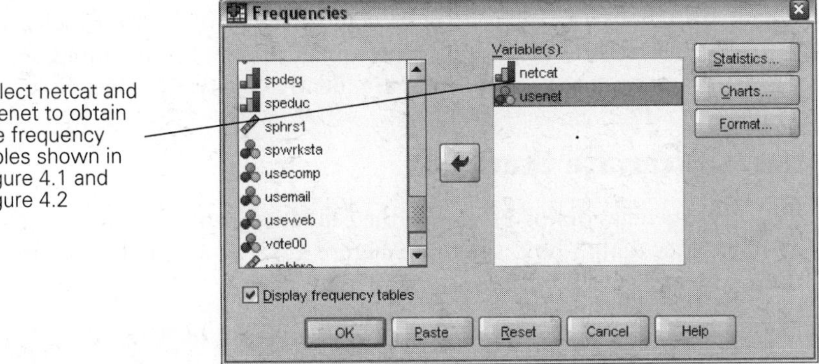

▶ In the Frequencies dialog box, select one or more variables and move them into the Variable(s) list. Make sure that the **Display frequency tables** option is selected, and click **OK**.

This produces frequency tables like those in Figure 4.1 and Figure 4.2, showing for each value the value label; the frequency, or count, of cases; the percentages of all the cases and all the valid (nonmissing) cases; and the cumulative percentages. The **Display frequency tables** option lets you display or suppress the actual frequency table. Deselect this option when you are interested only in the statistics or charts and not in the tabulation. For variables with a great many different values, such as age or weight, the detailed tabulation is long and not very interesting.

Format: Appearance of the Frequency Table

In the Frequencies dialog box, click the **Format** button to change the appearance of the frequency tables. The Frequencies Format dialog box, shown in Figure 4.14, controls the order in which values appear.

Figure 4.14 Frequencies Format dialog box

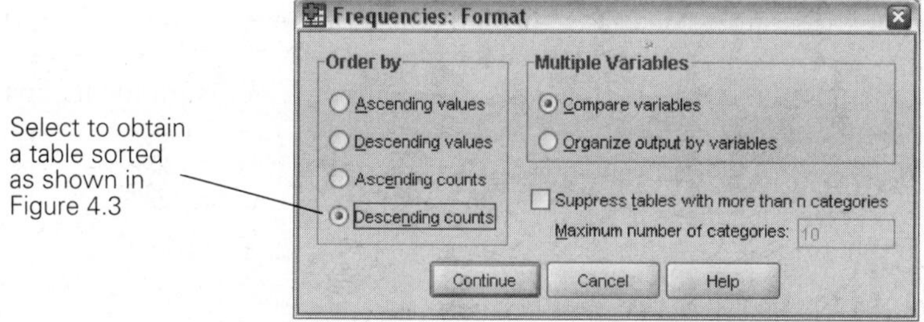

Order by. Choose an alternative to determine the order by which data values are sorted and displayed in a frequency table. For example, select Descending counts to see the most frequently occurring values first.

Statistics: Univariate Statistics

In the Frequencies dialog box, click the **Statistics** button. In the Frequencies Statistics dialog box, shown in Figure 4.15, you can choose any of the statistics discussed in this chapter.

Figure 4.15 Frequencies Statistics dialog box

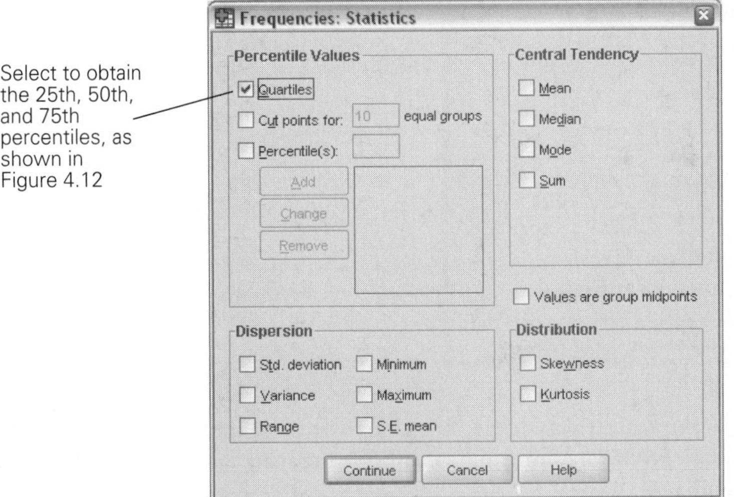

Some of the available options are described below:

Percentiles. To request percentiles, select Percentile(s), click in the text box beside it, type a percentile (from 1 to 99), and then click the Add button to add it to the list. Once the list has numbers in it, you can select one of them and either click Remove to remove it from the list or change the percentile number and click Change.

Values are group midpoints. Select if the data values represent ranges of values. For example, if the value 25 is used for all cases with values between 20 and 30, select this check box. This affects the computation of percentiles.

Charts: Bar Charts, Pie Charts, and Histograms

In the Frequencies dialog box, click the Charts button. In the Frequencies Charts dialog box, shown in Figure 4.16, you can request bar charts, pie charts, or histograms for the selected variables. If you select histograms, there's an option to sketch in a normal curve over the histogram, in case you want to compare your variable's distribution to the normal distribution discussed in Chapter 11.

Figure 4.16 Frequencies Charts dialog box

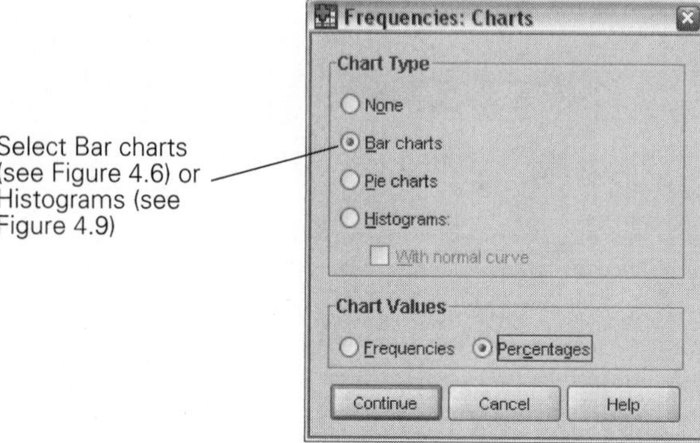

This is the fastest way to get bar charts or histograms for more than one variable. For more information about creating and editing charts, see Appendix A.

Exercises

Statistical Concepts

1. For which of the following variables would frequency tables be useful?

  a. Business miles driven per year

  b. Systolic blood pressure

  c. Income in dollars

  d. Happiness with marriage

  e. Square feet of office space

  f. CEO salaries in dollars

  g. Region of the country

  h. Number of cars owned

2. For which of the following variables would cumulative percentages be readily interpretable?

  a. Number of adults in a household

  b. Brand of car ownership

  c. College major

  d. Number of illnesses during the past year

3. The following data represent the number of periodicals read by 25 college students: 1, 1, 1, 1, 1, 1, 2, 2, 2, 3, 3, 3, 3, 3, 3, 4, 4, 5, 5, 5, 5, 8, 9, 9, 10.

a. Fill in the frequency table:

PERIOD Number of Periodicals

		Frequency	Percent	Valid Percent	Cumulative Percent
Valid	1				
	2				
	3				
	4				
	5				
	8				
	9				
	10				
	Total	25	100.0	100.0	

b. Using the same data, fill in the following bar chart:

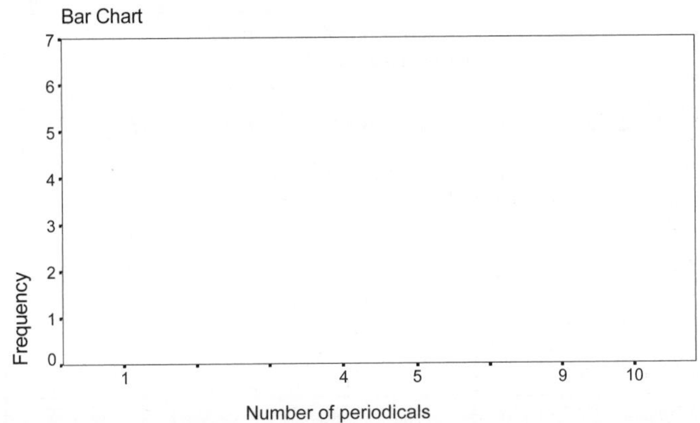

c. Why are there no bars for 6 and 7 periodicals?

d. Fill in the following histogram:

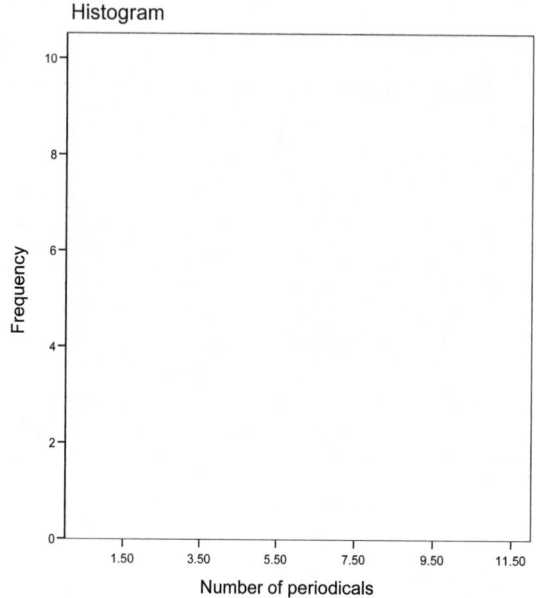

e. How does the histogram differ from the bar chart?

f. What is the mode for the number of periodicals read?

g. What is the median?

4. The following table contains the results of a Harris poll of 1254 adults (*Business Week*). The question asked was: How safe is your money in a savings and loan?

		Frequency	Percent	Valid Percent	Cumulative Percent
Valid	Very safe	351			
	Somewhat safe	414			
	Not very safe	251			
	Not at all safe	188			
	Total	1204			
Missing	Not sure	50			
Total		1254			

a. Fill in the missing entries.

b. For the previous question, explain whether *Not sure* should be considered a missing value. If it's not a missing value, where would you place it on the scale from *Very safe* to *Not at all safe*?

c. Sketch a pie chart.

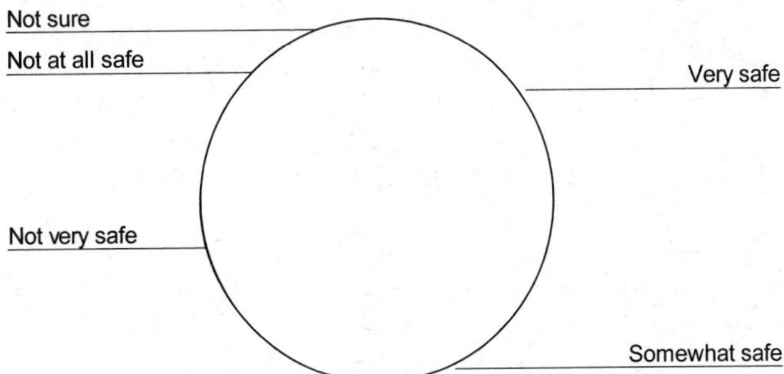

5. Without using the computer, construct a frequency table for the "gift" variable in Data Analysis question 1 (below).

Data Analysis

1. Twenty-five children were asked to sample each of three brands of cereals: Canary Crunch, Turtle Treats, and Ghostly Shadows. They were asked to choose their favorite and indicate how much they liked it (crazy about it, liked it, didn't particularly like it). They were also asked to select which of four "gifts" they'd like to find in it: a marble, a squirt gun, a whistle, or a magic ring. Here are their responses:

Child	Cereal	Like	Gift	Gender
1	Ghostly Shadows	crazy	squirt gun	M
2	Ghostly Shadows	like	squirt gun	M
3	Ghostly Shadows	not part	squirt gun	M
4	Canary Crunch	not part	ring	M
5	Turtle Treats	crazy	squirt gun	F
6	Turtle Treats	crazy	ring	F
7	Turtle Treats	crazy	squirt gun	F
8	Turtle Treats	like	ring	F
9	Ghostly Shadows	crazy	ring	M
10	Canary Crunch	not part	squirt gun	F
11	Turtle Treats	crazy	squirt gun	F
12	Ghostly Shadows	like	ring	F
13	Turtle Treats	crazy	squirt gun	M
14	Turtle Treats	like	ring	M
15	Ghostly Shadows	crazy	whistle	F
16	Canary Crunch	don't know	ring	M
17	Turtle Treats	crazy	whistle	F
18	Turtle Treats	like	ring	F
19	Ghostly Shadows	like	squirt gun	F
20	Turtle Treats	crazy	can't decide	M
21	Canary Crunch	like	ring	F
22	Turtle Treats	crazy	squirt gun	M
23	Ghostly Shadows	like	ring	F
24	Turtle Treats	like	ring	F
25	Turtle Treats	crazy	ring	F

Enter the data into SPSS and use the computer to answer the following questions:

a. What percentage of the sample are males and what percentage are females?

b. Which cereal was preferred by most children?

c. Based on the sample, which gift would you include in the cereal boxes? Explain the basis for your choice.

d. What percentage of the children were crazy about their favorite cereal? Why are the percentage and valid percentage columns in the frequency tables you obtained not the same?

Use the *gss.sav* data file to answer the following questions:

2. Write a short paper describing the General Social Survey (GSS) respondents in terms of their education (variable *degree*), whether they voted in the presidential election in 1996 (*vote96*), and for whom they voted in the 1996 presidential election (*pres96*). Include appropriate tables and charts. Don't just include the frequency tables generated by SPSS. Modify the tables so that they include only the appropriate information.

3. In the frequency table for *vote96*, there are two different codes for missing values. Do you think any additional codes should be considered missing? Which ones? Change them to missing and rerun the frequency table. Which of these missing value categories would concern you most if you want to analyze voting behavior?

4. How happy are GSS respondents? Write a short paper describing their responses to the following questions:

a. Taken all together, how would you say things are these days—would you say that you are very happy, pretty happy, or not too happy (variable *happy*)?

b. Taking things all together, how would you describe your marriage? Would you say that your marriage is very happy, pretty happy, or not too happy (*hapmar*)?

c. In general, do you find life exciting, pretty routine, or dull (*life*)?

5. Write a short report about the educational attainment of the GSS respondents. Examine the variables *educ* (years of education) and *degree* (highest degree earned).

6. Consider the variables *nethrs, webhrs, emailhrs*, and *hrs1*, the number of hours worked last week. Write a report, including graphs, that characterize how much time adults in the United States spend on each of these activities in 1996.

7. Make a frequency table of the variable *zodiac.* Are the number of people in each of the signs roughly the same? Would you expect them to be if births were equally distributed over the year? Would you use a bar chart or histogram to summarize birth signs in the sample?

8. Make a frequency table of *rincome* (respondent's income) and *income* (family income). What are the quartiles for each of the variables? Use the RECODE command to create new variables with values that correspond to the quartiles. That is, people with incomes less than or equal to the first quartile value are coded as 1, people with incomes in the second quartile are coded as 2, and so on. Make a frequency table of this new variable. Are approximately a quarter of the cases in each of the quartiles? Explain.

9. The variable *tvhours* tells you how many hours per day GSS respondents say they watch television.

a. Make a frequency table of the hours of television watched. Do any of the values strike you as strange? Explain. (Since you didn't conduct the survey yourself, there's no way for you to check whether the strange values are correct. You'll have to assume that they are.)

b. Based on the frequency table, answer the following questions: Of the people who answered the question, what percentage don't watch any television? What percentage watch two hours or less? Five hours or more? Of the people who watch television, what percentage watch one hour? What percentage watch four hours or less?

c. From the frequency table, estimate the 25th, 50th, 75th, and 95th percentiles. What is the value for the median? The mode?

d. Make a bar chart of the hours of television watched. What problems do you see with this display?

e. Make a histogram of the hours of television watched. What causes all of the values to be clumped together? Edit the histogram so that the smallest value displayed is 0 hours, the maximum is 24 hours, and the length of the interval is 1 hour. Compare this histogram to the bar chart you generated in question 4d. Which is a better display for these data?

10. The file *gss.sav* contains results from the year 2000 General Social Survey. Consider the Internet variables *nethrs, webhrs, emailhrs, useweb, usenet, and usemail.* Discuss how Internet usage has changed between the 2000 survey and the 2004 survey (*gssnet.sav*). Make appropriate charts to support your conclusions.

Use the *crimjust.sav* data file to answer the following questions:

11. Thirteen different crimes were described to a random sample of Vermont residents (variables *crime1* to *crime13.)* For each crime, respondents indicated whether the offender should go to prison or to a reparative board, under the supervision of the community. Write a report summarizing the citizens' opinions of how each crime should be handled. Classify the crimes into severity categories based on how often citizens think the offenders should go to prison.

12. If many people refuse to give an answer or claim not to know what answer to choose, you may have a problem drawing correct conclusions from your survey. For the crime scenarios, did many people fail to select one of the two answers? Do you think they present a serious problem for interpreting the results of the survey?

13. To see what actions Vermont citizens think the reparative boards should take, consider the variables *comserv* (perform community service), *essay* (write an essay), *classes* (attend classes), *apology* (apologize), and *restitut* (provide restitution). Prepare a briefing report to the governor outlining what steps people think the reparative boards should take. Include appropriate graphs and tables. (Don't just send a bunch of frequency tables; some governors have trouble reading output from software packages.) Include in your summary an analysis of how the citizens rate the criminal justice system overall (*cjsrate*) and whether they think crime (*crmcomp5*) and violent crime (*vlntcmp5*) have changed over the past five years.

14. Compare the age and education of the sample of Vermont residents to those of the sample of residents of the United States (*gssnet.sav*). Summarize and explain your findings. If you see large differences, speculate what might cause them.

Use the *manners.sav* data file to answer the following question. The file contains results from an ABC survey about manners in the United States. To figure out what's in the data file, from the Utilities menu choose File Info. Now you know what variables are in the data file.

15. Obtain frequency tables of the answers and write a short feature story on what people think about manners.

Use the *salary.sav* data file to answer the following questions:

16. Consider the distribution of employees within each job category (variable *jobcat*).

a. Describe the distribution of people in each employment category.

b. Make a pie chart that shows the percentages of employees within each job category. Modify the chart so that MBA trainees and technical employees are shown in a single slice.

c. Make a bar chart showing the distribution of employees within job categories. Have the bars represent the percentage of employees in each category.

17. Look at the *sexrace* variable.

a. What percentage of the bank's employees are males? White males? Minority females?

b. Make a pie chart of the distribution of the race/gender categories.

c. Select only clerical workers (variable *jobcat* equals 1; see "Case Selection" on p. 617 in Appendix B). Make a pie chart of the distribution of the race/gender categories for clerical workers only. Does the pie chart look similar to the one you obtained in question 16b? Comment on the similarities and differences.

18. Consider work experience (variable *work*).

a. Make a bar chart and a histogram of the values. Which do you think is the better summary of the data? Why?

b. What is the median number of years of work experience for the sample? The quartiles?

c. What percentage of the sample had no prior work experience? Of those who did have prior work experience, what is the median number of years of experience?

Use the *electric.sav* data file to answer the following questions:

19. Make a frequency table of the first coronary heart disease event (variable *firstchd*).

a. The sample in this file was selected in such a way that half of the 240 cases had not experienced coronary heart disease within 10 years, and half had. In the sample, what percentage of cases experienced sudden death? Either fatal or non-fatal myocardial infarction (MI)?

b. Do you agree or disagree with the statement "Half of the men in the study experienced coronary heart disease within 10 years." Explain your reasoning.

c. What percentage of the deaths in your sample were attributable to fatal MI? To sudden death?

d. What percentage of men who experienced coronary heart disease experienced fatal MI's?

e. What percentage of men who experienced coronary heart disease experienced sudden death?

f. Make a bar chart that shows the distribution of type of coronary heart disease events for men who experienced coronary heart disease. (To select only men who experienced coronary heart disease, in the Data Editor use **Select Firstchd ne 1**.) Write a paragraph describing your findings.

20. Make a frequency table for *firstchd* when only men who experienced coronary heart disease are selected. Explain why the percentages in this table differ from those in the table in question 18. Which percentages are more useful?

21. Make a frequency table of the number of men who died on each day of the week (variable *dayofwk*). (The day of death was obtained for all causes of death, not just for coronary heart disease.)

a. Of the men who died, what percentage died on Sunday?

b. What percentage of all men in your sample died on Sunday?

c. Explain why the percentages in questions 20a and 20b differ. Which percentage is more useful if you are studying the likelihood of deaths on different days of the week?

d. Make a pie chart of day of death. Summarize your findings.

e. What is the mode for day of death?

22. Generate the appropriate statistics and displays to summarize the distribution of cholesterol values and diastolic blood pressures in 1958 (variables *chol58* and *dbp58*). Briefly summarize your results.

a. Fifteen percent of the men had serum cholesterol values below what value? Ninety percent of men had serum cholesterol levels above what value?

b. What are the quartiles for diastolic blood pressure? Within what values did the middle 50% of the men fall?

5 Computing Descriptive Statistics

How can you summarize the values of a variable?

- What are scales of measurement, and why are they important?
- How does the arithmetic mean differ from the mode and the median?
- When is the median a better measure of central tendency than the mean?
- What does the variance tell you? The coefficient of variation?
- What are standardized scores, and why are they useful?

In the previous chapter, you used frequency tables, bar charts, pie charts, histograms, and percentiles to determine what percentage of your sample used the Internet, e-mail, and the World Wide Web and for how many hours a week. Frequency tables are essential for getting acquainted with the data. You should always display and visually examine your data before embarking on any more complicated analysis. Often, however, you want to summarize the information even further by computing summary statistics that describe the "typical" values, or the **central tendency**, as well as how the data spread out around this value, or the **variability.** In this chapter, you'll learn how to use the Frequencies and Descriptives procedures to compute the most commonly used summary statistics for central tendency and variability.

▶ This chapter continues to use the *gssnet.sav* data file. For instructions on how to obtain the Descriptives output discussed in the chapter, see "How to Obtain Univariate Descriptive Statistics" on p. 95.

? *What's a statistic?* Often when you collect data, you want to draw conclusions about a broader base of people or objects than are actually included in your study. For example, based on the responses of people included in the General Social Survey, you want to draw conclusions about the population of adults in the United States. The people you observe are called the **sample.** The people you want to draw conclusions about are called the **population.** A **statistic** is some characteristic of the sample. For example, the median age of people in the General Social Survey is a statistic. The term **parameter** is used to describe characteristics of the population. If you had the ages of all adults in the United States, the median age would be called a parameter value. Most of the time, population values, or parameters, are not known. You must estimate them based on statistics calculated from samples. ■■■

Summarizing Data

Consider again the data described in the previous chapter. Suppose you want to summarize the data values further. You want to know the typical time spent on the Internet for participants in the survey, their typical age or education, or the typical number of hours worked in a week for Internet users. A unique answer to these questions doesn't exist, since there are many different ways to define "typical." For example, you might define it as the value that occurs most often in the data (the mode), or as the middle value when the data are sorted from smallest to largest (the median), or as the sum of the data values divided by the number of cases (the arithmetic mean). To choose among the various measures of central tendency and variability, you must consider the characteristics of your data as well as the properties of the measures. Although the mode may be a plausible statistic to report for region of the country, it may be a poor selection for a variable like time per week on the Internet.

Scales of Measurement

One of the characteristics of your data that you must always consider is the scale on which they are measured. Scales are often classified as nominal, ordinal, interval, and ratio, based on a typology proposed by Stevens (1946). A nominal scale is used only for identification. Data measured on a **nominal scale** cannot be meaningfully ranked from smallest to largest. For example, status in the work force is measured on a nominal scale, since the codes assigned to the categories, although numeric, don't really mean anything. There is no order to *retired, in school, keeping house,* and *other*. Place of birth, hair color, and favorite statistician are all examples of variables measured on a nominal scale.

Variables whose values indicate only order or ranking are said to be measured on an **ordinal scale**. Job satisfaction and job importance are examples of variables measured on an ordinal scale. There are limitations on what you can say about data values measured on an ordinal scale. You can't say that someone who has a job satisfaction rating of 1 (*very satisfied*) is twice as satisfied as someone with a rating of 2 (*moderately satisfied*). All you can conclude is that one person claims to be more satisfied than the other. You can't tell how much more. The variable *netcat,* described in Chapter 4, is also measured on an ordinal scale. That's because time on the Internet is grouped into unequal categories. A person in category 2 doesn't spend twice the time on the Internet as a person in category 1. If you record people's actual annual incomes, you are measuring income on what is called a **ratio scale**. You can tell how much larger or smaller one value is compared with another. The distances between values are meaningful. For example, the distance between incomes of $20,000 and $30,000 is the same as the distance between incomes of $70,000 and $80,000. You can also legitimately compute ratios of two values. An income of $50,000 is twice as much as an income of $25,000. Age and years of education are both examples of variables measured on a ratio scale.

When the Data Editor is in Variable View, you can click on a cell in the Measure column to get a drop-down arrow to define the level of measurement. Select Scale for variables measured on an interval or ratio scale. In variable lists, SPSS displays an icon based on the level of measurement for a variable.

An **interval scale** is just like a ratio scale except that it doesn't have an absolute zero. You can't compute ratios between two values measured on an interval scale. The standard example of a variable measured on an interval scale is temperature. You can't say that a 40°F day is twice as warm as a 20°F day. Few variables are measured on an interval scale, and the distinction between interval and ratio scales is seldom, if ever, important in statistical analyses.

Although it is important to consider the scale on which a variable is measured, statisticians argue that Stevens' typology is too strict to apply to real-world data (Velleman and Wilkinson, 1993). For example, an identification number assigned to subjects as they enter a study might appear to be measured on a nominal scale. However, if the numbers are assigned sequentially from the first subject to enter the study to the last, the identification number is useful for seeing whether there is a relationship between some outcome of the study and the order of entry of the subjects. If the outcome is a variable like how long it takes a subject to master a particular task, it's certainly possible that instructions have improved during the course of a study and later participants fare better than earlier ones.

It's an oversimplification to conclude that the measurement scale dictates the statistical analyses you can perform. The questions that you want to be answered should direct the analyses. However, you should always make sure that your analysis is sensible. Using the computer, it's easy to calculate meaningless numbers, such as percentiles for place of birth or the median car color. In subsequent discussion, we'll occasionally refer to the scale of measurement of your data when describing various statistical techniques. These are not meant to be absolute rules but useful guidelines for performing analyses.

Mode, Median, and Arithmetic Average

The mode, median, and arithmetic average are the most commonly reported measures of central tendency. In Chapter 4, you saw how to compute the mode and median. You calculate the **mode** by finding the most frequently occurring value. The mode, since it does not require that the values of a variable have any meaning, is usually used for variables measured on a nominal scale. The mode is seldom reported alone. It's a useful statistic to report together with a frequency table or bar chart. You can easily find fault with the mode as a measure of what is typical. Even accompanied by the percentage of cases in the modal category or categories, it tells you very little.

If you are summarizing a variable whose values can be ranked from smallest to largest, the median is a more useful measure of central tendency. You calculate the **median** by sorting the values for all cases and then selecting the middle value. A problem with the median as a summary measure is that it ignores much of the available information. For example, the median for the five values 28, 29, 30, 31, and 32 is 30. For the five values 28, 29, 30, 98, and 190, it is also 30. The actual amounts by which the values fall above and below the median are ignored. The high values in the second example have no effect on the median.

The most commonly used measure of central tendency is the **arithmetic mean,** also known as the **average.** (For a sample, it's denoted as $\bar{X}$.) The mean uses the actual values of all of the cases. To compute the mean, add up the values of all the cases and then divide by the number of cases. For example, the arithmetic mean of the five values 28, 29, 30, 98, and 190 is

$$\text{Mean} = \frac{28 + 29 + 30 + 98 + 190}{5} = 75 \qquad \textbf{Equation 5.1}$$

Don't calculate the mean if the codes assigned to the values of a variable are arbitrary. For example, average car manufacturer and average religion don't make sense, since the codes are not meaningful.

? *Can I use the mean for variables that have only two values?* Many variables, such as responses to yes/no or agree/disagree questions, have two values. If a variable has only two values, coded as 0 or 1, the arithmetic mean tells you the proportion of cases coded 1. For example, if five out of 10 people answered *yes* to a question and the coding scheme used is 0=*no* and 1=*yes*, the arithmetic mean is 0.50. You know that 50% of the sample answered *yes*. ■■■

Comparing Mean and Median

Before you compute descriptive statistics for any variable, you should look at its histogram. A histogram shows the distribution of the values and suggests which summary measure may be useful. Look at Figure 5.1, which shows the histograms of age and years of education for people interviewed by the GSS.

The distribution of age has no small values. That's not unexpected, since only people over the age of 18 can be included in the GSS. The distribution of age has a "tail" toward larger values of age even though the GSS does put an upper limit on age. (In an odd quirk that dates back to technology in the 1970s, all ages greater than 89 are recorded as 89.) If you were asked to pick a "typical" age based on the histogram, you would probably chose a value somewhere in the range of 40 to 50 years, since that's where most of the values are.

Figure 5.1 Histograms of age and education

From the menus choose:

Graphs
Legacy Dialogs ▶
Histogram...

and select the variable age.
Double-click the chart to bring it into the Chart Editor, and then double-click the x axis. In the Properties dialog box, select X Axis Custom and set the interval width to 5.

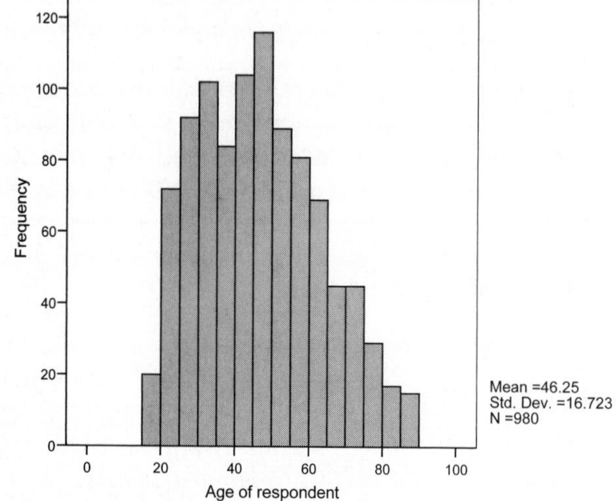

From the menus choose:

Graphs
Legacy Dialogs ▶
Histogram...

and select the variable educ.

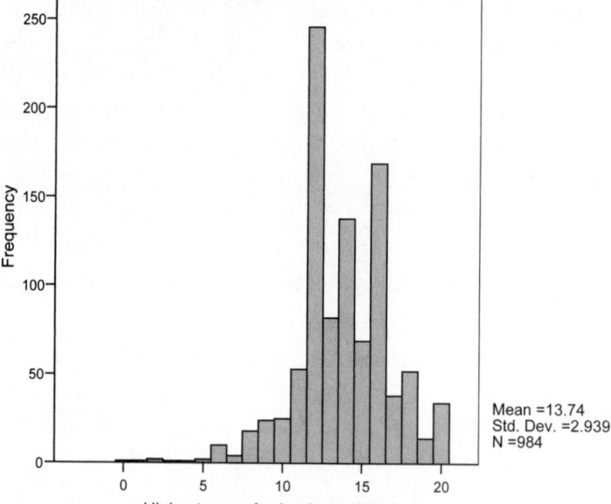

The histogram of highest year of school completed is somewhat different from the distribution of age. The smallest possible value for years of education is 0. The upper limit is 20 years. The GSS caps the years of education at 20, so if you spent 10 years working on your Ph.D. thesis, the GSS will credit you with only 20 years of education. The histogram shows two peaks that correspond to 12 years of school and 16 years of school. If you were asked to choose a typical number for years of school completed, it would probably be between 12 and 14 years.

Figure 5.2 contains descriptive statistics for age and education. The values are fairly close to what you would have guessed for "typical" based on the histograms.

Figure 5.2 Mean, median, and mode for age and education

In the Frequencies Statistics dialog box (see Figure 4.15), select Mean, Median, and Mode.

		AGE Age of respondent	EDUC Highest year of school completed
N	Valid	980	984
	Missing	4	0
Mean		46.25	13.74
Median		45.00	14.00
Mode		40	12

You see that the average age of the participants of the General Social Survey is 46.25 years. The median is somewhat lower, 45 years. The average number of years of school completed is 13.74, and the median is 14. The arithmetic mean is somewhat greater than the median for age and somewhat less than the median for years of education. The reason is that the age variable has a "tail" toward larger values, while the education variable has a tail toward smaller values. Although education has small values, the number of cases with many years of education outweighs those with small values. The older ages drive up the mean for age. They have no effect on the median, since it depends only on the values of the middle cases.

If the distribution of data values is exactly symmetric, the mean and median are always equal. If the distribution of values has a long tail (that is, the distribution is skewed toward larger values), the mean is larger than the median. If the tail extends toward smaller values, the mean will be smaller than the median. In this example, the differences between the mean and the median are not very large. This is not always true.

Consider the following example. You ask five employees of a company how much money they earned in the past year. You get the following replies: $45,000, $50,000, $60,000, $70,000, and $1,000,000. The average salary received by these five people is $245,000. The median is $60,000. The arithmetic mean doesn't really represent the data well. The CEO salary makes the employees appear much better compensated than they really are. The median better represents the employees' salaries.

Measures of central tendency that are less affected by extreme values are discussed in Chapter 7.

Whenever you have data values that are much smaller or larger than the others, the mean may not be a good measure of central tendency. It is unduly influenced by extreme values (called **outliers**). In such a situation, you should report the median and mention that some of the cases had extremely small or large values.

Summarizing Time Spent Online

In Chapter 3, you saw the distribution of hours spent on the Internet per week for people who use the Internet. The shape of the distribution was quite different from the distributions for age and education. There is a tall peak at small values of time and a pronounced tail toward larger values. Histograms for all three of the Internet time variables are shown in Figure 5.3. The shape of the distribution of values is similar for all three variables. In particular, note the presence of very large times for all three of the variables. There is even a person who claims to spend 100 hours per week online. The questions about time online do not differentiate between personal use and work use. It may be that the large values are for people who spend considerable parts of their workday online, working, of course, on job-related activities.

Figure 5.3 Histograms of Internet variables

From the menus choose:

Graphs
Legacy Dialogs ▸
Histogram...

Choose the variables nethrs, webhrs, and emailhrs, one at a time.

Edit the axes as described in Appendix A.

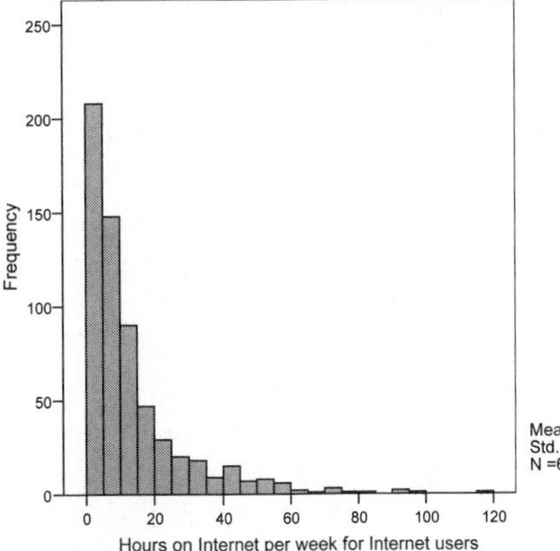

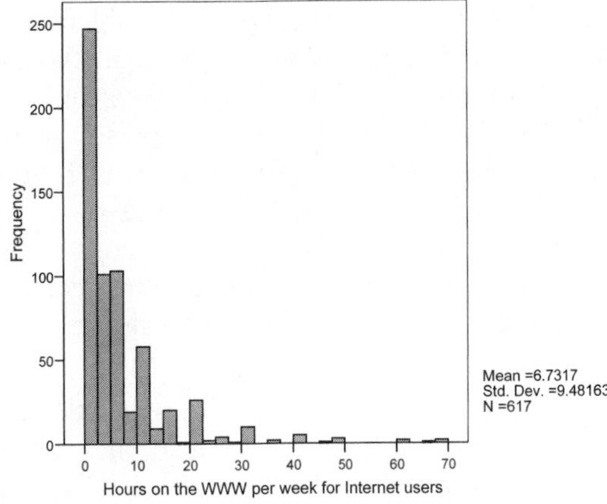

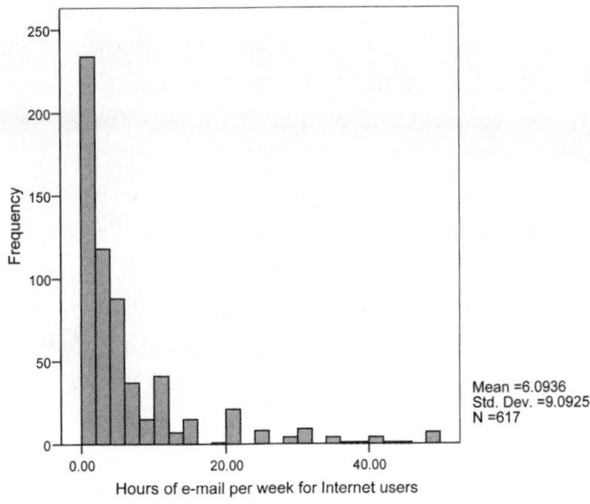

? *Is it possible that the large values for time are mistakes?*

Certainly it's possible. Errors in the data can be introduced in many different ways. People may give incorrect answers either deliberately or inadvertently. If you're ashamed to give the real answer or if you're annoyed with an interviewer, you may fabricate an answer. It's often very difficult even for a skilled interviewer to tell the difference between an honest answer and a lie. Sometimes people don't understand what the interviewer is really asking and give an answer that is correct but inappropriate to the question being asked. Interviewers can also be responsible for errors. They may fill out the form incorrectly. They may fail to ensure that people really understand the question. Even if a correct answer is given and correctly recorded on a survey form, it's possible to enter the information incorrectly into the database.

The first step you should take if you spot unusual values in a study that you have conducted is to go back to the original forms on which the information is recorded. Make sure the problem isn't one of data entry. If the unusual value appears in the original data, you must determine whether it is so outlandish that you know it must be wrong. The 160-year-old hospital patient is an error. The person who weighs 400 pounds is probably an error, but you can't be certain. If you know a value is in error, you should replace it with a missing value code so that it won't contaminate your analyses. If you have unusual, but correct, values in your data, you want to make sure to use statistical techniques that are not unduly affected by the presence of such values. ■■■

Figure 5.4 contains summary statistics for the Internet use variables. Notice the striking difference between the values for the mean and median. For all three variables, the mean is much larger than the median. The mean is hardly typical of most of the values. For all three of the variables, only about 30% of the cases have values greater than the mean. In this situation, the median is a much better summary measure since the very large values of time do not influence the median as much as they do the mean.

It's easy to calculate summary statistics that aren't typical of the data. For example, some studies that describe Internet use don't distinguish between people who use the Internet and people who don't. Summary measures of "typical" hours on the Internet are then impossible to interpret. If you have 100 people, 50 of whom don't use the Internet at all and 50 who use the Internet for five hours a week each, both the mean and median are 2.5 hours. What does this tell you? Nothing. The value of 2.5 hours is a meaningless statistic because it does not represent typical Internet usage by the people in your sample.

In the Frequencies Statistics dialog box (see Figure 4.15), select Mean, Median, and Mode.

Figure 5.4 Summary statistics for Internet variables

		NETHRS Hours on Internet per week for Internet users	WEBHRS Hours on the WWW per week for Internet users	EMAILHRS Hours of e-mail per week for Internet users
N	Valid	617	617	617
	Missing	367	367	367
Mean		12.8253	6.7317	6.0936
Median		7.0000	4.0000	2.0000
Mode		3.00	1.00[1]	1.00

1. Multiple modes exist. The smallest value is shown

Measures of Variability

Measures of central tendency don't tell you anything about how much the data values differ from each other. For example, the mean and median are both 50 for these two sets of ages: 50, 50, 50, 50, 50 and 10, 20, 50, 80, 90. However, the distribution of ages differs markedly between the two sets. **Measures of variability** attempt to quantify the spread of observations. We'll discuss the most common measures of variability in this chapter. Chapter 7 contains a discussion of additional measures.

From the menus choose:

Analyze
Descriptive Statistics ▸
Descriptives...

Select the variables age and educ, as shown in Figure 5.8. In the Descriptives Options dialog box, select Variance, as shown in Figure 5.9.

Figure 5.5 Measures of variability for age and education

	AGE Age of respondent	EDUC Highest year of school completed	Valid N (listwise)
N	980	984	980
Minimum	18	0	
Maximum	89	20	
Mean	46.25	13.74	
Std. Deviation	16.723	2.939	
Variance	279.654	8.636	

Range

The **range** is the simplest measure of variability. It's the difference between the largest and the smallest data values. Since the values for a nominal variable can't be meaningfully ordered from largest to smallest, it doesn't make sense to compute the range for a nominal variable such as region of the country. In Figure 5.5, you see that for the variable *age*, the smallest value (labeled *Minimum*) is 18. The largest value (labeled *Maximum*) is 89. The range is 71 years. A large value for the range tells you that the largest and smallest values differ substantially. It doesn't tell you anything about the variability of the values between the smallest and the largest.

You can use the Explore procedure, described in Chapter 7, to calculate the range and the interquartile range.

A better measure of variability is the **interquartile range.** It is the distance between the 75th and 25th percentile values. The interquartile range, unlike the ordinary range, is not easily affected by extreme values. The 25th percentile for the age variable is 33 years, and the 75th percentile is 58. The interquartile range is therefore 25, the difference between the two.

Variance and Standard Deviation

The most commonly used measure of variability is the **variance.** It is based on the squared distances between the values of the individual cases and the mean. To calculate the squared distance between a value and the mean, just subtract the mean from the value and then square the difference. (One reason you must use the squared distance instead of the distance is that the sum of distances around the mean is always 0.) To get the variance, sum up the squared distances from the mean for all cases and divide the sum by the number of cases minus 1.

The formula for computing the variance of a sample (denoted s^2) is

You can obtain the variance by clicking Options in the Descriptives dialog box. See Figure 5.9.

$$\text{Variance} = \frac{\text{sum of squared distances from the mean for all cases}}{(\text{number of cases} - 1)}$$

Equation 5.2

For example, to calculate the variance of the numbers 28, 29, 30, 98, and 190, first find the mean. It is 75. The sample variance is then

$$s^2 = \frac{(28-75)^2 + (29-75)^2 + (30-75)^2 + (98-75)^2 + (190-75)^2}{4}$$
$$= 5{,}026$$

Equation 5.3

If the variance is 0, all of the cases have the same value. The larger the variance, the more the values are spread out. In Figure 5.5, the variance for the age variable is 279.65 square years; for the education variable, it is 8.64 square years. To obtain a measure in the same units as the original data, you can take the square root of the variance and obtain what's known as the **standard deviation.** Again in Figure 5.5, the standard deviation (labeled *Std. Deviation*) for the age variable is 16.7 years; for the education variable, it is 2.9 years.

? *Why divide by the number of cases minus 1 when calculating the sample variance, rather than by the number of cases?* You want to know how much the data values vary around the population mean, but you don't know the value of the population mean. You have to use the sample mean in its place. This makes the sample values have less variability than they would if you used the population mean. Dividing by the number of cases minus 1 compensates for this. ■■■

The Coefficient of Variation

The magnitude of the standard deviation depends on the units used to measure a particular variable. For example, the standard deviation for age measured in days is larger than the standard deviation of the same ages measured in years. (In fact, the standard deviation for age in days is 365.25 times the standard deviation for age in years.) Similarly, a variable such as salary will usually have a larger standard deviation than a variable such as height.

The **coefficient of variation** expresses the standard deviation as a percentage of the mean value. This allows you to compare the variability of different variables. To compute the coefficient of variation, just divide the standard deviation by the mean and multiply by 100. (Take the absolute value of the mean if it is negative.)

$$\text{coefficient of variation} = \frac{\text{standard deviation}}{|\text{mean}|} \times 100$$

Equation 5.4

The coefficient of variation equals 100% if the standard deviation equals the mean. The coefficient of variation for the age variable is 36%. For the education variable, the coefficient of variation is 21%. Compared to their means, age varies more than education.

Standard Scores

The mean often serves as a convenient reference point to which individual observations are compared. Whenever you receive an examination back, the first question you ask is, How does my performance compare with the rest of the class? An initially dismal-looking score of 65% may turn stellar if that's the highest grade. Similarly, a usually respectable score of 80 loses its appeal if it places you in the bottom quarter of the class. If the instructor just tells you the mean score for the class, you can only tell if your score is less than, equal to, or greater than the mean. You can't say how far it is from the average unless you also know the standard deviation.

For example, if the average score is 70 and the standard deviation is 5, a score of 80 is quite a bit better than average. It is two standard deviations above the mean. If the standard deviation is 15, the same score is not very remarkable. It is less than one standard deviation above the mean. You can determine the position of a case in the distribution of observed values by calculating what's known as a **standard score,** or ***z* score.**

To calculate the standard score, first find the difference between the case's value and the mean and then divide this difference by the standard deviation.

$$\text{standard score} = \frac{\text{value} - \text{mean}}{\text{standard deviation}}$$

Equation 5.5

A standard score tells you how many standard deviation units a case is above or below the mean. If a case's standard score is 0, the value for that case is equal to the mean. If the standard score is 1, the value for the case is one standard deviation above the mean. If the standard score is –1, the value for the case is one standard deviation below the mean. (For many types of distributions, including the normal distribution discussed in Chapter 11, most of the observed values fall within plus or minus two standard deviations of the mean.) The mean of the standard scores for a variable is always 0, and their standard deviation is 1.

You can use the Descriptives procedure in SPSS to obtain standard scores for your cases and to save them as a new variable. Figure 5.6 shows the notes from the Descriptives procedure that indicate that a new variable, the standard score for age, has been created. In addition, a new variable, *zage*, has been saved in the Data Editor, containing the standard scores for age (see Figure 5.7).

Double-click Notes in the Output pane to display information on saved variables.

Figure 5.6 Descriptive statistics in the Viewer

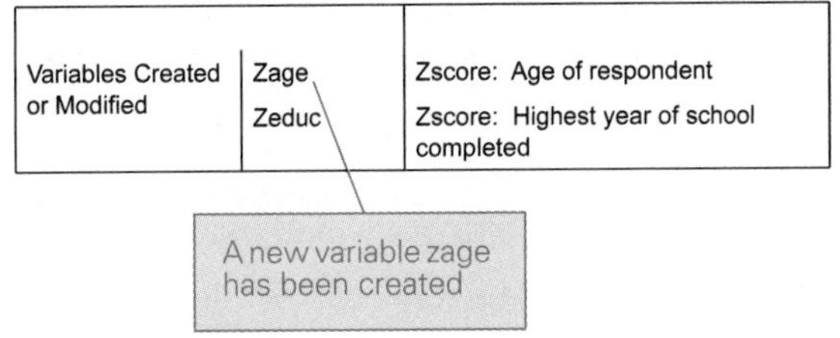

Variables Created or Modified	Zage	Zscore: Age of respondent
	Zeduc	Zscore: Highest year of school completed

To save standardized scores, select Save standardized values as variables in the Descriptives dialog box, as shown in Figure 5.8.

Figure 5.7 Data Editor with standard scores saved as a new variable

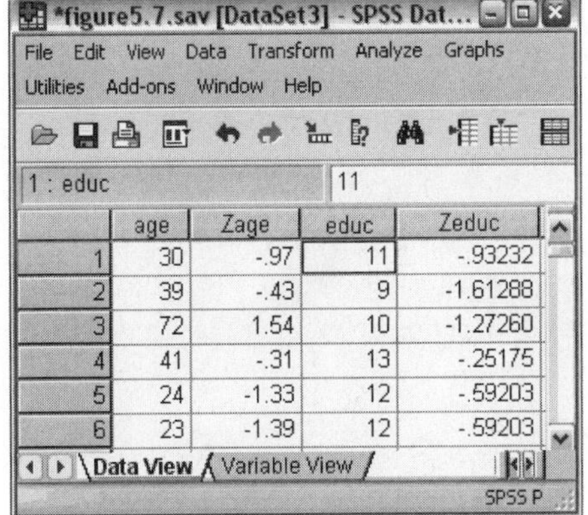

You see that the first case has an age of 30. From the standard score, you know that the case has an age somewhat less than average. The age for the case is about one standard deviation below the mean. The third case has an observed age of 72, which is 1.5 standard deviations above the mean. Always look at observations with large positive or negative scores. Make sure the data values are correct.

Standard scores allow you to compare relative values of several different variables for a case. For example, if a person has a standard score of 2 for income and a standard score of –1 for education, you know that the person has a larger income than most and somewhat fewer years of education. You couldn't meaningfully compare the original values, since the variables all have different units of measurement, different means, and different standard deviations.

? *Is it ever OK to just compute summary statistics without examining the data graphically?* No! Summary statistics can hide potentially serious problems with your data. Unless you look at a histogram, frequency table, or other display, you won't detect errors in data recording or entry, errors in setting up the data file (such as forgetting to identify missing values), or data values that may have been corrupted by the computer. ■■■

Summary

How can you summarize the values of a variable?

- Scales of measurement tell you about the properties of the values of a variable.
- The arithmetic mean is calculated by summing the values of a variable and dividing by the number of cases. Unlike the median and mode, the arithmetic mean uses all of the values of a variable.
- The median is a better measure of central tendency than the mean when there are data values that are far removed from the rest.
- The variance is a measure of the spread of data values around the mean. The coefficient of variation tells you the percentage the standard deviation is of the mean.
- A standardized score tells you how many standard deviation units above or below the mean an observation is.

What's Next?

In this chapter, you calculated summary statistics for all of the cases in your data file. In Chapter 6, you'll learn how to calculate summary statistics when the cases in the data file are subdivided into groups based on values of other variables.

How to Obtain Univariate Descriptive Statistics

You use the Descriptives procedure in SPSS to calculate basic univariate statistics for numeric variables. The Descriptives procedure also lets you save standardized scores (*z* scores) as new variables added to your data file. The Descriptives procedure calculates statistics very efficiently but without tabulating each individual value that occurs in the data. (Some statistics, such as the median and mode, require such tabulation and are not available from Descriptives.)

▶ To open the Descriptives dialog box, as shown in Figure 5.8, from the menus choose:

Analyze
  Descriptive Statistics ▸
    Descriptives...

Figure 5.8 Descriptives dialog box

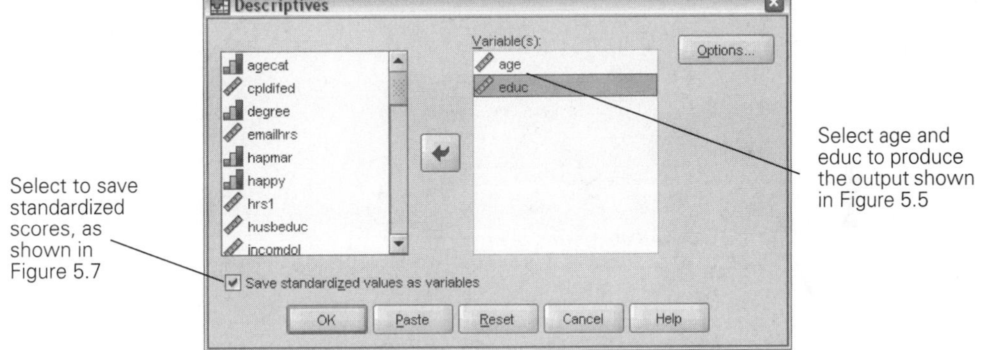

▶ In the Descriptives dialog box, select one or more variables and move them into the Variable(s) list. Only numeric variables appear in the source list, because you can't calculate the mean (for example) of a string variable. When you have the variables you want in the Variable(s) list, click OK.

This produces a default set of descriptive statistics for each of those variables, including the mean, standard deviation, minimum, and maximum. You can also choose the following:

Save standardized values as variables. Automatically creates standard scores (*z* scores) for the selected variables. The standard scores show up as new variables, so that you have the choice of using either the original values or the standardized values in later analysis. SPSS displays the names of the new variables.

Options: Choosing Statistics and Sorting Variables

In the Descriptives dialog box, click Options to obtain additional statistics or to specify the order in which the statistics for different variables are displayed (see Figure 5.9).

Figure 5.9 Descriptives Options dialog box

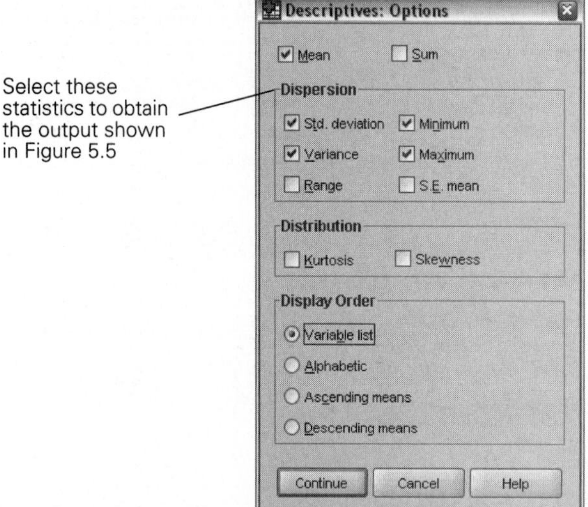

▶ Select the statistics you want.

You can also choose from the following group of options:

Display Order. This group lets you choose whether the statistics for different variables will appear in the order of the variables in the variable list, in alphabetical order of the variable names, or in the order of the variables' means (either ascending or descending).

Exercises

Statistical Concepts

1. Determine the level of measurement for each of these variables:

   a. Interest rate

   b. State in which company is incorporated

   c. Degree of satisfaction with a product

   d. Total family income

   e. Hours of television viewing

   f. Profit margin

   g. Birth order

   h. Preferred brand of gasoline

   i. Favorite music

2. A sample consists of 11 graduates of the University of Texas (coded 1), 10 graduates of the University of Michigan (coded 2), and 10 graduates of the University of Hawaii (coded 3). Which of the following statistics are appropriate for describing these data? Calculate the statistic if you think it is appropriate.

   a. Modal college attended

   b. Median college attended

   c. Mean college attended

   d. Variance of college attended

3. A sample contains five families who do not own a car (coded 0), 20 families who own one car (coded 1), and 10 families who own two cars (coded 2). Indicate which of the following statistics are appropriate and then calculate them.

   a. Modal number of cars owned

   b. Median number of cars owned

   c. Mean number of cars owned

   d. Variance of the number of cars owned

   e. Coefficient of variation

4. In a corporation, a very small group of employees has extremely high salaries while the majority of employees receive much lower salaries. If you were the bargaining agent for the employees, what statistic would you calculate to illustrate the low pay level for most employees, and why? If you were the employer, what statistic would you use to demonstrate a higher pay level for most employees, and why?

5. The number of dogs owned by 10 families are as follows: 0, 1, 1, 1, 2, 2, 2, 2, 2, 4. Fill in the following table based on these values:

	N	Range	Minimum	Maximum	Mean	Std. Deviation	Variance	Mode	Std. Error
DOGS Number of dogs owned	10					1.06			.34
Valid N (listwise)	10								

6. An absent-minded instructor calculated the following statistics for an examination: mean=50, range=50, number of cases=99, minimum=20, and maximum=70. She then found an additional examination with a score of 50. Recalculate the statistics, including the additional exam score.

7. Which measures of central tendency are appropriate for each of the following variables? If several can be calculated, indicate which makes most use of the available information.

   a. Number of siblings

   b. Political party affiliation

   c. Satisfaction with family

   d. Vacation days per year

   e. Type of car driven

   f. Weight of father

8. For each variable in question 7, would you make a bar chart or a histogram?

9. The number of pairs of shoes owned by seven college freshmen are 1, 2, 2, 3, 4, 4, and 5.

   a. Compute the mean, median, mode, range, and standard deviation.

   b. An eighth student, the heir to a shoe empire, is added to the sample. This student owns 50 pairs of shoes. Recompute the statistics.

   c. Which of the statistics are not much affected by the inclusion of an observation that is far removed from the rest?

   d. For all eight students, compute standardized scores for the number of shoes owned.

Data Analysis

Use the *gss.sav* data file to answer the following questions:

1. Compute descriptive statistics for the number of brothers and sisters (variable *sibs*), years of education (variable *educ*), and hours worked last week (variable *hrs1*). Obtain histograms for the variables as well. For each variable, compare the values of the different measures of central tendency. Indicate why and when you would prefer one measure over another.
2. Compute coefficients of variation for each variable. Which variable has the smallest coefficient of variation? Which has the largest?
3. Compute by hand the mean and median for variable *sibs* if people with 20 or more siblings are excluded. Comment on what change, if any, you see. Do you think the effect of outlying values would be different if you had a smaller sample?
4. Consider the variable *tvhours.*
   a. Compute standardized scores for the number of hours of television watched per day.
   b. Compute descriptive statistics and a histogram for the standardized variable.
   c. Does the shape of the distribution change when you compute standardized scores?
   d. In your data, what are the largest and smallest standardized scores?
   e. Compute the 5th, 25th, 50th, 75th, and 95th percentiles for the standardized scores. Does it appear that half of your cases have standardized scores greater than 0 and half less than 0? Explain why that's not the case.
   f. What's the standard score for someone who watches five hours of television a day? Someone who watches three hours? Someone who doesn't watch any television?
   g. How many hours of television does someone with a standardized score of 1 watch? With a standardized score of –1? –0.75?
5. Consider the variable *income.* (The variable is total family income in the year before the survey.)
   a. Make a frequency table for the variable. Does it make sense to make a histogram of the variable? A bar chart?
   b. What is the scale of measurement for the variable?
   c. What descriptive statistics are appropriate for describing this variable and why? Does it make sense to compute a mean?

d. Discuss the advantages and disadvantages of recording income in this manner. Would you record income in this way if you were doing a study? Describe other ways of recording income and the problems associated with each of them.

6. The Recode facility was used to compute new variables named *incomdol* and *rincdol* that are the midpoint of the income range for each value of *income* (family income) and *rincome* (respondent's income). For example, for all cases with *income* equal to 1, the value of *incomdol* is 500. Similarly, for cases with the value 3 for *income*, the value of *incomdol* is 3500. For cases with the value 23 for *income*, a value of 110,000 is used.

a. Perform appropriate analyses for the new variables. Make a histogram, if appropriate, and compute summary statistics. In the Frequencies procedure, you should select Values are group midpoints. This will change the calculation of some of the percentiles.

b. Do you think you know the exact income of people in your sample? Discuss the uncertainties introduced by this type of analysis.

7. When you write a paper using the results from surveys or experiments, you must describe the people in the sample. Write the introductory section of a paper in which you describe the GSS participants in terms of *age*, *educ*, *income*, and any other variables you think are important to characterize your sample.

Use the file *crimjust.sav* to answer the following question:

8. Describe the respondents to the Vermont criminal justice survey. Compare their characteristics to those of the Vermont population as a whole. You can find population information for Vermont on the Internet. Does your sample differ from the population in any important ways? Do you expect your sample to be a "miniature" of the population? Participation in any survey is voluntary, so people who agree to answer questions may be quite different from those who don't. In what ways do you think the people who respond to telephone surveys are different from those who don't? Discuss possible bias that this may introduce into survey results.

Use the *marathon.sav* file, or the *mar1500.sav* file if you are using the Student version, to answer the following questions:

9. Compare the age and sex distribution of Chicago marathon runners to that found in the GSS. Describe the differences. Explain.

10. Make a histogram of the running times.

a. Is the distribution approximately symmetric? Why not?

b. Describe the length of time it took people to run the marathon. Include appropriate graphs. Describe the merits of the mean, median, and mode for describing running times.

c. Calculate standard scores for the participants. How many runners have standard scores within 1 standard deviation of the mean? How many within two standard deviations?

Use the ***salary.sav*** data file to answer the following questions:

11. Consider the beginning salary for all employees (variable *salbeg*).

a. Make a histogram. Why are all of the values bunched together?

b. Edit the histogram by selecting Axis from the Chart menu. Restrict the maximum value for the horizontal axis to 20,000. Compare this histogram with the one in question 6a.

c. Compute the mean, median, and mode for beginning salary. Why do you think the mean is larger than the median? Which statistic do you think better summarizes the data values?

d. Compute the quartiles for beginning salary. Within what values do the middle 50% of the salaries fall?

e. Compute the coefficient of variation for beginning salary.

12. Compute standard scores for beginning salary.

a. What are the smallest and largest standard scores?

b. How many standard deviation units from the mean is the largest score?

c. Select females only. What is their average standard score for beginning salary?

d. On average, do women earn more or less than men in the sample?

Use the ***electric.sav*** data file to answer the following questions:

13. Calculate the mean, median, and mode for cholesterol values and diastolic blood pressures in 1958 (variables *chol58* and *dbp58*). For each variable, compare the values of these statistics. Which measure of central tendency do you think best summarizes each variable? Explain your selection.

14. Consider the number of cigarettes smoked in 1958 (variable *cgt58*). Describe the smoking habits of the men in 1958. Be sure to include what percentage of men in your sample were nonsmokers in 1958 as well as the mean and median number of cigarettes smoked by the smokers.

15. Compute the coefficient of variation for *chol58*, *dbp58*, and *cgt58*. Which has the largest? Which has the smallest?

16. Compute standardized scores for cholesterol values and diastolic blood pressure.

a. What is the smallest standardized score for diastolic blood pressure? The largest?

b. What is the smallest standardized score for cholesterol values? The largest?

c. Compute the means and standard deviations of the standardized scores. What are they? Compute the quartiles for the standardized scores. Compare the quartiles for the two variables.

d. Make histograms of the standardized and unstandardized values for diastolic blood pressure. Do the histograms differ? How?

e. What is the standard score for a person with a diastolic blood pressure of 120? 150? 80?

Use the *schools.sav* data file to answer the following question:

17. Prepare for the Chicago Board of Education a summary of the 1993 and 1994 performance of city high schools on standardized tests. In particular, focus on the changes from 1993 to 1994. Include a summary of the demographic characteristics (variables *loinc93* and *lep93*) and graduation rates of the schools (variables *grad93* and *grad94*). Include appropriate charts.

Comparing Groups

6

How can you determine if the values of the summary statistics for a variable differ for subgroups of cases?

- What are subgroups of cases?
- What can you learn from calculating summary statistics separately for subgroups of cases?
- How can you graph means for subgroups of cases?

In Chapter 4 and Chapter 5, you used the Frequencies and Descriptives procedures to calculate summary statistics for all people in the General Social Survey. You know how old they are, how much education they have, and how much time they spend each week on the Internet, the World Wide Web, and e-mail. You're probably wondering now how people who use the Internet differ from people who don't. Are they younger? Are they better educated? Are there differences between men and women? You want to calculate summary measures separately for different groups of cases. There's no easy way with the Frequencies or Descriptives procedure to produce such information. In this chapter, you'll use the Means procedure to calculate simple summary statistics for subgroups of cases. You'll examine who uses the Internet and who doesn't. (The Explore procedure described in Chapter 7 lets you examine the values of a variable for subgroups of cases in much greater detail.)

▶ This chapter uses the *gssnet.sav* data file. For instructions on how to obtain the SPSS output discussed in this chapter, see "How to Obtain Subgroup Means" on p. 109.

Age, Education, and Internet Use

In Figure 6.1, you see the mean and median years of age for people in each of five Internet usage categories.

Figure 6.1 Internet use by age

From the menus choose:

Analyze
Compare Means ▸
Means...

Select the variables age and netcat, as shown in Figure 6.6.

In the Means Options dialog box, select Mean, Median, and Number of Cases. Make sure that the code of –1 for netcat is not specified as missing.

Age of respondent

Weekly Internet use	Mean	Median	N
Not internet user	55.68	56.00	254
4 hours or less	41.78	42.00	201
4+ to 8 hours	45.82	45.00	135
8+ to 16 hours	42.91	43.00	135
16+hours	40.96	38.00	144
Total	46.51	45.00	869

Statistics for individual subgroups

Statistics for all cases analyzed

You see that the average age of all people is 46.51. (This differs slightly from the average shown in Figure 5.1 because only people who have valid answers for Internet usage and age are included in the computation.) From the last row of Figure 6.1, you see that 869 people are included in the computation of total age. These 869 people are assigned to one of five subgroups, based on their Internet usage. The first subgroup, *Not Internet user,* are nine years older than the group as a whole. The average age doesn't differ much across the other four categories of Internet use. People who use the Internet are younger than people who don't. There's no obvious relationship between the amount of time spent on the Internet and age, although those who use the Internet more than 16 hours a week are, on average, the youngest.

Plotting Means

A plot of the mean years of education for the five subgroups is shown in Figure 6.2. There is a bar for each of the subgroups. The height of the bar reflects the average years of education. You see that people who don't use the Internet have fewer years of education than people who use the Internet. The average years of education for everyone is 13.75, while for those who don't use the Internet, it is 11.55. The average years of education increase slightly across the Internet use categories. People who are online more than eight hours a week have the most education.

Figure 6.2 Bar chart of education by Internet use

From the menus choose:

Graphs
Legacy Dialogs ▸
Bar

In the Define Simple Bar Summaries for Groups of Cases dialog box, select Other statistic and select the variable educ. Select netcat for Category Axis.

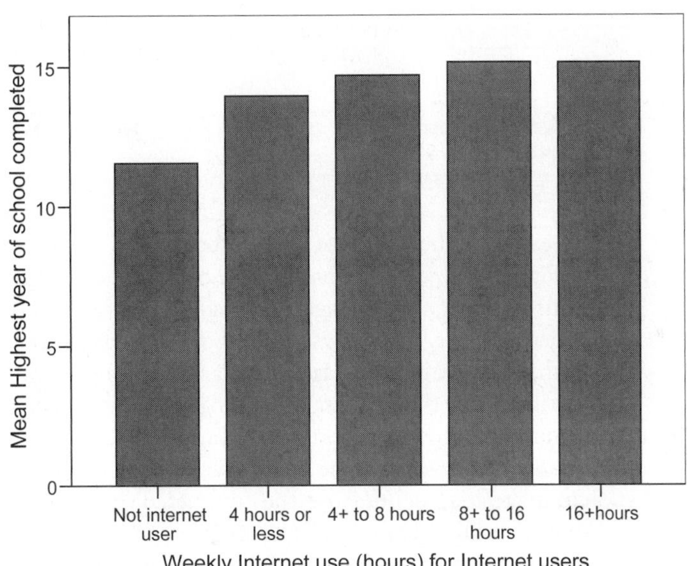

Layers: Defining Subgroups by More than One Variable

You have seen that Internet users are younger and better educated than those who don't use the Internet. Since age and education are related, the interpretation of this finding is tricky. If younger people are better educated, it may be that education alone is not related to Internet use.

Make the selections shown in Figure 6.7. Activate the table by double-clicking it. Use the Pivot Table Editor to rearrange the table.

Within each age category, separate statistics are shown for each degree category

Figure 6.3 Internet use for age and education groups

		Highest degree earned					
Age category		Less than high school	High school	Junior college	Bachelor	Graduate	Total
18-29	Mean	.60	.82	1.00	.97	1.00	.84
	N	25	101	12	36	9	183
30-39	Mean	.22	.83	.88	.96	1.00	.83
	N	18	84	17	48	18	185
40-49	Mean	.47	.76	1.00	.98	1.00	.83
	N	19	104	19	45	29	216
50-59	Mean	.32	.73	.71	.90	1.00	.74
	N	22	81	14	29	24	170
60-89	Mean	.07	.52	.83	.69	.85	.49
	N	45	123	6	26	20	220
Total	Mean	.29	.72	.90	.92	.97	.74
	N	129	493	68	184	100	974

To examine the effects of age and education together on Internet use, look at Figure 6.3, which contains means for the ***usenet*** variable for combinations of categories of age and highest degree received. (Since ***usenet*** is coded as 1 for Internet users and 0 for non-users, the mean of the variable is the proportion of cases that are Internet users.) There are a lot of numbers, but they tell an interesting story. For any age category, as education increases, so does the percentage of Internet users. From the first cell, you see that 60% of the 25 people in the age group 18–29 who did not complete high school use the Internet. From the *Total* column, you see that, overall, 84% of the 183 people aged 18–29 use the Internet. From the *Total* row, you see that 29% of the people with less than a high school diploma and 92% of those with a bachelors degree use the Internet. For any degree category, as age increases, the percentage of Internet users decreases. It's important to pay attention to the sample sizes, since some of the combinations of age and degree don't have very many cases. For example, there are only nine people with graduate degrees in the youngest age group. Means based on small numbers of cases are unreliable. If you choose another sample of nine people with graduate degrees who are younger than 30, you might get completely different results. (Though it's quite unlikely that any young recipient of a graduate degree would not use the Internet!)

Figure 6.4 is a bar chart of Internet use, age, and highest degree. Some of the education categories in Figure 6.3 have been combined so that the numbers of cases in some of the groups would be larger. From the bar chart, you can easily see that for each of the three education groups, Internet usage generally decreases with age. For each age category, Internet usage increases with education. It appears that both age and education are needed to predict Internet usage. (Boxplots, which are another way of comparing summary statistics for groups of cases, are described in Chapter 7.)

Figure 6.4 Bar chart of Internet usage by age and degree

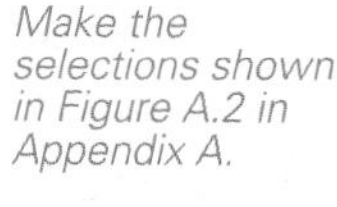

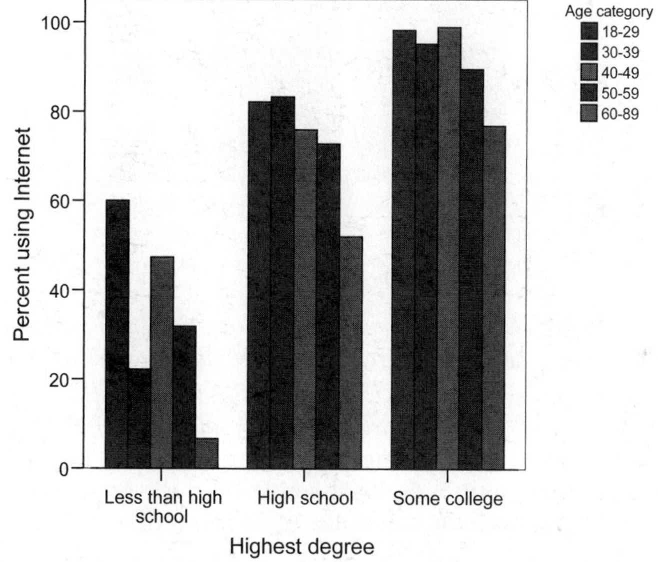

You can also group all of the bars for a particular age group together, as shown in Figure 6.5. This makes it easier to see the effect of education within each category of age. Note that education seems to have more effect for older age groups than for the youngest one.

To obtain this chart, interchange agecat and ndegree in Figure A.2 in Appendix A.

Figure 6.5 Bar chart of Internet use by degree and age

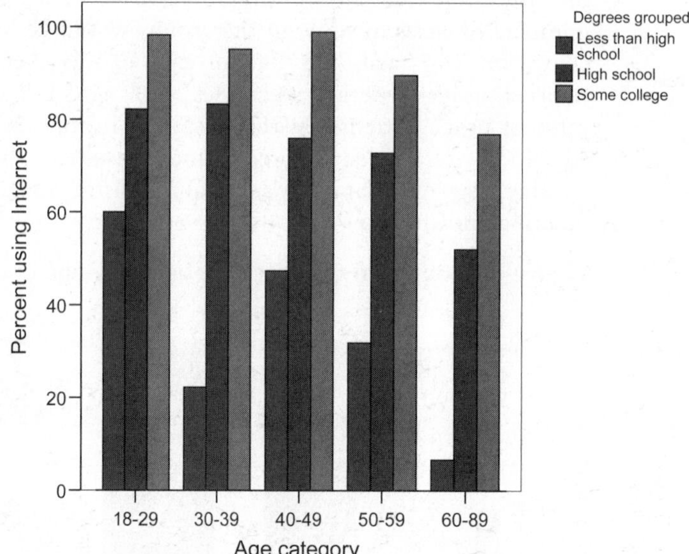

? *What problems are associated with calculating statistics for subgroups of cases?* As the number of subgroups that you want to compare increases, the sample size in each of the subgroups diminishes. When your means are based on a small number of cases, they are not very reliable. That is, the subgroup means can change substantially if you select another random sample from the same population. You'll learn more about the variability of sample means in Part 3.

Summary

How can you determine if the values of the summary statistics for a variable differ for subgroups of cases?

- Subgroups are formed when cases are subdivided into groups based on the values of one or more variables.
- By calculating summary statistics separately for subgroups of cases, you can see if there is a relationship between the summary statistics and the subgroups.
- You can make bar charts of the means of a variable for different subgroups.

What's Next?

By calculating summary statistics for subgroups of cases, you saw the striking effects of education and age on Internet usage. In Chapter 7, you'll examine the distributions of values for subgroups in more detail. You'll also learn about additional summary statistics and displays.

How to Obtain Subgroup Means

This section shows how to obtain subgroup means with SPSS. In addition, the Means command can:

- Use two or more variables simultaneously to define the subgroups.
- Display statistics other than the mean for the subgroups.

The Means procedure calculates statistics for groups defined by the categories of one or more categorical variables.

▶ To open the Means dialog box, from the menus choose:

Analyze
  Compare Means ▶
    Means...

Figure 6.6 Means dialog box

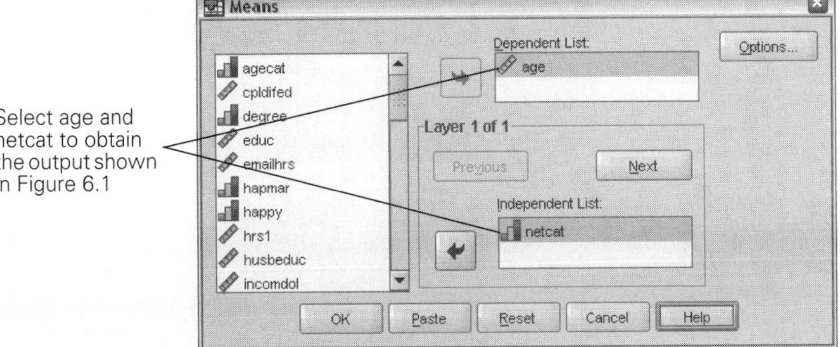

▶ In the Means dialog box, select the variable for which you want subgroup statistics from the source variable list and move it into the Dependent List.

You can move more than one variable into the Dependent List to analyze them all with a single command.

▶ Select a categorical variable in the source variable list and move it into the Independent List at the bottom of the dialog box.

The categories of this variable define the subgroups. You can move more than one variable into this list; SPSS analyzes each set of groups separately, one after another. For example, if you move both *degree* and *agecat* into the Independent List, you will obtain two separate analyses: one for the categories of highest degree received and one for the categories of age.

Layers: Defining Subgroups by More than One Variable

To use two categorical variables *simultaneously* to define the subgroups (as in Figure 6.3), move them into different layers. Notice the words *Layer 1 of 1* above the Independent List in Figure 6.6. This list is really a stack of lists. Once you move a variable into Layer 1, you can click Next to see the next layer, as shown in Figure 6.7. It will be empty until you move a variable into it; once you do, you can click Next again to build additional layers. Or you can click Previous to go back and add or remove variables in the previous layer. The layer controls are simple to operate if you just keep your eye on the Layer message so you don't lose your place. What's important is to understand what layers do, so you'll know when to use them.

Figure 6.7 Means dialog box displaying Layers 1 and 2

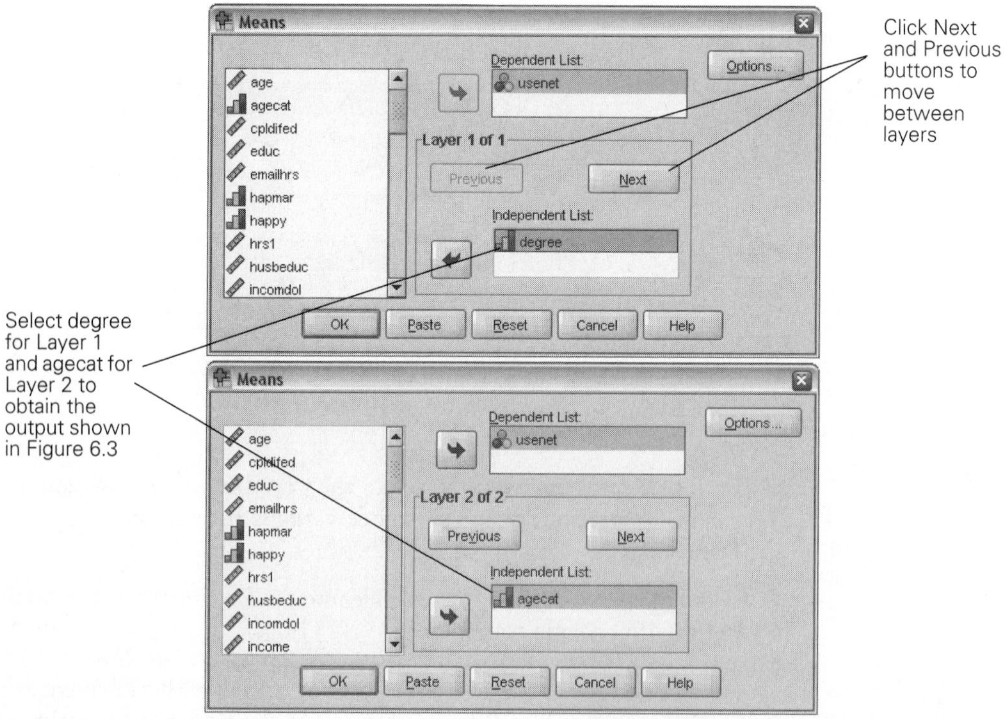

Click Next and Previous buttons to move between layers

Select degree for Layer 1 and agecat for Layer 2 to obtain the output shown in Figure 6.3

If Layer 1 of the Independent List contains one variable and Layer 2 contains another variable, SPSS uses both variables at the same time to define groups. For example, if you select *degree* and *agecat* as shown in Figure 6.7, SPSS displays statistics for each age group (the Layer 2 variable) within each degree category (the Layer 1 variable), as in Figure 6.3.

With more than one variable in any layer, SPSS analyzes the groups defined by each combination of the Layer 2 variable with the Layer 1 variable.

Options: Additional Statistics and Display of Labels

In the Means dialog box, click **Options**. In the Means Options dialog box, shown in Figure 6.8, you can add additional information to be displayed in each cell.

Figure 6.8 Means Options dialog box

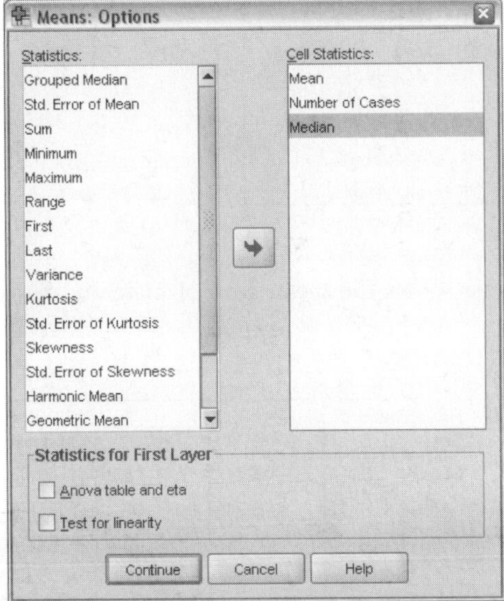

Some of the available options are:

Statistics. Allows you to select the subgroup statistics that are displayed. You can select any combination of one or more statistics. Select the statistics that you want and move them into the Cell Statistics list.

Statistics for First Layer. Lets you request an analysis-of-variance table for testing whether in the population subgroup means are equal. Only the categories of the variable in Layer 1 are used. (Other layers, if present, are ignored.) You can also request a test for whether the differences in subgroup means vary linearly across the subgroups. These concepts are explained in Chapter 15 and Chapter 16.

Exercises

Statistical Concepts

1. A market research company is trying to decide what color to make a new brand of breath mints. They ask 100 consumers to choose among the colors white, yellow, green striped, and red striped. A research analyst assigns the codes 1 through 4 to the possible choices and uses the Means procedure to find average color preferences for men and women in each of four income categories. How would you interpret the resulting table?

Mean

	Mint preferred	
	Respondent's sex	
TOTAL FAMILY INCOME	Male	Female
less than 15,000	1.29	3.37
15,000 - 25,000	3.66	1.10
25,000-50,000	1.72	1.77
50,000 +	3.11	2.35

2. The following table shows the mean ages of male and female Clinton, Bush, and Perot supporters.

Age of Respondent

VOTE FOR CLINTON, BUSH, PEROT	Respondent's Sex	Mean	N	Std. Deviation
CLINTON	Male	48.31	161	17.07
	Female	48.34	270	16.67
	Total	48.33	431	16.80
BUSH	Male	48.41	171	17.63
	Female	49.64	213	17.54
	Total	49.09	384	17.57
PEROT	Male	44.54	103	15.42
	Female	42.04	84	14.49
	Total	43.42	187	15.02
Total	Male	47.36	438	16.95
	Female	47.86	568	16.87
	Total	47.65	1006	16.90

a. Complete the following chart of mean ages:

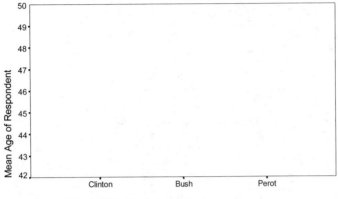

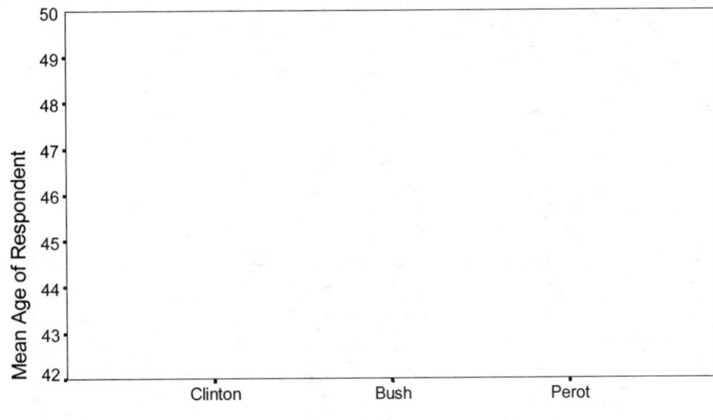

b. Complete the following clustered bar chart of mean ages:

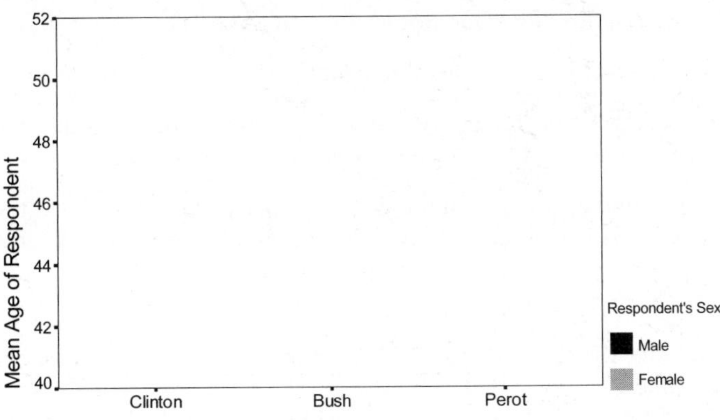

Data Analysis

Use the *gss.sav* data file to answer the following questions:

1. What kind of people are happy people? Write a short paper that compares people who are very happy, pretty happy, and not too happy (variable *happy*) on characteristics such as age, education, hours of television watched, and so on. Include appropriate charts. Don't submit all of the SPSS output that you produce. Instead, summarize the output that pertains to the points that you want to make. Be concise.

2. Many factors contribute to the happiness of a marriage (variable *hapmar*). Answer the question, "What distinguishes people who are very happy with their marriage from those who are less content?" Write a brief report outlining your findings.

3. In the 1996 election, how did Clinton, Dole, and Perot supporters differ (variable *pres96*)? Answer the following questions:

   a. Whose supporters were the oldest and by how much?

   b. Was there a relationship between years of education and candidate preference?

   c. Which candidate were wealthy people more likely to support?

4. What types of people claim that they would continue to work if they struck it rich (variable *richwork*)? Be sure to look at average education, age, income, and hours worked per week. The variables *rincdol* and *incomdol* contain incomes.

5. How do people who believe in life after death differ from those who don't? Use the variable *postlife* to answers the question.

6. Many factors are associated with satisfaction with one's job. Look at the variable *satjob.* (Respondents were asked, "On the whole, how satisfied are you with the work you do?" The question was asked of people who work or keep house.)

   a. Is job satisfaction related to education? Income? Age? Explain.

   b. Is the relationship between job satisfaction, education, and income similar for men and women? What do you base your answer on?

7. The variable *zodiac* contains the astrological signs of the GSS respondents. Examine differences in income (variable *rincdol*) for people born under different signs. Prepare appropriate graphical displays to support or refute the claim that people born under certain signs earn more money.

8. Describe the relationship between gender and years of education. Include appropriate graphics.

9. What is the average age and education of men and women who use computers (*usecomp*)? Use the Internet (*usenet*)? Use the World Wide Web (*useweb*)? Prepare an appropriate display.

10. Consider the variable *netcat.* Calculate the average age of men and women in each of the categories of Internet use. Describe the relationship. Does it differ for men and women? How?

Use the file *crimjust.sav* to answer the following questions:

11. Of people who think the criminal justice system is excellent (variable *cjsrate*), what percentage are men and what percentage are women? (Determine what coding scheme is used for sex. If necessary, use the Recode transformation to change the codes to 0 and 1. Then, you can use the Means facility to calculate the proportion of people who are coded as 1.)

12. Consider responses to the questions about trends in crime (variables *crmcomp5*, *vlntcmp5*, *drugcmp5*). Are men or women more likely to feel that crime is increasing? Is it the similar for all three?

Use the *manners.sav* data file to answer the following questions:

13. Consider the variable *tvdue*, which attributes bad manners to television. Is there a relationship between age and the responses? Is it the same for men and women?

14. Is there a relationship between age and attributing bad manners to teachers (variable *teachdue*)?

15. Swearing (variable *swear*), rudeness (variable *rude*), and getting and receiving obscene gestures (variables *gesture* and *gestothr*) may be related to age. Calculate the average age for people who admit to each of these activities and those who don't. Does the average age differ for men and women? Write a short summary describing the relationships. Make appropriate bar charts.

Use the *salary.sav* data file to answer the following questions:

16. Compute average beginning salaries (variable *salbeg*) for people in different job categories (variable *jobcat*).

a. Which job category has the largest average beginning salary? The smallest?

b. Compute the coefficient of variation for each job category. What can you say about the coefficients of variation for the different categories?

c. Make a bar chart of average beginning salary for each job category.

17. For each job category:

a. Find the average beginning salaries for males and females. Summarize your findings.

b. Make a bar chart of average beginning salary by job category and gender. Are the differences easier to see from the table or from the chart?

c. What possible explanations other than gender discrimination can you offer for the observed differences?

18. Repeat the analyses in questions 16 and 17, using years of education (variable *educ*) instead of beginning salary.

Use the *electric.sav* data file to answer the following questions:

19. Examine the differences in diastolic blood pressure, cholesterol, and years of education (variables *dbp59*, *chol58*, and *eduyr*) for those who were alive 10 years after the study started and those who were not (variable *vital10*). Generate appropriate charts. Write a summary of your results.

20. What is the average standardized score for diastolic blood pressure for those who were alive at 10 years and those who were not? Does the difference appear to be large? What does the average standardized score tell you?

21. Compute the average diastolic blood pressure for each category of first coronary heart disease event (variable *firstchd*).

a. Which group has the highest average diastolic blood pressure? How many cases is this mean based on? Does the number of cases influence how much confidence you have in the mean?

b. What is the average diastolic blood pressure for all cases? For those with sudden death? For those with no coronary heart disease?

c. Does systolic blood pressure appear to be related to coronary heart disease?

22. Look at the average number of cigarettes smoked per day in 1958 (variable *cgt58*) for each category of first coronary heart disease event. Summarize your findings.

23. Calculate the average diastolic blood pressure for those alive and not alive at 10 years, separately for those with a family history of coronary heart disease and for those without (variable *famhxcvr*). Summarize your findings. Make a bar chart of your results.

7 Looking at Distributions

What additional displays are useful for summarizing the distribution of a variable for several groups?

- What is a stem-and-leaf plot?
- How does a stem-and-leaf plot differ from a histogram?
- What is a boxplot?
- What can you tell from the length of a box?
- How is the median represented in a boxplot?

Out of concern for the health of college students and other readers who should heed the warnings of health professionals (and mothers) on the hazards of being "webheads" (obesity, social isolation, carpal tunnel syndrome, and so on), let's turn our attention to data from a healthier activity, the Chicago marathon. You'll apply some familiar techniques as well as learn new ones for looking at the distribution of data values.

SPSS contains many procedures that help you to examine and describe the distributions of variables. In Chapter 6, you used the Means procedure to look at the relationship between Internet use, age, and education. In this chapter, you'll use the Explore procedure, which contains additional descriptive statistics as well as plots to help you understand your data.

▶ This chapter uses the *marathon.sav* data file, which contains running times for all 28,764 people who completed the Chicago marathon in 2001. This file contains a variable, *time*, formatted to display as hours:minutes:seconds, and a variable, *hours*, formatted as decimal hours. In some circumstances, SPSS displays time-formatted values as a number of seconds. In those situations, we will use the *hours* variable so that it will be easier to understand. If you are running the Student version of SPSS or if you are running on a slow machine, you should use the file *mar1500.sav*. For instructions on how to obtain the Explore output shown in the chapter, see "How to Explore Distributions" on p. 131.

Marathon Completion Times

The Chicago Marathon has been run yearly since 1977. In 1993, almost 7,000 runners signed up. In 2001, nearly 29,000 people completed the race.

Histograms

Figure 7.1 is a histogram of the marathon running times in 2001. You see that times cluster around a central value. As you move further from the center, the number of people with those times decreases. The mean, 4.33 hours, falls nicely in the middle of the distribution. The distribution, not unexpectedly, has a tail toward larger times. Low marathon times are difficult to achieve. It's hard to break world records. High marathon times require much less effort. Since the distribution has a tail toward larger values, the median should be somewhat less than the mean.

From the menus choose:

Graphs
Legacy Dialogs ▶
Histogram...

Move hours into the Variable box. Double-click the chart to bring it into the Chart Editor, and then double-click the x axis. In the Properties dialog box, select X axis Custom and set the interval width to 0.25.

Figure 7.1 Histogram of marathon completion times

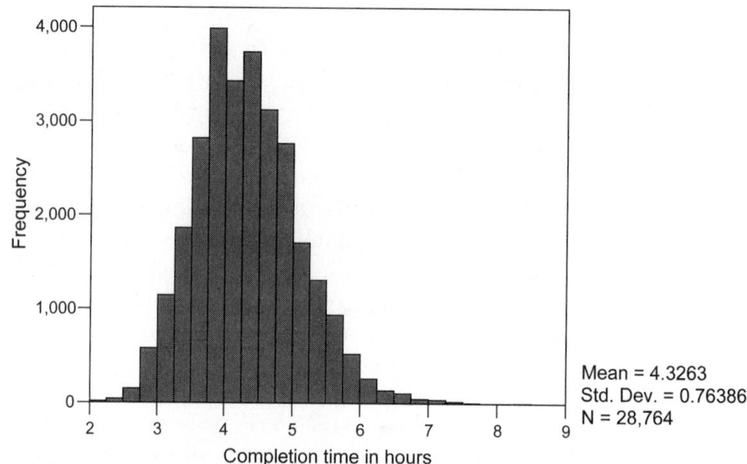

? *What kinds of things should I look for in a histogram?* You already know that you should look for cases with values very different from the rest. In fact, if there are such cases, they can cause most of your data values to bunch in one or two bars of the histogram, since the horizontal axis of the histogram is selected so that all data values can be shown. You should see also whether the distribution is symmetric, since many of the statistical procedures described in Part 3 require that the distribution be more or less symmetric.

You should also look for separate clumps of data values. For example, you may see a distribution with two peaks—one for men and one for women. In such a situation, you'd want to analyze the data for men and for women separately. ■■■

Descriptive statistics for the completion times are shown in Figure 7.2. As expected, the median time is just a little less than the mean time. Figure 7.2 contains another measure of central tendency known as the trimmed mean. As you've learned in Chapter 5, one of the shortcomings of the arithmetic mean is that very large or very small values in the data can change its value substantially. The trimmed mean avoids this problem. A **trimmed mean** is calculated just like the usual arithmetic mean, except that a designated percentage of the cases with the largest and smallest values are excluded. This makes the trimmed mean less sensitive to outlying values. The 5% trimmed mean excludes the 5% largest and the 5% smallest values. It's based on the 90% of cases in the middle. The trimmed mean provides an alternative to the median when you have some data values that are far removed from the rest.

From the menus choose:

Analyze
Descriptive Statistics ▸
Explore...

Move hours into the Dependent List and select Display Statistics.

Figure 7.2 Descriptive statistics for marathon completion times

			Statistic	Std. Error
HOURS Completion time in hours	Mean		4.3263	.00450
	95% Confidence Interval for Mean	Lower Bound	4.3175	
		Upper Bound	4.3352	
	5% Trimmed Mean		4.3054	
	Median		4.2751	
	Variance		.583	
	Std. Deviation		.76386	
	Minimum		2.15	
	Maximum		8.44	
	Range		6.29	
	Interquartile Range		1.0077	
	Skewness		.468	.014
	Kurtosis		.443	.029

In Figure 7.2, you see that the 5% trimmed mean doesn't differ much from the usual mean. That's not surprising because the distribution is not too far from being symmetric. (If a distribution is symmetric, cutting off the top and bottom of the distribution has no effect on the arithmetic mean since the two parts balance each other out. The number of marathon runners is also very large, so the effect of a few large times on the arithmetic mean is much less than if the number of runners were small.)

In Figure 7.2, you see that the standard deviation of the completion times is 0.76 hours. About two-thirds of the cases in Figure 7.1 are within one standard deviation of the mean. The fastest running time is 2:08:52; the slowest recorded, 8:26:06. (It's possible that someone is still running, very slowly.) The range—the difference between the largest and smallest times—is 6.29 hours. A single outlying value can have a large effect on the range. That's why the interquartile range is a better measure of variability. Unlike the ordinary range, the interquartile range is not easily affected by extreme values, since the bottom 25% and the top 25% of the data values are excluded from its computation. It is simply the difference between the 75th and the 25th percentile values. In Figure 7.2, you see that the interquartile range is one hour. That means that the middle 50% of the runners were within one hour of each other.

Age and Gender

The overall distribution of completion times is a mixture of times for men and women of different ages. It's a fact that, overall, male runners run faster than female runners. Look at Figure 7.3, which displays separate histograms for the completion times for men and women. If you compare these distributions to the overall distribution, you see that they have more pronounced peaks than the combined distribution. That's because the two distributions have different means, and combining them into a single histogram masks the individual distributions.

Figure 7.3 Histograms of completion times for men and women

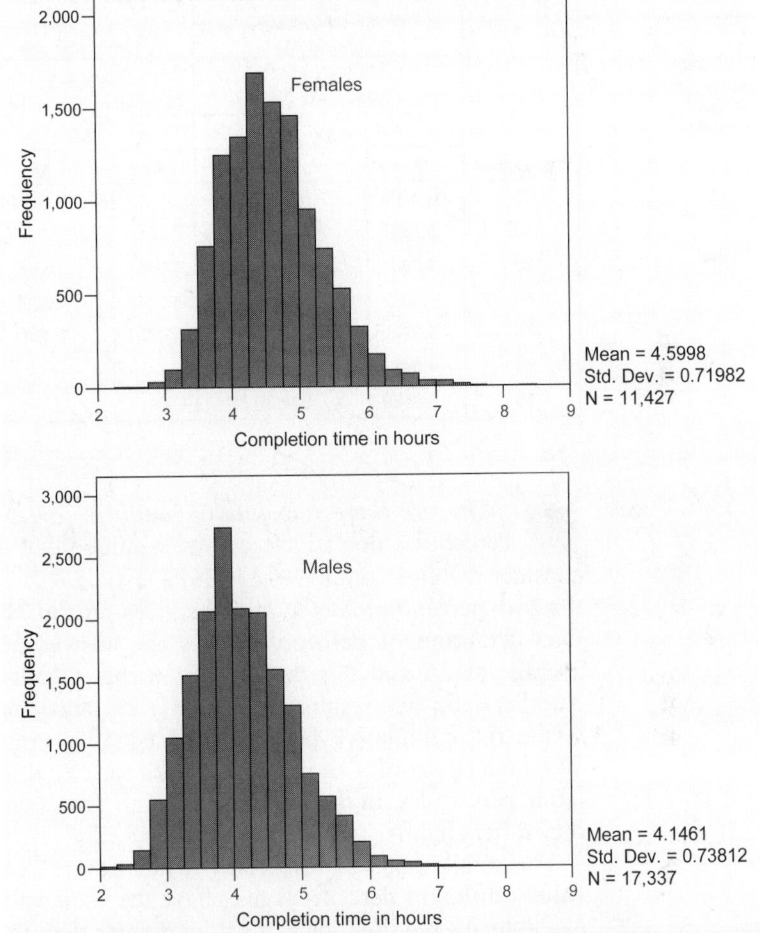

Make the selections shown in Figure 7.10. Click the Plots button and select Histogram. Edit the x axis. You can also make separate histograms for groups of cases, maintaining the same scales, by selecting Graphs, Legacy Dialogs, then Histogram, and moving sex into either Rows or Columns in Panel by.

Figure 7.4 shows percentiles of completion times for men and women. The difference in all of the percentile values between men and women is about half an hour. The fastest 5% of the women took an extra half-hour to complete the marathon compared to the fastest 5% of the men. Fifty percent of the men completed the marathon in 4.06 hours. Fifty percent of the women completed the marathon in 4.54 hours.

In the Explore Statistics dialog box, select Percentiles.

Figure 7.4 Percentiles for completion times for men and women

	Weighted Average (Definition 1)		Tukey's Hinges	
	HOURS Completion time in hours		HOURS Completion time in hours	
	SEX		SEX	
Percentiles	F	M	F	M
5	3.5503	3.0475		
10	3.7327	3.2406		
25	4.0861	3.6306	4.0861	3.6306
50	4.5347	4.0639	4.5347	4.0639
75	5.0222	4.6081	5.0221	4.6081
90	5.5453	5.1231		
95	5.8469	5.4825		

? *Why are there two sets of numbers for the same percentiles?* Percentiles don't have a single, unique definition. For example, consider the eight numbers, 25, 26, 27, 27, 27, 27, 30, and 31. What's the 25th percentile? Any number between 26 and 27 is a plausible value. One definition of percentiles gives the answer 26.5, since that's the average of 26 and 27, the interval within which the percentile falls. Another definition results in the answer 26, since that's the first value for which the cumulative percentage is equal t.o or greater than 25%. Weighted percentiles and Tukey's hinges are two different ways of calculating percentiles. In this example, they give the same results. Sometimes they differ slightly.

For small datasets, especially when several cases have the same values, different percentiles may have the same value. For the previous example, it's possible for percentiles greater than the 25th and less than the 75th to have the value 27. For small datasets, percentile values can vary a lot for samples from the same population, so you shouldn't place too much confidence in their exact values. (You also shouldn't worry about where the "equal" goes—that is, whether 25% of the cases have values less than the 25th percentile or whether 25% of the cases have values less than or *equal* to the 25th percentile. Statistical software packages implement arbitrary rules about where the "equal" goes. ■■■

Average completion times for men and women of different ages are shown in Figure 7.5. You see that for every age group, the average time for men is less than the average time for women. For men and women younger than 45, age doesn't seem to matter very much. The average time for people 25–39 years old is almost the same as the time for people 40–44 years of age. It's after that point that times start to increase.

In the Means dialog box, move time into the Dependent List, agecat8 into Layer 1, and sex into Layer 2.

In the Options dialog box, select Mean, Median, Standard Deviation, and Number of Cases.

Figure 7.5 Completion time by age and gender

AGECAT8 Age group	SEX	Mean	Median	Std. Deviation	N
24 or less	F	4:28:16	4:24:57	0:38:41	1731
	M	4:02:26	3:57:31	0:44:18	1375
	Total	4:16:50	4:15:06	0:43:12	3106
25-39	F	4:34:07	4:31:07	0:42:01	7100
	M	4:05:39	4:00:27	0:43:09	9462
	Total	4:17:51	4:15:01	0:44:56	16562
40-44	F	4:35:40	4:31:27	0:44:11	1272
	M	4:04:48	3:59:16	0:42:09	2616
	Total	4:14:54	4:11:34	0:45:12	3888
45-49	F	4:51:02	4:46:40	0:44:47	712
	M	4:12:36	4:08:59	0:42:35	1746
	Total	4:23:44	4:20:36	0:46:36	2458
50-54	F	4:59:38	4:57:21	0:49:17	375
	M	4:20:46	4:15:34	0:45:41	1209
	Total	4:29:58	4:24:38	0:49:24	1584
55-59	F	5:11:47	5:07:25	0:50:11	130
	M	4:32:09	4:28:36	0:44:25	532
	Total	4:39:56	4:37:42	0:48:13	662
60-64	F	5:25:25	5:28:48	0:46:18	28
	M	4:48:18	4:44:09	0:50:19	227
	Total	4:52:22	4:47:15	0:51:09	255
65+	F	6:02:01	5:56:16	0:47:02	15
	M	5:07:15	5:08:31	0:50:33	95
	Total	5:14:44	5:21:16	0:53:20	110
Total	F	4:35:59	4:32:04	0:43:12	11363
	M	4:08:45	4:03:47	0:44:18	17262
	Total	4:19:33	4:16:27	0:45:50	28625

If you look at the standard deviation column in Figure 7.5, you see that for both men and women, the variability of completion times increases slightly with age. At ages less than 25, the standard deviation for women's completion times is 39 minutes. By ages 50–54, the standard deviation is 49 minutes. It's often the case that the standard deviation increases with increasing values of the mean. (A simple example is children's heights. There's much less variability of heights at younger ages than there is at older ages. As average height increases, so does its spread.)

Figure 7.6 is a bar chart of completion times.

Figure 7.6 Bar chart of completion times by age and gender

In the Define Clustered Bar Summaries for Groups of Cases dialog box, select Other summary function and move hours into the Variable box. Select agecat6 for Category Axis and sex for Define Clusters by.

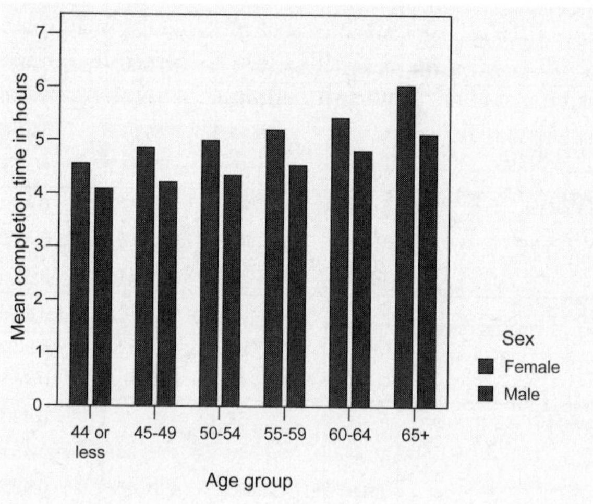

? *Why is the total standard deviation often larger than the standard deviation for males or females?* In Figure 7.5, for each age group there are three standard deviations. The standard deviation for males is computed by using only the values for men's completion times. Only female times are used for the women's standard deviation. To compute the total standard deviation, both men's and women's values are combined and a single standard deviation is calculated. Since the average completion times for men and women vary, when you put all of the values together, you get more variability than for the individual groups. If all 25 to 39-year-old men ran the marathon in exactly four hours and all 25 to 39-year-old women ran it in four and a half hours, the standard deviation for men would be 0 and the standard deviation for women women would be 0. When you computed the standard deviation for all

25 to 39-year-olds, the standard deviation would not be 0, since it would be calculated from a mixture of four-hour times and four-and-a-half-hour times.

Boxplots

Bar charts are convenient for displaying summary information about groups, but they provide very little information about anything other than the value of the measure that you are plotting. From the length of the bars in Figure 7.6, you can tell what the average completion times are. You can't tell anything else about the distribution of the values for the groups. A display that helps you to better visualize the distribution of a variable is the **boxplot**. It simultaneously displays the median, the interquartile range, and the smallest and largest values for a group of cases. A boxplot is more compact than a histogram but doesn't show as much detail. For example, you can't tell if your distribution has a single peak or if there are intervals that have no cases.

Figure 7.7 is an annotated boxplot of completion times for men and women in six age categories.

In the Define Clustered Boxplot Summaries for Groups of Cases dialog box, select hours as the variable, agecat6 as the category axis, and sex to define clusters.

From the Chart Editor menus choose:

Elements
Hide Data Labels

Figure 7.7 Boxplot of completion times by age and gender

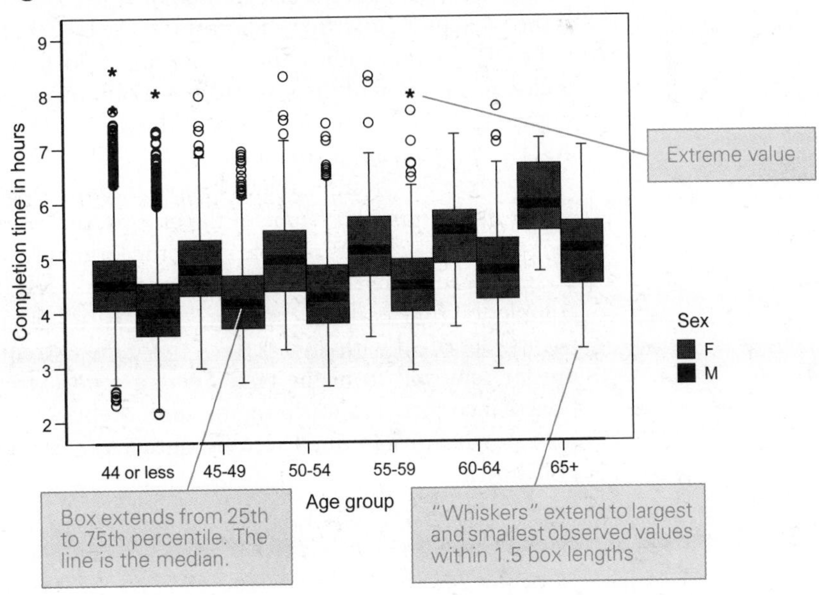

The lower boundary of each box represents the 25th percentile. The upper boundary represents the 75th percentile. (The percentile values known as Tukey's hinges are used to construct the box.) The vertical length of the box represents the interquartile range. Fifty percent of all cases have values within the box. The line inside the box represents the median. Note that the only meaningful scale in the boxplot is the vertical scale. All values are plotted on this scale. The width of a box doesn't represent anything.

In a boxplot, there are two categories of cases with outlying values. Cases with values between 1.5 and 3 box lengths from the upper or lower edge of the box are called **outliers** and are designated with an O. Cases with values of more than 3 box lengths from the upper or lower edge of the box are called **extreme values.** They are designated with asterisks (*). Lines are drawn from the edges of the box to the largest and smallest values that are outside the box but within 1.5 box lengths. (These lines are sometimes called **whiskers**, and the plot is sometimes called a **box-and-whiskers plot.**)

What can you tell about your data from a boxplot? From the median, you can get an idea of the typical value (the central tendency). From the length of the box, you can see how much the values vary (the spread or variability). If the line representing the median is not in the center of the box, you can tell that the distribution of your data values is not symmetric. If the median is closer to the bottom of the box than to the top, there is a tail toward larger values (this is also called **positive skewness**). If the line is closer to the top of the box, there is a tail toward small values (**negative skewness**). The length of the tail is shown by the length of the whiskers and the outlying and extreme points.

The Explore procedure also has more specialized charts and statistics for examining groups. These are discussed in Chapter 15 and Chapter 22.

In Figure 7.7, you see that for most of the groups the median is in the center of the box. (For some of the groups, the sample sizes are small, so the statistics are not as reliable as for the larger groups.) The length of the boxes (the interquartile range) is increasing slightly with age. The median completion time is also increasing with age for both men and women. The points identified with an asterisk (*) are the extremes—large values that are far removed from the rest. There are no extreme values below the boxes since there is a much tighter limit on how fast you can run than on how slowly you can run.There are outlier values on the plot, but they are not shown.

If you have a dataset with a small number of outlying points, you can identify the points on a boxplot by specifying a variable to be used for labeling. You can also identify outlying values by requesting a table of extreme values such as that shown in Figure 7.8. In the table, you see the five highest and lowest completion times for men and women. (To protect determined but slow participants, only the five fastest men and women runners are identified!)

In the Explore dialog box, select time for Dependent List, sex for Factor List, and name for Label Cases by.

In the Statistics dialog box, select Outliers.

Figure 7.8 Outliers for completion times by gender

	SEX			Case Number	Name	Value
TIME Time HR:MIN:SEC	F	Highest	1	28764		8:26:06
			2	28763		8:18:49
			3	28762		8:18:44
			4	28761		8:11:53
			5	28759		7:58:18
		Lowest	1	28	Ndereba (KEN)	2:18:47
			2	36	Alemu (ETH)	2:24:54
			3	39	McCann (AUS)	2:26:04
			4	40	Sobanska (POL)	2:26:08
			5	46	Curti (ITA)	2:28:59
	M	Highest	1	28760		8:01:20
			2	28758		7:58:17
			3	28757		7:44:39
			4	28754		7:40:10
			5	28749		7:26:45
		Lowest	1	1	Kimondiu (KEN)	2:08:52
			2	2	Tergat (KEN)	2:08:56
			3	3	Githuka (KEN)	2:09:00
			4	4	Ouaadi (FRA)	2:09:26
			5	5	Igarashi (JPN)	2:09:35

? *What should I do if I find outliers and extremes on my boxplots?* Use the case numbers to track down the data points and make sure that the values are correct. If these points are the results of data entry or coding errors, correct them.

Marathon Times for Mature Runners

It takes a lot more effort to run a marathon when you are 65 or older than when you are younger. Not many mature runners attempt marathons—the Chicago marathon had only 110 people in the 65-and-over category. To look at their running times in more detail than a histogram or boxplot allows, you can construct what's known as a stem-and-leaf plot.

Stem-and-Leaf Plots

A **stem-and-leaf plot** is a display very much like a histogram. However, more information about the actual data values is preserved. Consider Figure 7.9, which is a stem-and-leaf plot for completion times for runners age 65 and older. It looks like a histogram (turned sideways) because the length of each line corresponds to the number of cases in the interval. However, the cases are represented with different symbols. Each observed value is divided into two components—the leading digit or digits, called the **stem**, and a trailing digit, called the **leaf**. For example, the value 2.3 hours has a stem of 2 and a leaf of 3.

In a stem-and-leaf plot, each row corresponds to a stem and each case is represented by a leaf. More than one row can have the same stem. For example, in Figure 7.9, each stem is subdivided into two rows.

Figure 7.9 Stem-and-leaf plot for age 65+

In the Explore dialog box, move hours into the Dependent List and agecat6 into the Factor List.

In the Plots dialog box, select Stem-and-leaf.

```
completion time in hours Stem-and-Leaf Plot

 Frequency    Stem &  Leaf

     1.00        3 .  2
     9.00        3 .  567777899
    16.00        4 .  0122222333344444
    19.00        4 .  5555567777788889999
    21.00        5 .  001112223333444444444
    26.00        5 .  55555556666667777788899999
     4.00        6 .  1234
    11.00        6 .  55666677778
     3.00        7 .  011

 Stem width:       1.00
 Each leaf:        1 case(s)
```

Look at the first row with the stem of 6 in Figure 7.9. The four leaf values in that row are 1, 2, 3, and 4. What does this mean? In order to translate the stem-and-leaf values into actual numbers, you must look at the stem width given below the plot. In this case, it's 1.00. Leaf values always represent tenths of the stem width, so here they are multiples of one-tenth. You multiply the value in the stem column by the stem width and then add it to the leaf value to get the actual value represented by that leaf. The resulting running times are 6.1, 6.2, 6.3, and 6.4 hours. From the stem-and-leaf plot, you can tell the actual completion times to one decimal place. You couldn't do that from a histogram.

If the stem width were 100, you would multiply each stem by 100 and each leaf by 10 before adding them together. The values for the indicated row would be 610, 620, 630, and 640.

? *How would you make a stem-and-leaf plot of a variable such as income?* For a variable such as income, which has many digits, it's unwieldy and unnecessary to represent each case by the last digit. (Think of how many stems you would have!) Instead, you can look at income to the nearest thousand. For example, you can take a number such as 25,323 and divide it into a stem of 2 and a leaf of 5. In this case, the stem is the ten thousands and the leaf is the thousands. You no longer retain the entire value for the case, but that's not of concern, since annual income differences in the hundreds seldom matter very much. The Explore procedure always displays the stem width under the plot. ■■■

The number of rows used for each stem depends on the number of distinct stem values. If there are a lot of stem values, you'll see one row for each stem. When there are few values of the stem (for example, most completion times are in a small range), each stem can be subdivided into two or more rows, each one corresponding to a range of leaf values.

From the stem-and-leaf plot, you see that the older runners had longer running times than the younger runners. You also see that they did not contribute the outlying long values of eight hours or greater.

? *From this dataset, can you determine the average marathon running times for participants in the Chicago marathon?* Nothing is simple. Note that in this chapter, the term "completion times" was used. That's because the data file contained times only for people who actually completed the marathon. If everyone who started the marathon finished it, running times and completion times would be the same. It would be straightforward to talk about the average time required to run the marathon. However, not all people who start a marathon complete it. People get injured, sick, or decide that they just can't do it after all. You just don't know how fast or slow they would have run if they had made it to the finish line. As a simple example, think of a marathon where only people who could run the distance in under three hours crossed the finish line. Everyone else quit after 10 hours, before crossing the line. Would it be correct to say that it takes three hours to run the marathon? Of course not. If the slow runners had completed the marathon, the time to completion would be much larger.

If the number of dropouts is not large in comparison to the number who completed the marathon, it probably has little effect on the median time required to run the course. As the number of dropouts increases, so does its possible effect. When you looked at completion times for the various age groups, you couldn't tell what percentage of people in each group failed to complete the marathon and how long they had been running before they quit. Such information is very important if you're going to seriously study the effect of age on running marathons.

Eliminating cases from the computation of summary statistics can lead to nonsensical results. For example, if you are studying how long people live after a particular medical treatment and you exclude from your statistics people who have not yet died, your results will be meaningless. There are special statistical techniques, beyond the scope of this book, that deal with data in which some information, such as time to an event, is not available for everyone in the sample. ■■■

Summary

What additional displays are useful for summarizing the distribution of a variable for several groups?

- A stem-and-leaf plot, like a histogram, shows how many cases have various data values. A stem-and-leaf plot preserves more information than a histogram because it does not use the same symbol to represent all cases. Instead, the symbol depends on the actual value for a case.
- A boxplot is a display that shows both the central tendency and variability of the data.
- The length of the box in a boxplot is the distance between the 25th percentile and the 75th percentile. Fifty percent of the data values fall in this range.
- The median is represented by a line in a boxplot. If the median is not in the center of the box, the distribution of values is skewed.

What's Next?

Descriptive statistics such as the mean and variance are useful only for summarizing a variable that is measured on a meaningful scale. To summarize nominal variables or other variables with a small number of possible values, you can count how often various combinations of values occur. That's what the next chapter is about.

How to Explore Distributions

You can use SPSS to look at the distribution of values for a variable. The Explore procedure allows you to:

- Calculate descriptive statistics for all the cases in your data and for subgroups of cases.
- Identify extreme values. These are sometimes due to errors in collecting data or entering it into the computer. If they are correct, they can greatly influence statistical analysis, so you need to be aware of them.
- Calculate the percentiles of a variable's distribution. Again, you can do this both for all cases and for subgroups of cases.

- Generate plots. A variety of plots show graphically how data values are distributed.

▶ To open the Explore dialog box, from the menus choose:

Analyze
  Descriptive Statistics ▶
    Explore...

Figure 7.10 Explore dialog box

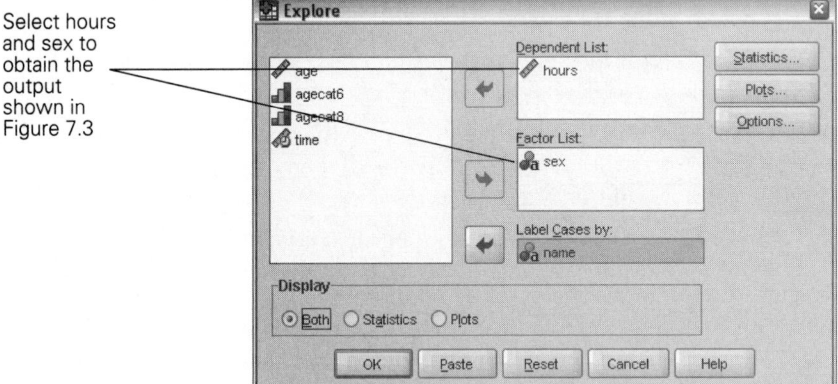

▶ In the Explore dialog box, select one or more numeric variables and move them into the Dependent List. Make sure that either Statistics or Both is selected in the Display group at the bottom left.

▶ If you want to calculate these statistics for subgroups of cases, such as men and women who completed the Chicago marathon, you must specify a factor variable.

A factor variable is simply a variable that distinguishes groups of cases. All of the cases in each group have the same value for the factor variable. If you want to specify a factor variable such as *sex*, select it and move it into the Factor List, as shown in Figure 7.10. You can move more than one variable into the Factor List. If you do, the calculations of subgroup statistics are done separately for the categories of each factor.

▶ If you want to label the five largest and smallest values on the outlier list or identify points on the boxplot, specify the variable used as the label in the Label Cases By box.

Explore Statistics

In the Explore dialog box, click **Statistics** to open the Explore Statistics dialog box (see Figure 7.11), where you can request outliers and additional statistics. (The **Descriptives** check box is selected by default. That's why SPSS calculated descriptive statistics, shown in Figure 7.2, even when you didn't change anything in the Explore Statistics dialog box.)

Figure 7.11 Explore Statistics dialog box

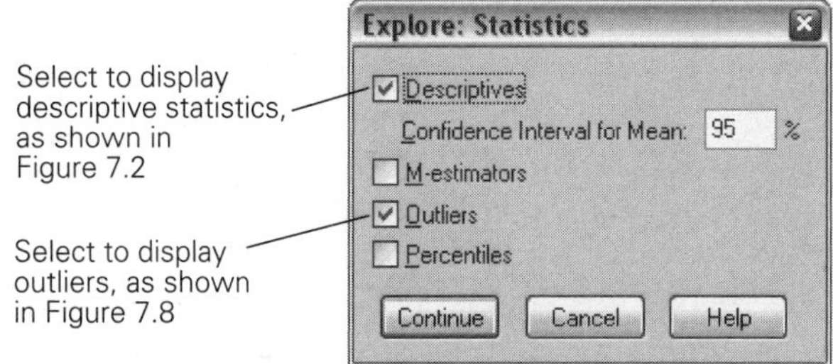

At least one statistic must be selected. In addition to descriptive statistics, you can select from the following:

M-estimators. Statistics that resemble the mean but give weights to observations depending on their distance from a central point.

Outliers. SPSS displays the five highest and five lowest values of the dependent variable. They are identified by case number or sequential position in the data file, as shown in Figure 7.8.

Percentiles. Displays the 5th, 10th, 25th, 50th, 75th, 90th, and 95th percentiles. The 25th, 50th, and 75th percentiles are called **quartiles**; they divide the cases into four equal groups based on the values of the dependent variable.

Graphical Displays

The Explore procedure offers several graphical ways of viewing the distribution of variables. By default, it displays boxplots and stem-and-leaf plots of all dependent variables for each category of each factor variable. To suppress the display of specific plots or to request additional ones, click **Plots** in the Explore dialog box. This opens the Explore Plots dialog box, as shown in Figure 7.12.

For more information about creating and editing charts in SPSS, see Appendix A.

Figure 7.12 Explore Plots dialog box

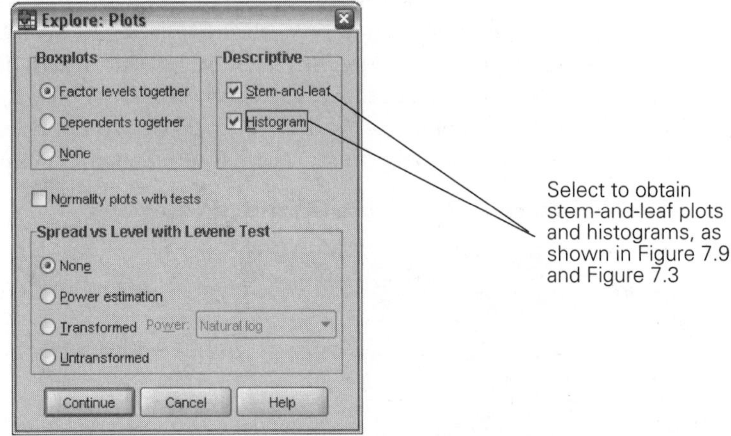

The following groups of options are available:

Boxplots. The Boxplots control group lets you rearrange the display of multiple boxplots or suppress them entirely. The first two alternatives are for situations in which you have more than one dependent variable and at least one factor variable. If you're interested primarily in comparing distributions across categories of the factors, select Factor levels together. If you're more concerned about whether the dependent variables are distributed differently within each factor category, select Dependents together. For example, if you have three variables that are IQ scores for the same children measured at ages 5, 7, and 10 and want to see the distributions of the three variables on the same plot, select Dependents together. On the other hand, if you want to look at differences between males and females at each age, select Factor levels together. This will produce for each variable male and female scores on the same display. If you don't want to see boxplots at all, select None in this group.

Descriptive. The Descriptive control group lets you select stem-and-leaf plots or histograms.

Spread vs. Level with Levene Test. Many statistical procedures, such as those discussed in Chapter 15 and Chapter 16, are dependent on the assumption that a dependent variable has about the same variance ("spread") for the cases with each level of a factor variable. If this assumption doesn't hold, you can sometimes transform the dependent variable so that the assumption does. When you have specified a factor variable, you can request Untransformed spread-versus-level plots to determine whether the dependent variable has the same variance within each factor level. If not, you can request Power estimation, which determines the best transformation to use. Once you know this, you can request Transformed spread-versus-level plots, using the Power drop-down list to indicate the desired transformation. After applying a transformation, you should make a stem-and-leaf plot of the data values to see what effect the transformation has.

The following check box is also available:

Normality plots with tests. Select this check box if you want to test whether the sample comes from a population that has a normal distribution. This causes SPSS to display the normality plots and tests shown in "Examining Normality" on p. 265 in Chapter 13, allowing you to test the null hypothesis that your data are from a normal distribution.

Options

The Explore Options dialog box lets you control the way in which missing data are handled by the Explore procedure. In the Explore dialog box, click Options to open the Explore Options dialog box, as shown in Figure 7.13.

Figure 7.13 Explore Options dialog box

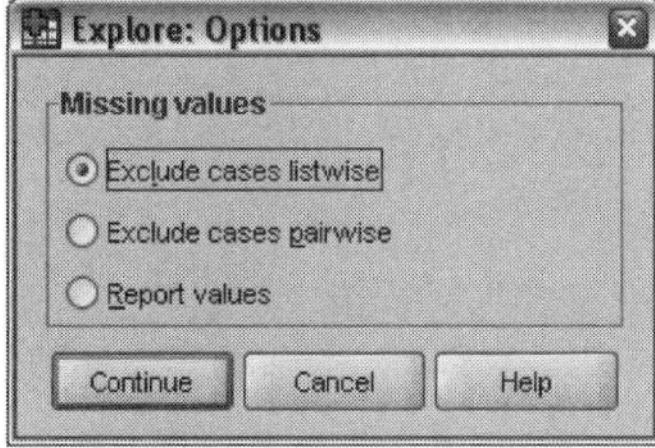

You can choose from the following alternatives:

Exclude cases listwise. When this is selected, cases with a missing value for any of the variables in the Dependent List or the Factor List in the Explore dialog box are omitted from all calculations and plots.

Exclude cases pairwise. When this is selected, each statistic or plot uses all of the cases that have nonmissing information for the variables actually needed for it. Cases with a missing value for one dependent variable are still used to calculate statistics for other dependent variables. This option uses all available data in the computations, but not all of the output is necessarily based on the same cases.

Report values. When this is selected, missing values for a factor variable are treated as a category of the factor. Statistics and plots contain a group of cases that have the factor variable declared missing. Missing values for a dependent variable are included in any frequency tables you request but not in the calculation of statistics.

Exercises

Statistical Concepts

1. Consider the following 20 ages: 21, 22, 22, 22, 26, 28, 30, 31, 32, 34, 36, 36, 36, 36, 38, 39, 40, 40, 41, and 80.

a. Complete the following stem-and-leaf plot for them:

```
AGE Stem-and-Leaf Plot

 Frequency     Stem &  Leaf
     4.00        2 .
     2.00        2 .
     4.00        3 .
     6.00        3 .
     3.00        4 .
     1.00 Extremes

 Stem width:         10
 Each leaf:        1 case(s)
```

b. Complete the following histogram:

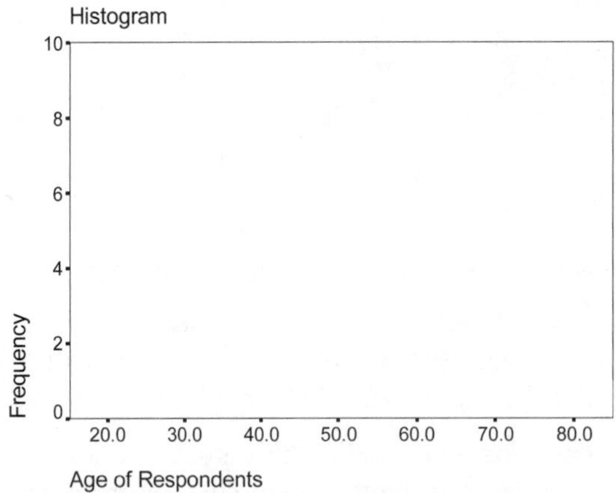

c. Compared to a histogram, what are the advantages of a stem-and-leaf plot?

2. Answer the following questions based on the boxplot below. The plots represent the time it took to ship products from three warehouses.

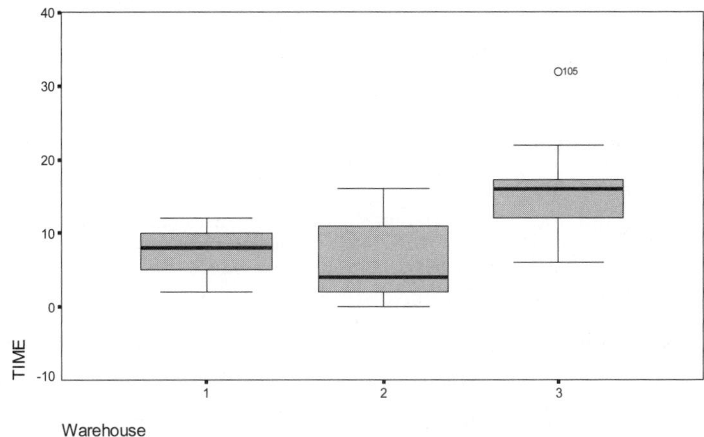

a. Estimate the median for warehouse 1.

b. Estimate the interquartile range for warehouse 2.

c. For warehouse 3, what is the largest value that is not an outlier?

d. Which warehouse has the most variability?

e. If you were to choose one of the warehouses to ship your product, which warehouse would you select and why?

3. Based on the summary statistics shown in the table below, sketch a boxplot of age.

	Age of Respondent												
		95% Confidence Interval for Mean											
	Mean	Lower Bound	Upper Bound	5% Trimmed Mean	Median	Variance	Std. Deviation	Minimum	Maximum	Range	Interquartile Range	Skewness	Kurtosis
Statistic	40.41	39.61	41.21	40.00	39.00	124.511	11.16	19	82	63	15.00	.529	.087
Std. Error	.41											.090	.179

		Percentiles						
		5	10	25	50	75	90	95
Weighted Average (Definition 1)	Age of Respondent	24.00	26.00	32.00	39.00	47.00	56.00	60.00
Tukey's Hinges	Age of Respondent			32.00	39.00	47.00		

Data Analysis

Use the *gss.sav* data file to answer to following questions:

1. Compute descriptive statistics and obtain a stem-and-leaf plot of age separately for people in each of the categories of variable *vote96*.
   a. Look at the stem-and-leaf plot for people who were ineligible. List all of their ages.
   b. Compare the shapes of the stem-and-leaf plots for people who voted and those who did not. What differences and similarities can you see?
   c. Look at the boxplot for voters and nonvoters. Based on the boxplot, summarize all that you can about the distribution of age in the two groups.
2. Look at the relationship between age and finding life exciting (variable *life*). Obtain boxplots of age for each group.
   a. Which group has the most variability? Indicate what criterion of variability you are using.
   b. Do you think the age distribution is symmetric for each group?
   c. For each group, below what value of age do 25% of the ages fall? Above what value do 25% of the ages fall?
   d. What are the medians for each group?
   e. What can you conclude about the relationship between finding life exciting and age?
3. In Chapter 4, you saw that the variable *tvhours* had some unusual values, including one person who claimed to watch television 24 hours per day. To assess the impact of the unusual values on the average hours of television watched, use the Explore procedure and compare the arithmetic mean, the 5% trimmed mean, and the median. Look at a histogram for the television variable again. Which of the measures of central tendency do you think best describe the television-viewing habits of the General Social Survey respondents?

Use the *gssft.sav* data file, which contains data for only full-time workers, to answer the following questions:

4. Use the Explore procedure to calculate descriptive statistics for age (*age*) and respondent's income (*rincdol*), education (*educ*), and hours worked (*hrs1*) for people in each of the job satisfaction categories (*satjob*). Compare the values of the arithmetic means, the medians, and the 5% trimmed means. Make boxplots for both variables. In the *very satisfied* group, identify the cases with the largest and smallest values of education. Summarize the relationship, if any, between job satisfaction, age, education, hours worked, and income.

5. Look at the responses to the questions about general happiness (variable *happy*). Identify characteristics that are associated with people who are *very happy*. Prepare a brief report with appropriate summaries, including graphs.

6. Summarize the relationship between hours worked last week (variable *hrs1*) and amount of Internet use (variable *netcat*).

Use the ***salary.sav*** data file to answer the following questions:

7. The job seniority variable indicates the length of time an employee has been on the job (variable *time*). Make a stem-and-leaf plot of this variable.

   a. How would you characterize the shape of the distribution? Does it have a single peak around which values are concentrated?

   b. What are the five shortest job seniorities? The five longest?

8. Make a boxplot of average time on the job for workers in the four gender/race categories (variable ***sexrace***). Write a brief description of the plot.

Use the *electric.sav* data file to answer the following questions:

9. Compute descriptive statistics and make a stem-and-leaf plot and a boxplot of average diastolic blood pressure (variable *dbp58*).

   a. Describe the distribution. Does it have a single peak? Does the distribution look symmetric? If it has a tail, is it toward large or small values?

   b. From the stem-and-leaf plot, what are the three largest diastolic blood pressure values? What are the values for the cases with the 10 smallest values?

   c. Look at the boxplot. Is the median in the middle of the box? What does that tell you? Are there outliers or extreme observations in the data values? Within what range do the middle 50% of the data values fall? What is the interquartile range for the data?

10. Select a variable from the *electric.sav* file and use the Explore procedure to study its distribution for men who were alive 10 years after the study started and in men who were not alive (variable *vital10*). Write a short summary of results, including appropriate graphs.

8 Counting Responses for Combinations of Variables

How can you study the relationship between two or more variables that have a small number of possible values?

- Why is a frequency table not enough?
- What is a crosstabulation?
- What kinds of percentages can you compute for a crosstabulation, and how do you choose among them?
- What is a dependent variable? An independent variable?
- What if you want to examine more than two variables together?
- How can you use a chart to display a crosstabulation?

Transportation by horse is obsolete. Radios and televisions coexist. Many institutions, commercial and private, are constantly assessing their roles in the face of rapidly changing technologies. Public libraries are no exception. Recently, a national random telephone survey of almost 3,100 people was conducted to gather detailed information about who uses the public library and why (Rodger et al., 2000). The role of the Internet was of particular interest.

In this chapter, you'll examine some of the data from the library survey. You already know that frequency tables of the responses to the individual questions is a good starting point. But if you want to look at the relationships between answers to several questions, you'll have to count the responses given to combinations of questions. When you want to look at the relationship between two or more variables that have a small number of values or categories (sometimes called **categorical variables**), you can use a **crosstabulation**—a table that contains counts of the number of times various combinations of values of two variables occur. For example, you can count how frequently people with different amounts of education use the public library or buy a particular product or service. A crosstabulation can also be used to look for errors in a data file. You should always look at simple relationships that you know should exist between variables to make sure that there are no errors in your data file. For example, if you crosstabulate gender and number of pregnancies and you find pregnant males in your data file, identify these cases and see what's going on. Cases that weren't unusual when you looked at variables individually may be unusual when the values of two or more variables are considered together.

▶ This chapter uses the *library.sav* data file. If you are using software that restricts the number of cases, you should use *lib1500.sav*. For instructions on how to obtain the crosstabulation output shown in this chapter, see "How to Obtain a Crosstabulation" on p. 157.

Library Use and Education

Figure 8.1 is a frequency table of reported library visits. You see that 39% of all people in the sample said they have not been to the public library in the last year. Slightly more than 25% visit the library less than once a month. Fewer than 10% use the library weekly. (Throughout this chapter, we'll assume that people actually use the library when they visit it.) Determining the characteristics of people who use the library is important for tailoring programs to users as well as for targeting programs to attract non-users.

In the Frequencies dialog box, select libfreq.

Figure 8.1 Frequency table for library use

		Frequency	Percent	Valid Percent	Cumulative Percent
Valid	Not in past year	1214	39.2	39.3	39.3
	Less than once a month	791	25.5	25.6	64.9
	Once a month	438	14.1	14.2	79.1
	2 or 3 times a month	358	11.6	11.6	90.7
	Once a week or more	288	9.3	9.3	100.0
	Total	3089	99.7	100.0	
Missing	Don't know	7	.2		
	Refused	1	.0		
	Total	8	.3		
Total		3097	100.0		

Differences in age, gender, education, and income are often associated with differences in everything from church attendance to voting behavior. It wouldn't be surprising if education is related to public library use as well. To examine the relationship between library use and highest degree earned, you want to take each row of the frequency table and subdivide it further based on education. Figure 8.2 is a crosstabulation of library use and highest degree. The rows of the table show library usage; the columns, education. A cell appears in the table for each combination of values of the two variables. The first cell, at the top left of the table, is for *people without a high school diploma who have not used the library in the past year*. You see that 208 people fall into this cell. The second cell in the first row of the table is for *people with only a high school diploma who have not used the library in the past year*. There are 499 people in this cell. Similarly, the very last cell in the table that's not a total tells you that 106 people with college or professional degrees use the library weekly.

To the right and at the bottom of the table are totals—often called **marginal totals** because they are in the table's margin. The marginal totals on the table show the same information as frequency tables for each of the two variables. In the right margin, in the column labeled *Total*, you have the total number of people within each usage category. In the last row, you have the number of people in each of the education categories. The first total of 334 is the number of people who reported having less than a high school education. The very last number—3047—is the total number of people in the table.

From the menus choose:

Analyze
Descriptive Statistics ▸
Crosstabs...

In the Crosstabs dialog box, select the variables libfreq and degree, as shown in Figure 8.14.

Figure 8.2 Crosstabulation of library use and degree

		DEGREE highest degree				
		Less than high school	High school	Some college	College	Total
LIBFREQ frequency of use	Not in past year	208	499	270	215	1192
	Less than once a month	64	229	246	243	782
	Once a month	34	123	144	133	434
	2 or 3 times a month	14	122	106	114	356
	Once a week or more	14	82	81	106	283
Total		334	1055	847	811	3047

123 high school graduates used the library once a month

? *Will the marginal totals that I get in a crosstabulation table always be the same as those I would get from frequency tables for the variables individually?* Not if you have missing values for either of the two variables in the crosstabulation. For example, the crosstabulation in Figure 8.2 includes only cases that have nonmissing values for both education and for library use. The marginal totals for library use are therefore based on cases that have nonmissing values for *both* library use and education. When you make a frequency table for library use, the only cases excluded are those with missing values for library use. That's why there are 3047 cases in the crosstabulation table of education and library use and 3089 cases with valid values in the frequency table for library use alone. ■■■

In Figure 8.2, you see that 208 people without high school diplomas, 499 with high school diplomas, 270 people with some college, and 215 people with college degrees did not use the library. Can you tell from the counts just what the relationship is between education and library use? Of course not, since you cannot simply compare the counts when there are different numbers of people in the four education groups. To compare the groups, you must look at percentages instead of counts. That is, you must look at the percentage of people in each of the education groups who gave each of the library use responses.

Row and Column Percentages

Figure 8.3 contains both the counts and the percentages within columns. From the totals for the first row (at the far right), you see that, overall, 39% of the sample did not visit the library in the last year. Reading across that first row, you also see that 62% of those without high school diplomas, 47% of those with only high school diplomas, 32% of those with some college, and 27% of those with college degrees did not use the public library in the last year. Phrasing the results positively, 73% of college graduates use a public library at least once a year, while only 38% of those who have not graduated from high school use the public library at least once a year. About 4% of those without a high school diploma use the library weekly compared to 13% of those with college degrees. It appears that as education increases so does library use.

In the Crosstabs dialog box, click Cells. Then select Column, as shown in Figure 8.16.

Use the Pivot Table Editor to change labels.

Figure 8.3 Crosstabulation showing column percentages

			DEGREE highest degree				
			Less than high school	High school	Some college	College	Total
LIBFREQ frequency of use	Not in past year	Count	208	499	270	215	1192
		Column %	62.3%	47.3%	31.9%	26.5%	39.1%
	Less than once a month	Count	64	229	246	243	782
		Column %	19.2%	21.7%	29.0%	30.0%	25.7%
	Once a month	Count	34	123	144	133	434
		Column %	10.2%	11.7%	17.0%	16.4%	14.2%
	2 or 3 times a month	Count	14	122	106	114	356
		Column %	4.2%	11.6%	12.5%	14.1%	11.7%
	Once a week or more	Count	14	82	81	106	283
		Column %	4.2%	7.8%	9.6%	13.1%	9.3%
Total		Count	334	1055	847	811	3047
		Column %	100.0%	100.0%	100.0%	100.0%	100.0%

You can change the label to indicate that the column percentages are shown

Column percentages sum to 100% in each column

The percentages that you used to make comparisons are known as **column percentages**, since they express the number of cases in each cell of the table as a percentage of its column total. That is, for each education level, they tell you the distribution of library use. The column percentages sum up to 100% for each of the columns.

You can also calculate row percentages for the table. **Row percentages** tell you what percentage of the total cases in a row fall into each of the columns. For each library use category, they tell you the percentage of cases in each education group. (You can also compute what are called **total percentages**. The count in each cell of the table is expressed as a percentage of the total number of cases in the table.) Figure 8.4 contains counts and row percentages for our example.

Figure 8.4 Crosstabulation showing row percentages

In the Crosstabs dialog box, click Cells. Then select Row.

			DEGREE highest degree				
			Less than high school	High school	Some college	College	Total
LIBFREQ frequency of use	Not in past year	Count	208	499	270	215	1192
		Row %	17.4%	41.9%	22.7%	18.0%	100.0%
	Less than once a month	Count	64	229	246	243	782
		Row %	8.2%	29.3%	31.5%	31.1%	100.0%
	Once a month	Count	34	123	144	133	434
		Row %	7.8%	28.3%	33.2%	30.6%	100.0%
	2 or 3 times a month	Count	14	122	106	114	356
		Row %	3.9%	34.3%	29.8%	32.0%	100.0%
	Once a week or more	Count	14	82	81	106	283
		Row %	4.9%	29.0%	28.6%	37.5%	100.0%
Total		Count	334	1055	847	811	3047
		Row %	11.0%	34.6%	27.8%	26.6%	100.0%

Default label changed

Row percentages sum to 100% across each row

From the row percentages, you see that 17% of people who did not use the library in the past year did not graduate from high school, 42% had only a high school diploma, 23% had some college, and 18% were college graduates. In this example, the row percentages aren't very helpful, since you can't make much sense of them without taking into account the overall percentages of cases in each of the education categories. You can't tell whether the large percentage of high school graduates in the *non-user* category is due to a large number of high school graduates in your sample or to low library usage in this group.

? *How can I tell whether a table contains row or column percentages?* If the column labeled *Total* shows all 100%, the table contains row percentages, which necessarily sum to 100 for each row. If the row labeled *Total* contains all 100%, the table contains column percentages.

For a particular table, you must determine whether the row or column percentages answer the question of interest. This can be done easily if one of the variables can be thought of as an independent variable and the other as a dependent variable. An **independent variable** is a variable that is thought to influence another variable, the **dependent variable**. For example, if you are studying the incidence of lung cancer in smokers and nonsmokers, smoking is the independent variable. Smoking influences whether people get cancer, the dependent variable. Similarly, if you are studying the income categories of men and women, gender is the independent variable because it might influence how much you get paid.

If you can identify one of your variables as independent and the other as dependent, then you should compute percentages so that they sum to 100 for each category of the independent variable. In other words, what you want to see is the same number of people in each of the categories of the independent variable. Having the percentages sum to 100 for each category of the independent variable is the equivalent of having 100 cases in each category. For example, you want 100 smokers and 100 nonsmokers. Then you can compare the incidence of lung cancer in the two groups. In the current example, *education category* is the independent variable and *library use* is the dependent variable. That means you'd like to see 100 people in each of the education categories. Since education is the column variable in Figure 8.3, you use column percentages that sum to 100 for each category of education.

? *Are you sure that college graduates really use the public library more than those without a college degree?* No. The results in this chapter are based on a survey in which people volunteered how much they use the public library. It's always possible that college graduates feel guiltier than nongraduates about their failure to use the public library and hence report inflated usage. ■■■

Bar Charts

You can display the results of a crosstabulation in a clustered bar chart. Consider Figure 8.5, which is a bar chart of library use by education. The length of a bar tells you the number of cases in a category.

In the Define Clustered Bar Summaries for Groups of Cases legacy dialog box, select degree as the category axis and libfreq to define clusters.

Figure 8.5 Bar chart of library use by education

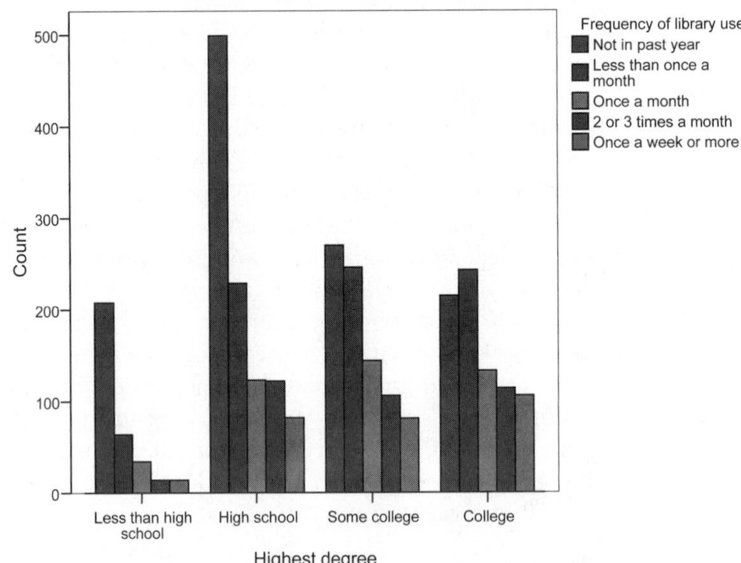

There is a cluster of bars for each of the four education categories. Within each cluster, there is a bar for each of the five library usage categories. Since there are unequal numbers of people in the education categories, comparing bar lengths across education categories presents the same problem as looking at simple counts in a crosstabulation. All you can really do with this bar chart is compare bar lengths within a cluster and see whether the patterns are the same across clusters.

From the Chart Editor menus choose:

Edit
Select Chart

In the Properties dialog box, click the Variables tab. For libfreq, select Stack and then click Apply.

Figure 8.6 Stacked bar chart

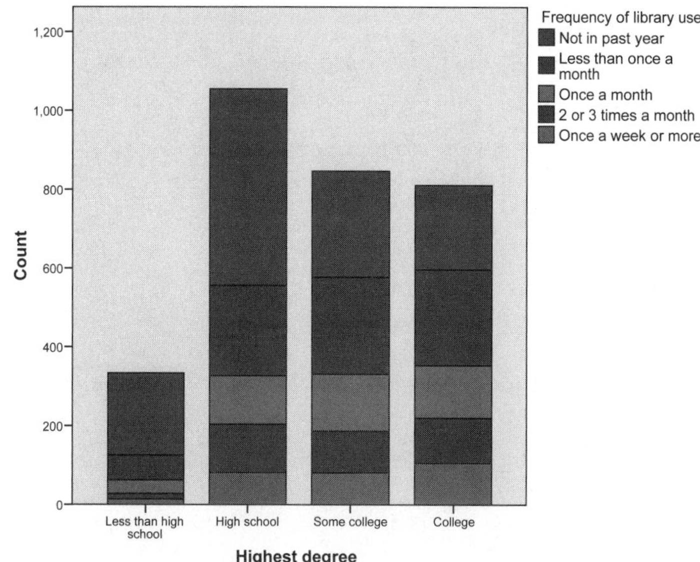

Stacked Bar Charts

You can stack the bars in a clustered bar chart one on top of the other. The result is the stacked bar chart shown in Figure 8.6. Now it's easier to see for each education category the percentage of people in each of the library use categories. However, the lengths of the bars aren't equal for the four education categories, so that still gets in the way.

Ideally, you want each of the bars to be of the same length, so you can easily compare the areas across bars. What you'd really like to see is a plot of the column percentages from Figure 8.3. You can do this by turning the counts in each bar into percentages, as shown in Figure 8.7. Now each of the bars has the same length, and you can easily compare the library use distributions across education. You see that people in the lowest education group are the least likely to use the library. People with college degrees use the library most frequently.

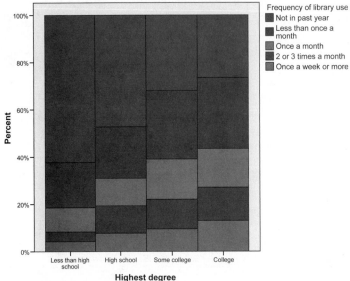

In the Chart Editor, activate the chart in Figure 8.6. From the Options menu, choose Scale to 100%.

Figure 8.7 Stacked bar chart with percentage scale

Adding Control Variables

So far, you've considered the relationship between education and library use. It's possible that if you consider additional variables, the relationship you've seen between the two variables may change. For example, it may be that the relationship between education and library use is different for people with and without children under the age of 17. Children's assignments may compel parents at all educational levels to visit the library more frequently. To test this, you can make separate tables of library use and education for people with and without children under the age of 17. The presence of children is then called a **control variable**, since its effect is removed, or "controlled for." Figure 8.8 shows crosstabulations for those with and without children. Instead of showing two separate tables, one for people with school-age children and one for people without school-age children, the two tables are interleaved so that you can easily see the effect of children for each education category.

In the Crosstabs dialog box, select kids for the row variable, libfreq for the column variable, and degree for the layer variable. Display row percentages.

Figure 8.8 Library use by education and presence of children

DEGREE highest degree	KIDS kids lt 17		LIBFREQ frequency of use: Not in past year	Less than once a month	Once a month	2 or 3 times a month	Once a week or more	Total
Less than high school	None	Count	136	33	13	3	3	188
		Row %	72.3%	17.6%	6.9%	1.6%	1.6%	100.0%
	One or more	Count	72	31	21	11	11	146
		Row %	49.3%	21.2%	14.4%	7.5%	7.5%	100.0%
High school	None	Count	332	124	56	62	37	611
		Row %	54.3%	20.3%	9.2%	10.1%	6.1%	100.0%
	One or more	Count	167	104	67	60	43	441
		Row %	37.9%	23.6%	15.2%	13.6%	9.8%	100.0%
Some college	None	Count	174	166	75	57	44	516
		Row %	33.7%	32.2%	14.5%	11.0%	8.5%	100.0%
	One or more	Count	96	80	69	48	36	329
		Row %	29.2%	24.3%	21.0%	14.6%	10.9%	100.0%
College	None	Count	160	153	72	65	57	507
		Row %	31.6%	30.2%	14.2%	12.8%	11.2%	100.0%
	One or more	Count	55	88	61	49	48	301
		Row %	18.3%	29.2%	20.3%	16.3%	15.9%	100.0%

Each cell contains counts and row percentages. You can see that for each education group, the presence of children increases the likelihood of library use.

In the Define Clustered Bar Summaries for Groups of Cases legacy dialog box, select Other statistic and select the variable libuseyr. Select degree for Category Axis and kids for Define Clusters by.

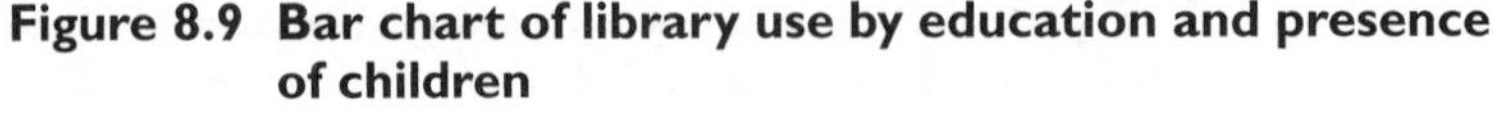
Figure 8.9 Bar chart of library use by education and presence of children

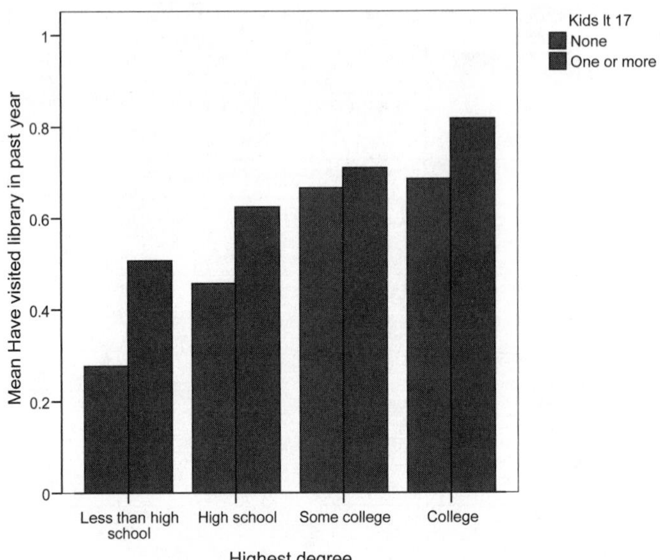

Figure 8.9 presents this result graphically. To simplify the bar chart, only the percentage *using* the library in the last year is plotted. Note that the increase in library use when children are present is largest for those who did not attend college. Children needing a ride to the library account for a large increase in library visits by adults. (The question in the survey was, "In the past year, have you, yourself, visited a public library?" It's possible that during library visits, adults with children don't personally use the library.)

Library Use and the Internet

You've looked at the relationship between library use, education, and the presence of young children. The possible effect of Internet use has been ignored. To look at the effect of the Internet, start simply. Figure 8.10 is a stacked bar chart of library use for those who use the Internet and those who don't have access to the Internet. (The study grouped people into three categories: Internet users, those who don't have Internet access, and those who have access but don't use the Internet. The last group had few people, so we will exclude it from our analyses. As an exercise, you can determine whether those who had access but didn't use the Internet behaved more like non-users or users.)

Figure 8.10 Stacked bar chart of frequency of library use by Internet use

In the Define Stacked Bar Summaries for Groups of Cases legacy dialog box, select internet as the category axis and libfreq to define stacks. In the Chart Editor, select either bar. From the Options menu, choose Scale to 100%.

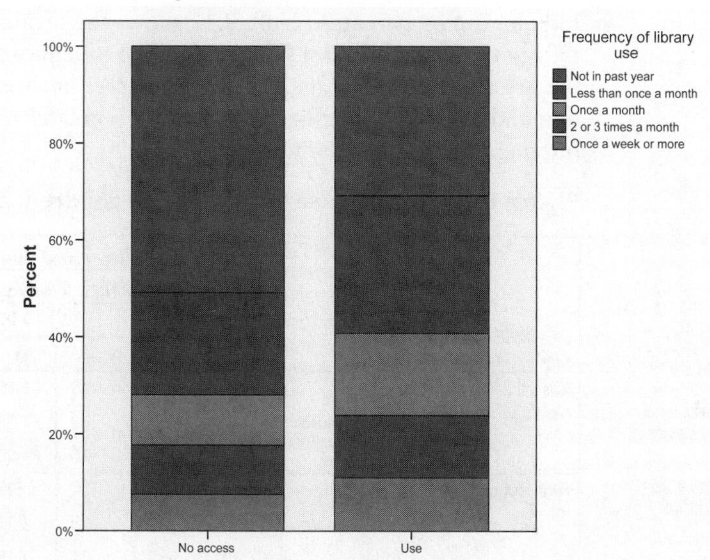

You see that, overall, Internet users are more likely to use the library than non-users. About 30% of Internet users haven't used the library in the last year, while close to 50% of those without Internet access have not used the library. From Figure 8.11, which contains total percentages, you see that almost 40% of the sample used both the Internet and the library. More than 20% didn't use either. Almost 22% used the library but not the Internet, and 17% used the Internet but not the library.

Figure 8.11 Internet and library use with total percentages

In the Crosstabs dialog box, select internet as the row variable and libuseyr as the column variable.

In the Cell Display dialog box, select Total in the Percentages group.

			LIBUSEYR Have visited library in past year		
			No	Yes	Total
INTERNET internet status	No access	Count	646	629	1275
		% of Total	22.2%	21.6%	43.8%
	Use	Count	502	1132	1634
		% of Total	17.3%	38.9%	56.2%
Total		Count	1148	1761	2909
		% of Total	39.5%	60.5%	100.0%

You've already seen that school-age children and education are associated with library visits. Does this relationship still hold when controlling for Internet use by parents? Figure 8.12 shows the proportion of people using the library during the past year, controlling for education, Internet use, and the presence of school-age children. You see that for almost every education and child combination, library usage is greater for people who use the Internet than for those who do not.

In the Means dialog box, select libuseyr as the dependent variable, internet as the Layer 1 variable, kids as the Layer 2 variable, and degree as the Layer 3 variable.

Use the Pivot Table Editor to rearrange the table.

Figure 8.12 Library use by education, children, and Internet use

		LIBUSEYR Have visited library in past year					
		INTERNET internet status					
		No access		Use		Total	
DEGREE highest degree	KIDS kids lt 17	Mean	N	Mean	N	Mean	N
Less than high school	None	.22	152	.79	19	.28	171
	One or more	.40	98	.73	37	.49	135
	Total	.29	250	.75	56	.37	306
High school	None	.40	380	.54	193	.45	573
	One or more	.58	179	.68	222	.63	401
	Total	.46	559	.61	415	.52	974
Some college	None	.63	205	.67	281	.65	486
	One or more	.60	89	.74	223	.70	312
	Total	.62	294	.70	504	.67	798
College	None	.69	112	.69	375	.69	487
	One or more	.85	33	.81	262	.82	295
	Total	.72	145	.74	637	.74	782
Total	None	.46	849	.65	868	.56	1717
	One or more	.56	399	.74	744	.68	1143
	Total	.49	1248	.69	1612	.61	2860

? *How can you get means into cells of a crosstabulation?* The output in Figure 8.12 is from the Means procedure. Since the variable *libuseyr* is coded as 0 if a library has not been visited in the past year and 1 if it has, the mean of the variable is the proportion of people visiting a library in the past year. ■■■

The relationship between Internet use and library use is most noticeable at lower education levels. For example, 75% of Internet users without a high school diploma use the library, compared to 29% of those who are not Internet users. (Note, however, that there are only 56 people without a high school diploma who use the Internet, so you can't put too much faith in the observed percentage.) For those with only a high school diploma, you see the same effect. More than 60% of those who use the Internet also use the library, compared to 46% of those who don't use the Internet. Internet use had little effect on library use in college graduates. The results are furthered summarized in Figure 8.13.

Figure 8.13 Bar chart of education, children, and Internet use

In the Define Clustered Bar Summaries for Groups of Cases legacy dialog box, select Other statistic and select the variable libuseyr. Select degree for Category Axis, internet for Define Clusters by, and kids for Panel by Columns.

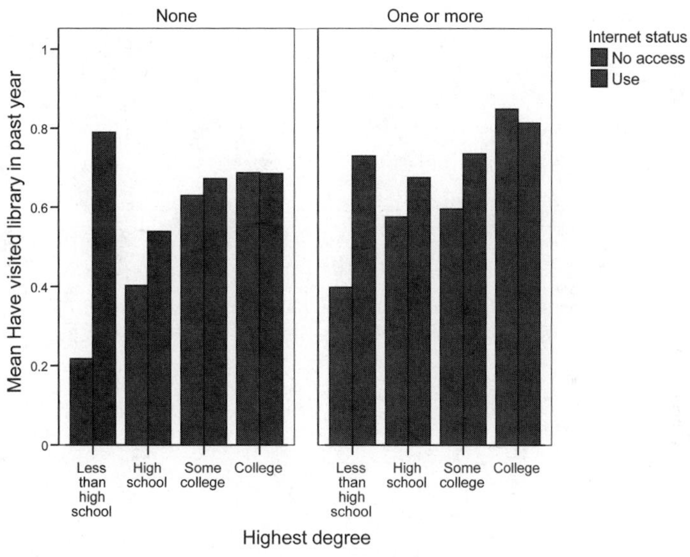

The chart tells an interesting story. Library use increases with education, the presence of children under age 17, and use of the Internet. By analyzing the effects of education, children, and the Internet simultaneously, you were able to determine the effects of each of the variables while "controlling" for the other variables. If you had analyzed the effect of Internet use alone, ignoring education and children, you couldn't tell if differences that you found between Internet users and non-users were really due to differences in education or the presence of school-age children. Your comparison of Internet use would have been **confounded** by the possible effects of education and presence of school-age children.

? *Are people going to the library to use the Internet?* That would certainly be an easy explanation for the observed association between library use and Internet use. Fortunately, the survey was well designed and people were asked where they most often access the Internet. Only 19 people who used the Internet reported that the public library was where they accessed the Internet most often. ■■■

Summary

How can you study the relationship between two or more variables that have a small number of possible values?

- A crosstabulation shows the number of cases that have particular combinations of values for two or more variables.
- The number of cases in each cell of a crosstabulation can be expressed as the percentage of all cases in that row (the row percentage) or the percentage of all cases in that column (the column percentage).
- A variable that is thought to influence the values of another variable is called an independent variable.
- The variable that is influenced is called the dependent variable.
- If there is an independent variable, percentages should be calculated so that they sum to 100% for each category of the independent variable.
- When you have more than two variables, you can make separate crosstabulations for each of the combinations of values of the other variables.
- Bar charts can be used to display a crosstabulation graphically.

What's Next?

So far, you've used crosstabulations only to summarize the relationship between two variables. In Chapter 17, you'll learn how to compute statistical tests to determine whether the two variables in a crosstabulation are related. In Chapter 19, you'll compute statistics that measure the strength of the relationship between the two variables in a crosstabulation. In Chapter 9, you'll see how scatterplots can be used to display the values of two variables that are measured on a meaningful numeric scale.

How to Obtain a Crosstabulation

This section shows how to obtain crosstabulations of two or more variables with SPSS. The Crosstabs procedure tabulates the different combinations of values that occur for two or more variables. You should use crosstabulations only if your variables have a small number of distinct values. In addition, the Crosstabs command can:

- Display percentages, expected counts, and residuals within each cell.
- Calculate tests of statistical independence and measures of association for pairs of variables ("bivariate" statistics).
- Control some aspects of the table format.
- Display clustered bar charts.

▶ To open the Crosstabs dialog box, from the menus choose:

Analyze
 Descriptive Statistics ▶
  Crosstabs...

Figure 8.14 Crosstabs dialog box

Select libfreq and degree to obtain the crosstabulation shown in Figure 8.2

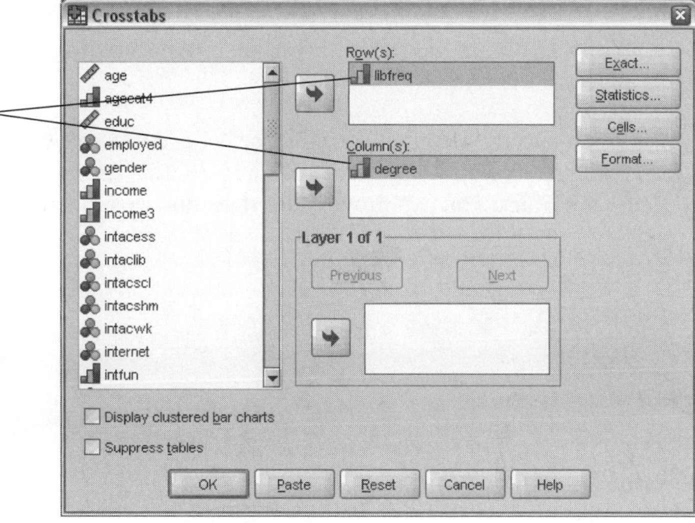

▶ In the Crosstabs dialog box, select a categorical variable and move it into the Row(s) list. Select another categorical variable and move it into the Column(s) list.

This produces a single table with the categories of the first variable down the left side, defining the rows, and the categories of the second variable across the top, defining the columns. Each cell contains the number of cases with the corresponding values for the two variables.

▶ Move several variables into the Row(s) list or the Column(s) list or both to obtain multiple bivariate crosstabulations.

SPSS displays a table for each combination of row and column variables. If one list has two variables and the other list has three, six tables are displayed. In addition, you can display a clustered bar chart. For each row in the crosstabulation, you will see a set of bars representing the counts in each cell in the row. Bars from the same row are clustered together.

The following option is also available:

Suppress tables. Lets you suppress the display of the actual tables in case you are interested only in the statistics (see "Bivariate Statistics" on p. 161).

Layers: Three or More Variables at Once

You can obtain a different bivariate crosstabulation for each category of a control variable.

▶ To specify a control variable, select the variable and move it into the Layer 1 list, as shown in Figure 8.15. As with row and column variables, you can move more than one variable into this list.

Figure 8.15 Selecting a layer variable

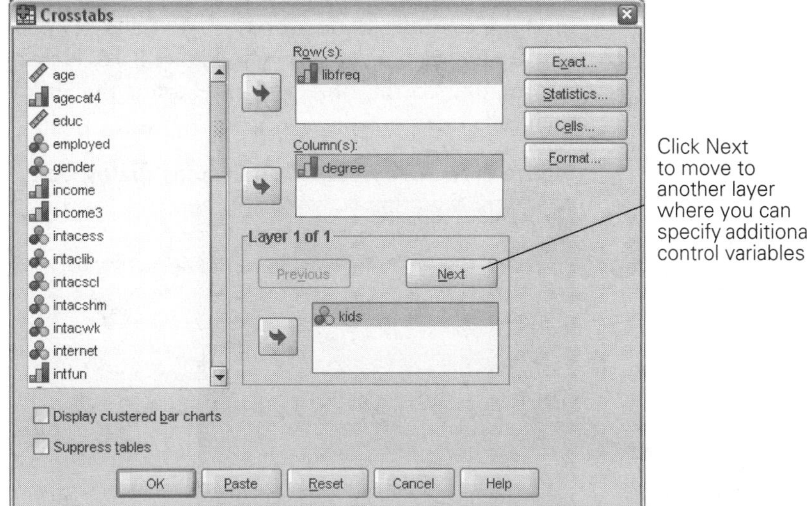

SPSS displays all of the bivariate crosstabulations for each category of the first control variable, then for each category of the second control variable, and so on, through the list.

▶ Optionally, you can click Next to specify additional layers of control variables.

If you move one or more variables into the Layer 2 list, SPSS displays all of the bivariate crosstabulations for each category of the Layer 1 variables, combined with each category of the Layer 2 variables. After you have a Layer 2 list, you can continue to click Next and build additional layers. Or you can click Previous to go back and add or remove variables in the previous layer.

Cells: Percentages, Expected Counts, and Residuals

In the Crosstabs dialog box, click **Cells** to open the Crosstabs Cell Display dialog box, as shown in Figure 8.16. Here you can add additional information to the observed case count that is displayed by default in each cell.

Figure 8.16 Crosstabs Cell Display dialog box

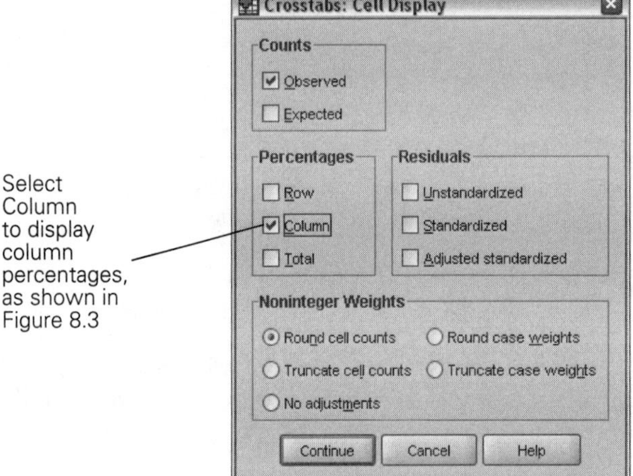

The following groups of options are available:

Counts. The observed count is the number of cases in the cell. The expected count is the number of cases that would be in the cell if the row and column variables were statistically independent.

Percentages. Row percentages sum to 100% across each row of the table. Column percentages sum to 100% down each column. Total percentages sum to 100% over the table as a whole.

Residuals. Residuals are the difference between observed and expected cell counts.

Bivariate Statistics

In the Crosstabs dialog box, click **Statistics** to open the Crosstabs Statistics dialog box, which allows you to choose statistics to be displayed along with your crosstabulation. (You will learn more about these statistics in Chapter 17 and Chapter 19.)

Figure 8.17 Crosstabs Statistics dialog box

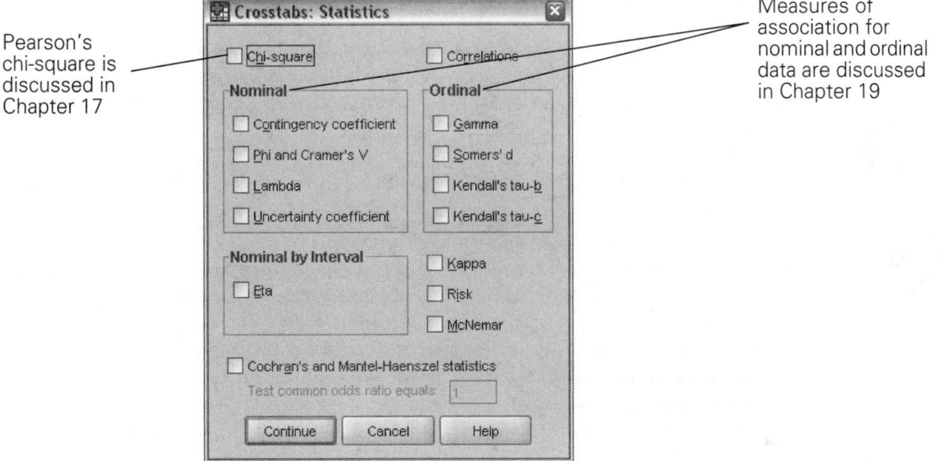

Format: Adjusting the Table Format

In the Crosstabs dialog box, click **Format** to open the Crosstabs Table Format dialog box, which allows you to modify the format of the tables.

Figure 8.18 Crosstabs Table Format dialog box

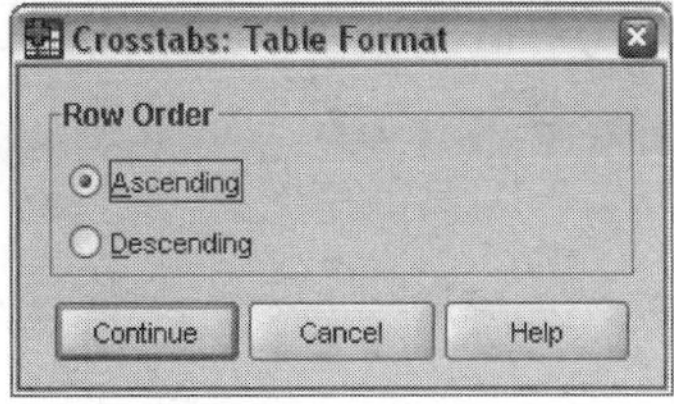

Row Order. Determines whether the rows of the table are listed in ascending or descending order of the categories of the row variable.

Exercises

Statistical Concepts

1. Indicate whether you would use the Frequencies, Crosstabs, or Means procedure to find the following:

    a. The average age of purchasers of different brands of a product

    b. The number of married, single, widowed, divorced, and never-married purchasers of different brands of a product

    c. The number of purchasers of each brand

    d. The average age of male and female purchasers of each brand

    e. The number of men and women in each of the marital categories who purchase each brand

2. Identify the dependent and independent variables, if possible, for each of the following pairs of variables:

    a. Satisfaction with job and race

    b. Belief in life after death and gender

    c. Astrological sign and excitement with life

    d. Mother's highest degree and daughter's highest degree

    e. Happiness with one's marriage and belief in life after death

3. If you construct a crosstabulation for each of the pairs of variables in the previous question, with the first variable forming the rows of the table and the second variable forming the columns, should you calculate row or column percentages? Answer for all five pairs.

4. The following table indicates whether each of 20 people owns or rents a home (1=*own*, 2=*rents*) and how satisfied they are with city services (1=*not satisfied*, 2=*satisfied*, 3=*very satisfied*).

Person	Owner	Satisfied
1	1	1
2	1	1
3	1	1
4	1	1
5	1	1
6	1	2
7	1	3
8	1	3
9	2	2
10	2	2
11	2	3
12	2	3
13	2	3
14	2	3
15	2	3
16	2	3
17	2	3
18	2	3
19	2	3
20	2	3

a. Summarize the data by filling in the values for the cell counts and marginals of the following crosstabulation:

Count

		SATISFY			
		not satisfied	satisfied	very satisfied	Total
OWNER	own				
	rents				
Total					

b. Calculate the row and column percentages.

c. What percentage of the sample are homeowners?

d. What percentage of the sample are very satisfied with city services?

e. What percentage of homeowners are very satisfied with city services? Of non-owners?

5. Fill in the missing information in the following table:

			DEPTH			Total
			Small	Medium	Large	
CURE	No	Count	29		30	69
		% within CURE				
		% within DEPTH				
	Yes	Count	15	10	20	
		% within CURE				
		% within DEPTH				
Total		Count	44	20	50	114
		% within CURE				
		% within DEPTH				

If you want to show the relationship between the cure rate and depth of tumor, which percentages would you look at?

6. Your local newspaper's Sunday supplement contains an article on job satisfaction and marital status. The article contains the following table. The authors conclude that marriage makes people more satisfied with their jobs, since 62% of the very satisfied people are married, while only 17% have never been married. Comment on the conclusions from the study.

			Job or Housework				Total
			Very satisfied	Mod satisfied	A little dissatisfied	Very dissatisfied	
MARITAL STATUS	MARRIED	Count	302	257	58	28	645
		% within Job or Housework	61.6%	54.4%	45.7%	52.8%	56.5%
	WIDOWED	Count	26	19	5	1	51
		% within Job or Housework	5.3%	4.0%	3.9%	1.9%	4.5%
	DIVORCED	Count	66	77	20	11	174
		% within Job or Housework	13.5%	16.3%	15.7%	20.8%	15.2%
	SEPARATED	Count	12	14	7	1	34
		% within Job or Housework	2.4%	3.0%	5.5%	1.9%	3.0%
	NEVER MARRIED	Count	84	105	37	12	238
		% within Job or Housework	17.1%	22.2%	29.1%	22.6%	20.8%
Total		Count	490	472	127	53	1142
		% within Job or Housework	100.0%	100.0%	100.0%	100.0%	100.0%

7. Below is a crosstabulation of belief in life after death and highest degree earned. Calculate the appropriate percentages, and write a few sentences summarizing the table.

BELIEF IN LIFE AFTER DEATH * RS Highest Degree Crosstabulation

			RS Highest Degree					
			Less than HS	High school	Junior college	Bachelor	Graduate	Total
BELIEF IN LIFE AFTER DEATH	YES	Count	132	392	45	110	52	731
		% within RS Highest Degree						80.7%
	NO	Count	37	82	12	29	15	175
		% within RS Highest Degree						19.3%
Total		Count	169	474	57	139	67	906
		% within RS Highest Degree	100.0%	100.0%	100.0%	100.0%	100.0%	100.0%

8. For which of the following pairs of variables do you think a crosstabulation would be appropriate?

a. Weight in pounds and daily intake in calories

b. Number of cars and highest degree earned

c. Body temperature in degrees and survival after an operation

d. Eye color and undergraduate grade point average

9. You are interested in studying the relationship between highest degree earned by a person in school, the person's marital status, and several other variables. Which SPSS procedure would you use to investigate the relationship of these two variables and the following:

a. The number of hours of television watched per week

b. Religious affiliation

c. Job satisfaction

d. Number of siblings

e. Zodiac sign

Data Analysis

Use the *gss.sav* data file to answer the following questions:

1. What percentage of men and what percentage of women use computers (*usecomp)*? Use the Internet (*usenet*)? Use the World Wide Web (*useweb*)? Use e-mail (*usemail*)?

2. Describe the relationship between time on the Internet (variable *netcat*) and gender. Does the relationship change when you control for education (variable *degree*)?

3. Make a crosstabulation showing the relationship between age (variable *agecat*), gender, and use of the Internet.

   a. What percentage of males under 30 years of age use the Internet? What percentage of females under 30 years of age use the Internet? Is the relationship between age and Internet use similar for men and women? Prepare a chart to support your response.

   b. Make a stacked bar chart showing the distribution of time on the Internet (variable *netcat*) for men and women. What does it show?

4. Make a crosstabulation to examine the relationship between highest degree earned (variable *degree*) and a person's perception of life (variable *life*).

   a. How many people without a high school diploma find life exciting?

   b. What percentage of people without a high school diploma find life exciting?

   c. Of the people who find life exciting, what percentage do not have a high school diploma?

   d. If *degree* is the column variable and *life* is the row variable, what kind of percentages should you compute for the table?

   e. Summarize the relationship, if any, between perception of life and highest degree earned.

5. What kinds of people claim that they will continue to keep working if they strike it rich (variable *richwork*)? Use the Crosstabs procedure to examine the relationship between *richwork* and characteristics such as gender, education, job satisfaction, and perception of life (variables *sex*, *degree*, *satjob*, and *life*). Write a brief report summarizing your findings. Use graphical displays when appropriate.

6. Belief in life after death (variable *postlife*) may be related to characteristics such as age, education, income, marital status, and so on. How many characteristics can you identify that are associated with belief in life after death? Write a short report about your findings.

7. Look at the relationship between a person's highest degree (*degree*) and his or her general happiness (*happy*). Does there seem to be a relationship between the two variables?

8. Repeat the analysis you did in question 4 separately for men and women. Does the relationship between the two variables appear to be similar for males and for females? Explain how you reached your conclusion.

9. How many characteristics can you identify that distinguish people who voted in 1996 from those who did not (variable *vote96*)? For example, are degree, gender, and marital status related to the likelihood of voting?

10. You've been hired to do a "postmortem" on the election of 1996. (That's the last election covered by the GSS.) Write a short paper describing what types of people voted for each of the candidates (variable *pres96*).

11. For total family income (variable *income*), compute a new variable (*income4*) that has the value of 1 if a person is in the first quartile of family income, 2 if in the second, and so on. (See Appendix B.)

a. Crosstabulate family income in quartiles and job satisfaction.

b. Would you consider family income or job satisfaction to be the dependent variable? Explain.

c. Compute the appropriate percentages for studying the relationship between family income and job satisfaction. How many people in each of the income quartiles are very satisfied with their jobs?

d. Make a bar chart showing the relationship between average family income and job satisfaction. Use the variable *incomdol*.

e. Is the relationship between income and job satisfaction the same for men and women? Support your answer.

f. Make a stacked bar chart showing job satisfaction for each of the quartiles of total family income. Summarize the results.

g. Repeat the question using respondent's income (variables *rincome* and *rincdol*) instead of total family income. Do your conclusions about the relationship of income and job satisfaction change? How?

12. Are people in all of the zodiac signs (variable *zodiac*) equally happy (variable *happy*)? Support your conclusion.

Use the *library.sav* file to answer the following questions:

13. In the sample, what percentage of men and what percentage of women have visited the library in the past year (*libuseyr*)? What percentage of men and what percentage of women use the Internet (variable *Internet*)? How does Internet use in the library survey compare to the Internet use in the General Social Survey?

14. What percentage of men and what percentage of women strongly agree with the statement that they enjoy surfing the Internet (*surfnet*)? What percentage strongly disagree?

15. Consider responses to the variable *surflib*. What percentage of men and what percentage of women say they strongly agree with the statement that they enjoy browsing a library's collection?

16. Examine the relationship between age (variable *agecat4*) and income (variable *income3*) and library use in the past year (variable *libuseyr*). Is the relationship with library use similar for age and income? Prepare bar charts showing the percentage using the library in the past year for people classified by age and by income.

Use the *manners.sav* data file to answer the following question:

17. Write a paper describing the relationship between manners, their importance and practice, and gender, age, education, and income.

Use the *crimjust.sav* data file to answer the following questions:

18. Age, education, and gender may affect peoples' views toward crime and the criminal justice system. Support or refute the following statements with appropriate summaries from the data file:

a. Women rate the criminal justice system (variable *cjstrate*) as poorer than do men.

b. Older people think that crime has increased more in the past five years than do younger people.

c. Men think that requiring a letter of apology from the offender (variable *apology*) is more important than do women.

d. Compared to older people, younger people think that essays (variable *essay*) and letters of apology to the victim (variable *apology*) are very important.

e. College graduates think that letters of apology and essays are less important than do people without college degrees.

19. If someone in your household is a victim of crime (variable *hhcrime*), how does that influence your evaluation of the criminal justice system? Write a short paper comparing the responses of people with and without crime in their households to questions about the criminal justice system (variable *cjsrate*) and the performance of judges (variable *judgrate*), prosecutors (variable *proscrat*), and police (variable *policrat*).

Use the *salary.sav* data file to answer the following questions:

20. Consider the relationship between the gender/race categories (variable *sexrace*) and job category (variable *jobcat*).

a. How many white males are employed as clerical workers? What percentage of white males are clerical workers? Of clerical workers, what percentage are white males?

b. If you make a crosstabulation table in which *sexrace* is the column variable and *jobcat* is the row variable, would you compute row or column percentages? Why? Make such a table and write a short paragraph summarizing your results.

21. On the Transform menu, use the Recode facility to create a new variable named *degree*. Compute it in the following way from years of education (variable *edlevel*):

edlevel	degree	Value label
0–11	1	Less than high school
12	2	High school diploma
13–15	3	Some college
16+	4	College degree

You can assign the value labels in the Data Editor after the variable is computed.

a. What percentage of white males have college degrees? White females? Nonwhite males and females?

b. Write a paragraph summarizing the relationship between education and the gender/race categories. Do the categories seem to differ in educational attainment?

Use the *electric.sav* data file to answer the following questions:

22. Use a crosstabulation to look at the relationship between vital status at 10 years (variable *vital10*) and family history of coronary heart disease (variable *famhxcvr*).

a. What percentage of men without a family history were alive 10 years after the study started? What percentage of those with a family history were alive?

b. If *famhxcvr* is the column variable, would you compute row or column percentages? If *famhxcvr* is the row variable, would you compute row or column percentages?

23. Using the *cgt58* variable (number of cigarettes smoked per day in 1958), compute a new variable called *smoke* that has the value 0 if the person did not smoke at the start of the study and a value of 1 if he or she did. (Use the Recode facility on the Transform menu to do this.) Make a crosstabulation that shows the relationship between smoking status and vital status after 10 years (variable *vital10*). What percentage of smokers are alive after 10 years? What percentage of nonsmokers?

24. Using the *eduyr* variable, create a new variable called *degree* that has the following coding scheme:

eduyr	degree	Value label
0–8	1	Grammar school or less
9–11	2	Some high school
12	3	High school graduate
13–15	4	Some college
16	5	College graduate

You can assign the value labels in the Data Editor after the variable is computed.

a. Use a crosstabulation to examine the relationship between *degree* and *vital10*. What percentage of people with less than a high school diploma are alive? What percentage of people with a high school diploma or more are alive?

b. Look at the relationship between *degree* and *smoke*. Write a short paragraph summarizing your results.

Plotting Data

How can you display the relationship between two variables that are measured on a scale with meaningful numeric values?

- What is a scatterplot, and why is it useful?
- What is a scatterplot matrix?
- How can a scatterplot be used to identify unusual observations?
- What can you learn from a three-dimensional plot?

In previous chapters, you've used a variety of graphical displays to summarize your data. You've made pie charts, bar charts, histograms, and boxplots. All of these plots show the distribution of the values of a single variable. In this chapter, you'll learn how to display the values of two variables that are measured on a meaningful scale. Using **scatterplots**, you can look at the relationships of pairs of variables, such as beginning salary and college GPA, weight and cholesterol, or the compensation of CEOs and corporate profits.

A scatterplot is one of the best ways to look for relationships and patterns among variables. It is simple to understand, yet it conveys much information about the data. In Part 4 of this book, you'll learn methods for summarizing and describing relationships, but these methods are no substitute for plots. You should always plot the data first and then think about appropriate methods for summarizing them.

▶ This chapter uses demographic data from a sample of 122 nations (The World Almanac and Book of Facts). To work along in SPSS, use the *country.sav* data file. For specific instructions on how to create scatterplots, see "How to Obtain a Scatterplot" on p. 186. For an overview of creating and modifying charts in SPSS, see Appendix A.

Examining Population Indicators

You'll learn about scatterplots by looking at the relationships between female life expectancy, birthrate per 1000 population, and percentage of population living in urban areas for a sample of 122 countries. Birthrates as well as life expectancies have traditionally been related to economic prosperity. Better education, better health, and increasing urbanization led to declining birthrates in Western nations. Studies of developing nations (Robey, Rutstein, and Morris, 1993) suggest, however, that birthrates are declining in the absence of economic improvement. Changes in cultural values and availability of family planning have resulted in declining fertility rates, although not necessarily in increasing life expectancy.

Simple Scatterplots

Figure 9.1 is a scatterplot of female life expectancy and birthrate. Life expectancy, the average number of years a newborn female is expected to live, is plotted on the vertical axis (the y axis), and the birthrate per 1000 population is plotted on the horizontal axis (x axis). Consider the circled point. It represents Indonesia, with a female life expectancy at birth of 64 years and a birthrate of 26 births per 1000 population. (Dotted lines are drawn from the point to the axes to help you see what the values for the point are.)

To create this scatterplot, from the menus choose:

Graphs
Legacy Dialogs
Scatter/Dot...

In the Simple Scatterplot dialog box, select lifeexpf for Y Axis and birthrate for X Axis.

Figure 9.1 Scatterplot of life expectancy with birthrate

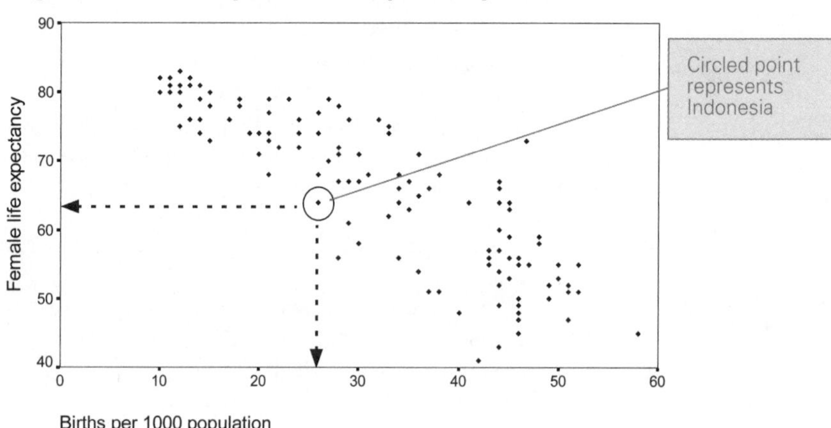

? *Does it matter which variable I plot on the* y *axis and which variable I plot on the* x *axis?* If one of the two variables that you're plotting is considered a dependent variable and the other an independent variable, it's traditional to plot the dependent variable on the *y* axis. For example, if you're plotting salary and years of education, plot the salary on the *y* axis, since salary may well depend on years of education. ■■■

What can you tell from the scatterplot shown in Figure 9.1? First of all, you see that the points are not randomly scattered over the grid. There seems to be a pattern. The points are concentrated in a band from the top left of the plot to the bottom right. As the birthrate increases, life expectancy decreases. In fact, the relationship appears to be more or less linear. That is, a straight line might be a reasonable summary of the data.

In a scatterplot, you can also determine whether there are cases that have unusual combinations of values for the two variables. In this plot, there aren't any cases that are really far removed from the overall pattern. However, if there were a country with a birthrate of 10 per 1000 and a female life expectancy of 50, that would be an unusual point that should be scrutinized to ensure that it is correct. Individually, the values for the two variables are not unusual. Neither would stand out on a stem-and-leaf plot or a histogram. There are many countries with a birthrate of around 10. There are also many countries with a life expectancy of around 50 years. What would be unusual is the combination of these two values.

Labeling the Points

In Figure 9.1, you can't tell which point represents which country. Identifying points can be important. If you find unusual points, you can investigate them more easily. If points represent entities such as car brands, companies, or countries, by examining the labels, you can identify types of cars or companies on the plot. For example, you can see if all of the European countries or African countries cluster together on the plot.

Figure 9.2 is the same plot as Figure 9.1, with some of the countries labeled. You see that Niger has the highest birthrate of all the countries in the plot and one of the lowest female life expectancies. The United States (and Canada) are in the upper left corner of the plot. They have low birthrates and relatively high life expectancies. Jordan sticks out somewhat in the plot. It has a higher life expectancy than other countries with similar birthrates.

? *Won't labeling the points cause the plot to be unreadable, since there will be labels all over?* SPSS has a special point-identification capability that lets you click on a particular point for the label to appear (see "Changing the Markers to Represent Number of Cases" on p. 195). If you click on the point again, the label disappears. That means that you can look at the points as much as you want and not worry about the plot becoming unreadable.

To display country labels, select the variable country for Label Cases by, as shown in Figure 9.15. You can then display case labels, as described in "Displaying Labels for Selected Points" on p. 194.

Figure 9.2 Life expectancy with birthrate showing labels

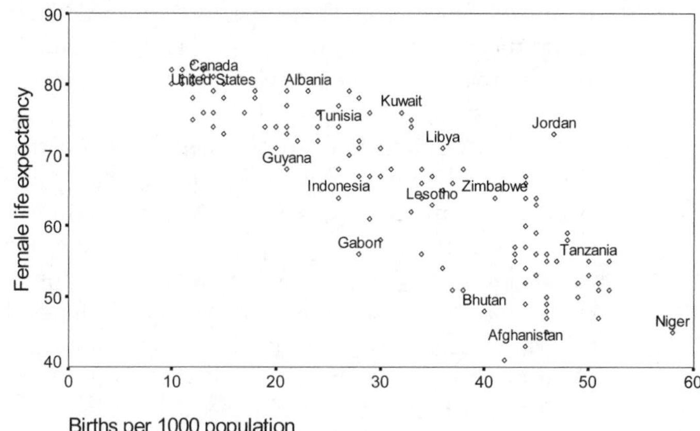

Identifying Different Groups

Often, when you look at the relationship between two variables, you are interested in how other factors affect the relationship. For example, if you're looking at the relationship between salary and years of experience, you may want to know if it's similar for males and females. Or, if you're looking at the relationship between blood pressure and weight, you may want to see if it's similar for smokers and nonsmokers. An easy way to determine whether different groups of cases behave similarly is to identify each point with a marker that indicates what group it's in. Sometimes the variable whose values determine the marker is called a **control variable** because when you identify the point with its value, you "control" for the effect of the variable.

For example, if you want to see whether the relationship between birthrate and life expectancy is similar for developing nations and for those considered developed, you can identify each of the points in Figure 9.1 as belonging to a developing or developed country.

To identify points by development status, select develop for Set Markers by, as shown in Figure 9.15.

Figure 9.3 Life expectancy with birthrate by development status

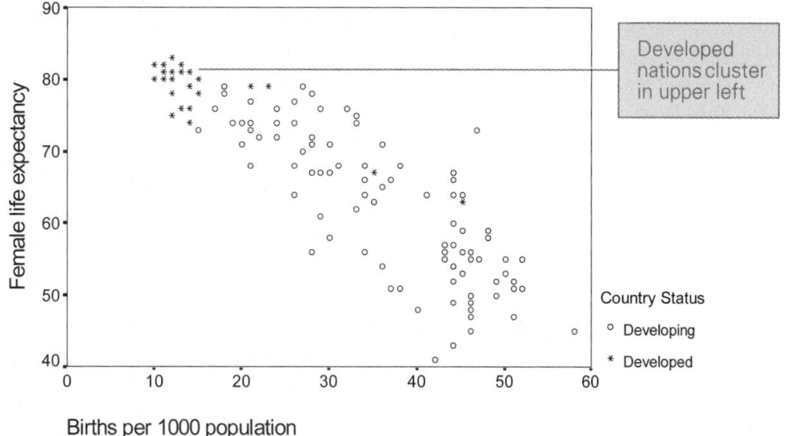

Figure 9.3 is the same plot as Figure 9.1. The only difference is that points are identified with two different markers, depending on whether a country is considered developed or developing. (Membership in the Group of 77, an organization of developing countries, was the primary criterion for determining development status. A couple of exceptions were made to this rule for clearly developing countries that are not members.)

You see that most of the developed nations cluster in the upper left corner of the plot. They have low birthrates and high life expectancies. The developing countries have a large spread in values for birthrate and life expectancy. For example, their birthrates range from 15 per 1000 all the way to almost 60 per 1000. The developing countries take up most of the plot.

? *What happens if I have two points with the same values for life expectancy and birthrate, but one is developed and one is developing?* If you have data in which points overlap (cases have identical or very similar values for the two variables) but the cases have different values for the variable that determines the marker used, only one of the markers will appear. If that happens often, you'll have difficulty interpreting the plot. This can be a serious problem for large datasets. It's not much of a problem in Figure 9.3, since the developed and developing countries are well-separated on the plot.

Representing Multiple Points

In Figure 9.3, you see a lot of points that are close to one another. In fact, some of them are overlapping, but you can't tell because when two points overlap, only one point is visible. (Figure 9.3 has 112 visible points to represent 122 countries, since 10 points overlap.)

When you have many points, it's helpful to have symbols whose size indicates how many cases are represented by each symbol. For example, consider Figure 9.4. The size of the point depends on the number of cases at that location on the plot. Note that the largest points are in the upper left corner of the plot. That's because most of the developed nations have very similar birthrates and life expectancies.

Figure 9.4 Life expectancy with birthrate showing multiple points

In the Chart Editor, activate Figure 9.2. From the Options menu, choose Bin Element. In the Properties dialog box, click the Binning tab. Under Count Indicator, select Marker Size and then click Apply.

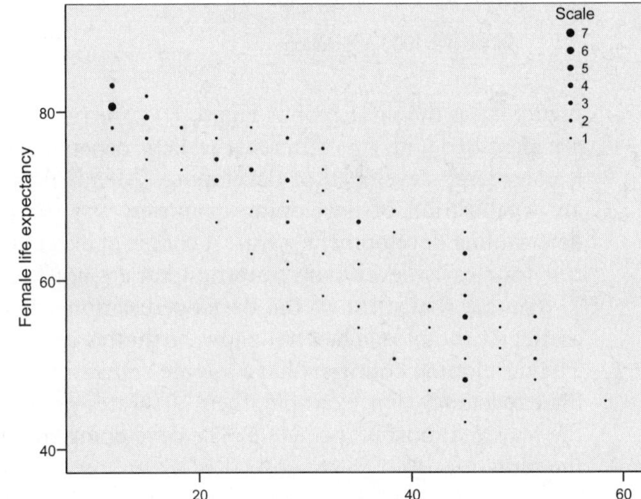

Scatterplot Matrices

So far, you've looked at the relationship between life expectancy and birthrate. What if you want to see how these variables relate to another variable, say percentage of urban population? You can certainly make additional plots of birthrate against percentage urban and of life expectancy against percentage urban. However, when you are interested in the relationships between several pairs of variables, you can make a scatterplot matrix of all of the variables of interest. A **scatterplot matrix** is a display that contains scatterplots for all possible pairs of variables.

Figure 9.5 Birthrate, life expectancy, and percentage urban

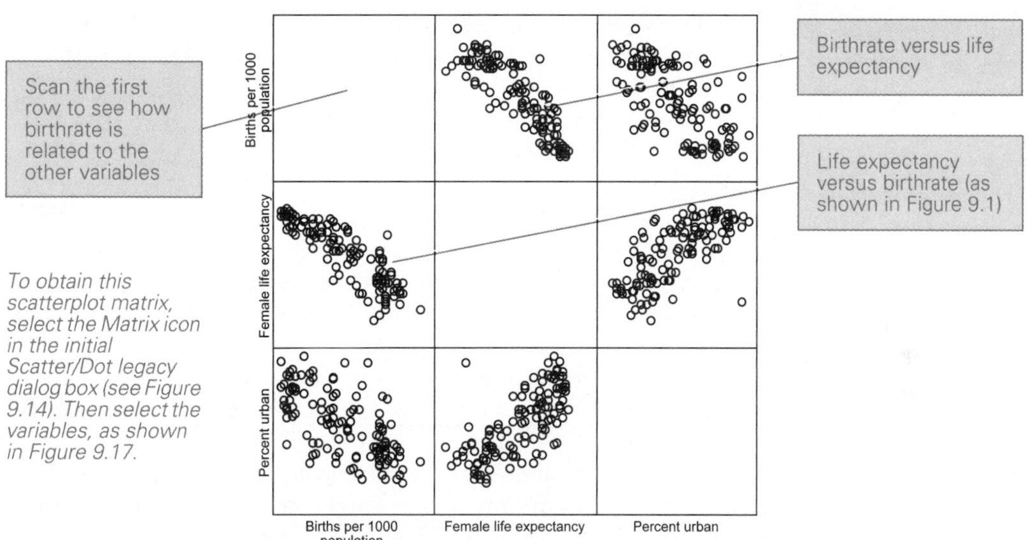

To obtain this scatterplot matrix, select the Matrix icon in the initial Scatter/Dot legacy dialog box (see Figure 9.14). Then select the variables, as shown in Figure 9.17.

Figure 9.5 is a scatterplot matrix of birthrate, female life expectancy, and percentage urban. A scatterplot matrix has the same number of rows and columns as there are variables. In this example, you see three rows and three columns. Each cell of the matrix, except for cells on the diagonal, is a plot of a pair of variables. The labels below the plot and on the left side of the plot tell you which variables are plotted in each cell. For example, the first row is labeled *Births per 1000 population.* This tells you that for all plots in the first row, the birthrate per 1000 population is plotted on the *y* axis (the vertical axis). For all plots in the first column, the birthrate is plotted on the *x* axis (the horizontal axis), since *Births per 1000 population* is the label for the first column. Similarly, the label *Female life expectancy* tells you that the life expectancy is plotted on the vertical axis for all plots in the second row and on the horizontal axis for all plots in the second column.

Look at the first plot in the first row. It's a plot of birthrate with life expectancy. This plot differs from the one you see in Figure 9.1, since female life expectancy is now on the horizontal axis instead of the vertical axis. (The first plot in the second row is the same as Figure 9.1.) In a scatterplot matrix, plots for all possible pairs of variables are displayed. The plots above the diagonal are the same as the plots below the diagonal. The only difference is that the variables are flipped. That is, the variables that are on the horizontal axis above the diagonal are on the vertical axis below the diagonal, and vice versa.

You can identify points in a scatterplot matrix in the same way as in a scatterplot. The point is labeled in all of the plots.

? *What is the easiest way to read a scatterplot matrix?* Try to scan across an entire row or column. For example, by reading across the first row, you see how birthrate relates first to life expectancy and then to urbanization. Similarly, the last row tells you how urbanization relates to birthrate and then to life expectancy at birth. The easiest way to identify an individual plot in a scatterplot matrix is to look up or down to find out which variable is on the horizontal axis, and look right or left to find out which variable is on the vertical axis. ■■■

In the scatterplot matrix, you see that the strongest relationship appears to be between life expectancy and birthrate. There is a negative relationship between the two variables. As the birthrate decreases, the life expectancy increases. The birthrate also decreases with increasing urbanization, but not as strongly. That's not surprising, since urban dwellers are usually better educated than their rural counterparts. They also have less of an economic need for children than do people in agricultural areas. Life expectancy and urbanization are positively related. As urbanization increases, so does life expectancy. There's a strange point in the urbanization and life expectancy plots that you'll examine in the section "Identifying Unusual Points" on p. 184.

Overlay Plots

The scatterplots you've looked at so far show the values of a single pair of variables. One of the variables is plotted on the horizontal axis, and the other is plotted on the vertical axis. Sometimes it's informative to plot several pairs of variables on the same axes. For example, if you want to study the relationship between high school GPA and the percentile ranks on standardized tests of math and verbal skills, you can make two plots: math percentile against GPA and verbal percentile against GPA. Since the scales for the variable pairs are the same for both plots, you can literally put one plot on top of the other. That's what an **overlay plot** does.

Figure 9.6 Overlay scatterplot of birth and death versus urbanization

To create this overlay scatterplot, select the Overlay Scatter icon in the initial Scatter/Dot legacy dialog box (see Figure 9.14). Then select the variables, as shown in Figure 9.16.

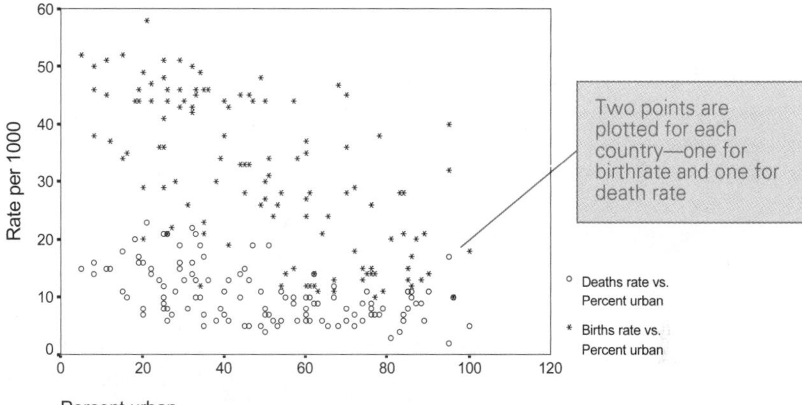

Figure 9.6 is an overlay plot of birthrate against percentage urban and death rate against percentage urban. The marker tells you whether the point is a birthrate or a death rate. Each country appears twice on the plot, once with its birthrate plotted against percentage urban and once with its death rate plotted against percentage urban.

? *How did this plot differ from the one in which you identify members of different groups by different markers?* Although the plots might look similar, since several different markers are used to identify the points, the plots are completely different. Figure 9.3 plots two variables, with a third variable used to classify each country as developing or developed. Figure 9.6, the overlay scatterplot, plots two separate *pairs* of variables—really it's two scatterplots on top of one another. That means each country appears not once, but twice, in the overlay plot. The marker tells you which pair of variables is being plotted. ■■■

Figure 9.6 has a lot of points. You may have some difficulty seeing the relationships between the two pairs of variables. To make it easier to sort out what's going on, you can draw summary curves for each pair of variables. Look at Figure 9.7. It's the same as Figure 9.6, but there are now two curves, called **loess smooths**, on the plot.

Figure 9.7 Overlay plot with loess smooths

To display loess smooths, see "Adding a Fit Line" on p. 196.

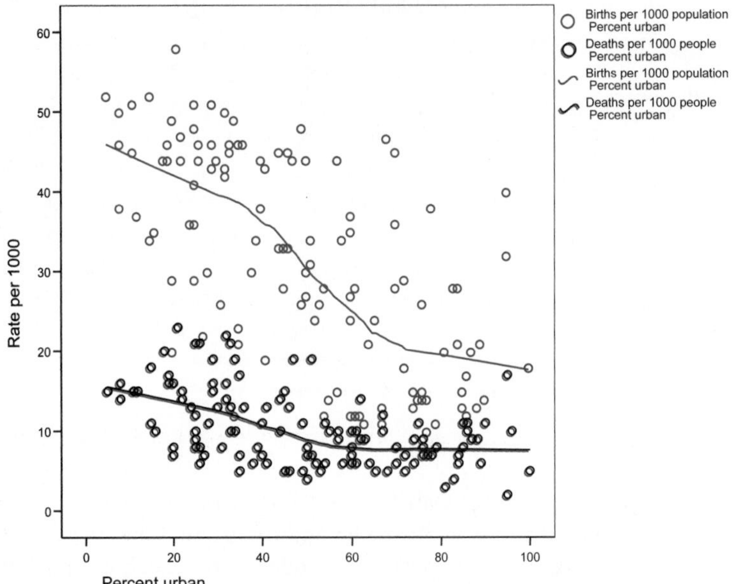

From the legend, you can tell which curve is for birthrate and which is for death rate. Now it's easier to see that both death rates and birthrates decrease as urbanization increases. The highest birthrates and death rates are for countries with little urbanization. You also see that birthrates decline more steeply than death rates. Although the curves on the plot are not straight lines, they're not too far from being linear. That is, you could straighten out the curves and they would still be fairly good summaries of the relationships. In Part 4 of this book, you'll see how to fit straight lines that summarize points in a scatterplot.

? *Where did the curves come from?* The curves drawn on the plots are called loess smooths. A loess smooth doesn't make assumptions about the mathematical form of the relationship between the two variables. It just looks at cases that have similar values for the *x* variable and, using an intricate mathematical algorithm, figures out where a reasonable average *y* value for them might be. By looking at a loess smooth of a plot, you can get an idea of the relationship between the two variables. If your plot has areas with few points, the loess smooth may not be good for them, since it depends on having points with similar *x* values. ■■■

You should use an overlay plot only when the plots that are being superimposed have similar scales. You don't want to overlay plots of per capita domestic product against urbanization and birthrate against urbanization. Per capita domestic product and birthrate have very different values, and plotting both of them on the same axis will result in a useless plot. The axis that shows both per capita GDP and birthrate will have a wide range, since it must accommodate numbers in the thousands. You'll be unable to see differences in birthrates on such a scale, since they will be bunched together. (If you want to plot variables with different scales on the same plot, you may want to standardize the variables. Then they will both have a mean of 0 and a standard deviation of 1.)

Three-Dimensional Plots

All of the scatterplots you've looked at so far plotted the values of cases on two axes, the horizontal (x) axis and the vertical (y) axis. The cases were represented in two dimensions. You can also plot data in three dimensions. For example, you can make a scatterplot that simultaneously shows the values of female life expectancy, urbanization, and birthrate.

Figure 9.8 3-D plot of life expectancy, birthrate, and urbanization

To create this 3-D scatterplot, select the 3-D icon in the Scatter/Dot dialog box (see Figure 9.14). Then select the variables, as shown in Figure 9.18.

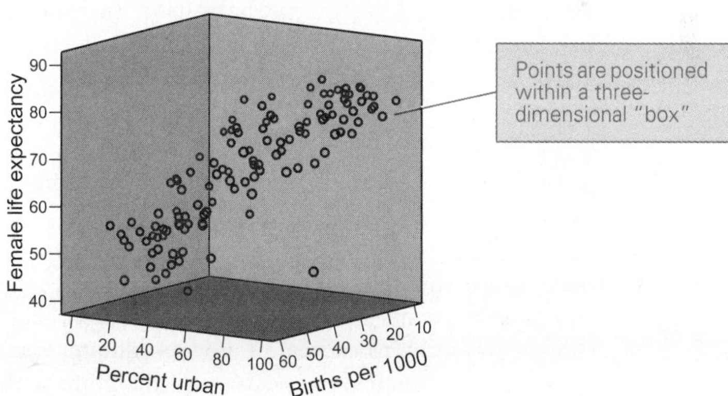

In Figure 9.8, instead of a plot that looks flat, like a sheet of paper, you see what looks like a box. You now have three axes, instead of two, on which to plot the values of the variables. (The additional axis is, not unexpectedly, called the z axis.) Life expectancy is plotted on the y axis, percentage urban is plotted on the x axis, and birthrate is plotted on the z axis. (In this plot, you see that there are several points that are far removed from the rest. We'll examine them later.)

To draw spikes to the floor, in the Chart Editor double-click on a point. In the Properties dialog box, click the Spikes tab. Select Floor.

Figure 9.9 3-D plot with spikes

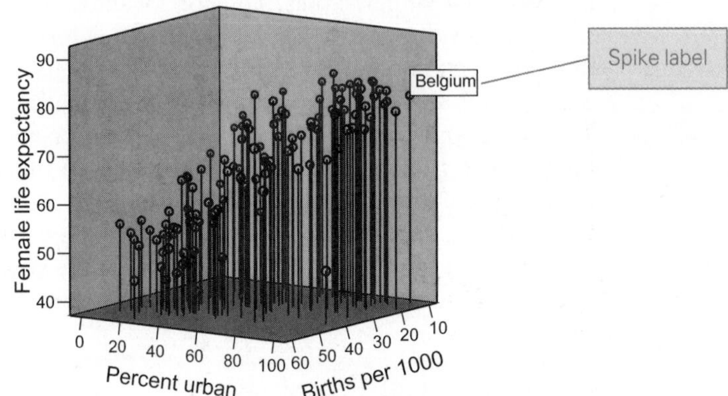

To make it easier to tell where the points are located, you can draw what are called **spikes**. Figure 9.9 is the same plot as Figure 9.8 with spikes drawn to the floor of the plot. The height of the spike tells you a point's value for the *y* variable. By looking at the position of the bottom of the spike, you can see the values for the *x* and *z* variables.

Look at the spike labeled *Belgium* in Figure 9.9. If you compare its height to the *y* axis, you see that the value for life expectancy is around 80 years. Now look at the bottom of the spike. It's position on the percentage urban axis is about 95. On the birthrate axis, its position is about 10.

? *How did you get those numbers?* Reading the numbers off a 3-D plot is tricky. Look at Figure 9.10. The point represents a case with values $y = 63$, $x = 25$, and $z = 42$. To read the values for the point from the axes, you have to follow the arrows. The bottom of the spike indicates the position of the point in the xz plane; to tell what the point's values are, you must draw lines perpendicular to the axes that define that plane, as shown. Similarly, to estimate the y coordinate, you must first draw your own spike to visualize the position of the point within the xy plane (also shown). Remember, everything is in three dimensions. ■■■

Figure 9.10 Reading a 3-D scatterplot

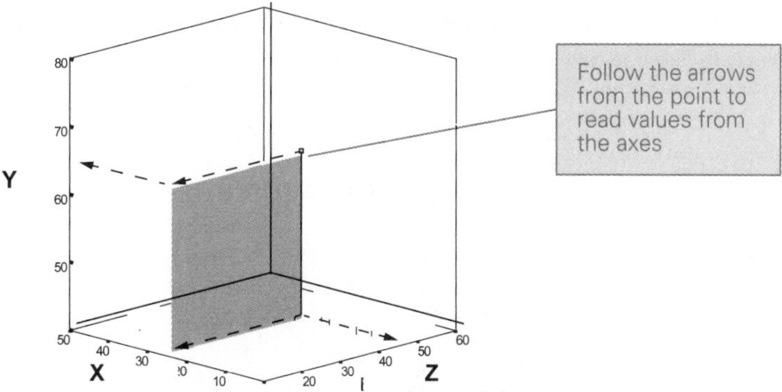

If you look at the heights of the spikes in Figure 9.9, you see that the tall spikes, which indicate long life expectancy, are for countries with low birthrates and high urbanization. Similarly, the shortest spikes are for countries with high birthrates and low urbanization. There is a transition from high to low spikes over the range of birthrates and urbanization.

Figure 9.11 3-D plot with countries identified by status

To identify countries by status, in the 3-D Scatterplot dialog box, specify develop for Set Markers by, as shown in Figure 9.18.

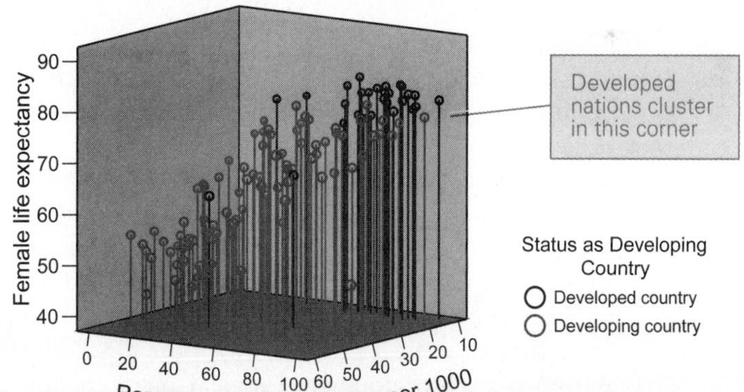

You can identify the points in Figure 9.8 as corresponding to developed or developing nations. This scatterplot is shown in Figure 9.11. Note how the developed countries cluster together in the corner that corresponds to high urbanization and low birthrates.

Identifying Unusual Points

You can identify unusual points in Figure 9.11 by looking for short spikes in the midst of tall ones. However, it's easier to spot unusual points when they're not obscured by a bunch of spikes. Figure 9.12 shows the original 3-D scatterplot with some of the points labeled.

Figure 9.12 3-D scatterplot with points labeled

To display country labels, select the country variable for Label Cases by, as shown in Figure 9.18. You can then display case labels as described in "Editing a Scatterplot" on p. 192.

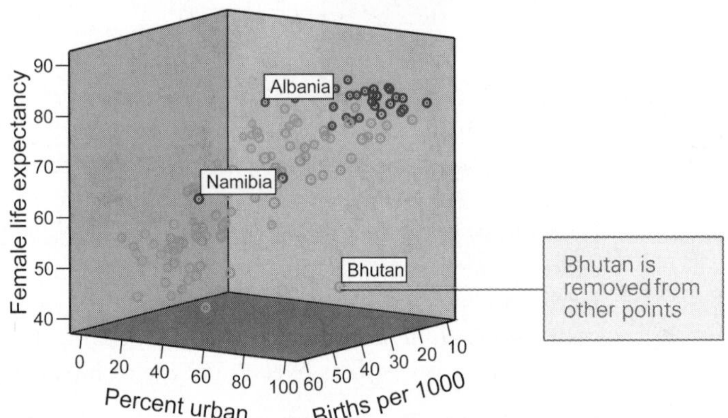

The point farthest removed from the rest is *Bhutan*. In the Data Editor, you see that the female life expectancy for Bhutan is 48 years. The birthrate is 40 per 1000 population, and the percentage urban is 95. Individually, none of these values is strange. There are a lot of countries with birthrates in the 40s, life expectancies in the 40s, or urban percentages in the 90s. What's strange about Bhutan is the combination of 95 for percentage urban and 48 for life expectancy. (You can also see the point as an outlier in the scatterplot matrix of life expectancy and percentage urban.)

Since you know that Bhutan is a tiny Himalayan kingdom, you should immediately suspect problems with the data. It's unlikely that the explanation is an intriguing sociological phenomenon at play. Sure enough, Bhutan is 95% rural, not 95% urban. So the value for percentage urban should be only 5.

? *What should I do now that I've found a problem with the data?* Fix it. Go into the Data Editor, and change the percentage urban value for Bhutan from 95 to 5. Then save the data file. If you don't do this, you won't get the correct answers in the subsequent chapters that use the *country.sav* data file.

Rotating 3-D Scatterplots

It's easy to examine a two-dimensional plot. You see the exact position of all the points. There aren't any surprises. Examining a three-dimensional plot is considerably more complicated. You can rotate it in all directions and get different views of the same plot. You can see if, in some views, the points cluster together, indicating a relationship among the three variables. SPSS has extensive facilities for rotating three-dimensional plots.

Figure 9.13 Rotated 3-D scatterplot

To rotate a 3-D scatterplot, bring it into the Chart Editor and from the Edit menu, select 3-D Rotation. Hold down the left mouse button to rotate the axes.

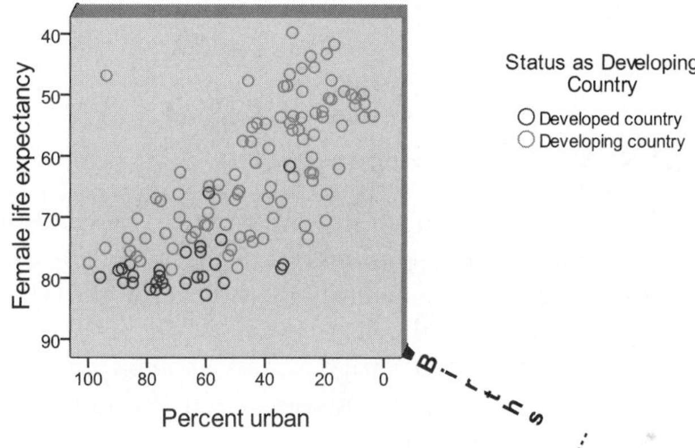

For a simple example of a 3-D rotation, look at Figure 9.13. The axes have been rotated so that you get an almost two-dimensional view of life expectancy and percentage urban. That's the same plot you see in the scatterplot matrix.

Summary

How can you display the relationship between two variables that are measured on a scale with meaningful numeric values?

- A scatterplot displays the values of two variables for each case.
- A scatterplot matrix displays scatterplots of all possible pairs of variables.
- You can identify unusual points in a scatterplot by looking for points that are far removed from the rest.
- A 3-D scatterplot shows the values of three variables at once.

What's Next?

In this chapter, you constructed a variety of scatterplots that are essential for examining the relationships between variables. In Part 4 of this book, you'll learn how to quantify the strength of the relationship between two variables. You'll also learn ways to predict the values of one variable from those of other variables. In Chapter 10, which starts Part 3, you'll learn how to test hypotheses about the population based on results observed in samples.

How to Obtain a Scatterplot

The Scatterplot procedure displays two-dimensional plots or three-dimensional plots (as seen in perspective) of the actual values of two or more variables. You can easily produce bivariate scatterplots (with or without control variables), overlay scatterplots, scatterplot matrices, and 3-D scatterplots. Once you have generated a plot, you can use the Chart Editor to enhance it with such things as labels or a best-fitting line or curve.

Note: This section provides information specific to creating and modifying scatterplots. For a general overview of creating SPSS charts, see Appendix A.

Obtaining a Simple Scatterplot

▶ To open the Scatter/Dot dialog box (see Figure 9.14), from the menus choose:

Graphs
 Legacy Dialogs ▶
  Scatter/Dot...

Figure 9.14 Scatter/Dot dialog box

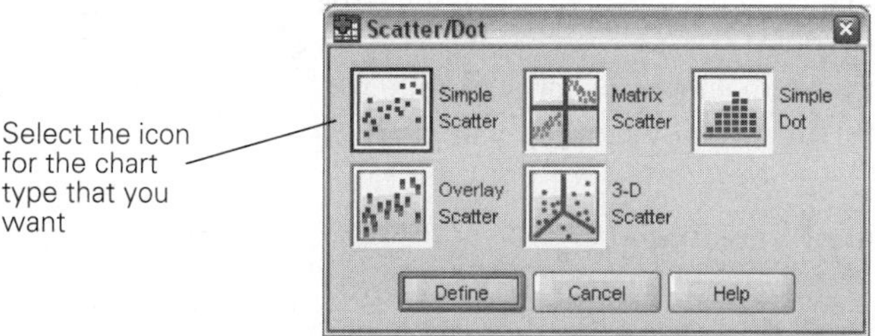

▶ Select the icon for the chart type that you want, and then click Define.

This opens a dialog box in which you can choose variables and options for the scatterplot (see Figure 9.15).

Figure 9.15 Simple Scatterplot dialog box

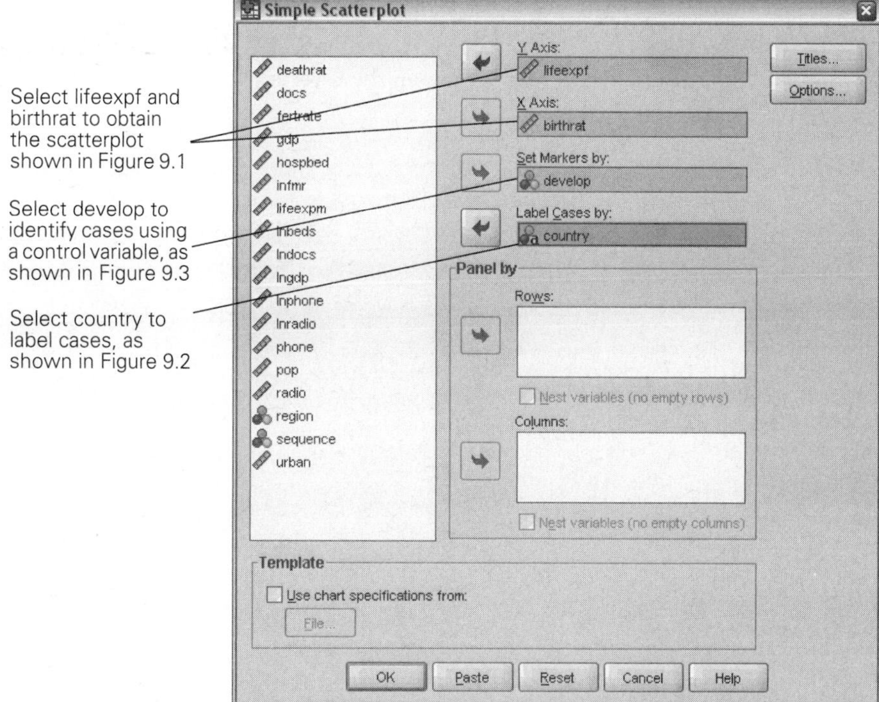

▶ Select numeric variables for the vertical (y) and horizontal (x) axes and move them into the appropriate boxes. You can request only one simple scatterplot at a time.

▶ Click **OK** to produce the plot.

Optionally, you can move a case-label variable into the Label Cases By box in the Simple Scatterplot dialog box. The values of this variable can then be displayed as case labels in the plot. (See "Changing the Markers to Represent Number of Cases" on p. 195.)

Optionally, you can choose a chart template or specify titles and options (for more information, see Appendix A).

Scatterplot with Control Variable

In a simple, matrix, or 3-D scatterplot, you can specify a control variable whose value determines the plot symbol that is used for each case (see Figure 9.3).

▶ To specify a control variable, move a variable into the Set Markers By box, as shown in Figure 9.15.

Obtaining an Overlay Scatterplot

In the initial Scatter/Dot dialog box (see Figure 9.14), select the **Overlay Scatter** icon and click **Define** to open the Overlay Scatterplot dialog box, as shown in Figure 9.16.

Figure 9.16 Overlay Scatterplot dialog box

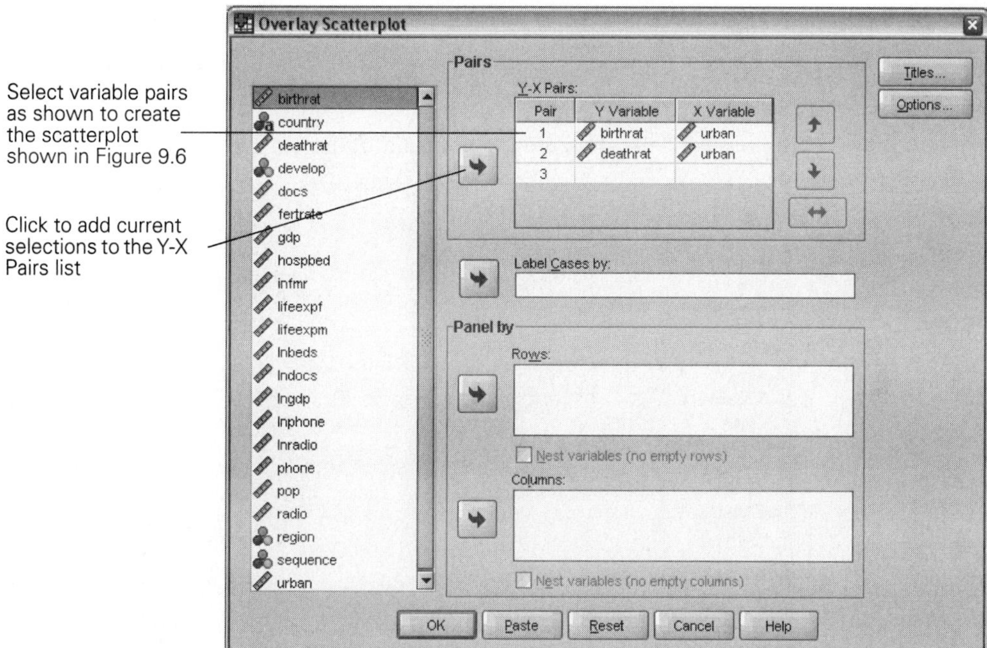

- Select two numeric variables in the source variable list and move the pair into the Y-X Pairs list.

- You can move a variable into the Label Cases By box, as discussed in "Simple Scatterplots" on p. 172. Optionally, you can choose a chart template or specify titles and options (for more information, see Appendix A). Click OK to produce the plot.

Obtaining a Scatterplot Matrix

In the initial Scatter/Dot dialog box (see Figure 9.14), select the **Matrix Scatter** icon and click **Define** to open the Scatterplot Matrix dialog box, as shown in Figure 9.17.

Figure 9.17 Scatterplot Matrix dialog box

Select birthrat, lifeexpf, and urban as shown to create the scatterplot matrix shown in Figure 9.5

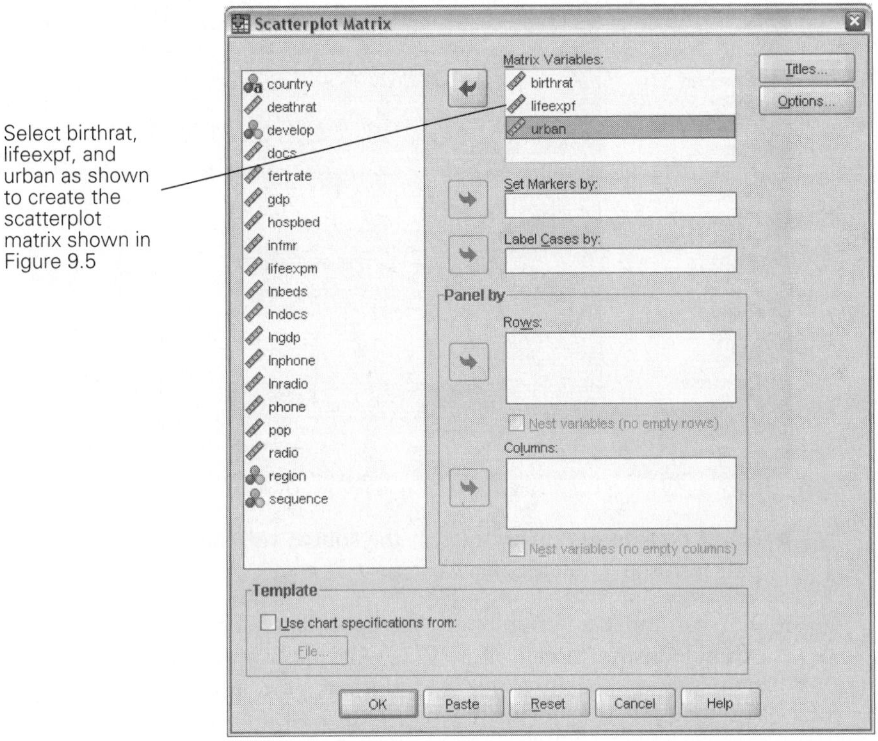

- Select three or more numeric variables and move them into the Matrix Variables list. (Remember that you are requesting a matrix of plots; if you move six variables into the matrix, you will get 36 tiny plots on your screen.)

You can move a variable into the Label Cases By box, and you can move a control variable into the Set Markers By box. (These specifications apply to all plots in the matrix.) Optionally, you can choose a chart template or specify titles and options (for more information, see Appendix A). Click **OK** to produce the plot.

Obtaining a 3-D Scatterplot

In the initial Scatter/Dot dialog box (see Figure 9.14), select the **3-D Scatter** icon and click **Define** to open the 3-D Scatterplot dialog box, as shown in Figure 9.18.

Figure 9.18 3-D Scatterplot dialog box

Select y, x, and z variables as shown to obtain the plot shown in Figure 9.8

Select develop to identify cases by status, as shown in Figure 9.11

Select country to label cases, as shown in Figure 9.8

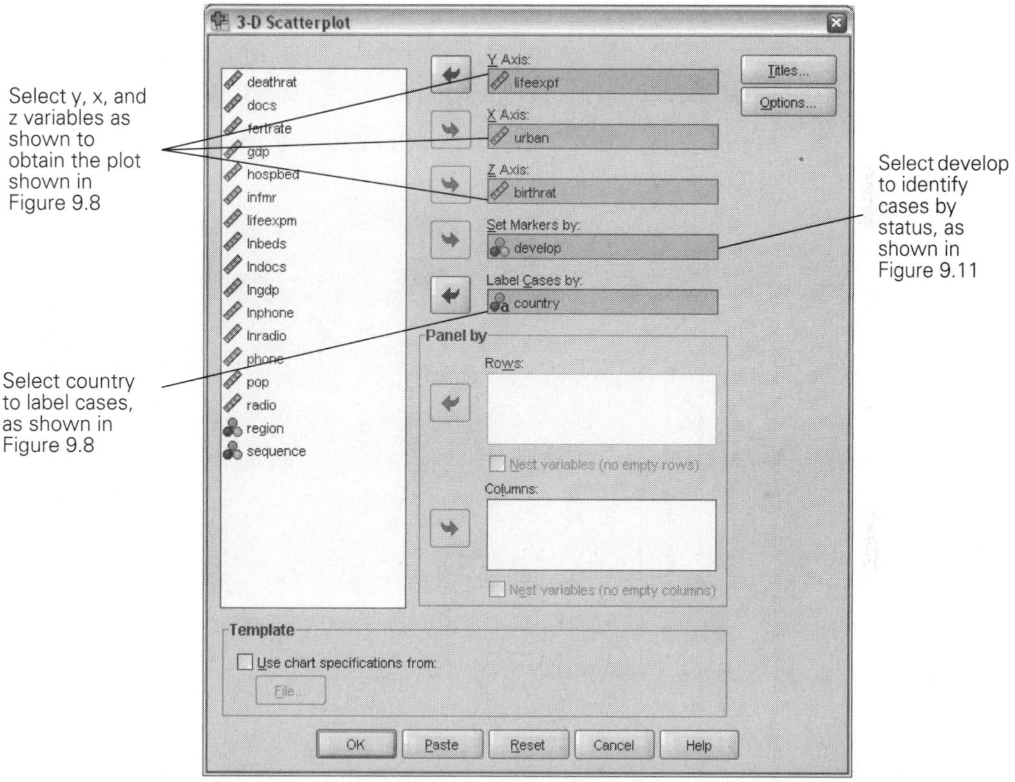

- Select three or more numeric variables and move them into the Y, X, and Z Axis boxes. The *y* axis is vertical in the initial orientation of the plot, although you can rotate it as you please in the Chart Editor.

You can move a variable into the Label Cases By box, and you can move a control variable into the Set Markers By box. Optionally, you can choose a chart template or specify titles and options (for more information, see Appendix A). Click **OK** to produce the plot.

Editing a Scatterplot

In the Chart Editor, you can add or remove labels from points, select cases to represent multiple cases plotted at the same point, connect the points using interpolation, and draw lines or curves that summarize the relationship between the variables. You can also rotate a 3-D plot.

▶ To bring the chart into the Chart Editor, double-click the chart in the Viewer.

The Chart Editor menus and toolbar are displayed, as shown in Figure 9.19.

Figure 9.19 Editing a chart in the Chart Editor

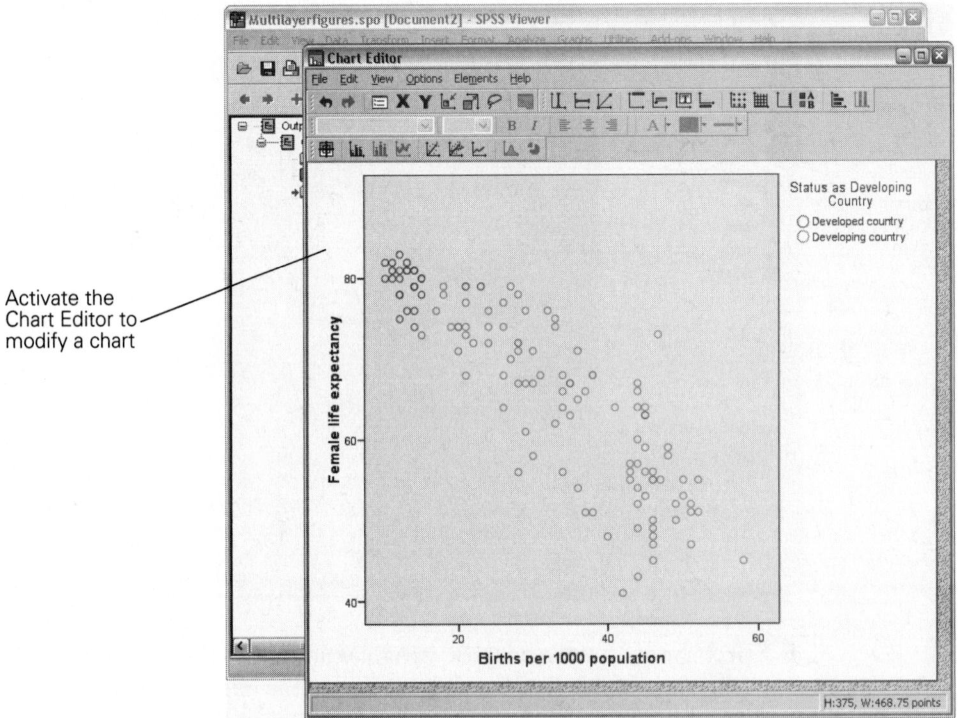

Labeling All Points

▶ To label all of the points on the scatterplot, from the Elements menu, choose Show Data Labels, as shown in Figure 9.20.

Figure 9.20 Show Data Labels

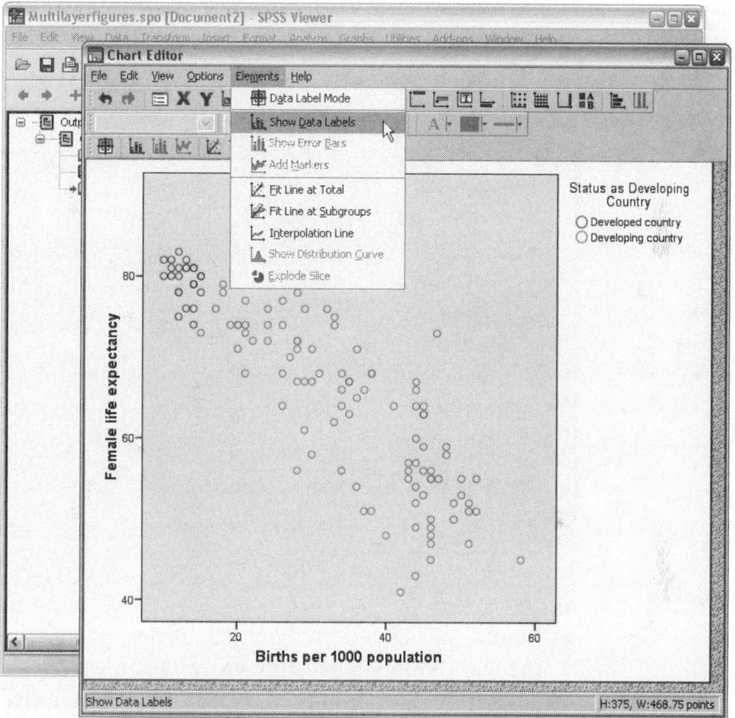

When you choose Show Data Labels, the points are labeled by the value of the variable specified in the Label Cases By box in the Scatterplot dialog box. A Properties dialog box also opens, as shown in Figure 9.21.

Figure 9.21 Properties dialog box for labeling data points

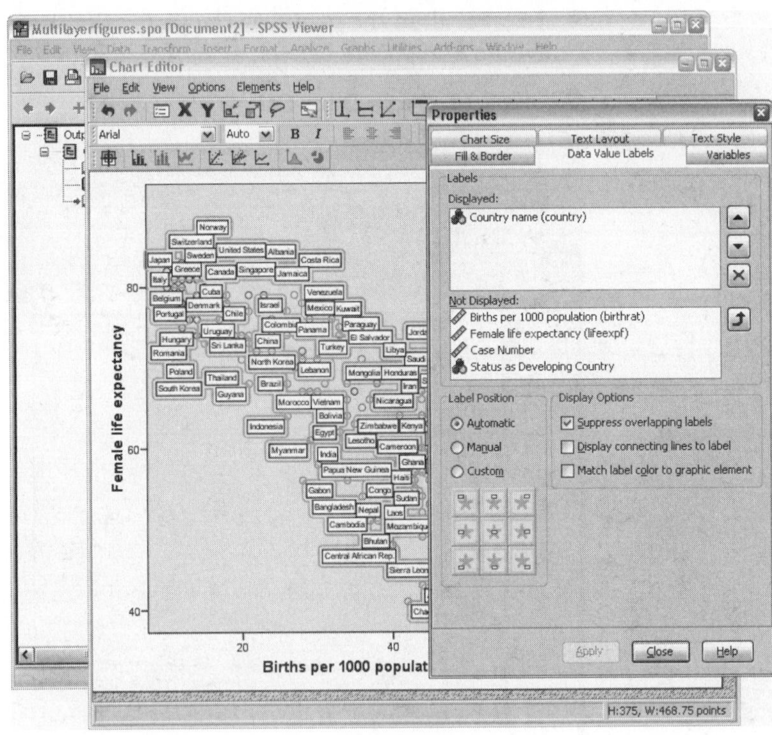

Points are labeled by the value of the variable in the Displayed list on the Data Value Labels tab. Only variables that you have selected in the Scatterplots dialog box appear here. To move a variable into the Displayed list, select it in the Not Displayed list, and then click the up arrow button. To remove a variable, select it in the Displayed list, and then click the red X button. When you have finished using the Properties dialog box, click **Close**.

Displaying Labels for Selected Points

When you have a large dataset or points that are close together, you might not want to label all of the points. You might want to identify only points that are of interest. To do this, you must be in Data Label Mode.

- ▶ To turn on Data Label Mode, click the Data Label Mode tool on the Chart Editor toolbar, as shown in Figure 9.22. The cursor changes to the Data Label Mode cursor.

▶ Click on any point in the chart to display a label for that point. To hide the label, click on the point again.

▶ To turn off Data Label Mode, click the Data Label Mode tool again.

Figure 9.22 Chart Editor in Data Label Mode

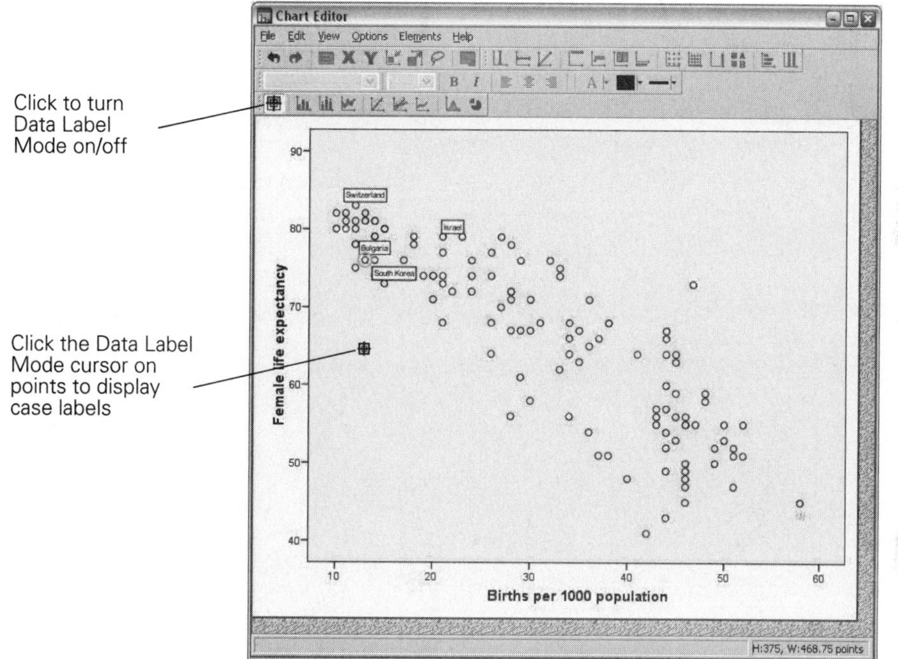

Changing the Markers to Represent Number of Cases

If you have a large dataset in which many points overlap, you can count the number of cases that fall approximately in the same location of the plot (the bin) and represent the number of points with a symbol or color that varies in size and intensity.

▶ To change the markers to represent the number of cases, double-click on the plot to bring it into the Chart Editor.

▶ From the Options menu, choose Bin Element. Click the Binning tab. For Count Indicator, specify whether you want the size of the marker or the intensity of its color to vary depending on the number of points in that bin.

▶ Click Apply.

Adding a Fit Line

- ▶ To draw a line or curve through the data points
- ▶ From the Elements menu choose:

  Fit Line at Total

A Properties dialog box opens, as shown in Figure 9.23.

Figure 9.23 Fit Line tab

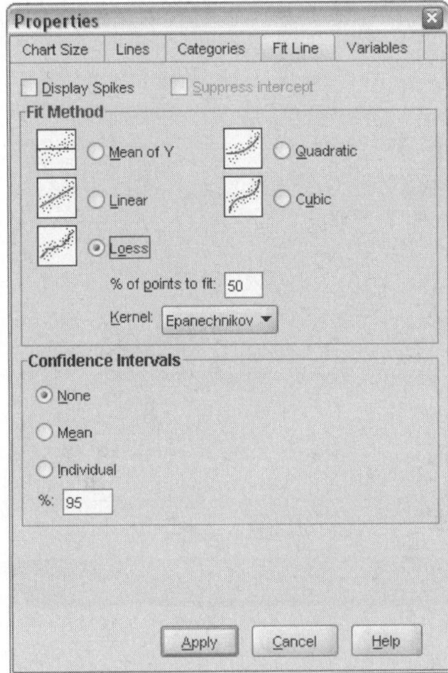

- ▶ To draw a linear regression line, select Linear.
- ▶ To draw confidence intervals, select Mean or Individual.
- ▶ Click Apply.

Rotating a 3-D Scatterplot

▶ To rotate a 3-D scatterplot, double-click it in the Viewer to bring it into the Chart Editor.

▶ From the Edit menu, choose 3-D Rotation.

▶ Hold down the left mouse button and move the mouse to achieve the desired rotation. For more precise positions, you can enter specific values in the 3-D Rotation dialog box.

▶ The Properties dialog box, shown in Figure 9.24, controls displays of the wireframe and backplanes.

Figure 9.24 3-D Elements tab

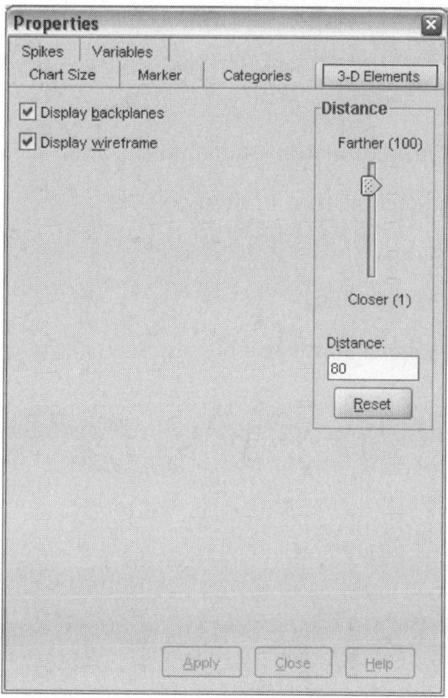

Exercises

Statistical Concepts

1. Indicate whether you would use the Crosstabs procedure, the Means procedure, or scatterplots to display the relationships between the following pairs of variables:

a. Job satisfaction (measured on a scale of 1–4) and income in dollars

b. Likelihood of buying a product and marital status

c. Current stock price and profit-to-earning ratio

d. Car color preference of husband and car color preference of wife

e. Miles per gallon and weight in pounds for a car

f. Hours studied for an examination and letter grade on the exam

g. Job satisfaction and income measured in dollars

h. Race and marital status

i. Systolic blood pressure and age

j. Husband's highest degree and wife's highest degree

2. The following table contains the age at first marriage, years of education, and gender for five people. Plot these values, identifying whether each is for a male or female.

agewed	educ	sex
18	12	M
22	13	F
30	16	M
16	10	M
25	18	F

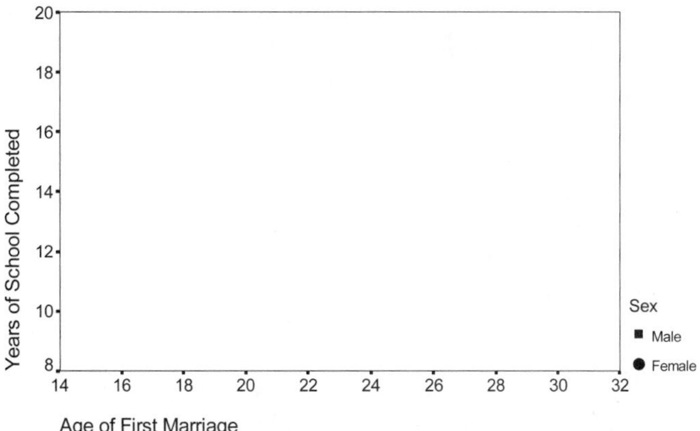

Data Analysis

Use the *gss.sav* data file to answer the following questions:

1. Make a scatterplot of time on the Internet (*nethrs*) and time on e-mail (*emailhrs*). Describe the relationship between the two variables. Explain why it occurs.

2. Make a scatterplot of time on the Web (*webhrs*) and time on e-mail (*emailhrs*). Is the relationship between these two variables similar to that for Internet hours and e-mail hours? Why?

3. Make a scatterplot matrix of all three variables. Identify the variables plotted in each graph. Indicate which variable is on the *y* axis and which variable is on the *x* axis.

4. Make a scatterplot of respondents' grouped income (*rincome*) and respondents' income recoded to the midpoint of the income value (*rincdol*). Why don't the points fall on a straight line? Under what circumstances would the points fall on a straight line?

5. For the married couples in the General Social Survey, make a scatterplot of years of husband's education (*husbeduc*) against years of wife's education (*wifeduc*).

a. Write a sentence describing the relationship.

b. Edit the scatterplot so that a symbol represents the number of cases. Do you think this makes it easier to see what's going on?

c. Identify each of the points on the scatterplot by how exciting the respondent perceives life to be (variable *life*). What problems do you see with this plot?

d. Using Data Label Mode, identify by ID number the cases that have the value of less than five years for wife's education.

6. Make a scatterplot matrix of the respondent's education (*educ*), the respondent's mother's education (*maeduc*), and the respondent's father's education (*paeduc*).

a. What can you tell from the plot about the relationships of the variables?

b. Do the variables appear to be positively or negatively related?

c. By default, points that do not have valid values for all variables in the scatterplot matrix are excluded from the plot. Does that exclude any particular group of people from your plot? What type of people are excluded if you make a scatterplot of *husbeduc*, *wifeduc*, *paeduc*, and *maeduc*?

7. Make an overlay plot of the following two plots: hours worked (variable *hrs1*) and respondent's income (variable *rincmdol*), and hours of Internet use (variable *nethrs*) and respondent's income. Describe the problems you see with this plot.

8. Make a three-dimensional plot of *educ*, *maeduc*, and *paeduc*. From this plot, identify three cases that are outliers. Click on one of the points so that the ID is displayed. Then rotate the plot. Pay attention to how the identified point moves as you spin the plot in different directions. Do the points appear randomly scattered or do you see a pattern?

Use the *salary.sav* data file to answer the following questions:

9. Make a scatterplot with current salary (variable *salnow*) on the vertical axis and beginning salary (variable *salbeg*) on the horizontal axis.

a. Describe the relationship between the two variables.

b. Identify each of the points by gender. Does this help you see anything interesting in the data?

c. Turn the plot into a plot in which symbol size depends on the number of cases. Comment on the useful features of this kind of plot.

10. Make a scatterplot matrix of beginning salary, current salary, and educational level (variable *edlevel*). Comment on the relationships between pairs of these variables.

Use the *electric.sav* data file to answer the following questions:

11. Make a scatterplot of weight (variable *wt58*) on the vertical axis and height (variable *ht58*) on the horizontal axis.

a. Does there appear to be a relationship between the two variables? Describe it.

b. Edit the chart so that symbols are used to represent the number of points. What advantage, if any, do you see to this type of plot?

12. Plot weight against height, this time identifying each point by the respondent's vital status at 10 years (variable *vital10*). Does the relationship between height and weight appear to be different for those who are alive and for those who are not?

13. Make a scatterplot matrix of height, weight, and diastolic blood pressure (variable *dbp58*). Write a brief summary of what you see.

Use the *schools.sav* data file to answer the following question:

14. Using the plots available on the Graphs menu, examine the relationship between percentage of low income students (*loinc93*), percentage of limited English proficiency students (*lep93*), graduation rates (*grad93*), and the school performance variables. Write a report based on your observations. Include appropriate graphs.

Use the *country.sav* data file to answer the following questions:

15. Plot birthrate (variable *birthrat*) against infant mortality rate (variable *infmr*), identifying countries by their development status (variable *develop*). Edit the plot to draw a loess smooth. Use Data Label Mode to identify the countries that are different from the rest. Summarize the relationship between the two variables.

16. Make a scatterplot matrix of birthrate, infant mortality rate, and the number of doctors per 10,000 population (variable *docs*). Identify the country with the largest number of doctors per 10,000 population. Which country has the smallest? Describe the different relationships between the three pairs of variables.

17. Make a three-dimensional plot of birthrate, infant mortality rate, and the number of doctors. Click to identify one of the points that is removed from the rest. Spin the plot, looking at how the labeled point moves. Are there views in which the points cluster together more tightly than in others? Do different points appear to be outliers as you rotate the plot?

Use the *buying.sav* data file to answer the following questions:

18. Make a scatterplot of wives' buying scores and husbands' buying scores (variables *wsumbuy* and *hsumbuy*). Does there appear to be a relationship between the two variables?

19. Make a plot of wives' influence scores and husbands' influence scores (variables *wsuminf* and *hsuminf*). Does there seem to be a relationship between the scores?

Use the file *bodyfat.sav* (Johnson, 1996) to answer the following questions:

20. Make a plot of percentage of body fat (*fatpct1*) against weight in pounds (*weight*) for the 252 men in the study.

   a. Are there any points that are far removed from the rest? Use Data Label Mode to identify them.

   b. Does there appear to be a relationship between the two variables? How would you summarize it?

   c. Add a linear fit line to the plot. Does a straight line appear to be a reasonable summary of the relationship between the two variables?

   d. Fit a loess smooth to the data. Does the smooth function look like a straight line?

21. Make a scatterplot matrix of percentage of body fat (*fatpct1*), height (*height*), weight (*weight*), and abdominal circumference (*abdomcirc*).

   a. Do all pairs of variables appear to be strongly related to each other? Which variables appear to be the least strongly related to each other?

   b. Are the same cases outlier in all of the plots?

   c. If you had to predict the values of one variable from another, which pair of variables would be the best candidates?

22. Describe the relationships between abdominal circumference (*abdomcirc*), hip circumference (*hipcirc*), knee circumference (*kneecirc*), and neck circumference (*neckcirc*). Include appropriate plots.

10 Evaluating Results from Samples

What can you say about a population, based on the results observed in a random sample?

- Are the results you observe in a sample identical to the results you would observe from the entire population?
- What is the sampling distribution of a statistic?
- How is it used to test a hypothesis about the population?
- What factors determine how much sample means vary from sample to sample?
- What is an observed significance level?
- What is the binomial test, and when do you use it?

In previous chapters, you've answered questions like "What percentage of adults in the United States use the Internet?" or "What is the relationship between education and public library use?" All you did was report what you found in your sample.

In this section of the book, you'll begin to look at the problems you face when you want to draw conclusions about a larger number of people or objects than those actually included in your study. You'll learn how to draw conclusions about the population based on the results observed in a sample.

▶ This chapter uses computer-generated data in the file *simul.sav*. For information on how to obtain the binomial test results shown in the chapter, see "Binomial Test" on p. 404 in Chapter 18.

From Sample to Population

In the General Social Survey, you found that almost 47% of the sample said that they used the Internet. Unless errors have been made while recording or entering the data, you know this for a fact. Similarly, you know exactly how old the people in the sample are, how much education they have, and so on. You can describe in great detail and with much certainty the results observed in this sample. Unfortunately, that's not really what's of interest. What you really want to do is draw conclusions about the larger group that the people in the GSS represent, the **population.**

The participants in the GSS are a sample from the population of adults in the United States. Based on the results you observe from the participants, you want to draw conclusions about *all* adults in the United States. You want to be able to say, for example, that in the United States, younger people are more likely to use the Internet than older people.

On first thought, that might not seem too complicated. Why not assume that what's true for the sample is also true for the population? That would certainly be simple. But would it always be correct? Do you really believe that since 47% of your sample use the Internet, that's exactly the percentage of Internet users in the adult population? Common sense tells you that it's very unlikely that the results you see in a sample are identical to those you would obtain if you made measurements or inquiries of the entire population of interest. If that were the case, one quick poll before an election would eliminate the need to even hold elections.

What's true instead is that different samples give different results, and it's highly unlikely that any one sample will hit the population results on the nose. To see what you can conclude about the population based on a sample, you must consider what results are possible when you select a sample from a population.

A Computer Model

Although we could use mathematical arguments to derive the properties of samples and populations, it's less intimidating and more fun to discover them for yourself. You can use the computer to keep drawing random samples from the same population and see how much the results change from sample to sample. This process is known as a **computer simulation.**

? *What's a random sample?* A **random sample** gives every member of the population (animal, vegetable, mineral, or whatever) the same chance of being included in the sample. No particular type of creature or thing is systematically excluded from the sample, and no particular type is more likely to be included than any other. Each member is also selected independently; including one particular member doesn't alter the chance of including another.

A sample is **biased** if, for example, rich people have a better chance of being included than poor people, or healthier people are more likely to be selected than sick people. You can't draw correct conclusions about the population based on the results from such a sample. ■■■

Let's use the computer to solve the following problem. A noted physician claims that she has a better treatment for the disease of interest. Of 10 patients who received her new treatment, 70% were cured. Extensive literature on the topic indicates that nationwide, only 50% of patients with this disease are cured. Based on the results of her experiment, can you tell if the physician has really made inroads into the treatment of this disease?

Are the Observed Results Unlikely?

To evaluate the physician's claim, you have to ask yourself the question, are the results she observed (7 out of 10 cures) unlikely if the true population cure rate is 50%? You know that if half of all people with a disease can be cured, that doesn't mean that any time you select 10 patients, exactly 5 will be cured by the treatment. Consider a coin-tossing analogy. You know that if a coin is fair, heads and tails are equally likely. If you flip a fair coin 10 times, however, you don't expect to see exactly 5 heads every 10 flips. Sometimes you get more heads and sometimes, more tails. (Try flipping a coin 10 times and see how many heads—cures—you get. Record your results. Repeat this as many times as you have the patience for and then make a stem-and-leaf plot of the results. You can compare your results with those you'll see in this chapter.)

To evaluate the physician's claim, instead of spending the afternoon flipping a coin, you can use the computer to construct a population in which half of the patients are cured and half are not. That's the situation if the physician's claim is not true. Then you can have the computer take a random sample of 10 patients and record the percentage that are cured. Have it repeat this procedure 500 times.

The reason you're doing this is to see what kind of sample results are possible if the new treatment is not different from the standard one. You can then determine whether finding 70% cured in a sample of 10 patients is an unusual finding when the true cure rate is 50%.

A stem-and-leaf plot of the results of the 500 experiments is shown in Figure 10.1.

You can obtain stem-and-leaf plots using the Explore procedure, as described in Chapter 7. Select the variable cured10 in the Explore dialog box.

Figure 10.1 Stem-and-leaf plot of percentage cured for sample size 10

```
Frequency    Stem &  Leaf

    3.00 Extremes    (=<10)
   24.00        2 .  00000000
   70.00        3 .  00000000000000000000000
   98.00        4 .  000000000000000000000000000000000
  114.00        5 .  00000000000000000000000000000000000000
   95.00        6 .  0000000000000000000000000000000
   66.00        7 .  0000000000000000000000
   24.00        8 .  00000000
    6.00 Extremes    (>=90)

 Stem width:      10.00
 Each leaf:       3 case(s)
```

For most samples, the cure rate is close to 50%

From this plot, you can tell approximately how often you would expect to see various outcomes in samples of size 10. The distribution of all possible sample outcomes for a statistic (such as the percentage cured) is called the **sampling distribution** of the statistic.

? *Exactly what is a statistic anyhow?* A **statistic** is some characteristic of a sample. The sample mean and variance are both examples of statistics. The term **parameter** is used to describe the characteristics of the population. For example, the average height of people in your sample is a statistic. If you measured the heights of all people in the population of interest, that would be called a parameter of the population. Parameters are usually designated (by statisticians, at least) with Greek symbols. For example, the mean of a population is called μ (mu), while the mean of a sample is called $\bar{X}$. Similarly, the standard deviation of the population is called σ (sigma), while the value for a sample is called s. Most of the time, population values, or parameters, are not known. You must estimate them based on statistics calculated from samples. ■■■

The sampling distribution is usually calculated mathematically. In this case, you're using a computer to give you some idea of what it looks like. In Figure 10.1, you see that for most samples, the percentage of cures is close to 50%. In fact, 307 out of the 500 experiments resulted in cure rates of 40%, 50%, or 60%. The further you move from 50%, in either direction, the fewer samples you see. Although various outcomes are possible, the outcomes are not equally likely. For example, only 6 experiments out of 500 resulted in a cure rate of 90% or greater.

You can calculate descriptive statistics for the data summarized in Figure 10.1. These summary statistics are shown in Figure 10.2. The values range from a minimum of 10% to a maximum of 90%, but the mean is very close to 50%. (In fact, for the mathematically computed sampling distribution, the mean value is exactly 50%, the mean of the population from which the samples are being drawn.) The standard deviation of the percentages, labeled *Std. Deviation* in Figure 10.2, is 16.22%. The standard deviation tells you how much the percentage cured varies in samples of size 10. (The standard deviation of the distribution of all possible values of a statistic is called the **standard error** of the statistic. For example, the standard deviation of all possible values of a sample mean is called the standard error of the mean.)

You can obtain these statistics using the Descriptives procedure, as described in Chapter 5.

Figure 10.2 Descriptive statistics for samples of size 10

	N	Minimum	Maximum	Mean	Std. Deviation
CURED10	500	10.00	90.00	50.0200	16.2212
Valid N (listwise)	500				

? *What's the difference between a standard deviation and a standard error?* Standard deviation refers to the variability of the observations in a sample. The term standard error is used when you are talking about the variability of a statistic. For example, if you have a sample of 10 systolic blood pressures, you can calculate their mean, variance, and standard deviation in the usual way. From the standard deviation of the 10 blood pressure measurements, you can also estimate how much *average* blood pressures calculated from samples of 10 people vary. That's the standard error of the mean for samples of this size. Figure 10.2 contains descriptive statistics for 500 means for samples of size 10. The standard deviation of these 500 means is an estimate of the standard error of the mean for samples of size 10. ■■■

Using Figure 10.1 as a guideline, you can estimate whether the physician's results are unusual if the true cure rate is 50%. You see that 96 out of 500 simulated experiments (19.2%) resulted in cure rates of 70% or more. That indicates that even if the new treatment is no better than the standard, you would expect to see cure rates at least as large as those observed by the physician almost 1 out of every 5 times you repeated the experiment. (In fact, it is possible to calculate mathematically that the probability of obtaining 7 or more cures in a sample of 10 is close to 17% when the true cure rate is 50%.)

Of course, it's always possible that the new treatment is really *less* effective than the usual treatment. So if you want to test the hypothesis that the new treatment is not different from the standard treatment, you must evaluate the probability of results as extreme as the one observed in either direction—increasing or decreasing the cure rate. You can estimate from Figure 10.1 that the probability of 30% or fewer cures and the probability of 70% or more cures is $(96 + 97)/500 = 38.6\%$.

Based on this, you have little reason to believe that the physician is really onto something. Her results are certainly not incompatible with samples selected from a population in which the true cure rate is 50%.

? *Why look at cure rates of 70% or more and cure rates of 30% or less?* Consider the following analogy. Your friend gives you a coin and claims that it is not fair. That is, heads and tails are not equally likely. Your friend wants your opinion. What outcomes will make you suspicious of the coin? Obviously, too many or too few heads (or tails) will cause you to be suspicious. You have to consider both possibilities if you don't know whether the coin is biased in favor of heads or tails. On the other hand, if you know that the coin would be rigged only in favor of heads, because that's what the coin's owner always bets on, you can ignore the possibility of getting too few heads.

Returning to the noted physician example, you are interested in both possibilities—too few and too many cures. That's because it's possible that the new treatment may work worse than the standard, and you want to know that. If there is a reason why the new treatment can't be worse—for example, if it involves adding meditation to the standard treatment—you can restrict your attention to cure rates at least as large as the one observed. ■■■

The Effect of Sample Size

As you saw above, when the true cure rate is 50%, there's a good chance that anywhere from 3 to 7 patients could be cured in a sample of 10. Most of the outcomes that can occur would not be considered unusual, because they could reasonably occur if the true cure rate is 50%. What's more, if the new treatment results in a cure rate of 60% or 70%, you probably would not detect the improvement, since many sample rates that are compatible with true rates of 60 or 70% are also compatible with the 50% rate. That means that based on a sample of only 10 patients, it's very difficult to evaluate a new treatment.

? *Can you ever tell from a sample of just 10 patients that a new treatment is better?* Yes. Since the existence of one little green man could convince you that there's life on Mars, similarly, 10 cures of a previously incurable disease could convince you that it's worth pursuing your treatment. It all depends on how unlikely your results are. ■■■

To see what effect sample size has on your ability to evaluate the physician's claim, consider what happens if you take samples of 40 patients, instead of just 10, from the same population with a cure rate of 50%. The results of this computer experiment are shown in Figure 10.3. (Note that each stem in the plot is now divided into two rows.) When you compare Figure 10.3 with Figure 10.1, you see that the values are much closer to 50% than before. Values greater than 60% or less than 40% are now noticeably less likely. These rates were not particularly unusual when you had samples of 10 patients. Based on Figure 10.3, you would estimate your chance of finding a sample rate of 70% or more or 30% or less when the true rate is 50% to be about 3 in 500, 0.6%. That means that only about 1 in 200 times would such a cure rate occur if the new treatment doesn't differ from the standard treatment.

In summary, when you have samples of 40 cases, an observed rate of 70% or more, or 30% or less, is possible, but not very likely when the true population rate is 50%. If the physician sees the same cure rate of 70% based on a sample of 40 patients, you would be more likely to believe that perhaps she's onto something. Her results really would be unusual when the true cure rate is 50%.

? *Just how unusual does "unusual" need to be?* The rule of thumb that is usually used to characterize results as unusual is a probability of 5% or less. That is, if results as extreme or more extreme than those observed are expected to occur in 5 (or fewer) samples out of 100, the results are considered unusual, or statistically significant. ■■■

Figure 10.3 Stem-and-leaf plot of percentage cured for sample size 40

To obtain this stem-and-leaf plot, select the variable cured40 in the Explore dialog box.

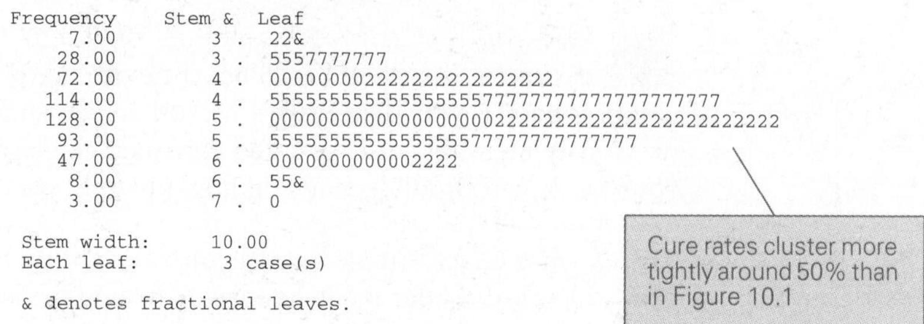

```
Frequency    Stem &  Leaf
     7.00       3 .  22&
    28.00       3 .  5557777777
    72.00       4 .  000000002222222222222222
   114.00       4 .  55555555555555555555777777777777777777
   128.00       5 .  000000000000000000002222222222222222222222
    93.00       5 .  5555555555555555557777777777777
    47.00       6 .  0000000000002222
     8.00       6 .  55&
     3.00       7 .  0

 Stem width:      10.00
 Each leaf:        3 case(s)

 & denotes fractional leaves.
```

Cure rates cluster more tightly around 50% than in Figure 10.1

Larger samples improve your chances of detecting a difference in the cure rates (if in fact there is one) because there is less variability in the possible outcomes. Consider Figure 10.4, which contains descriptive statistics for the distribution shown in Figure 10.3. The mean value is again close to 50%. The standard deviation, however, is much smaller than for samples of size 10. It is now 7.29%, compared to the standard deviation of 16.22% in Figure 10.2. There's a pattern in the way that sample size affects the variance of the sampling distribution of means. If you increase the sample size by a factor of four, the variance decreases by a factor of four. Since the standard deviation is the square root of the variance, it decreases by a factor of two.

Figure 10.4 Descriptive statistics for samples of size 40

	N	Minimum	Maximum	Mean	Std. Deviation
CURED40	500	30.00	70.00	49.8350	**7.2864**
Valid N (listwise)	500				

The standard error of the mean is much smaller than for samples of size 10

The Binomial Test

In the previous example, you estimated the probability of various outcomes of an experiment from a stem-and-leaf plot obtained by repeated samples from the same population. The reason for doing it this way is to show you that when you take a sample from a population, the value you calculate for a statistic such as the mean is one of many possible values you can obtain. The possible values have a distribution—the sampling distribution of the statistic. Results vary from sample to sample, and you must take this variability into account when drawing conclusions about the population based on results observed from a sample.

Fortunately, in most situations, you don't personally need to determine the possible outcomes and their likelihoods by performing computer experiments. These can be mathematically calculated for you by SPSS. For example, you can use the **binomial test** to determine whether an observed cure rate is unlikely if the true rate is 50%. Your goal is to compare your experiment's success rate to a standard or usual rate. You observe the outcome of interest for a sample of subjects or objects.

To use the binomial test, your experiment or study must have only two possible outcomes, such as cured/not cured, pass/fail, buy/not buy, defective/not defective, and so on. All of the observations must be independent, and the probability of success must be the same for each member of the sample population.

? *What do you mean by independent?* For observations to be **independent**, one subject's response can't influence that of another. For example, if students collaborate on an exam, their scores are not independent. One student's results influence those of another. If you make multiple observations on the same subject, the observations are similarly not independent. Curing the same patient from 10 bouts of a disease is not equivalent to curing 10 patients from 1 bout. The 10 observations from a single patient are not independent. ■■■

Figure 10.5 shows the results of the binomial test for the 10-subject experiment. You see that there are 10 cases, 7 of which are coded 1, indicating a cure, and 3 of which are coded 0, indicating no cure. The population value that you want to test against (0.5) is labeled *Test Prop.* The proportion of successes in the sample, 0.7, is labeled *Observed Prop.* The probability of obtaining results as extreme or more extreme than the ones you observe in your sample, when the true probability of a cure is 0.5, is labeled *Exact Sig. (2-tailed).*

The observed significance level tells you that the probability of obtaining a cure rate of 70% or greater or 30% or less, when the true cure rate is 50%, is 0.34. (Note how close this exact probability is to your estimated probability of 0.386 from Figure 10.1.) Since the observed significance level is larger than 0.05, the usual frame of reference, you don't have enough evidence to believe that the physician has achieved a cure rate different from 50%. The sample with an observed cure rate of 70% is not particularly unusual if the true population cure rate is 50%. In fact, more than 34% of samples from this population are as unusual as the one sample that the physician observed.

For instructions on how to obtain a binomial test, see "Binomial Test" on p. 404 in Chapter 18.

Figure 10.5 Binomial test: Sample size 10

		Category	N	Observed Prop.	Test Prop.	Exact Sig. (2-tailed)
CURE	Group 1	1	7	.70	.50	**.344**
	Group 2	0	3	.30		
	Total		10	1.00		

There is a 34% chance of observing a cure rate as extreme as 70% when the true rate is 50%

The results from the 40-patient experiment are shown in Figure 10.6. There are now 28 cases with the response of 1, and 12 cases with the response of 0, giving the same observed proportion of 0.70. The test proportion is unchanged at 0.50. The observed significance level is 0.017. That means that, with samples of size 40, you would expect to see samples as unusual as the one observed less than 2% of the time. (Again, this value is reasonably close to the empirical estimate of 0.6% from Figure 10.3. The observed significance level is not calculated exactly but is based on an approximation that is quite close to the exact value for large samples.) If the physician finds a 70% cure rate based on 40 patients, you're much more likely to believe that the physician is doing better than the usual 50%.

Figure 10.6 Binomial test: Sample size 40

		Category	N	Observed Prop.	Test Prop.	Asymp. Sig. (2-tailed)
binom40	Group 1	1.00	28	.70	.50	.017[1]
	Group 2	.00	12	.30		
	Total		40	1.00		

1. Based on Z Approximation.

Probability of results this extreme decreases to less than 2%

? *Would you embrace her cure based only on these results?* Of course not. A statistical analysis is useless if a study is poorly designed. Here are some important concerns: How were patients selected for inclusion in her study? Is there something about them that would make them more likely to be cured than those in the population at large? Were there objective criteria for establishing a cure, or was it a subjective judgment? Did the evaluator and/or the patient know that a new drug was being used?

The correct way to conduct an evaluation of a new treatment is to allocate patients randomly to two treatment groups. One receives the standard treatment, and the other receives the new one. Ideally, neither the patient nor the physician knows which treatment the patient is receiving. Evaluation is done, based on well-established criteria, by physicians who are unaware of which patients received which treatment. These precautions help to ensure that the results of the study measure what they were intended to measure. ■■■

Summary

What can you say about a population, based on the results observed in a random sample?

- When you take a sample from a population, you won't get the same results as you would if you had data for the entire population.
- The sampling distribution of a statistic tells you, for a particular sample size, about the distribution of all possible sample values of that statistic.
- From the sampling distribution of a statistic, you can tell if observed sample results are unusual under particular circumstances.
- As the sample size increases, the variability of statistics calculated from the sample decreases.
- The observed significance level is the probability of observing a sample difference at least as the large as the one observed, when there is no difference in the population.
- A binomial test is used to test the hypothesis that a variable comes from a binomial population with a specified probability of an event occurring. The variable can have only two values.

What's Next?

In this chapter, you saw that the results you observe when you perform an experiment or conduct a survey are only one of many possible outcomes. Different samples from the same population give different results. You also saw how sampling distributions can be used to determine how likely or unlikely various sample results are. In Chapter 11, you'll learn more about testing hypotheses about a population, based on results observed in a sample. You'll also learn about the importance of the normal distribution in hypothesis testing.

Exercises

Statistical Concepts

1. In a large university, there is a proposal to drop statistics as a requirement for graduation. Each of the 500 professors at the university commissions a survey to gauge student support for the proposal. Each survey contains a random sample of 50 students. The results are summarized in the following table and stem-and-leaf plot of the percentage of students favoring the proposal:

	VOTE												
		95% Confidence Interval for Mean											
	Mean	Lower Bound	Upper Bound	5% Trimmed Mean	Median	Variance	Std. Deviation	Minimum	Maximum	Range	Interquartile Range	Skewness	Kurtosis
Statistic	39.9240	39.3101	40.5379	39.8800	40.0000	48.820	6.98712	20.00	64.00	44.00	8.0000	.082	-.035
Std. Error	.31247											.109	.218

```
VOTE Stem-and-Leaf Plot

 Frequency    Stem &  Leaf

     5.00 Extremes    (=<24)
      .00        2 .
     9.00        2 .  6666
    15.00        2 .  8888888
    26.00        3 .  0000000000000
    23.00        3 .  22222222222
    43.00        3 .  444444444444444444444
    46.00        3 .  66666666666666666666666
    61.00        3 .  888888888888888888888888888888
    55.00        4 .  000000000000000000000000000
    55.00        4 .  222222222222222222222222222
    45.00        4 .  4444444444444444444444
    38.00        4 .  6666666666666666666
    30.00        4 .  888888888888888
    22.00        5 .  00000000000
    10.00        5 .  22222
     9.00        5 .  4444
     8.00 Extremes    (>=56)

 Stem width:      10.00
 Each leaf:        2 case(s)
```

Based on the stem-and-leaf plot and summary statistics:

a. What is your best guess for the percentage of students favoring the proposal?

b. When the 500 professors presented their results to the president, she was aghast that the results of all of the surveys were not similar. She is considering censuring the professors whose polls were far removed from the average value. She thinks that they "rigged" their polls to support their own viewpoints. How would you defend the professors at their hearing?

c. Based on the stem-and-leaf plot, if the true percentage favoring the proposal is 40%, what's the probability that a poll will estimate the value to be 25% or greater? 55% or greater? Less than 35%?

2. As superintendent of Chicago schools, you are interested in seeing whether ACT scores have improved between 1993 and 1994. You obtain a sample of 56 schools and find that scores have improved in 19 and worsened in 37.

   a. How can you tell if your observed results are plausible if there has been no change in ACT scores?

   b. From the binomial test, you find that the observed significance level is 0.0231. Does this support the claim that ACT scores have improved?

   c. Explain to the mayor whether it's possible that ACT scores have not really changed?

   d. What's the probability that you would see results as extreme as the ones observed when average ACT scores have not changed?

Data Analysis

Use the *gss.sav* data file to answer the following questions:

1. A social science researcher claims that half of the adults in the United States are male and half are female. Assume that the General Social Survey respondents are a random sample of the United States population.

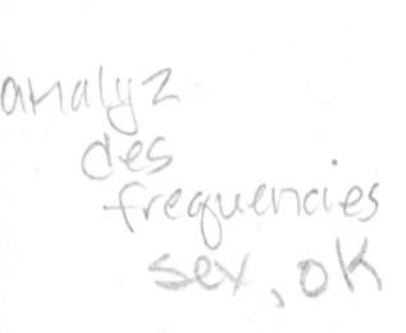

   a. What percentage of the sample are males? Females?

   b. If the adult population is really half male and half female, are your observed results unusual?

   c. What do you think about the researcher's claim in light of the General Social Survey data?

2. A sociologist claims that half of adults in the United States believe in a life after this one.

   a. Analyze the variable ***postlife*** and write a brief commentary on the sociologist's claim.

   b. What if the sociologist claimed that 3 out of 4 adults believe in life after death? Based on the data, would you believe his claim?

11 The Normal Distribution

What is the normal distribution, and why is it important for data analysis?

- What does a normal distribution look like?
- What is a standard normal distribution?
- What is the Central Limit Theorem, and why is it important?

In Chapter 10, you learned how to evaluate a claim about the mean of a variable that has two possible values. Using the binomial test, you calculated the probabilities of getting various sample results when the probability of a success was assumed to be known. In this chapter, you'll learn how to test claims about the mean of a variable that has more than two values. You'll also learn about the normal distribution and the important role it plays in statistics.

▶ This chapter examines data on serum cholesterol levels from the *electric.sav* data file. In addition, some figures use simulated datasets included in the file *simul.sav*. The histograms and output shown can be obtained using the SPSS Graphs menu (see Appendix A) and the Descriptives procedure (see Chapter 5).

The Normal Distribution

You may have noticed that the shapes of the two stem-and-leaf plots in Chapter 10 are similar. They look like bells (on their sides). The same data are displayed as histograms in Figure 11.1 and Figure 11.2, where a bell-shaped distribution with the same mean and variance as the data is superimposed. You can see that most of the values are bunched in the center. The farther you move from the center, in either direction, the fewer the number of observations. The distributions are also more or less symmetric. That is, if you divide the distribution into two pieces at the peak, the two halves of the distribution are very similar in shape, but mirror images of each other. (The theoretical bell distribution is perfectly symmetric.)

You can obtain histograms using the Graphs menu, as described in Appendix A.

In the Histogram dialog box, select the variables cured10 and cured40.

Figure 11.1 Simulated experiments: Sample size 10

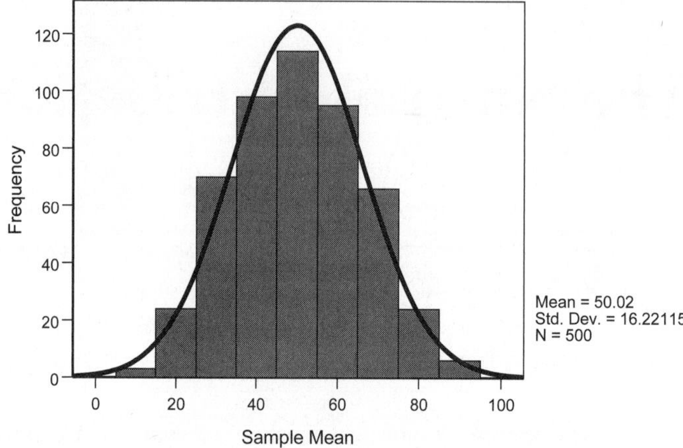

Figure 11.2 Simulated experiments: Sample size 40

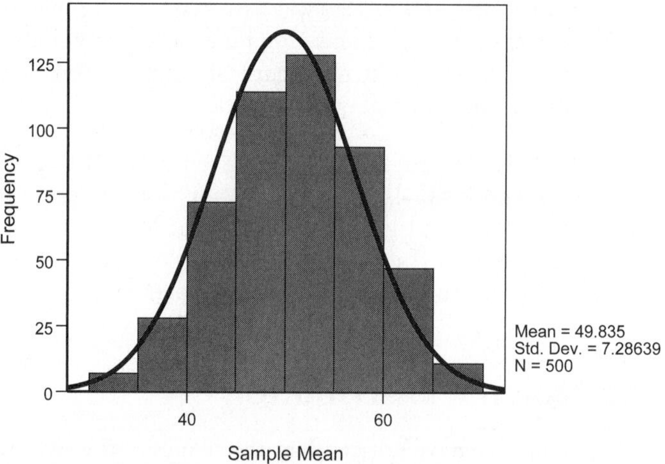

Many variables—such as blood pressure, weight, and scores on standardized tests—turn out to have distributions that are bell-shaped. For example, look at Figure 11.3, which is a histogram of cholesterol levels for a sample of 239 men enrolled in the Western Electric study (Paul et al., 1963). Note that the shape of the distribution is very similar to that in Figure 11.2. That's a pretty remarkable coincidence, since Figure 11.2 is a plot of many sample means from a distribution that has only two values (1=*cured*, 0=*not cured*), while Figure 11.3 is a plot of actual cholesterol values.

Figure 11.3 Histogram of cholesterol values

To obtain this histogram, open the electric.sav data file and select chol58 in the Histogram dialog box.

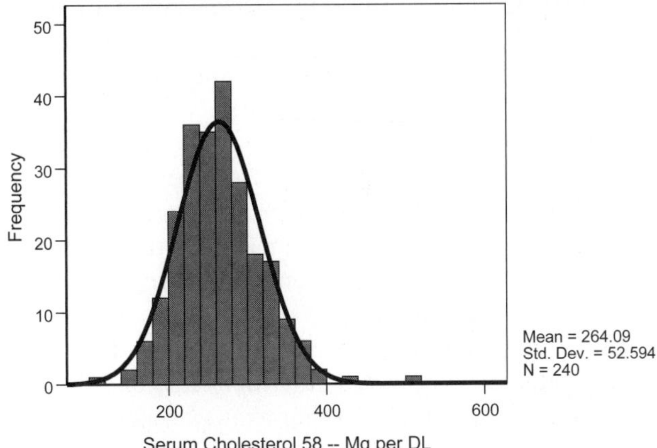

The bell distribution that is superimposed on Figure 11.1, Figure 11.2, and Figure 11.3 is called the **normal distribution.** A mathematical equation specifies exactly the distribution of values for a variable that has a normal distribution. Consider Figure 11.4, which is a picture of a normal distribution that has a mean of 100 and a standard deviation of 15. The center of the distribution is at the mean. The mean of a normal distribution has the same value as the most frequently occurring value (the mode), and as the median, the value that splits the distribution into two equal parts.

Figure 11.4 A normal distribution

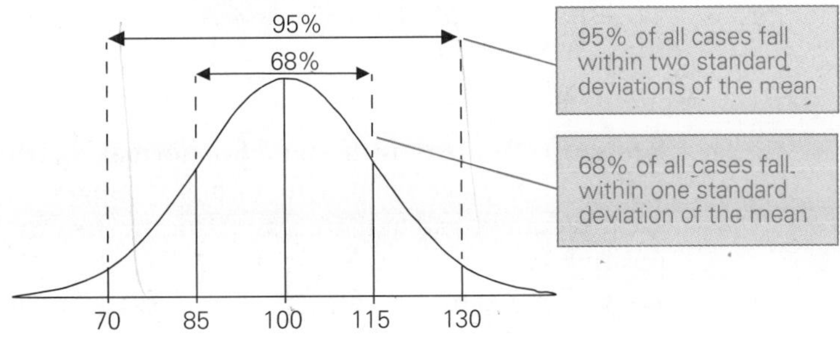

If a variable has exactly a normal distribution, you can calculate the percentage of cases falling within any interval. All you have to know are the mean and the standard deviation. Suppose that scores on IQ tests are normally distributed, with a mean of 100 and a standard deviation of 15, as was once thought to be true. In a normal distribution, 68% of all values fall within one standard deviation of the mean, so you would expect 68% of the population to have IQ scores between 85 (one standard deviation below the mean) and 115 (one standard deviation above the mean). Similarly, 95% of the values in a normal distribution fall within two standard deviations of the mean, so you would expect 95% of the population to have IQ scores between 70 and 130.

Since a normal distribution can have any mean and standard deviation, the location of a case within the distribution is usually given by the number of standard deviations it is above or below the mean. (Recall from Chapter 5 that this is called a **standard score,** or ***z* score.**) A normal distribution in which all values are given as standard scores is called a **standard normal distribution.** A standard normal distribution has a mean of 0 and a standard deviation of 1. For example, a person with an IQ of 100 would have a standard score of 0, since 100 is the mean of the distribution. Similarly, a person with an IQ of 115 would have a standard score of +1, since the score is one standard deviation (15 points) above the mean, while a person with an IQ of 70 would have a standard score of –2, since the score is two standard deviation units (30 points) below the mean.

Figure 11.5 The standard normal distribution

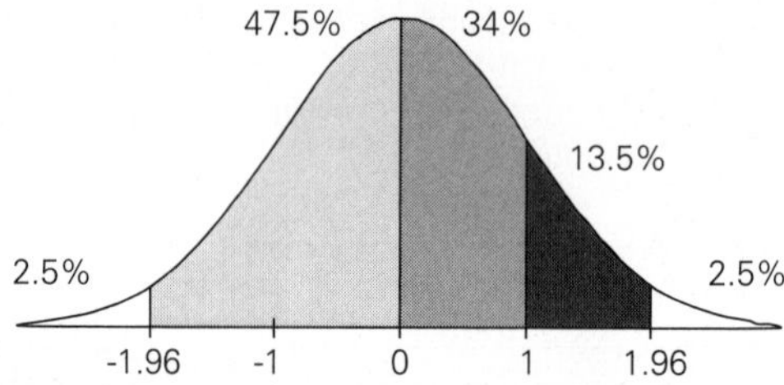

Some of the areas in a standard normal distribution are shown in Figure 11.5. Since the distribution is symmetric, half of the values are greater than 0, and half are less. Also, the area to the right of any given positive score is the same as the area to the left of the same negative score. For example, 16% of cases have standardized scores greater than +1, and 16% of cases have standardized scores less than –1. Appendix D gives areas of the normal distribution for various standard scores. The exercises show you how to use SPSS to calculate areas in a normal distribution.

? *If you're more than two standard deviations from the mean on some characteristic, does that mean you're abnormal?* Not necessarily. For example, pediatricians often evaluate a child's size by finding percentile values. They may tell the parents that their child is at the 2.5th percentile or 97.5th percentile for height. (For a normal distribution, these percentiles correspond to standardized scores of –2 and +2.) The small or large percentile values don't necessarily indicate that something is wrong. Even if you took a group of healthy children and looked at their height distribution, some of them would be more than two standard deviations from the mean. Somebody has to fall into the tails of the normal distribution. This also leads to a convincing argument against grading on the curve. Even in a brilliant, hard-working class, some students will receive scores more than two standard deviations below the mean. Does that make their performance unacceptable? Not necessarily. ■■■

Samples from a Normal Distribution

If you look again at Figure 11.3, you'll see that the normal distribution that is superimposed on the cholesterol data doesn't fit the data values exactly. The observed data are not perfectly normal. Instead, the distribution of the data values can be described as approximately normal. That's not surprising. Even if you assume that cholesterol values have a perfect normal distribution in the population, you wouldn't expect a sample from this distribution to be exactly normal. You know that a sample is not a perfect picture of the population. You expect that samples from a normal population would appear to be more or less bell-shaped, but it would be unrealistic to expect that every sample is exactly normal. In fact, even the population distribution of most variables is not exactly normal. Instead, it's usually the case that the normal distribution is a good approximation. Slight departures from the normal distribution have little effect on statistical analyses that assume that the distribution of data values is normal.

Means from a Normal Population

Since we've established that the normal distribution is a reasonable representation of the distribution of data values for many variables, we can use this information in testing statistical hypotheses about such variables. For example, suppose you want to test whether highly paid CEOs have average cholesterol levels that are different from the population as a whole. In 1991, *Forbes* sent out a survey to the 200 most highly compensated CEOs requesting their cholesterol levels. The 21 CEOs who responded had an average cholesterol of 193 mg/dL. Assume that in the population, cholesterol levels are normally distributed with a mean of 205 and a standard deviation of 35. Based on this information, how would you determine if the CEOs differ from the rest of us not only in their net worth but in average cholesterol as well?

To answer this question, you need to know whether 193 is an unlikely sample value for the mean when the true population value is 205. To arrive at this information, you'll follow the same procedure as you did in Chapter 10. However, instead of taking samples from a population in which only two values can occur, you'll take repeated samples from a normal population.

Figure 11.6 Distribution of 500 sample means

To obtain this histogram, open the simul.sav file and select the variable normal21 in the Histogram dialog box.

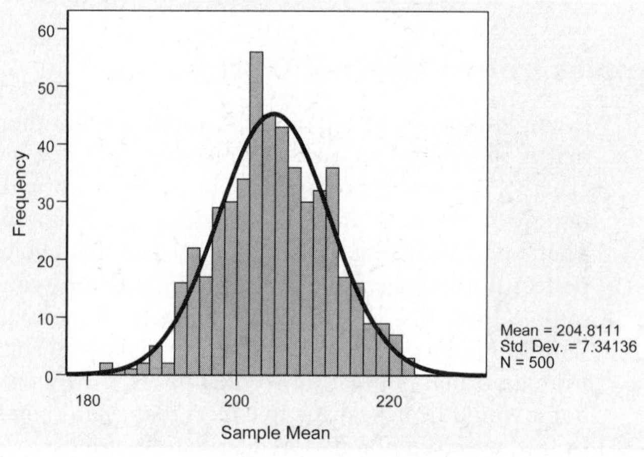

Figure 11.6 shows the distribution of 500 sample means from a normal distribution with a mean of 205 and a standard deviation of 35. Each mean is based on 21 cases. As you can see, the distribution of sample means is also approximately normal. That's always the case when you calculate sample means for data from a normal population. The mean of the sample means is very close to 205, the population value. In fact, for the theoretical sampling distribution of the means, the value is exactly 205. (Remember, the theoretical distribution of sample means is mathematically derived and tells you precisely what the distribution of the sample means is for all possible samples of a particular size.) In Figure 11.6, the standard deviation of the means, also known as the **standard error of the mean**, is 7.34.

Standard Error of the Mean

You saw in Chapter 10 that the standard error of the mean tells you how much sample means from the same population vary. It depends on two things: how large a sample you take (that is, the number of cases used to compute the mean) and how much variability there is in the population. Means based on large numbers of cases vary less than means based on small numbers of cases. Means calculated from populations with little variability vary less than means calculated from populations with large variability.

If you know the population standard deviation (or variance) and the number of cases in the sample, you can calculate the standard error of the mean by dividing the standard deviation by the square root of the number of cases. In this example, the population standard deviation is 35 and the number of cases is 21, so the standard error of the mean is:

$$\frac{35}{\sqrt{21}} = 7.64$$ **Equation 11.1**

Note that the value we calculated based on the 500 samples with 21 hypothetical CEOs in each sample was not exactly 7.64 but was very close. What we obtained was an *estimate* of the true value. That's because we did not take all possible samples from the population but restricted our attention to 500.

? *Will the standard error of the mean always be smaller than the standard deviation of the data values?* Yes. It's always the case that the standard error of the mean is smaller than the standard deviation of the data values. That makes sense if you think about it. When you calculate a mean, it falls in between the smallest and largest sample values. It's not as extreme as the actual data values in your sample. Thus, the mean has less variability than the original observations. The larger the sample that you take, the more you smooth out the variability of the individual data values when you calculate the mean. ■■■

Are the Sample Results Unlikely?

Now that you know about the important properties of the sampling distribution of the mean from a normal population, let's return to the cholesterol levels of the CEOs. Figure 11.6 gives you a rough idea of how often you can expect various values for sample means when cholesterol is normally distributed in the population, with a mean of 205 and a standard deviation of 35. It's easy to see that the observed sample value of 193 is not a particularly unusual value.

You can use the characteristics of the normal distribution to calculate *exactly* how often you would expect to see, based on 21 cases, a sample mean of 193 or less (12 less than the population mean), or 217 or greater (12 more than the population mean). You're interested in both large and small cholesterol values, since you don't know in advance whether CEO values will be larger or smaller than those of the general population. It may be that the *foie gras* on Parisian business trips raises their cholesterol levels, or that exercising in swanky health clubs while the rest of us work decreases their cholesterol levels.

First, you must calculate a standard score for the observed mean. You calculate it in the usual way: subtract the population mean from the observed mean and then divide by the standard deviation. The only trick to remember is that since you're dealing with a distribution of means, you must use the standard deviation of the means (the standard error of the mean), not the standard deviation of the sample values themselves. In our example, the standard score is

$$Z = \frac{193 - 205}{7.64} = -1.57 \qquad \textbf{Equation 11.2}$$

Look at Figure 11.7 for a summary of the situation.

Figure 11.7 How unlikely is a sample mean of 193?

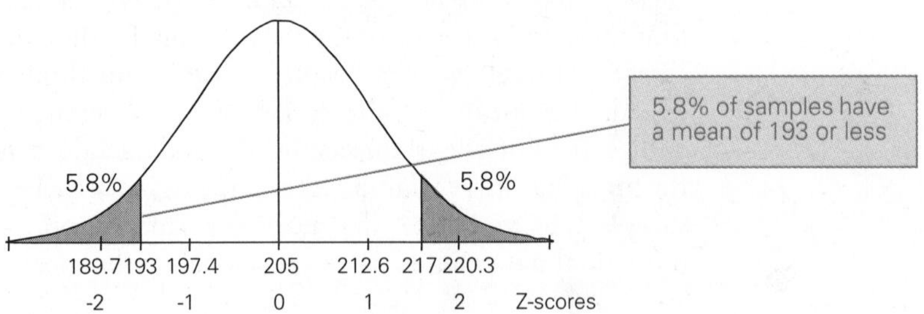

You see that the distribution of all possible sample means of 21 cases is normal, with a mean of 205 mg/dL and a standard error of 7.64 mg/dL. The observed sample mean of 193 has a standard score of –1.57. In a normal distribution, 11.6% of the cases have standardized values less than –1.57 or greater than +1.57. Based on this, you don't have enough evidence to conclude that CEO cholesterol levels are different from those in the general population. (The observed significance level of 0.116 is larger than 0.05, the usual criterion for unusual.)

? *Since only 21 out of 200 CEOs responded, shouldn't you be concerned about the results from a survey that has a response rate of 10.5%?* Absolutely. There are many reasons why those who responded to the survey may differ from those who did not. It may be that CEOs with low cholesterol levels are more likely to volunteer this information than CEOs with high cholesterol levels. Or it may be that CEOs who have experienced medical problems are more likely to know, and perhaps to volunteer, their cholesterol levels than CEOs who are healthy. Our analysis was based on the rather shaky assumption that CEOs who responded don't differ from those who didn't. We also made the simplifying assumption that middle-aged males have the same cholesterol distribution as the general population. If this isn't the case, we'd have to compare CEO values to those for middle-aged males. (Unfortunately, all results were from middle-aged males.) Our analysis also assumes that the CEOs reported their correct current cholesterol values. Anyone who has read an annual report to shareholders knows that CEOs can cast any kind of data in the best possible light. ■■■

Testing a Hypothesis

In the previous example, you used statistical methods to test a hypothesis about the population based on results observed in a sample. Here's a summary of the procedure you followed:

1. You wanted to see if the average cholesterol levels of highly paid CEOs differ from those of the general population. You obtained a sample of cholesterol values from 21 such highly paid CEOs.
2. You calculated the average cholesterol value for the 21 CEOs in your sample to be 193 mg/dL.
3. You used the normal distribution with a mean of 205 and a standard error of 7.64 to determine how often you would expect to see average cholesterol values less than 193 or greater than 217, when the population mean is 205.
4. You found that sample means as unusual as the one you observed are expected to occur in about 11.6% of samples from the population, so you didn't have enough evidence to conclude that average cholesterol levels for CEOs are different from those of the population.

Means from Non-Normal Distributions

You probably weren't too surprised that the distribution of sample means from a normal population is also normal. That makes a certain amount of sense. But it is surprising that the distributions of means shown in Figure 11.1 and Figure 11.2 at the beginning of this chapter also appear to be normal.

Remember that these are not means of a variable that has a normal distribution. The cure variable has only two equally likely values—0 for *not cured* and 1 for *cured.* This remarkable finding is explained by what's called the Central Limit Theorem. The **Central Limit Theorem** says that for samples of a sufficiently large size, the distribution of sample means is approximately normal. The original variable can have any kind of distribution. It doesn't have to be bell-shaped at all.

? *Sufficiently large size? What does that mean?* How large a sample you need before the distribution of sample means is approximately normal depends on the distribution of the original values of a variable. For a variable that has a distribution not too different from normal, sample means will have a normal distribution even if they're based on small sample sizes. If the distribution of the variable is very far from normal, larger sample sizes will be needed for the distribution of sample means to be normal. The important point is that the distribution of means gets closer and closer to normal as the sample size gets larger and larger—regardless of what the distribution of the original variable looks like. ■■■

Means from a Uniform Distribution

As an example of the Central Limit Theorem, let's see what the distribution of sample means looks like if cholesterol values had a uniform distribution in the population. In a uniform distribution, all values of a variable are equally likely. Figure 11.8 shows a histogram of 5,000 values from a uniform distribution with a range of 135 to 275. All of the bars representing values from 135 to 275 are of approximately equal length.

Figure 11.8 A uniform distribution

300
200
100
0
137 157 177 197 217 237 257
Value
Std. Dev = 40.24
Mean = 205
N = 5000.00

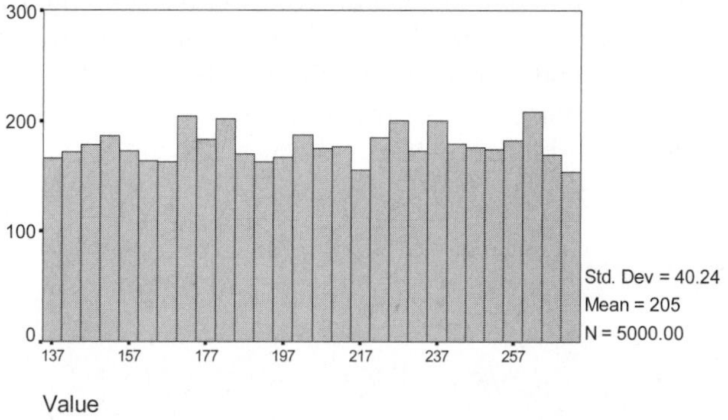

Let's see what happens if we take a sample of 10 cases from the distribution and compute their mean. Figure 11.9 shows the histogram of 500 such sample means.

Figure 11.9 500 samples of 10 from a uniform distribution

To obtain this histogram, open the simul.sav file and select the variable unif10 in the Histogram dialog box.

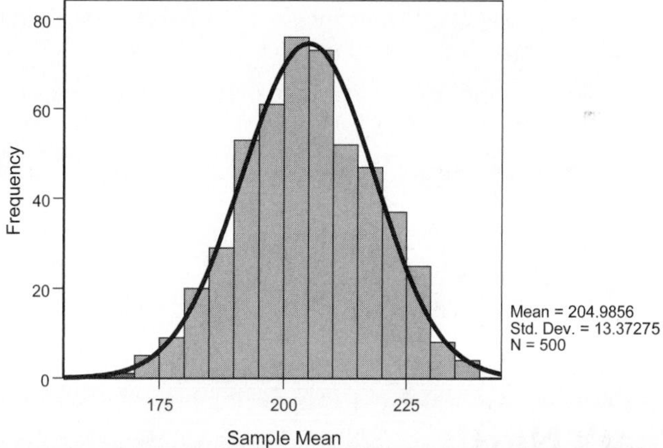

What is amazing is that the distribution of sample means looks nothing like the original distribution of values. The distribution of means is approximately normal, even when the distribution of a variable is not, provided that the sample size is large enough. This remarkable fact explains why the normal distribution is so important in data analysis. If the variable you're studying does have a normal distribution, then the distribution of sample means will be normal for samples of any size. The more unlike the normal distribution the distribution of your variable is, the larger the samples have to be for the distribution of means to be approximately normal. You'll be able to use the properties of the normal distribution to test a variety of hypotheses about population means based on the results observed in samples.

Summary

What is the normal distribution, and why is it important for data analysis?

- A normal distribution is bell-shaped. It is a symmetric distribution in which the mean, median, and mode all coincide. In the population, many variables, such as height and weight, have distributions that are approximately normal.
- Although normal distributions can have different means and variances, the proportional distribution of the cases about the mean is always the same.
- A standard normal distribution has a mean of 0 and a standard deviation of 1.
- The Central Limit Theorem states that for samples of a sufficiently large size, the distribution of sample means is approximately normal. (That's why the normal distribution is so important for data analysis.)

What's Next?

In this chapter, you performed a very simple statistical test. You tested whether a sample might be coming from a population with a known mean and standard deviation. You used the properties of the normal distribution to help you evaluate whether your sample results were unusual. In the chapters that follow, you will learn how to test a variety of hypotheses. The basic idea will not change, just some of the details.

Exercises

Statistical Concepts

1. You are interested in estimating the mean vacancy rate on Saturday night at hotels in major metropolitan areas. Let's assume that in the population of all hotels in major metropolitan areas, the distribution of vacancy rates is approximately normal, with a mean of 50 and a standard deviation of 15. The following stem-and-leaf plot is a sample of values from a normal distribution with a mean of 50 and a standard deviation of 15:

```
Frequency    Stem &  Leaf

    6.00 Extremes    (0), (6), (9), (10), (12)
    4.00        1 .  6&
    9.00        2 .  0124&
   22.00        2 .  5667788899
   23.00        3 .  0012223444
   43.00        3 .  5556666677788888899999
   67.00        4 .  0000001111112222222223333333444444
   67.00        4 .  5555555566666666677788888899999999
   78.00        5 .  0000001111112222222222333333344444444
   61.00        5 .  5555566666677777888888899999
   42.00        6 .  00000011112222233334
   27.00        6 .  5555666777789
   25.00        7 .  00011112344
   14.00        7 .  55668&
   10.00        8 .  1113&
    1.00        8 .  &
    1.00 Extremes    (88)

Stem width:     10.00
Each leaf:       2 case(s)


& denotes fractional leaves.
```

a. Based on the figure, approximately what percentage of all hotels have vacancy rates within one standard deviation of the mean? Within two standard deviations?

b. How do the values you estimated from the figure compare to the values shown in Appendix D?

c. The stem-and-leaf plot above is based on 500 values from a normal distribution with a mean of 50 and a standard deviation of 15. How would you expect the distribution of values to change if you took 1000 values from the same normal distribution?

2. Consider what would happen if you took a random sample of five hotels and calculated the average vacancy rate. Would you expect the rate to be 50? Would you expect the standard deviation to be 15? Explain your answer.

3. If you repeatedly took a random sample of five hotels and computed their average vacancy rate 500 times, you would get a distribution of means that looks like the following:

```
Frequency    Stem &  Leaf

   5.00 Extremes     (29), (31), (32)
   7.00        3 .  555&
  13.00        3 .  666777
  15.00        3 .  8888999
  21.00        4 .  0000111111
  37.00        4 .  222222222333333333
  33.00        4 .  4444444555555555
  55.00        4 .  666666666666667777777777777
  55.00        4 .  888888888888888999999999999
  69.00        5 .  0000000000000000000011111111111111
  59.00        5 .  22222222222222223333333333333
  50.00        5 .  444444444444444555555555
  35.00        5 .  66666666667777777
  20.00        5 .  8888888999
  11.00        6 .  00011
   8.00        6 .  223
   3.00        6 .  4&
   2.00        6 .  6
   2.00 Extremes     (68)

Stem width:      10.00
Each leaf:        2 case(s)


& denotes fractional leaves.
```

   a. What's similar and what's different about the above distribution and the distribution of individual values shown in question 1? Be sure to comment on the means and standard deviations of the two distributions.

   b. Estimate the standard deviation of the above distribution based on the relationship between the standard deviation and the standard error of the mean.

4. You're interested in buying a hotel. The seller assures you that the average vacancy rate for Saturday night is 50%, just like that for all hotels in the area. You take a sample of five Saturdays and compute the average vacancy rate to be 75%. What would you conclude about the seller's claim? How often would you expect to see a rate of 75% or more if the seller's claim is correct?

   a. The seller is unhappy with your statistics. He tells you that five Saturdays are too few for you to draw meaningful conclusions. He recommends that you examine a random sample of 40 Saturdays instead. In this situation, what is the hypothesis that you are interested in testing?

b. Sketch what you would expect the sampling distribution of the mean to be, based on samples of size 40, if the null hypothesis is correct. Be sure to indicate what the mean and standard deviation are for the distribution.

c. On the basis of the sample of size 40, you find the average vacancy rate to be 60%. The seller is noticeably relieved, since he is sure that a value of 60% is close enough to 50% for you to believe his claim. Do you? Explain your answer.

5. Explain why you agree or disagree with each of the following statements:

a. It's better to include a small number of subjects in a study than a large number.

b. All samples from the same population give the same results.

c. How much the mean varies from sample to sample depends on both the size of the sample and the variability in the population.

d. Both variables and statistics have distributions.

6. Assume you are told that in the population of adults in the United States, nostril width is normally distributed, with a mean of 0.9 inches and a standard deviation of 0.2 inches. List all of the facts about nostril width that you can deduce from the statement.

7. If grades on an examination are approximately normally distributed, with an average of 70 and a standard deviation of 10, what percentage of the students:

a. Received grades less than 70?

b. Received grades greater than 70?

c. Received grades less than 60?

d. Received grades less than 50?

e. Received grades less than 50 or greater than 90?

f. Received grades less than the median?

8. Two researchers are studying the effect of positive thinking on recovery time after surgery. Both take a sample of 25 persons about to undergo surgery, teach them how to think positively, and then examine how long they stay in the hospital.

a. The first researcher calculates the average stay to be 12 days and the standard deviation to be 3 days. He reports these results in the *Journal of Positive Living*. The second researcher calculates the average stay to be 12.5 days. He reports the standard error of the mean to be 0.6 days. When he submits his results to the same journal, the editors question his findings. They want to know why his measure of variability is so much less than the first researcher's. Explain to the editors of the journal the difference between the two statistics. Indicate the relationship of the two statistics as well.

b. What would be the standard error of the mean if the standard deviation remained at 3 but the sample size were increased to 50? What if it were decreased to 10?

9. In this exercise, you will use SPSS to compute areas under the normal curve. Assume that scores on a test are normally distributed in the population with a mean of 100 and a standard deviation of 15. You have seven people with scores of 55, 70, 85, 100, 115, 130, and 145. You want to tell them what proportion of people in the population have scores less than or equal to theirs.

a. First, use the Data Editor to enter the seven scores into a data file. Call the variable *score*.

b. Next, compute standardized scores (z scores) for the seven data values. You can't use the Descriptives procedure to do this, since Descriptives will standardize the scores using the mean and standard deviation of the seven scores in your file. You want to standardize the scores using the population mean of 100 and the population standard deviation of 15. To do this, you must use the Compute facility on the Transform menu. Compute a variable named *zscore* equal to the following expression:

(score – 100)/15

Then look at the Data Editor to make sure you have the correct z score for each value of *score*.

c. In the Compute Variable dialog box, there is a function called CDFNORM(zvalue). You supply a standardized score (z value), and it tells you what proportion of cases in a standard normal distribution have values less than equal to your standardized score. Compute a variable *cumprob* equal to the following expression:

cdfnorm(zscore)

The variable *cumprob* is the proportion of cases in a normal distribution with standardized scores less than or equal to *zscore*.

d. In the Data Editor, you now have three variables: *score*, *zscore*, and *cumprob*. In Variable View, select *cumprob* and change the number of decimal places to 4.

e. Based on the values of *cumprob*, indicate the areas in the following drawings:

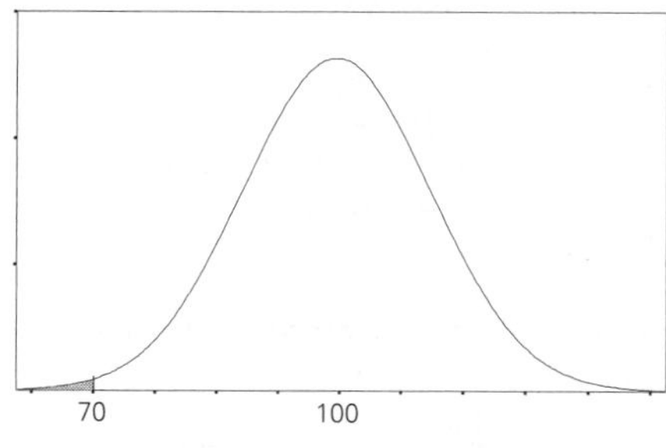

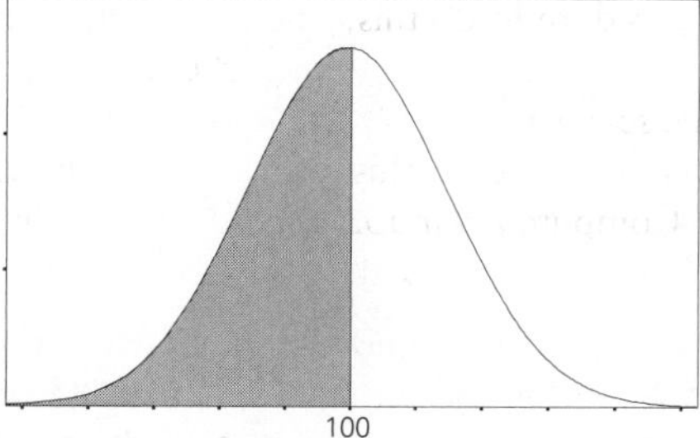

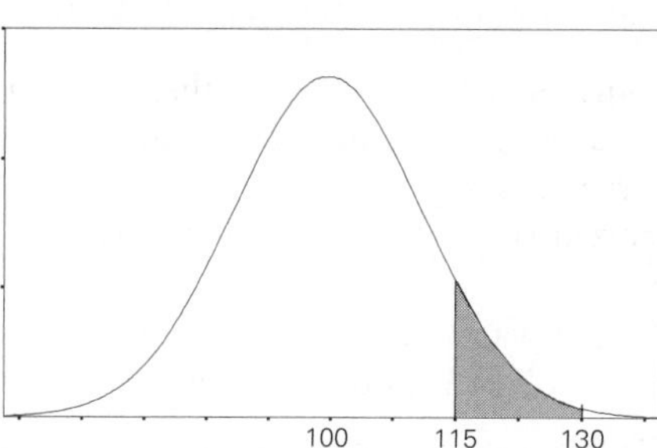

10. Using the CDFNORM function described above and simple arithmetic, you can calculate any area in the normal distribution. For example, if you want to know the proportion of cases in a standard normal distribution with standardized scores in absolute value at least as large as the one you observed, you would compute a variable (call it *twotailp*) using the following expression:

$2 \times (1 - \text{cdfnorm}(\text{abs}(z\text{score})))$

Here's what the expression does. Consider a standard score of +2. To calculate the proportion of cases with standardized scores greater than +2, you have to subtract the proportion of cases with values less than 2 from the total area under the curve, which equals 1. For example, in a normal distribution, 97.72% of the cases have standardized scores less than or equal to 2. The percentage with standardized scores greater than 2 is then

$$100\% - 97.72\% = 2.28\%$$

Since the normal distribution is symmetric, scores less than –2 and greater than +2 are equally likely. You can therefore get the probability of obtaining a score less than –2 or greater than +2 by doubling the probability of a score greater than +2. Thus, the probability of obtaining a standard score at least 2 in absolute value is 2 times 0.0228, or 0.0456. The absolute value function abs(zscore) ensures that you'll get the same two-tailed probability for negative z scores (such as –2) and positive z scores (such as +2).

a. Using the Compute facility, calculate the variable *twotailp* for the z scores in your data file. Increase the number of decimal places displayed to 4.

b. Based on the results you obtained in question 10a, answer the following: What's the probability that a person has a test score less than 85 or greater than 115? What's the probability that a person has a test score greater than 145? Less than 100? Less than 55 or greater than 145? Between 55 and 145?

11. Modify the expression in question 10 to compute the probability of a standardized score at least as large as the one observed and of the same sign. Call the new variable *onetailp*.

a. What's the relationship between *onetailp* and *twotailp*?

b. If a person asks you, "What percentage of people in the population scored at least as well as I did?", which probability would you quote?

c. If a person asks you, "What percentage of people in the population scored as 'weird' as I did?", which probability would you quote?

d. What percentage of people have standardized scores greater than –2? Greater than +3?

Data Analysis

Use the *gssnet.sav* file to answer the following questions:

1. Make histograms of *age, educ,* and *hrs1*. Superimpose a normal curve on them. Do any of these distributions appear approximately normal? Describe how each of the distributions is different from the normal distribution.

2. Generate a fake, normally distributed IQ score for each case in the file. To do this, use the Compute facility and place the following expression into the dialog boxes:

   IQ = norm(15) + 100

   These instructions generate a random sample from a normal distribution with a mean of 100 and a standard deviation of 15.

   a. Make a histogram of IQ and superimpose the normal curve on it.

   b. If IQ is exactly normally distributed, what percentage of cases should have values between 70 and 130? What percentage of cases in your sample have IQ's in this range?

   c. What percentage of the cases would you expect to have IQ's of 115 or more if IQ is exactly normally distributed? What percentage of cases in your sample have IQ's greater than 115?

   d. What percentage of cases in your sample have IQ's less than 85? What would you expect if the distribution is exactly normal?

3. Compute standard scores for the IQ variable. Make a histogram with a normal curve superimposed.

   a. What is the mean of this distribution? The standard deviation?

   b. From Appendix D, what percentage of the cases would you expect to have standard scores between –1 and +1? Between 0 and 1.5? Greater than +2? Less than –2?

   c. Compute the quartiles of the standardized variable. How do your observed quartiles compare to the quartiles of a normal distribution?

Use the *electric.sav* data file to answer the following questions:

4. Make histograms of *chol58, dbp58, ht58, cgt58,* and *wt58*. Superimpose a normal distribution with the same mean and variance on each one. Does the distribution of any of these variables appear to be approximately normal?

5. Standardize each of the variables in question 4. For the standardized variables, compute the 5th, 16th, 50th, 84th, and 95th percentiles. How do your observed percentiles compare to those that would be expected if the distributions were exactly normal?

12 Testing a Hypothesis about a Single Mean

Sometimes it's hard to tell the difference between an urban legend and a "fact." Everyone knows that the average IQ is 100 points, that the normal work week is 40 hours, and that the early bird catches the worm. Most people have little recourse but to believe what they hear. You, as a student of statistical methods, can actually evaluate the veracity of such claims. You just need good data. The General Social Survey can be used to test whether the 40-hour work week is a myth. (You'll have to gather your own data on early morning worm-seeking behavior in birds.)

In Chapter 11, you learned how to test whether a sample comes from a population with a known mean. Your test was based on the fact that if you select samples from a population that has a normal distribution, the distribution of the sample means is normal as well. You also saw that for sufficiently large sample sizes, the distribution of sample means will be normal even if the population from which you select your sample is not normal.

To use the normal distribution to test whether your sample comes from a population with a particular mean, you have to know both the mean and standard deviation *of the population*. For example, to see if the cholesterol levels of CEOs appear to be unusual, you had to know the mean and standard deviation of the cholesterol values in the general population. Often, however, you don't know the standard deviation of the population values and instead must estimate it from your sample. In this chapter, you'll learn how to use a distribution closely related to the normal, the **t distribution**, to test whether a sample comes from a population with a specified mean when you don't know the population standard deviation.

▶ This chapter uses the *gssft.sav* data file, which includes data for full-time workers only. You must select people with college degrees to run the analysis. For instructions on how to obtain the *t* test output shown in this chapter, see “How to Obtain a One-Sample T Test” on p. 251. For instructions on how to restrict your analysis to people with college degrees, see “Example: Selecting College Graduates” on p. 619 in Appendix B.

Examining the Data

Since many readers of this book aspire to join the workforce with a college degree in hand, we’ll analyze the length of the work week for college graduates who are employed full time. (Later chapters look at the relationship of education to hours worked.)

The first step in analyzing data is examining a plot of the data values. From a frequency table, stem-and-leaf plot, or histogram, you can see if there are any strange data values. You can also see whether the distribution of data values looks approximately normal. If the distribution of data values looks not too far from normal, you can be confident that the distribution of means will be normal, even for small sample sizes.

Figure 12.1 Histogram of hours worked for college graduates

From the Graphs menu, choose Legacy Dialogs, and then Histogram. In the Histogram dialog box, select hrs1.

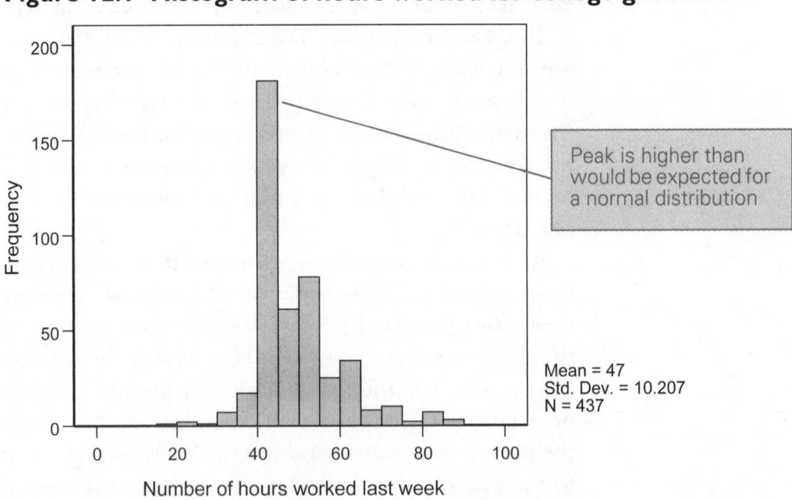

Figure 12.1 is a histogram of the hours worked the previous week for the 437 college graduates in your sample. You see that the distribution of the data values is not quite normal. The peak at 40 hours is higher than you would expect for a normal distribution. There is also a tail toward larger values of hours worked. It appears that people are more likely to work a long week than a short week. Since the sample size is quite large and the

distribution is somewhat bell-shaped, you can count on the Central Limit Theorem to ensure that the sampling distribution of the means is approximately normal.

In Figure 12.2, you see that the average work week is 47 hours, and the standard deviation is 10.2 hours.

You obtain these descriptive statistics as part of the one-sample t test output. See Figure 12.4.

Figure 12.2 Descriptive statistics for hours worked

	N	Mean	Std. Deviation	Std. Error Mean
hrs1 Number of hours worked last week	437	47.00	10.207	.488

You know that even if college graduates work an average of 40 hours a week, you don't expect a random sample of 437 of them to hit the norm on the head. You know that means calculated from samples from the same population vary. How much they vary depends on the size of the sample and the standard deviation of the values in the population.

If you know the population value for the standard deviation of the number of hours worked per week, you can use the procedure described in Chapter 11 to determine whether the observed sample mean of 47 is unlikely if the population mean for number of hours worked is 40.

? *How would you go about determining if 47 is an unlikely value if you know the standard deviation?* First, find the difference between the observed sample mean and the hypothetical population mean. The difference is $47 - 40$, or 7 hours. Then you calculate the standard error of the mean, which is the population standard deviation divided by the square root of the sample size. Let's assume that you know that the population standard deviation is 10.2 hours. The standard error is then $10.2/(\sqrt{437}) = 0.49$.

Next, you have to figure out the standard score for your observed mean. You do this by dividing the difference between the observed and hypothetical mean by the standard error. For this example, the standard score is $7/0.49 = 14.3$. Since 99% of the cases in a normal distribution have standardized values between –2.6 and +2.6, you know that a standard score of 14.3 is extremely unusual. ■■■

The T Distribution

It may seem reasonable that if you don't know the value for the population standard deviation, you should just substitute the sample standard deviation and proceed as before to base your test on the normal distribution. For small sample sizes, that's not a good idea. Here's why. You've already seen that when you take a sample from a population and calculate the sample mean, it's very unlikely that the sample mean will be the same as the population mean. The same is true for the sample variance. If you take a sample from a population and calculate the variance, it is very unlikely that the sample variance will be the same as the population variance. Sample variances, just like sample means, have sampling distributions. That is, if you take repeated samples of the same size from a population and calculate their variances, these variances will spread out into a distribution. (The distribution will not be normal, however. If the samples are from a normally distributed population, the distribution of sample variances has what's called an *F* distribution. This distribution is discussed in Chapter 15.)

If you use the sample standard deviation instead of a known population value in the computation of the standard score, you introduce additional uncertainty into the result. For example, if your sample standard deviation is smaller than the population value, the resulting standard score will be too large. If the observed sample standard deviation is too large, the standard score will be too small. That's why, when you don't know the population standard deviation but estimate it from the sample, the distribution of standard scores is no longer normal. Instead, it follows what's called the *t* distribution. The *t* distribution takes into account the fact that, by using the sample standard deviation instead of the population standard deviation, you're introducing error into the computation of the standard score. The *t* distribution looks like a normal distribution but it has more area in the tails. That's because large standard scores can result not only from sample means that are far from the population mean but also from poor estimates of the population standard deviation.

Another way that the *t* distribution differs from the normal distribution is that its shape depends on the number of cases in your sample. On first reflection, this may seem odd, but it's not. You know that if you estimate a population standard deviation based on a sample of 4 cases, the possible results will have much more variability than if you estimate the population standard deviation based on 4000 cases. You're much more confident that the estimate based on the larger sample size is closer to the true value than the estimate based on the smaller sample size. If you have a large sample size, the fact that you don't know the population standard deviation becomes much less important than if you have a small sample size.

Look at Figure 12.3, which shows the *t* distribution for degrees of freedom of 3, 10, and 50. (It is customary to identify a *t* distribution not by the actual number of cases in the sample, but by what's called the degrees of freedom. In this example, it's just the number of cases in the sample minus 1.)

Figure 12.3 T distribution for 3, 10, and 50 degrees of freedom

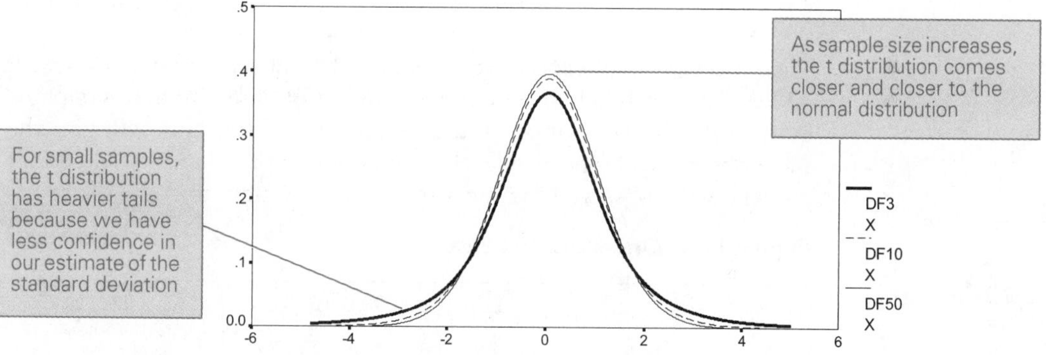
For small samples, the t distribution has heavier tails because we have less confidence in our estimate of the standard deviation
As sample size increases, the t distribution comes closer and closer to the normal distribution
.5
.4
.3
.2
.1
0.0
-6
-4
-2
0
2
4
6
DF3 X
DF10 X
DF50 X

In Figure 12.3, you see that as the degrees of freedom increase, the *t* distribution becomes more peaked. That is, the percentage of the total area in the tails decreases. For 50 degrees of freedom, the *t* distribution and the normal distribution are indistinguishable. For small degrees of freedom, the tails of the *t* distribution are heavier than those of the normal distribution. When you have to estimate the population standard deviation, you'll find more seemingly large deviations from the population mean than you would if you knew the standard deviation.

? *What do the Guinness Brewery in Dublin and the* t *distribution have in common?* W. S. Gosset, of course. Gosset was a chemist who worked for Guinness in the early 1900's. Since he carried out experiments based on small sample sizes, he worried about the consequences of using a sample value for the standard deviation when testing hypotheses about the mean. Standard practice was to just use the normal distribution. Gosset derived the *t* distribution and published his results under the pseudonym Student. That's why the *t* distribution is often called Student's *t*.

Calculating the T Statistic

To test the hypothesis that a sample comes from a population with a known mean but an unknown standard deviation, you calculate what's called a **t statistic.** The calculations are exactly the same as for the standard score, except that the value of the sample standard deviation is used in place of the population value in calculating the standard error of the mean.

For this example, the value for the t statistic is

$$t = (47 - 40)/0.49 = 14.3 \qquad \textbf{Equation 12.1}$$

The observed significance level is obtained from the t distribution with 436 degrees of freedom. (Since the number of cases in your sample is quite large, the t distribution and the normal distribution will give the same observed significance levels.) The observed t value is so large that you know the observed significance level is very close to 0.

Figure 12.4 One-sample t test

From the menus choose:

Analyze
Compare Means ▶
One-Sample T Test...

Select hrs1 and specify 40 for Test Value, as shown in Figure 12.8.

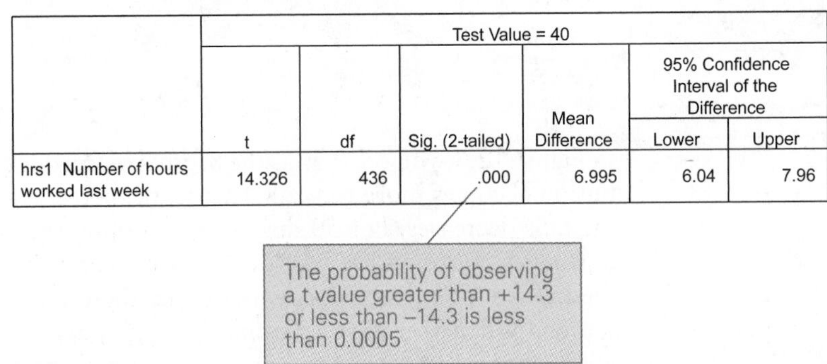

	Test Value = 40					
					95% Confidence Interval of the Difference	
	t	df	Sig. (2-tailed)	Mean Difference	Lower	Upper
hrs1 Number of hours worked last week	14.326	436	.000	6.995	6.04	7.96

The probability of observing a t value greater than +14.3 or less than −14.3 is less than 0.0005

Figure 12.4 contains the value for the t statistic as well as additional output from the SPSS One-Sample T Test procedure. You see that the difference between the observed mean and the hypothetical population mean of 40 (labeled *Test Value*) is 7.0. For your sample, the average work week was 7.0 hours longer than the hypothesized 40 hours.

The probability of observing a sample t value greater than +14.3 or less than −14.3 is given by the entry labeled *Sig. (2-tailed)*. Since the observed significance level is less than 0.0005, SPSS displays it as 0.000. This does not mean that the probability is 0. It is less than 0.0005. Based on the observed significance level, you can conclude that it's quite unlikely that college graduates work a 40-hour week on average. They seem to work much more.

The probability given on the output is called "two-tailed" because it is the sum of the areas in both tails of the *t* distribution—the area less than –14.3 and the area greater than +14.3. You're interested in both of these areas since the average number of hours worked by college graduates can be either less than 40 hours or greater than 40 hours. Both alternatives are possible and of interest.

? *When should I use the One-Sample T Test procedure?* You should use the One-Sample T Test procedure if you have a single sample of data and want to test whether your sample comes from a population with a known mean. For example, you can use the One-Sample T Test procedure to see whether 16-ounce boxes of cereal really weigh 16 ounces on average, or to test whether children who are born prematurely have an average IQ of 100. In both of these examples, you have a single set of data—cereal boxes that you weighed and premature babies whom you tested. You want to compare your sample means to known population values. Your test values are not estimated from another set of data. They are known values. If you have two samples of data, for example, CEOs and non-CEOs, or premature and full-term infants, and you want to compare their means, you should *not* use the one-sample *t* test. Instead, you'll probably want to use the independent-samples *t* test described in Chapter 14. ■■■

Confidence Intervals

From the results of the *t* test, you're reasonably confident that the average work week for college graduates is not 40 hours. What do you think it is? Based on your sample, your best guess is 47 hours, but you know that it's most unlikely that the true population value for all college graduates is actually equal to the value found in your sample. Let's see how, based on the normal distribution, you can obtain useful information about a plausible range of values for the population mean.

Figure 12.5 Sampling distribution of means

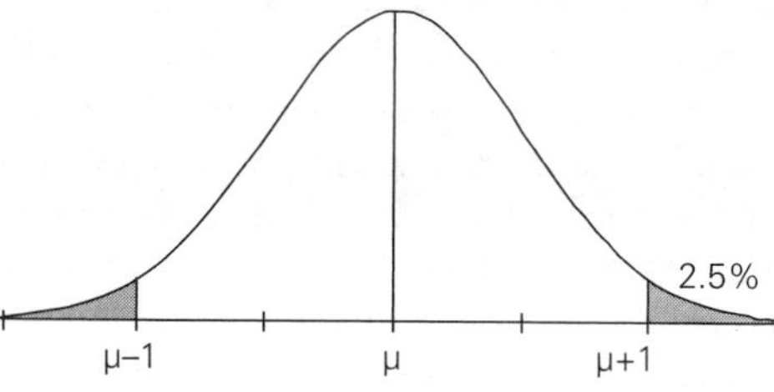

First, consider Figure 12.5, which is the distribution of all possible sample means for samples of 437 college graduates. (For simplicity, we've assumed that the population standard deviation is 10.2 hours, so the standard error of the mean is about 0.5 hours.) The first thing that should strike you about this distribution is that you don't know its mean. You know only the average hours worked for one sample of 437 college graduates. In Figure 12.5, the unknown population mean is identified by the Greek symbol μ (mu), the common abbreviation for the population mean (well, common if you read statistics books...).

Although you don't know the value for the mean of the sampling distribution of means, we're supposing that you do know the standard deviation of the distribution—the standard error. It's about 0.5 hours. If the sample means are normally distributed, you also know that approximately 95% of the sample means should be within about two standard errors, or 1 hour, of the unknown population mean. (More precisely, within about 1.96 standard errors.) Only 5% of the sample means fall in the shaded region of Figure 12.5.

Based on the previous information, you can calculate a range of values—an interval—that should include the population mean 95% of the time. You calculate the lower limit of this interval by subtracting 1.96 times the standard error from your sample mean. You calculate the upper interval by adding 1.96 times the standard error to the sample mean. For this example, the interval is from 47 – 1 to 47 + 1; that is, from 46 to 48 hours. This interval has a special name. It's called a **95% confidence interval** for the population mean.

? *If the population standard deviation isn't known—if it must be estimated from the sample—how does that change the computation of the confidence interval?* The only difference is that you must use values from the *t* distribution instead of the normal distribution. For example, if you have a sample of 10 cases, instead of using the value 1.96 in the computation of the 95% confidence interval, you must use the value 2.26 because 95% of the values in a *t* distribution with 9 degrees of freedom are between –2.26 and 2.26. ■■■

It may help you to understand the idea behind confidence intervals if you identify a possible sample mean in Figure 12.5 and calculate the confidence interval. Consider first a sample mean that is 1.5 standard errors above the population mean. It's shown in Figure 12.6, as is the 95% confidence interval based on it.

Figure 12.6 Sample mean 1.5 standard errors above population mean

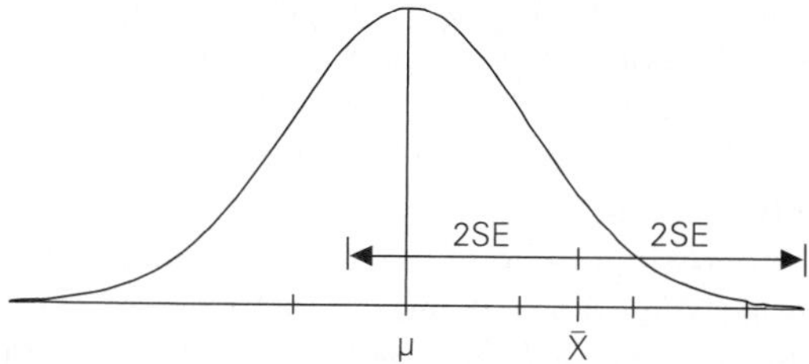

Does the confidence interval include the unknown population value? Sure it does. The confidence interval extends for two standard errors below the sample mean, while the population mean is only 1.5 standard errors less than the sample mean. Now look at Figure 12.7, which shows a sample mean 1 standard error below the population mean. Again, the 95% confidence interval includes the population value.

Figure 12.7 Sample mean 1 standard error below population mean

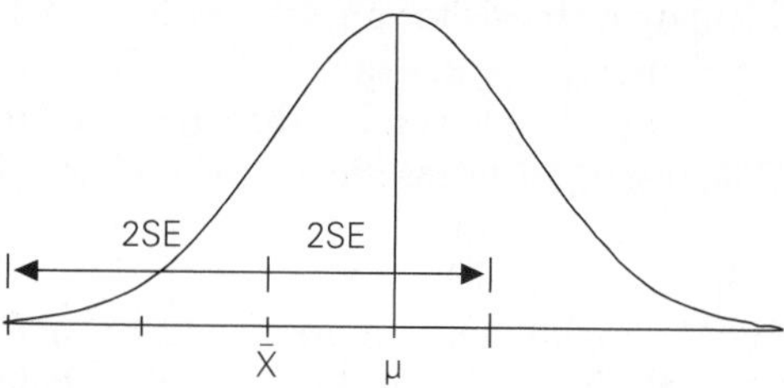

The only time the confidence interval won't include the population value is when your sample mean falls into the shaded region of Figure 12.5. The shaded region corresponds to the 5% of the distribution that is more than two standard error units from the population mean. Unfortunately, you can never tell whether your particular sample mean is one of the unlucky ones in the shaded region. All you can do is calculate the 95% confidence interval and hope that your sample is one of the 95-out-of-100 for which the confidence interval includes the population value.

? *Am I correct in saying that the probability is 95% that the population mean for the average hours worked by college graduates is in the range of 46 hours to 48 hours?* Not quite. Once you actually compute the confidence interval based on a sample mean, the resulting interval either contains the population mean or it doesn't. The correct statement is that you're 95% *confident* that the interval contains the population mean.

The following analogy may make the idea clearer (and might make this book a *New York Times* best-seller). Before a baby is conceived, the probability that it will be a girl is roughly 50%. Once the deed is done, the baby is either a boy or a girl. The same is true for confidence intervals. Before you conduct your survey or experiment, you know that 95 times out of 100, the 95% confidence interval based on your sample mean will include the population mean. Once you've calculated the interval, it either does or it doesn't include the population value. ■■■

Other Confidence Levels

Although the 95% confidence interval is the most commonly reported, you can calculate intervals for any confidence level. What changes is the width of the interval. For example, to construct a 99% confidence interval, for a large sample you use the value 2.57 instead of 1.96. That's because 99% of sample means are within 2.57 standard error units of the unknown population mean. The 99% confidence interval for the average number of hours worked by college graduates is from 45.73 hours to 48.26 hours. Although you're more confident that this interval contains the population mean, the interval is wider than before. A wide interval is less useful than a narrower one. For example, if you are confident that a drug prolongs life by 5 to 6 years, that's more useful than knowing that the drug prolongs life from 1 month to 15 years. Whenever you compute a confidence interval, you trade off the degree of confidence and the interval width—the higher the confidence level, the wider the interval.

? *How can I get SPSS to compute confidence intervals for me?* The easiest way to get a confidence interval for a mean is to run the Explore procedure. A 95% confidence interval is part of the default output. You can change the confidence level as described in "Explore Statistics" on p. 133 in Chapter 7. SPSS also computes confidence intervals as part of the output in other procedures. ■■■

Confidence Interval for a Difference

You can specify the confidence level in the One-Sample T Test Options dialog box.

You can easily convert the observed confidence interval for the average hours worked in a week by college graduates into a confidence interval for the difference between the population mean and the hypothesized value of 40. All you have to do is subtract the value 40 from the lower and upper bounds of the confidence interval for the population mean. This gives you a 95% confidence interval from 6 to 8 hours. You are 95% confident that the difference between the number of hours actually worked by the average college graduate and the mythical 40-hour work week is between 6 and 8 hours. It appears likely that the average college graduate works quite a bit more than 40 hours (or inflates the hours worked when asked). The 95% confidence interval for the population mean difference is shown in Figure 12.4.

When you run the one-sample *t* test, you can calculate a confidence interval at whatever level you want. For example, the 99% confidence interval for the difference from 40 hours extends from 5.73 to 8.26 hours. As you expect, it is wider than the 95% confidence interval. That's because the only way you can be more certain that you're trapping the population value is to increase the range of values included in the interval.

Confidence Intervals and Hypothesis Tests

If you look at the 95% confidence interval for the difference between the hours actually worked and 40 hours, you will notice that the value 0 is not included in the interval (the interval ranges from 6 to 8 hours). That tells you that 0 is not a plausible population value. It appears unlikely that the true difference between the average hours worked and 40 hours is 0. There's a close relationship between hypothesis tests and confidence intervals: if a value is not included in a 95% confidence interval for the difference, you can reject the hypothesis that it's a plausible value for the population difference, using a 5% criterion for unusual.

From the confidence interval, you can conclude that the true difference is unlikely to be smaller than 6 hours or larger than 8 hours. If you did a one-sample *t* test and set your hypothetical week to be less than 46 hours or greater than 48 hours, the two-tailed probability level for that *t* test would be less than 0.05. If you set the hypothetical value to be between 46 and 48 hours, the observed significance level would be greater than 0.05. Similarly, if you calculate a 99% confidence interval and it does not contain a particular value, then the corresponding *t* test for that hypothetical value will have an observed significance level of less than 0.01.

Null Hypotheses and Alternative Hypotheses

In the last few chapters, you've learned how to test a hypothesis statistically. You tested whether a new drug has the same cure rate as the standard treatment, whether CEOs have the same average cholesterol levels as the general population, and whether college graduates work a 40-hour week. You followed the same steps for testing all three hypotheses, although the statistic on which you based your conclusion differed. (For the cure-rate example, you used a binomial test; for the CEO cholesterol levels, a test based on the normal distribution; and for the work week, a test based on the *t* distribution.) The statistical test you use depends on the hypothesis of interest and the type of data available.

In each of the three situations, there are two hypotheses or claims of interest. The first is that there is nothing going on: the new drug is as effective as the standard, CEOs have the same average cholesterol levels as the general population, and college graduates really do work a 40-hour week. In statistical terms, these are called **null hypotheses.** (Notice that each null hypothesis is precise. It describes a hypothetical but exact state of affairs.) The **alternative hypothesis** describes the situation when the null hypothesis is false. The following are all alternative hypotheses: the new treatment changes the cure rate, CEOs have different cholesterol levels than the general population, and college graduates don't work a 40-hour week.

When you statistically test a hypothesis, you assume that the null hypothesis correctly describes the state of affairs. The null hypothesis is the frame of reference against which you'll judge your sample results. You assume that the cure rate is the population value of 50%. You assume that the CEOs have average cholesterol levels of 205. You assume that college graduates work a 40-hour week. The null hypothesis describes a well-defined situation. If the population cure rate is 50%, you can determine how often you expect to see various possible sample outcomes, such as 12 cures or more in a sample of size 20. If the true cholesterol value is 205 mg/dL, you can calculate how often you would expect to see samples of size 21 with average cholesterol levels of 193 mg/dL or less. If the true work week is 40 hours, you can determine how often a sample of 437 people would have an average work week as long as 47 hours.

You must state the null hypothesis so that it perfectly describes a single situation. The null hypothesis cannot state that college graduates don't work a 40-hour week. That statement cannot serve as a frame of reference for evaluating sample results, since it describes many possible outcomes. (On the other hand, college graduates working the same number of hours as people who haven't graduated from college is a perfectly acceptable null hypothesis. It describes a situation that can be used as a frame of reference for evaluating your observed sample results.)

Most of the time when you perform an experiment or conduct a study, the null hypothesis claims the opposite of what you would like to be true. If you've synthesized a new compound that you think improves memory, the null hypothesis would state that it does not. If you think that men and women are not equally satisfied with their jobs, the null hypothesis would state that they are.

An alternative hypothesis can specify the direction of the difference that you expect to observe. If you know that the cure you are touting cannot be worse than the standard, your alternative hypothesis can claim that your cure rate is better than the standard. However, the direction of your alternative hypothesis must be stated in advance. You can't look at the data values and then decide on the direction. The reason for this is that if you know the direction in advance, your observed significance level can be restricted to include possible outcomes only in the direction of interest. That will make your observed significance level smaller than if you consider both alternatives. For example, if you know that your treatment can only be better, you can calculate the probability of getting sample results as extreme as the ones you observed in a positive direction. If your treatment can help or hinder, then you must consider differences at least as large as the ones observed in either direction.

? *How can I get the correct observed significance level for an alternative hypothesis that specifies a direction?* If you are using a statistical test that calculates the observed significance level from a symmetric distribution, such as the normal or the *t* distribution, you can divide the two-tailed observed significance level by 2. That will give you a one-tailed significance level. It's called one-tailed, since it's based on the area in only one tail of the distribution. ■■■

Rejecting the Null Hypothesis

Since the null hypothesis serves as the frame of reference against which sample results are evaluated, if your sample results appear to be unlikely when the null hypothesis is true, you reject the null hypothesis. That is, if the probability of obtaining sample results as extreme as the ones you've observed (the observed significance level) is small, usually less than 0.05, you are entitled to reject the null hypothesis. In the previous example, you rejected the null hypothesis that college graduates work a 40-hour week.

You did not reject the null hypothesis that CEOs have an average cholesterol value that is the same as that of the general population. In the chapters that follow, you'll learn more about what you can and can't conclude when you reject or don't reject the null hypothesis. In particular, in Chapter 14 you'll learn about factors that affect your ability to reject the null hypothesis when it is false.

Summary

How can you test the hypothesis that a sample comes from a population with a known mean?

- A one-sample *t* test is used to test the null hypothesis that a sample comes from a population with a particular mean.
- A confidence interval is a range of values that, with a designated likelihood, includes the unknown population value.
- The null hypothesis is the frame of reference used to evaluate a claim about a population.
- The alternative hypothesis specifies the situation if the null hypothesis is false.

What's Next?

In this chapter, you used the one-sample *t* test to test the null hypothesis that a sample comes from a population with a given mean. You identified the null and alternative hypotheses. You assumed that the null hypothesis was true and calculated the observed significance level, which told you how often you would expect to see sample results as extreme as the ones you observed, if in fact the null hypothesis is true. If this observed significance level was small, you rejected the null hypothesis. You concluded that as a college graduate, you probably won't be working a 40-hour week.

In this chapter, you were interested in drawing conclusions based on a single sample of data. In Chapter 13, you'll learn how to apply the techniques described in this chapter to testing hypotheses about two related samples.

How to Obtain a One-Sample T Test

This procedure tests the null hypothesis that a sample comes from a population with specified mean. It also displays a confidence interval for the difference.

▶ To open the One-Sample T Test dialog box, from the menus choose:

Analyze
  Compare Means ▸
    One-Sample T Test...

Figure 12.8 One-Sample T Test dialog box

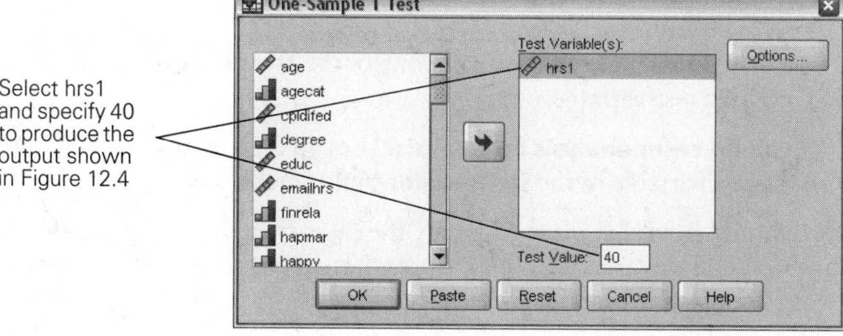

▶ In the One-Sample T Test dialog box, select in the source variable list the variable you want to test and move it into the Test Variable(s) list. You can move more than one variable into the list to test all of them against the specified test value.

▶ Enter a number into the Test Value text box and click OK.

For each variable selected, SPSS calculates the *t* statistic and its observed significance level.

Options: Confidence Level and Missing Data

To change the confidence level for which SPSS displays the confidence interval for the difference between the population mean and the test value, or to control the handling of cases with missing values, click Options in the One-Sample T Test dialog box. This opens the One-Sample T Test Options dialog box, as shown in Figure 12.9.

Figure 12.9 One-Sample T Test Options dialog box

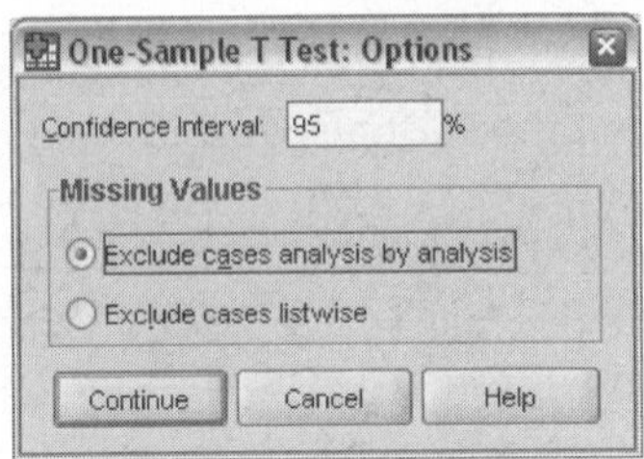

Available options include:

Confidence Interval. Allows you to specify a confidence level between 1 and 99.

Missing Values. Two alternatives control the treatment of missing data for multiple test variables:

Exclude cases analysis by analysis. Uses all cases that have valid data for each variable in the statistics for that variable.

Exclude cases listwise. Uses only the cases that have valid data for all specified test variables. This ensures that all of the tests are performed using the same cases.

Exercises

Statistical Concepts

1. You suspect that your favorite candy bar manufacturer's eight-ounce candy bars weigh less than advertised. You go out and buy 200 candy bars from different stores. You find that their average weight is 7.75 ounces, with a standard deviation of 0.5 ounces. Do you have enough evidence to believe that you are being short-changed? Explain.

2. Your local pizza chain claims that the delivery time of their pizzas is normally distributed with a mean of 30 minutes and a standard deviation of 10 minutes.

   a. You order a single pizza and it arrives in 42 minutes. Do you have reason to disbelieve the chain's claim?

   b. Twenty of your friends in different locations order pizzas. The average delivery time is 42 minutes. Do you have reason to disbelieve the chain's claim? How often would you expect an average delivery time of 42 if the chain's claim is correct?

c. Compute a 95% confidence interval for the true average delivery time based on the results in question 2b.

d. Compute a 95% confidence interval for the true difference based on the results in question 2b.

3. For which of the following situations is a one-sample *t* test appropriate:

a. You want to know if the average salary for males is the same as the average salary for females. You have available a sample of male salaries and female salaries.

b. You want to know if the average difference in systolic blood pressure in a standing and reclining position is 0. You have values for differences for 54 people.

c. You want to know if the average ACT score for your school is 18.

d. You want to know if two schools have the same average ACT scores.

Data Analysis

Use the *gssft.sav* file, which contains data from the General Social Survey for full-time employees only, to answer the following questions:

1. In this chapter, you tested the hypothesis that college graduates who work full time work a 40-hour work week. Now test the hypothesis that for all full-time workers, the population value for average hours worked is 40 hours (variable *hrs1*).

a. What assumptions do you need to use the one-sample *t* test? Do you think the data meet the assumptions?

b. What is the null hypothesis that you want to test? The alternative hypothesis?

c. Test the hypothesis and write a brief summary of your conclusions.

d. Explain the difference between 10.63, the standard deviation of your sample, and 0.275, the standard error of the mean.

e. If your sample size were doubled, how would you expect the value of the standard deviation to change? How would the value of the standard error of the mean change? Estimate both the standard deviation and the standard error for a sample twice as large.

f. What is the 95% confidence interval for the average number of hours per week by full-time workers? How does it differ from the 95% confidence interval for the difference?

g. Based on the 95% confidence interval for the mean difference, can you reject the null hypothesis that the average population value for hours worked is 43 hours? Explain.

h. Based on the confidence interval for the mean, what is a plausible range of population values for the average hours worked?

2. Repeat the analysis in question 1 for women who work full time (use the Select Cases facility to analyze only cases where variable *sex* equals 2). Summarize your conclusions.

Use the *gss.sav* file to answer the following questions:

3. The variable *sibs* is the respondent's number of siblings.

a. Make a histogram of the variable. Do you think its distribution is normal? Is it symmetric? Explain why it looks the way it does.

b. Assume that the sample size is large enough for the Central Limit Theorem to hold. Test the null hypothesis that the average number of siblings is 2.5. Summarize your conclusions.

c. Without running the One-Sample T Test procedure, test the null hypothesis that the average number of siblings is 3. Indicate what procedure you followed.

4. Since the General Social Survey contains people of all ages, you can't conclude anything about the average size of today's family. To get a better idea of current family size, use the Select Cases facility to restrict your analysis only to respondents who are 21 years of age or younger. Test the null hypothesis that the average number of siblings they have is 2.5. Write a short summary of your results.

5. Based on the output you generated for question 4, answer the following questions:

a. Can you determine the probability that the null hypothesis is true? If so, what is it?

b. Can you determine the probability that the null hypothesis is false? If so, what is it?

c. Have you proved that the average number of siblings for people 21 or younger is 2.5?

d. What does the two-tailed significance level tell you?

6. For married couples, *cpldifed* is the difference between years of education for husbands and wives.

a. Make a histogram of the differences. Is the distribution of differences reasonably symmetric? Do you see any outliers? What is the largest positive difference and the largest negative difference?

b. Use the Explore procedure to calculate the 5% trimmed mean for the average difference in years of education for a couple. How much is the difference between the arithmetic mean and the trimmed mean?

c. Perform a one-sample *t* test to test the null hypothesis that the average difference in education is 0.

d. Use the Select Cases facility to select only cases with differences in education of less than 17 years. Rerun the one-sample t test without the outlier. Summarize your conclusions. Include discussion of the effect, if any, of the outlier.

e. The difference between a respondent's father's and mother's education is in *prtdifed.* Can you reject the null hypothesis that the average education for a person's father and mother is equal? What is the 95% confidence interval for the difference?

Use the *electric.sav* data file to answer the following questions:

7. Use the Select Cases facility to select only men with coronary heart disease (variable *chd* equals 1). Test the hypothesis that they come from a population in which the average serum cholesterol is 205 mg/dl (variable *chol58*).

a. State the null and alternative hypotheses.

b. What do you conclude about the null hypothesis based on the t test?

c. What is the difference between your sample mean and the hypothetical population value?

d. How often would you expect to see a sample difference at least this large in absolute value if the null hypothesis is true?

e. Give a range of values that you are 95% confident include the population value for the mean cholesterol of men with coronary heart disease. Does that interval include your test value of 205?

f. What is the range of values that you are 95% confident include the true difference between 205 and the average cholesterol for the population of men with coronary heart disease?

8. Select only men without coronary heart disease (variable *chd* equals 0).

a. Is it plausible that they are a sample from a population in which the average weight is 175 pounds (variable *wt58*)? Explain your reasoning.

b. What is the 99% confidence interval for the population value for average weight for men without coronary heart disease?

c. On the basis of the confidence interval you computed in question 8b, can you reject the null hypothesis that the population value is 180 pounds?

Use the *schools.sav* data file to answer the following question:

9. The leader of the Chicago schools claims that dramatic improvements have occurred between 1993 and 1994. Look at the variables that show the change in the percentage of schools meeting or exceeding state standards (*mathch94*, *readch94*, and *scich94*). Test the hypothesis that the true change in the percentage meeting state standards is 0. Write a short report to the mayor detailing your findings.

13

Testing a Hypothesis about Two Related Means

How can you test the null hypothesis that the average difference between a pair of measurements is 0?

- What are paired experimental designs, and what are their advantages?
- What types of problems can occur when you use paired designs?
- What is a paired *t* test?
- What are Type 1 and Type 2 errors?
- Why do you use a normal probability plot?

In Chapter 12, you used the one-sample *t* test to test whether the average work week for college graduates is 40 hours. You were interested in drawing conclusions about one group of people only—college graduates who work full time. For each person in your analysis, you had a single measurement: the number of hours worked the previous week. In this chapter, you'll learn about a closely related test—the paired-samples *t* test. You can use the **paired-samples t test** to analyze the results of experiments when the same person or animal is observed under two different conditions, or studies in which you have a pair of subjects (or measurements) that are matched in some way. One type of such study is the "before and after" design. For example, you might obtain a student's pulse rate before and after completing an exam, or you might record the blood pressure of a patient before and after a treatment.

▶ This chapter uses the *endorph.sav* data file. For instructions on how to obtain the paired-sample *t* test output shown, see "How to Obtain a Paired-Samples T Test" on p. 268.

Marathon Runners in Paired Designs

Dale, Fleetwood, Weddell, and Ellis (1987) investigated the possible role of β-endorphins in the collapse of runners. (β-endorphins are morphine-like substances manufactured in the body.) They measured plasma β-endorphin concentrations for 11 runners before and after they participated in a half-marathon run. The question of interest was whether average β-endorphin levels changed during a run. The authors postulated that runners were able to continue running despite pain and discomfort because β-endorphin levels increased and produced a sense of well-being. Since the same variable, β-endorphin level, was measured twice on each subject, this study is an example of a paired design.

The advantage of a paired design is that it makes it easier to detect true differences when they exist. When the same person is measured before and after a marathon, observed differences in β-endorphin levels are more easily attributable to running. If you obtain values for two separate groups of people, one group before the race and another group after, some of the observed difference between the two means might be the result of inherent differences between people in the two groups. For example, the "before" people might have naturally lower levels of β-endorphin than the "after" people.

Paired designs are not restricted to situations in which the same person or object is measured under two different conditions. If you are interested in whether sons are taller than their fathers, you can create father-son pairs. Or, if you are interested in whether wives spend more time on household chores than husbands, you can form wife-husband pairs. The important consideration is that the two members of a pair are matched in some way. In the housework example, if you obtain values from spouses, you're controlling for some of the factors that might be associated with time spent working around the house. For example, the number and ages of children, socioeconomic class, and the size of the house are the same for spouses. You can rule them out as possible explanations for any observed differences between men and women.

? *What if my matched pairs of cases don't really match?* If it turns out that the pairs of subjects are not really similar, then a paired design will actually make it harder for you to detect true differences than if you didn't pair the subjects but used two independent groups of subjects. For example, if you arbitrarily create pairs of students and then assign each member of a pair to one of two teaching programs, you'll be worse off than if you randomly assigned students to the two teaching programs. The same is true if the characteristics you use to create pairs aren't related to the variable being measured. For example, if you match students on the basis of height when you're studying methods of teaching reading, the pairing will not do you any good. Analyzing the data with a paired t test will hinder your ability to detect a true difference between the teaching methods. ■■■

Looking at Differences

Whenever you have a paired design, you are primarily interested in the difference between the two measurements for the same individual or for the matched pair. The sign of the difference is important because it tells you the direction of the change. If you subtract the *before* value from the *after* value and the result is positive, that means the values after some event are larger than the values before the event. If the result is negative, the values after the event are smaller than those before the event. If there has been no change, then you expect to have roughly the same number of positive and negative signs.

Figure 13.1 β-endorphin levels for 11 runners

You can obtain a listing like this one by choosing:

Analyze
Reports ▸
Case Summaries...

		AFTER	BEFORE	DIFF
1		29.60	4.30	25.30
2		25.10	4.60	20.50
3		15.50	5.20	10.30
4		29.60	5.20	24.40
5		24.10	6.60	17.50
6		37.80	7.20	30.60
7		20.20	8.40	11.80
8		21.90	9.00	12.90
9		14.20	10.40	3.80
10		34.60	14.00	20.60
11		46.20	17.80	28.40
Total	N	11	11	11

Figure 13.1 shows the data values and the difference between the *after* and *before* values for the 11 runners. You can see the stem-and-leaf plot of the difference in Figure 13.2.

Figure 13.2 Stem-and-leaf plot of differences in β–endorphin levels

You can obtain stem-and-leaf plots using the Explore procedure, as described in Chapter 7.

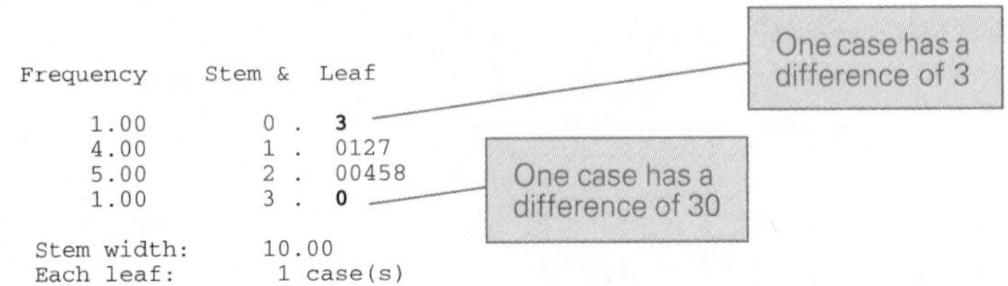

Frequency Stem & Leaf
1.00 0 . 3
4.00 1 . 0127
5.00 2 . 00458
1.00 3 . 0
Stem width: 10.00
Each leaf: 1 case(s)
One case has a difference of 3
One case has a difference of 30

See "Examining Normality" on p. 265 for tests that can be used to check for normality.

The first row of the display in Figure 13.2 represents a case with a difference of 3, while the last row represents a case with a difference of 30. (The stem width is 10, which means that the stem values must be multiplied by 10 before adding them to the leaf values.) As you can see in both figures, all of the differences are positive. That is, the *after* values are always greater than the *before* values. The stem-and-leaf plot doesn't suggest any obvious departures from normality, although when you have so few observations, it's hard to tell if the data come from a population that has a normal distribution. (In the section titled "Examining Normality" on p. 265, you'll learn about statistical tests that can be used to check for normality.) The paired-samples t test, like the one-sample t test to which it's closely related, requires the differences to be a random sample from a normal population, or the sample size to be large enough so that you can rely on the Central Limit Theorem to make the distribution of sample mean differences normal.

? *Isn't it usually impractical to get a random sample from a population?* Yes, in practice that's seldom possible. A large-scale survey like the General Social Survey can draw a random sample of most of the population of the United States, while an investigator who wants to study β–endorphins probably can't take a random sample of American marathon runners. What's possible and important is to make sure that the sample you select is not in some way biased. That is, it should not differ from the population in any important way. For example, if you're studying a new treatment, your patients should not be healthier or sicker than the diseased population of interest. All statistical procedures are based on the assumption that the selected sample is fair. ■■■

Is the Mean Difference Zero?

The null hypothesis for a paired design is that there is no difference between the average values for the two members of a pair in the population. In other words, the average population difference is 0. The alternative hypothesis is that there is a difference in the average values.

How would you go about statistically testing the hypothesis that the sample comes from a population where the average difference is 0? Just as in the previous chapters, you have to determine if your sample results are unlikely if the null hypothesis is true. That is, you want to know how often you would expect to see a mean difference at least as large as the one you've observed in your sample if the real population difference is 0.

? *What if I want to test whether the two means are equal?* Testing whether the average difference is 0 is the same as testing whether the two means are equal. For example, testing whether the average blood pressure before treatment is the same as the average blood pressure after treatment is the same as testing whether the average difference in the before and after blood pressure values is 0. ■■■

Two Approaches

If you think about it, you'll realize that you already know how to solve this problem. It's identical to the one you solved in Chapter 12. Once you've calculated the difference between the pair of measurements, you have one sample of differences, and you want to know if it comes from a population with a mean of 0. Instead of testing whether the average *work week* is 40 hours, you want to test whether the average *difference* between the two measurements is 0. You can obtain the answer from SPSS in one of two ways: you can compute the differences and run the One-Sample T Test procedure, or you can run the Paired-Samples T Test procedure, which automatically computes the differences for you.

? *When does it makes sense to use the paired-samples* t *test?* You should use the paired-samples *t* test only when you have measurements for the same variable on two different occasions for the same subject, or when you have values for the same variable for matched pairs of cases. You can't use a paired-samples *t* test to compare average height and weight, for example, because those are entirely different variables. ■■■

Computing the One-Sample T Test

See Appendix B for information on computing variables in SPSS.

To use the One-Sample T Test procedure in SPSS to solve this problem, you first compute a new variable that is the difference between the *after* value and the *before* value. (You don't have to do this if you have a variable that is the difference.)

The output from the One-Sample T Test procedure is shown in Figure 13.3. You see that the average difference between the *after* and *before* marathon values is 18.74 picomoles per liter. The standard deviation of the difference is 8.33 pmol/l. The 95% confidence interval for the average difference is from 13.14 to 24.33 pmol/l. Since the confidence interval does not include the value of 0, you can reject the null hypothesis that the average difference between the two measurements in the population is 0. It's unlikely that you would see a sample difference at least as large as 18.74 (for samples of 11 pairs) when the true difference is 0. An equivalent way of testing the hypothesis is to look at the t value and its associated two-tailed significance level. As you expect, the significance level is small ($p < 0.0005$), leading you to reject the null hypothesis. It appears that β-endorphin levels rise during a marathon run. (In fact, the authors found that the median β-endorphin level for a sample of people who collapsed during the run was 110 pmol/l. That was significantly different from the median levels for the runners who did not collapse.)

Figure 13.3 Output from the One-Sample T Test procedure

You can obtain this output using the One-Sample T Test procedure, as described in Chapter 12.

	N	Mean	Std. Deviation	Std. Error Mean
DIFF	11	18.7364	8.3297	2.5115

	Test Value = 0					
					95% Confidence Interval of the Difference	
	t	df	Sig. (2-tailed)	Mean Difference	Lower	Upper
DIFF	7.460	10	.000	18.7364	**13.1404**	24.3324

Since the confidence interval does not include 0, you can reject the null hypothesis that the average difference is 0

? *How do I fill in the One-Sample T Test dialog box for this kind of test?* It's easy. The Test Variable is the variable that contains the differences between the two values of a variable for a pair. The Test Value remains at 0, its default, since you're interested in testing whether the test variable comes from a population with a mean of 0. ■■■

The Paired-Samples T Test

To use the one-sample *t* test to test a hypothesis about the mean difference between pairs of observations, you have to compute the differences between the pair of values for each case. You can skip that step if you use the Paired-Samples T Test procedure. This procedure automatically calculates the differences. Figure 13.4 is output from the Paired-Samples T Test procedure for the same problem.

From the menus choose:

Analyze
Compare Means ▸
Paired-Samples...

Select the before and after variables, as shown in Figure 13.7.

Figure 13.4 Output from the Paired-Samples T Test procedure

	Pair 1	
	AFTER	BEFORE
Mean	27.1636	8.4273
N	11	11
Std. Deviation	9.6779	4.2483
Std. Error Mean	2.9180	1.2809

	Pair 1
	AFTER & BEFORE
N	11
Correlation	.515
Sig.	.105

			Pair 1
			AFTER - BEFORE
Paired Differences	Mean		**18.7364**
	Std. Deviation		8.3297
	Std. Error Mean		2.5115
	95% Confidence Interval of the Difference	Lower	13.1404
		Upper	24.3324
t			7.460
df			10
Sig. (2-tailed)			.000

Results in the lower part of table are the same as in Figure 13.3

You see that there are separate summary statistics for the *after* and *before* measurements. The average value for the β-endorphins after running is 27.16 pmol/l, while before running it is 8.43. (The statistic labeled *Correlation* measures how strongly the *before* and *after* values are related. This statistic is discussed further in Chapter 20.)

Statistics for the differences between the two measures are shown in the second part of the table. You see that the statistics in this part of the table are identical to those in Figure 13.3. That's because the same analysis is being performed with both procedures.

Are You Positive?

Based on the paired-samples *t* test, you concluded that β-endorphin levels appear to change during a half-marathon run. Can you be absolutely certain of this conclusion? The answer is no. Whenever you reject the null hypothesis, there is a chance that you are wrong. That's because you reject the null hypothesis when the observed sample results appear to be unlikely, not impossible. The observed significance level even tells you the probability that you would see results as extreme as the ones you observed *when the null hypothesis is true*. If your null hypothesis is true, and you decide to reject it whenever the observed significance level for a sample is less than 0.05, you will be rejecting a perfectly good null hypothesis 5% of the time.

Statisticians creatively call this type of error—rejecting the null hypothesis when it is true—a **Type 1 error**. The other error—failing to reject the null hypothesis when it is false—is called a **Type 2 error**. You would make a Type 2 error if you conclude that running doesn't change β-endorphin levels when in fact it does. Table 13.1 summarizes what can happen when you test a null hypothesis.

Table 13.1 Testing a null hypothesis

	The null hypothesis is:	
Your action	True	False
Reject	Type 1 error	You are correct
Not reject	You are correct	Type 2 error

In Chapter 14, you'll learn more about the factors that contribute to Type 1 and Type 2 errors.

Some Possible Problems

When you have a paired design in which the same person is studied under two different conditions, poor experimental design may also cause you to reject the null hypothesis when it is true. When you use a paired design, you should keep in mind the following caveats:

- If you want to compare two treatments on the same person, you must make sure that enough time passes between treatments so that one treatment wears off before the other begins. Otherwise, you won't be able to separate the effects of the two interventions. For example, if you want to study the effect of two different tasks on pulse rate, you must make sure the subject's pulse returns to normal before the second task is begun.
- If you have subjects repeat the same task or test twice, they may do better the second time because of the learning effect. That's why you should be wary of experiments in which the same (or a similar) test is administered before and after an intervention. Even if the intervention is ineffective, the learning effect may cause a change in the two sets of scores.

Examining Normality

See Chapter 18 for information about nonparametric tests, which do not require the assumption of normality.

Since a paired-samples *t* test is fundamentally the same as a one-sample *t* test, the same assumptions are required for its use. The differences should come from a normal population, or the sample size should be large enough so that the distribution of sample mean differences is approximately normal. In Figure 13.2, you see that there are no strange or outlying values. Although the sample size is small, the observed data may well be from a normal distribution.

Although you can examine a histogram or a stem-and-leaf plot to see if the data appear to come from a normal distribution, there are special plots and statistics that make it easier for you to assess normality. One such plot, called a **normal probability plot**, or **Q-Q plot**, is shown in Figure 13.5. For each data point, the Q-Q plot shows the observed value and the value that's expected if the data are a sample from a normal distribution. The points should cluster around a straight line if the data are from a normal distribution. The normal probability plot of the difference variable is more or less linear, so the assumption of normality appears to be reasonable.

Figure 13.5 Q-Q plot for β-endorphin differences

From the menus choose
Analyze
Descriptive Statistics ▶
Q-Q Plots...
Select the variable diff.

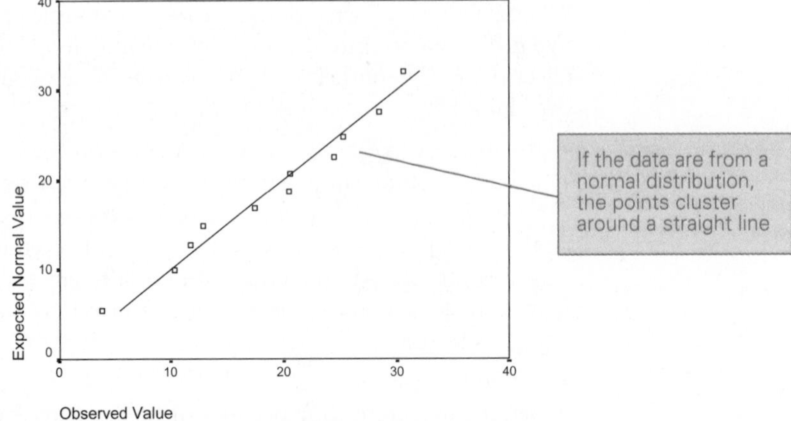

There are also formal statistical tests that you can use to test the null hypothesis that the data are a sample from a normal distribution. If the observed significance levels for the tests are small, you have reason to doubt the assumption of normality. Two commonly used tests are shown in Figure 13.6.

Figure 13.6 Shapiro-Wilk's and K-S Lilliefors tests for normality

Normality tests can be obtained using the Explore procedure, as described in Chapter 7.

	diff		
	Statistic	df	Sig.
Kolmogorov-Smirnov[1]	.129	11	.200*
Shapiro-Wilk	.969	11	.872

*. This is a lower bound of the true significance.

1. Lilliefors Significance Correction

Significance levels are large enough that normality assumption is not unreasonable

Their observed significance levels are reasonably large, indicating that normality is not an unreasonable assumption. Of course, for small sample sizes, you may be unable to reject the normality assumption even if it's wrong. If the assumption of normality appears suspect and the sample size is small, you should consider transforming the data values—for example, by taking logs or square roots—to make the distribution more normal. If this is unsuccessful, you can use a statistical test that does not require the assumption of normality. Such tests are described in Chapter 18.

? *Should I not use the* t *test if the significance level for the test of normality is small?* It depends upon your sample size. If you have a large sample, even small differences from normality may result in a small observed significance level for the tests of normality. This is just another example of the fact that when samples are large, even small differences may be statistically significant, although not necessarily of any practical importance. So, if your sample is large and the distribution of values is not extremely far from normal, you don't really have to worry. The *t* test does not by any means require that the sample come from a perfectly normal population. In general, a graphical examination of assumptions is more informative than a statistical test. ■■■

Summary

How can you test the null hypothesis that the average difference between a pair of measurements is 0?

- In a paired design, the same subject is observed under two conditions, or data are obtained from a pair of subjects that are matched in some way.
- Paired designs help to make the two groups being compared more similar. Some of the differences between subjects are eliminated.
- If you observe the same subject under two conditions, you must make sure that the effect of one treatment has worn off before the other one is given.
- You can compare two related means using a paired-samples *t* test.
- A Type 1 error is made when you reject a null hypothesis that is true. A Type 2 error is made when you don't reject a null hypothesis that is false.
- The normal probability plot is used to examine the assumption of normality.

What's Next?

In this chapter, you learned how to test hypotheses about two means when the data are obtained from matched pairs of observations. You saw that the paired *t* test is the same as a one-sample *t* test of the differences. In Chapter 14, you'll learn how to test hypotheses about two independent means.

How to Obtain a Paired-Samples T Test

The Paired-Samples T Test procedure tests the null hypothesis that the difference in means of two related variables is 0. It also displays a confidence interval for the difference between the population means of the two variables.

▶ To open the Paired-Samples T Test dialog box (see Figure 13.7), from the menus choose:

Analyze
  Compare Means ▶
    Paired-Samples T Test...

Figure 13.7 Paired-Samples T Test dialog box

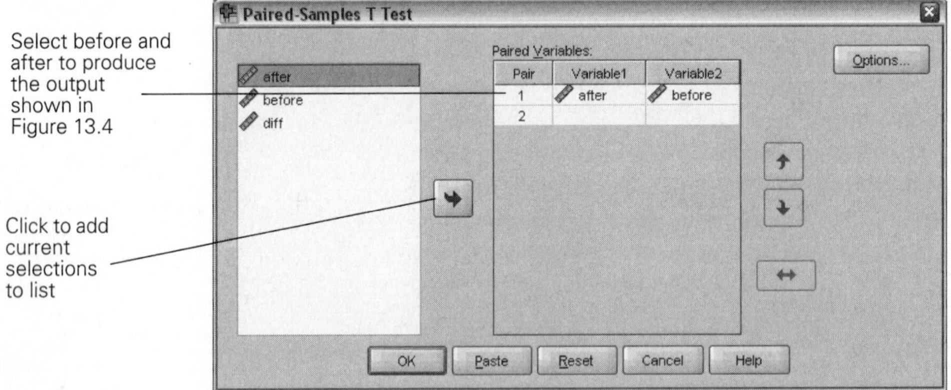

To compare means between two groups of cases for one or more variables, see Chapter 14.

▶ In the Paired-Samples T Test dialog box, move each of the two variables whose means you want to compare to the Paired Variables list.

▶ Click OK.

You can select more than one pair of variables if desired. For each pair of variables, SPSS calculates a t statistic and its observed significance level.

Options: Confidence Level and Missing Data

In the Paired-Samples T Test Options dialog box (see Figure 13.8), you can change the confidence level for which SPSS displays the confidence interval for the difference between the population means of the pair of variables.

▶ In the Paired-Samples T Test dialog box, click Options to open the Paired-Samples T Test Options dialog box.

Figure 13.8 Paired-Samples T Test Options dialog box

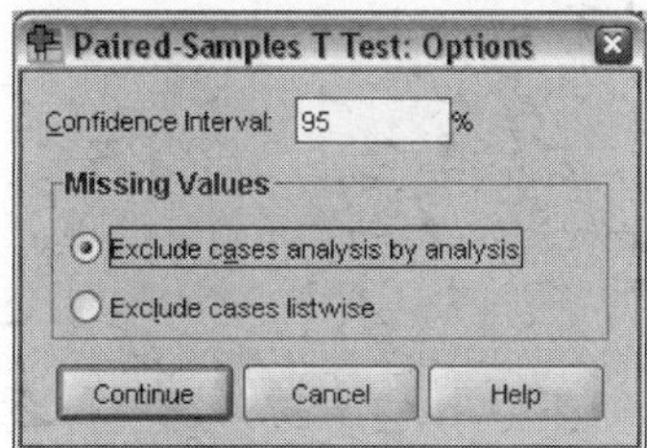

Confidence Interval. Specify a value (such as 90 or 99) in this text box.

Missing Values. Two alternatives control the treatment of missing data for multiple pairs of variables:

Exclude cases analysis by analysis. Uses all cases that have valid data for the two variables in a pair in the test for that pair.

Exclude cases listwise. Uses only the cases that have valid data for all variables in any of the pairs. This ensures that all of the tests are performed using the same cases.

Exercises

Statistical Concepts

1. An investigator wants to test the hypothesis that children who drink orange juice before class will be more attentive than children who drink milk. He selects a classroom of children and obtains an alphabetical list of the students. He assigns the first child to orange juice therapy, the next to milk therapy, and so on down the list. He wants to analyze the experiment using the paired *t* test, since he has formed pairs of children based on the alphabetic list. Suggest to him how he might analyze his data. Do you think this is a paired experiment? If not, give an example of a paired design for this question.

2. Studies sometimes use twins who have been raised separately to investigate questions like "What are the roles of parental influence and genetic heritage in children's intellectual development?" Discuss the advantages and disadvantages that you see in using twins for studies of this nature.

3. Discuss any problems you see in the following studies:

a. Anxiety often affects performance on tests. A psychologist has developed a new method for reducing stress during statistics exams. To evaluate the new method, he tests each of 50 students under two conditions. He gives each student the final exam before stress-reduction training and then again after the training. He then compares the two sets of scores.

b. A market researcher wants to study consumer preferences for five brands of pizza. He invites 250 people to a pizza party. Each person is instructed to make sure that he or she eats a piece of all five brands. As they are leaving, each participant fills out a questionnaire evaluating the five brands of pizza.

c. As a drug manufacturer, you're interested in studying the effectiveness of a new drug for headache relief. You place an ad in the newspaper recruiting "headache sufferers" who want to volunteer for the study. At the beginning of the study, you question each participant about the frequency and duration of headaches. Then you send the sufferers home with a week's supply of new medicine. A week later, you ask each participant the same questions about their headaches.

d. You want to compare two methods for weight reduction. You recruit 123 people who are interested in losing weight. You instruct everyone to use the first method until they have lost 10 pounds and then use the second method until they have lost 10 more. You then compare the length of time it takes to lose the first 10 pounds to the length of time it takes to lose the next 10 pounds.

Data Analysis

Use the *gssft.sav* data file to answer the following questions:

1. Use the Compute facility (discussed in Appendix B) to create a new variable that is the difference between a husband's and wife's hours worked last week (variables *husbhr* and *wifehr*).

   a. Make a histogram of the difference. What should you look for in the histogram?

   b. Make a Q-Q plot of the differences. How does the distribution differ from the normal distribution?

   c. Do you think it's reasonable to believe that the distribution of mean differences is normal? Why? Perform a paired t test using the Paired-Samples T Test procedure. Write a brief summary of your results. Be sure to state your null and alternative hypotheses.

   d. Run a one-sample t test on the difference variable. Compare your results to those from the paired t test. In what ways are the two tests different?

2. You are studying the work habits of spouses. You want to know whether husbands and wives who are both employed full time work the same average number of hours per week (variables *husbhr* and *wifehr*). Use the Select Cases facility to choose cases where *husbft* = 1 and *wifeft* = 1 (both spouses work full time). Perform the appropriate analyses and write a short summary of your results.

3. Based on the analyses you performed in question 2, is it reasonable to conclude that women who work full time on average work fewer hours than men who work full time? Explain your reasoning.

Use the *siqss.sav* data file to answer the following questions:

4. Consider the difference between the number of e-mails a person sends in a day and the number of e-mails he or she receives.

   a. Compute a variable that is the difference between the number of e-mails a person sends (*emsend*) and the number of e-mails a person receives (*emrec*). Make a histogram. Describe the histogram. Are there outliers?

   b. Test the null hypothesis that the average number of e-mails sent is equal to the average number of e-mails received. What do you conclude?

   c. Use the Select Cases facility to remove outliers. Do your conclusions change?

5. Repeat the analyses in question 4 for the number of personal e-mails sent (*emsendpr*) and the number of personal e-mails received (*emgetprs*).

6. Repeat the analyses for the number of work e-mails (*emsentw*, *emrecw*).

Use the *electric.sav* data file to answer the following question:

7. Make histograms and Q-Q plots of *dbp58*, *cgt58*, *wt58*, *chol58*, and *ht58*. Describe for each variable the types of departures from normality that you see.

Use the *schools.sav* data file to answer the following question:

8. Look at the changes between 1993 and 1994 in graduation rates (variables *grad93* and *grad94*), ACT scores (variables *act93* and *act94*), and percentages of students taking ACT tests (variables *pctact93* and *pctact94*). Does it look like the Chicago school system is improving? Which schools appear to be "outliers"?

Use the *renal.sav* data file to answer the following question:

9. For patients who developed acute renal failure (variable *type* equals 1), determine if there was a statistically significant change in average BUN and creatinine at admission (variables *admbun* and *admcreat*) and average BUN and creatinine at discharge (variables *finbun* and *fincreat*). For all patients, see if there was a statistically significant change in creatinine between the time of admission (variable *admcreat*) and the time of surgery (variable *precreat*).

Use the *country.sav* data file to answer the following question:

10. Test the null hypothesis that the average life expectancy for males is the same as the average life expectancy for females (variables *lifeexpm* and *lifeexpf*). Look at the distribution of the differences. Summarize your conclusions.

Use the *buying.sav* data file to answer the following question:

11. Test the following hypotheses: husbands' and wives' average buying scores are equal (variables *hsumbuy* and *wsumbuy*); wives' average buying scores and their husbands' prediction of them are equal (variables *wsumbuy* and *hpredsum*); husbands' average buying scores and their wives' prediction of them are equal (variables *hsumbuy* and *wpredsum*); average influence scores assigned by husbands and wives are equal (variables *hsuminf* and *wsuminf*). Be sure to look at the distribution of differences. Summarize your results.

14 Testing a Hypothesis about Two Independent Means

How can you test the null hypothesis that two population means are equal, based on the results observed in two independent samples?

- Why can't you use a one-sample *t* test?
- What assumptions are needed for the two-independent-samples *t* test?
- Can you prove that the null hypothesis is true?
- What is power, and why is it important?

All flavors of social scientists are agonizing over the effects of Internet use. One day you're told that e-mail helps you to connect to friends and family and makes you a happy, social person. Several days later the news is bad. Internet users don't spend time with their families, they're depressed and addicted. Evaluating the effects of the Internet on society will, no doubt, keep faculty and graduate students occupied for many years to come. (Note that medical researchers still argue over whether chocolate, cheese, and red wine, which have been consumed for centuries, are good or bad for you, in moderation of course.)

You, too, can participate in Internet research by using the General Social Survey to test hypotheses about differences between those who use the Internet and those who don't. You have already found that Internet users appear to be better educated and younger. In this chapter, you'll test hypotheses about television-viewing behavior in Internet users and non-users. You'll determine whether Internet use is related to hours of television viewing.

You'll learn how to test whether two population means are equal, based on the results observed in two independent samples—one from each of the populations of interest. The statistical technique you'll use is called the **two-independent-samples *t* test**. You can use the two-independent-samples *t* test to see if in the population men and women have the same scores on a test of physical dexterity or if two treatments for high cholesterol result in the same mean cholesterol levels.

▶ This chapter uses the *gss.sav* data file. For instructions on how to obtain the independent-samples *t* test output shown in this chapter, see "How to Obtain an Independent-Samples T Test" on p. 296.

Examining Television Viewing

The first step of any statistical analysis is to examine the data carefully. You want to make sure that the values are plausible. You also want to examine the shape of the distribution so that you can select an appropriate statistical test for testing hypotheses of interest. Figure 14.1 contains histograms of the number of hours of television viewing per day reported by Internet users and non-users. (The GSS question is, "On the average day, about how many hours do you personally watch television?".) You see that both distributions have a tail toward large values, indicating that there are people who report watching television for many hours each day. Some of these values raise statistical concerns as well as concerns about the sanity of some of our fellow citizens. There are people who report watching television for 24 hours a day. You know that isn't possible. It may be that people are reporting how many hours they have the television turned on. "Watch television" is not a very well-defined term. If you have the television on while you're doing homework, are you studying or watching television? It's probably the case that you're doing some of both. When you tally the number of hours you've spent studying for a test, the television time will probably be counted as study time. To an interviewer from the General Social Survey, you might more honestly report it as television time.

Figure 14.1 Histograms of hours spent watching television

From the Graphs menu, choose Legacy Dialogs, and then Histogram. In the Histogram dialog box, select tvhours as the dependent variable and usenet as the Panel by Rows variable.

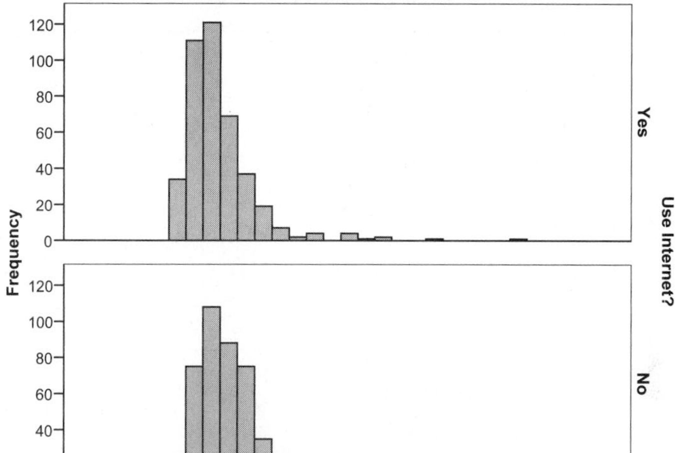

We'll proceed with our analyses, assuming that the reported television times are correct. However, we will examine the impact of the outlying points on the results of the analyses. If our conclusions change when the suspect data values are removed, we'll have to consider other approaches to analyzing the data.

From the descriptive statistics in Figure 14.2, you see that Internet users reported an average of 2.42 hours of television viewing per day compared to 3.52 hours for those who don't use the Internet. Internet users, on average, report watching television for about an hour a day less than those who don't use the Internet. Notice that the 5% trimmed means, which are calculated by removing the top and bottom 5% of the values, are 0.2 hours less for both groups than the arithmetic means. Removing those very large values makes both means smaller.

Figure 14.2 Descriptive statistics for hours spent watching television

			Statistic	
			USENET Use Internet?	
			0 No	1 Yes
TVHOURS Hours per day watching TV	Mean		3.52	2.42
	95% Confidence Interval for Mean	Lower Bound	3.26	2.22
		Upper Bound	3.77	2.63
	5% Trimmed Mean		3.22	2.18
	Median		3.00	2.00
	Variance		7.801	4.604
	Std. Deviation		2.793	2.146
	Minimum		0	0
	Maximum		24	20
	Range		24	20
	Interquartile Range		2.00	2.00

? *Why is the number of Internet users and non-users different than in earlier chapters?* The General Social Survey doesn't ask all people all questions. Everyone is asked core questions (about age, education, and income, for example) and a certain number of specialized questions. This is done so that the interviews don't become unbearably long. The people who are asked any particular question are still a random sample from the United States adult population. In 2004 the people who were asked about Internet use were not asked questions about television viewing. You'll use data from an earlier GSS survey to look at the relationship of TV viewing and Internet use. ■■■

You know that even if the average hours of television viewing in the population are the same for Internet users and non-users, the sample means will not be equal. Different samples from the same population result in different sample means and standard deviations. To determine if observed sample differences among groups might reflect differences in the population, instead of just sample-to-sample variability, you need to determine if the observed sample means are unusual if the population means are equal. You need to figure out how often you would see a difference of at least 1.1 hours between the two independent groups of Internet users and non-users when there is no difference between the two groups in the population.

? *What do you mean by independent groups?* Samples from different groups are called **independent** if there is no relationship between the people or objects in the different groups. For example, if you select a random sample of males and a random sample of females from a population, the two samples are independent. That's because selecting a person for one group in no way influences the selection of a person for another group. The two groups in a paired design are not independent, since either the same people or closely matched people are in both groups. ■■■

Since you have means from two independent groups, you can't use the one-sample *t* test to test the null hypothesis that two population means are equal. That's because you now have to cope with the variability of two sample means: the mean for Internet users and the mean for those who don't use the Internet. When you test whether a single sample comes from a population with a known mean, you have to worry only about how much individual means from the same population vary. The population value to which you compare your sample mean is a fixed, known number. It doesn't vary from sample to sample. You assumed that the value of 205 mg/dL for the cholesterol of the general population is an established norm based on large-scale studies. Similarly, the value of 40 hours for a work week is a commonly held belief.

The two-independent-samples *t* test is basically a modification of the one-sample *t* test that incorporates information about the variability of the two independent-sample means. The standard error of the mean difference is no longer estimated from the variance and number of cases in a single group. Instead, it is estimated from the variances and sample sizes of the two independent groups.

Distribution of Differences

In the one-sample *t* test, you looked at the distribution of all possible sample means from a population. You saw that the amount in which sample means vary depends on the standard deviation of the values and on the sample size. For the same population, means calculated from large samples vary less than means calculated from small samples. For the same sample size, means calculated from a population with a lot of variability will vary more than means calculated from a population with less variability.

When you want to test hypotheses about two independent population means, you have to look at the distribution of all possible *differences* between the two sample means. Fortunately, the Central Limit Theorem works for differences of sample means as well as for the sample means themselves. So, if your data are samples from approximately normal populations or your sample size is large enough so that the Central Limit Theorem holds, the distribution of differences between two sample means is also approximately normal.

Standard Error of the Mean Difference

If two samples come from populations with the same mean, the mean of the distribution of differences is 0. However, that's not enough information to determine if the observed sample results are unusual. You also need to know how much the sample differences vary. The standard deviation of the difference between two sample means, the **standard error of the mean difference**, tells you that. When you have two independent groups, you must estimate the standard error of the mean difference from the standard deviations and the sample sizes in each of the two groups.

? *How do I estimate the standard error of the difference?*
The formula is

$$S_{\bar{X}_1 - \bar{X}_2} = \sqrt{\frac{S_1^2}{n_1} + \frac{S_2^2}{n_2}}$$

where S_1^2 is the variance for the first sample and S_2^2 is the variance for the second sample. The sample sizes for the two samples are n_1 and n_2. If you look carefully at the formula, you'll see that the standard error of the mean difference depends on the standard errors of the two sample means. You square the standard error of the mean for each of the two groups. Next you sum them, and then take the square root. ■■■

Computing the T Statistic

Once you've estimated the standard error of the mean difference, you can compute the *t* statistic the same way as in the previous chapters. You divide the observed mean difference by the standard error of the difference. This tells you how many standard error units from the population mean of 0 your observed difference falls. That is,

$$t = \frac{(\bar{X}_1 - \bar{X}_2) - 0}{S_{\bar{X}_1 - \bar{X}_2}}$$

Equation 14.1

If your observed difference is unlikely when the null hypothesis is true, you can reject the null hypothesis.

? *How is this different from the one-sample* t *test?* The idea is exactly the same. What differs is that you now have two independent-sample means, not one. So you estimate the standard error of the mean difference based on two sample variances and two sample sizes.

Output from the Two-Independent-Samples T Test

Figure 14.3 shows the results from SPSS of testing the null hypothesis that the average hours of daily television viewing is the same in the population for Internet users and non-users.

From the menus choose:

Analyze
Compare Means ▶
Independent-Samples T Test...

Select the variables tvhours and usenet, as shown in Figure 14.10.

Figure 14.3 T test output

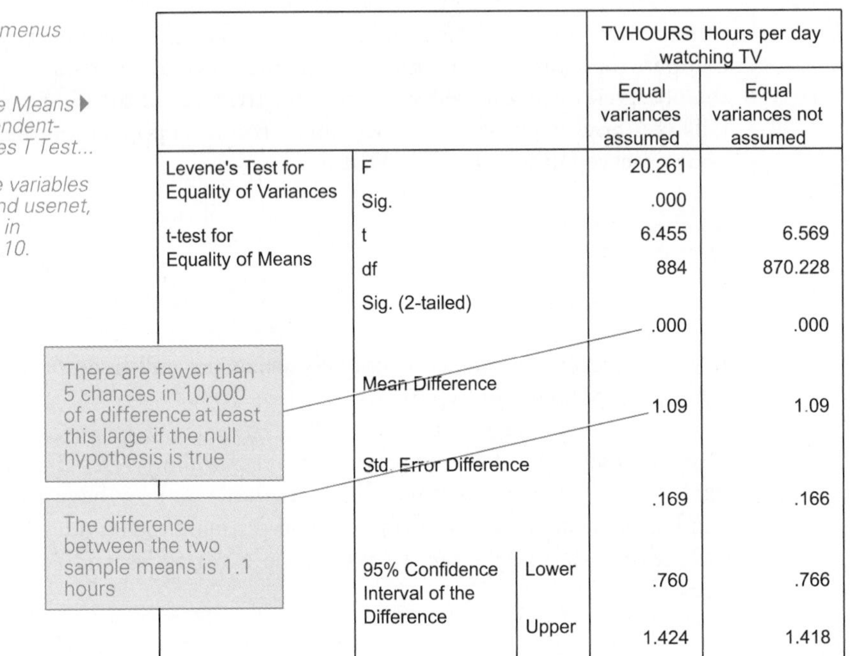

			TVHOURS Hours per day watching TV	
			Equal variances assumed	Equal variances not assumed
Levene's Test for Equality of Variances	F		20.261	
	Sig.		.000	
t-test for Equality of Means	t		6.455	6.569
	df		884	870.228
	Sig. (2-tailed)		.000	.000
	Mean Difference		1.09	1.09
	Std. Error Difference		.169	.166
	95% Confidence Interval of the Difference	Lower	.760	.766
		Upper	1.424	1.418

There are fewer than 5 chances in 10,000 of a difference at least this large if the null hypothesis is true
The difference between the two sample means is 1.1 hours

In the output, there are two slightly different versions of the *t* test. One makes the assumption that the variances in the two populations are equal; the other does not. This assumption affects how the standard error of the mean difference is calculated. You'll learn more about this distinction later in this chapter.

Consider the column labeled *Equal variances not assumed*. You see that for the observed difference of 1.1 hours, the *t* statistic is 6.57. (To calculate the *t* statistic, divide the observed difference of 1.1 hours by 0.17, the standard error of the difference estimate when the two population variances are not assumed to be equal.) The degrees of freedom for the *t* statistic are 870.

The observed two-tailed significance level is less than 0.0005. This tells you that fewer than 5 times in 10,000 would you expect to see a sample difference of 1.1 hours or larger when the two population means are equal. Since this is less than 5%, you reject the null hypothesis that Internet users and non-users watch the same average number of hours of television each night. Your observed results are very unusual if the null hypothesis is true.

Confidence Intervals for the Mean Difference

Take another look at Figure 14.3. The 95% confidence interval for the true difference is from 0.77 hours to 1.42 hours. This tells you it's likely that the true mean difference is anywhere from 45 minutes to 85 minutes. Since your observed significance level for the test that the two population means are equal was less than 5%, you already knew that the 95% confidence interval does not contain the value of 0. (Remember, only likely values are included in a confidence interval. Since you found 0 to be an unlikely value, it won't be included in the confidence interval.)

To calculate a 99% confidence interval, specify 99 in the T Test Options dialog box (see Figure 14.12).

? *If I compute a 99% confidence interval for the true mean difference, will it also not include 0?* Since the observed significance level is less than 0.01, you know that the 99% confidence interval will not include the value of 0. The 99% confidence interval for the mean difference extends from to 0.66 hours to 1.53 hours. ■■■

Testing the Equality of Variances

There are two slightly different *t* values in Figure 14.3. That's because there are two different ways to estimate the standard error of the difference. One of them assumes that the variances are equal in the two populations from which you are taking samples, and the other one does not.

In Figure 14.2, you saw that the observed standard deviation for Internet users was somewhat smaller than the standard deviation for nonusers. You can test the null hypothesis that the two samples come from populations with the same variances using the Levene test, shown in Figure 14.4. If the observed significance level for the Levene test is small, you reject the null hypothesis that the two population variances are equal.

In the Independent-Samples T Test dialog box, select tvhours and usenet, as shown in Figure 14.10.

Figure 14.4 Levene test for equality of variances

			TVHOURS Hours per day watching TV	
			Equal variances assumed	Equal variances not assumed
Levene's Test for Equality of Variances	F		20.261	
	Sig.		.000	
t-test for Equality of Means	t		6.455	6.569
	df		884	870.228
	Sig. (2-tailed)		.000	.000
	Mean Difference		1.09	1.09
	Std. Error Difference		.169	.166
	95% Confidence Interval of the Difference	Lower	.760	.766
		Upper	1.424	1.418

You reject the hypothesis that the two population variances are equal based on the Levene test

For this example, you reject the equal variances hypothesis, since the observed significance level for the Levene test is less than 0.005. That means you should use the results labeled *Equal variances not assumed* in Figure 14.4. Notice that the estimate of the standard error of the difference is not the same in the two columns. This affects the t value and confidence interval. When you use the estimate of the standard error of the difference that does not assume that the two variances are equal, the degrees of freedom for the t statistic are calculated based on both the sample sizes and the standard deviations in each of the groups. This is an approximation, and the result is usually not an integer. If the equal variance t test is used, the degrees of freedom are just the sum of the two sample sizes minus 2. In this example, both t tests give very similar results, but that's not always the case.

? *Why do you get different numbers for the standard error of the mean difference depending on the assumptions you make about the population variances?* If you assume that the two population variances are equal, you can compute what's called a pooled estimate of the variance. The idea is similar to that of averaging the variances in the two groups, taking into account the sample size. The formula for the pooled variance is

$$S^2 = \frac{(n_1 - 1)S_1^{\ 2} + (n_2 - 1)S_2^{\ 2}}{(n_1 - 1) + (n_2 - 1)}$$

It is this pooled value that is substituted for both $S_1^{\ 2}$ and $S_2^{\ 2}$ in Equation 14.1. If you do not assume that the two population variances are equal, the individual sample variances are used in Equation 14.1. ■■■

Effect of Outliers

You know from Figure 14.1 that some people reported watching television for very long periods of time, including 24 hours a day. Although you know that a person can't actually watch television for 24 hours a day, some of the other large values are possible, although not particularly believable.

? *Why does the GSS report values that appear to be suspect?* General Social Survey interviewers are trained to conduct interviews in accordance with good survey practice. Their role is to record answers, not to influence or challenge them. Imagine the bias that would be introduced if interviewers were allowed to use their personal judgment to determine which answers they felt were plausible and which were not. With all that gray hair, are you sure you're 35? You graduated from college even though you don't understand a simple question? Do you really earn that much and live in this dump?! Besides creating a hostile environment, challenging answers would result in data that are hopelessly contaminated by interviewer styles and prejudices. The General Social Survey reports the responses given under well-controlled circumstances; users of the data must decide how they will deal with questionable answers and inconsistencies. ■■■

Since the arithmetic mean is affected by data values that are far removed from the rest, you want to make sure that your analysis of differences between the two means is not unduly influenced by the outlying points.

This is particularly troublesome for small datasets because a single case can make a big difference in the mean. If removal of the people who watch television 24 hours a day makes the significant difference between Internet users and non-users disappear, you've got a problem. There is no single correct solution for dealing with outliers. A variety of strategies may be useful, such as using statistical techniques that aren't affected by strange data points (some are discussed in Chapter 18), analyzing the data with and without the strange values and seeing whether the results change, or capping the outlying values to decrease their influence.

Figure 14.5 shows the results of the *t* test when people who watch television for more than 12 hours a day are removed from the analysis. The average difference between the two groups has changed only slightly. It went from 1.09 hours to 1.05 hours. You can still reject the null hypothesis that average television-viewing time is the same for the two groups. It's reassuring that your conclusions don't change.

Figure 14.5 T test output when television hours greater than 12 are removed

From the Data Editor menus choose:

Data
Select Cases...

Select If condition is satisfied. Enter tvhours <= 12 in the Numeric Expression box. Repeat the analysis as shown in Figure 14.10. (For more information, see Appendix B.)

TVHOURS
Hours per day watching TV

USENET Use Internet?	N	Mean	Std. Deviation	Std. Error Mean
0 No	469	3.40	2.491	.115
1 Yes	411	2.35	1.866	.092

TVHOURS
Hours per day watching TV

			Equal variances assumed	Equal variances not assumed
Levene's Test for Equality of Variances	F		25.449	
	Sig.		.000	
t-test for Equality of Means	t		7.013	7.145
	df		878	857.737
	Sig. (2-tailed)		.000	.000
	Mean Difference		1.05	1.05
	Std. Error Difference		.150	.147
	95% Confidence Interval of the Difference	Lower	.758	.763
		Upper	1.347	1.342

Introducing Education

From the previous analysis, you can conclude that Internet users, on average, watch fewer hours of television per day than non-users. You are 95% confident that the true difference is between 0.8 and 1.4 hours. You may be tempted to rush out and publish the finding that Internet use decreases television-viewing time. But, as a skilled researcher, you know that you must draw conclusions carefully.

You can't say that Internet use *causes* people to watch less television. Causation is very difficult to show in a non-experimental setting. Just because two variables are related doesn't mean that one causes the other. You didn't randomly assign people to be Internet users or non-users, so the two groups may differ in many important ways besides Internet use. This is a serious problem in many observational studies. For example, if you find that people who exercise have lower cholesterol levels, you can't conclude that exercise decreases cholesterol. You know that people who exercise are different from people who don't. They may have healthier diets, smoke less, and be exemplary in other ways. You can't attribute the lower cholesterol to exercise, since it might be due to any or all of the other uncontrolled differences between the groups. If you randomly assign people to exercise or no-exercise programs, you stand a better chance of isolating the impact of exercise.

? *Isn't it misguided to classify people into just two groups: Internet users and non-users?* Absolutely. People use the Internet in different locations, for different purposes, and for different amounts of time. People who have Internet access only at work may be quite different from people with cable modems or DSL service at home. People who use the Internet for one hour a week should not be grouped together with those who use it for long periods of time. You're analyzing only two groups in this chapter, since you're learning about statistical tests for two independent samples. Many different, and better, criteria for forming groups can be considered (for example, heavy Internet use at home versus light Internet use at home versus no Internet use at home). ■■■

In previous chapters, you saw that in the GSS sample, Internet users are younger and better educated than non-users. That may explain some of the observed differences in television-viewing habits. For example, if people with more education watch less television in general and are more likely to use the Internet, you'll find that Internet use and television viewing are related. Their relationship is explained by education. If you don't include education in your analysis of Internet use, it becomes a *lurking* variable.

? *A lurking variable?!* That's standard statistical jargon for a variable that affects the response you are studying but is not included in your analysis. Age, education, gender, and income are all likely to be lurking variables if you ignore them when looking at differences between Internet users and non-users. ■■■

You can start your investigation of the possible effects of age, education, and hours worked per week on television viewing by using the two-independent-samples *t* test to test whether the population values for average age, years of education, and hours worked last week by the respondent and respondent's spouse differ for the two groups. Figure 14.6 shows descriptive statistics for the two groups. Based on the *t* tests in Figure 14.7, you reject the null hypothesis that in the population the two groups have the same average age, education, and hours worked. Internet users are significantly younger, better educated, and work more hours per week. You don't reject the null hypothesis that the average hours worked by the spouses of Internet users and non-users is the same, since the observed significance level is greater than 0.05.

In the Independent-Samples T Test dialog box, select age, educ, hrs1, and sphrs1.

Figure 14.6 Descriptive statistics for age, education, and hours worked

	USENET Use Internet?	N	Mean	Std. Deviation	Std. Error Mean
AGE Age of respondent	0 No	734	51.75	18.857	.696
	1 Yes	653	40.79	13.212	.517
EDUC Highest year of school completed	0 No	733	12.05	2.702	.100
	1 Yes	652	14.55	2.522	.099
HRS1 Number of hours worked last week	0 No	356	40.80	13.960	.740
	1 Yes	532	43.74	13.481	.584
SPHRS1 Number of hours spouse worked last week	0 No	171	40.98	11.990	.917
	1 Yes	238	43.38	12.498	.810

Figure 14.7 T tests for age, education, and hours worked

		Levene's Test for Equality of Variances		t-test for Equality of Means						
									95% Confidence Interval of the Difference	
		F	Sig.	t	df	Sig. (2-tailed)	Mean Difference	Std. Error Difference	Lower	Upper
AGE Age of respondent	Equal variances assumed	131.217	.000	12.388	1385	.000	10.96	.885	9.222	12.692
	Equal variances not assumed			12.637	1314.977	.000	10.96	.867	9.256	12.658
EDUC Highest year of school completed	Equal variances assumed	7.327	.007	-17.752	1383	.000	-2.50	.141	-2.779	-2.226
	Equal variances not assumed			-17.823	1379.733	.000	-2.50	.140	-2.778	-2.227
HRS1 Number of hours worked last week	Equal variances assumed	.441	.507	-3.136	886	.002	-2.94	.936	-4.774	-1.099
	Equal variances not assumed			-3.114	742.904	.002	-2.94	.943	-4.787	-1.085
SPHRS1 Number of hours spouse worked last week	Equal variances assumed	1.050	.306	-1.948	407	.052	-2.40	1.232	-4.822	.022
	Equal variances not assumed			-1.961	375.077	.051	-2.40	1.224	-4.806	.006

? *Why is the difference in average hours worked by the spouses, 2.4 hours, not statistically significant when the difference in average hours worked by respondents, 2.9 hours, is so highly significant ($p < 0.002$)?* Whether a difference is statistically significant depends on the magnitude of the difference, the variability in the two groups, and the sample sizes in the groups. The difference between 2.9 hours and 2.4 hours is not very great, and the standard deviations of the spouses are actually smaller than the standard deviations for the respondents. The sample sizes are the reason that one of the differences is highly significant and the other is not. Only married people with working spouses are asked how many hours the spouse worked. There are only 238 working Internet spouses and 171 working non-Internet spouses, compared to 532 Internet users and 356 non-users. If the sample included more spouses who worked, a difference of 2.4 hours might well be significant. Remember that when you don't reject the null hypothesis, you haven't shown that the hours worked by spouses are equal. You just didn't have enough evidence to reject the null hypothesis. ■■■

Figure 14.8 is a bar chart of the average hours of television watched when groups are formed by age and education. You see that for most age groups, as education increases, television viewing decreases. In fact, for four out of the five age groups, those without a high school diploma report the largest number of hours of television viewing. For all age groups except those 18–29, people with graduate degrees report the least television viewing. (There are only six people with graduate degrees in the age group 18–29, so you don't have much confidence in the estimate of the average hours of television viewing for such a small group.) Within an education category, there is no clear age effect, although the oldest groups report watching the most hours of television.

In the Define Clustered Bar Summaries for Groups of Cases legacy dialog box, select Other statistic and move tvhours into the Variable box. Select agecat for Category Axis and degree for Define Clusters by.

Figure 14.8 Bar chart of television hours for education and age groups

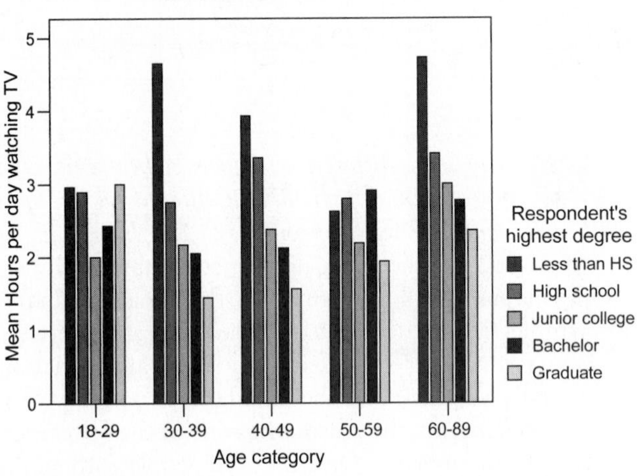

Since television hours are related to education and education is related to Internet use, it's very likely that some of the differences in television viewing between Internet users and non-users is due to differences in education. You can't look at overall differences between the two groups; you must look at differences within an educational category. That is, you must control for the effects of education. For example, you must compare television viewing for college graduates who are Internet users and non-users.

Figure 14.9 is the two-independent-samples *t* test when the comparison of average hours of television viewing is restricted to those with at least a college degree. You see that the average hours of television viewing for those who don't use the Internet is 2.36 hours compared to 2.16 hours for those who do. The difference is 0.2 hours, or about 12 minutes a day.

That's quite a change from the 1.1 hours you saw when all respondents were included in the analysis. The observed significance level for this difference is 0.5, so you can no longer reject the null hypothesis that average hours of television viewing is the same for the two groups. Controlling the average hours of television use for college education has greatly decreased the difference between the two groups.

Figure 14.9 Comparison of television hours for college graduates

Select cases with degree greater than 2. Repeat the analysis as shown in Figure 14.10. Activate the pivot table and choose:

Pivot
Transpose Rows and Columns

TVHOURS
Hours per day watching TV

USENET Use Internet?	N	Mean	Std. Deviation	Std. Error Mean
0 No	45	2.36	1.401	.209
1 Yes	161	2.16	1.981	.156

TVHOURS
Hours per day watching TV

			Equal variances assumed	Equal variances not assumed
Levene's Test for Equality of Variances	F		1.209	
	Sig.		.273	
t-test for Equality of Means	t		.615	.744
	df		204	98.487
	Sig. (2-tailed)		.539	.458
	Mean Difference		.19	.19
	Std. Error Difference		.315	.261
	95% Confidence Interval of the Difference	Lower	-.428	-.323
		Upper	.816	.711

Based on these latest findings, what can you conclude? Can you conclude that Internet use and television viewing are not related? Of course not. You've looked at only one of several education groups and ignored possible differences in age between the two groups. What you can conclude is that the study of Internet usage and activities such as television viewing, spending time with one's family, and the like, is a very complex undertaking. There are many factors that influence these behaviors, and chances are excellent that the distribution of these factors for Internet users and non-users is different.

? *So what should I do, just give up my promising research?* Of course not. Just because a problem is complex doesn't mean it can't be solved. There are many statistical techniques that help you untangle relationships between variables. You'll learn about some of them in subsequent chapters. Just remember, fancy statistical techniques are not a substitute for careful thought or good experimental design. ■■■

Can You Prove the Null Hypothesis?

You may have noticed that when phrasing the results of a hypothesis test, we've always been careful to say that you either rejected the null hypothesis or did not reject it. The phrase "You've proved the null hypothesis" has never been used. The reason is that you can't prove the null hypothesis. Think about it. Do you really believe that you've shown that the average hours of television viewing are exactly the same for college graduates who use the Internet and those who do not? Of course not. Your sample results are compatible with many values besides 0 for the population difference. Even if you observed a sample difference of 0, that wouldn't tell you that the true difference is 0. Sample differences of 0 are compatible with values other than 0 for the true population difference.

From the 95% confidence interval, you see that values anywhere in the range of –0.43 to 0.82 years are plausible for the average difference in hours of television viewing between the two groups. If you increase your sample size, your confidence interval will become shorter; that is, it will include fewer "plausible" values, but it will never exclude all values except 0. What all this means is that based on the results of a hypothesis test, you should never claim that you've shown that the null hypothesis is really true. All you can claim is that your sample results are not unusual if the null hypothesis is true.

Sometimes a legal analogy is drawn to statistical hypothesis testing. The null hypothesis is compared to the presumption of innocence in a legal case. Failure to find a defendant guilty doesn't prove innocence. All it says is that there was not enough evidence to establish guilt.

Interpreting the Observed Significance Level

When the observed significance level is small, you reject the null hypothesis and conclude that the two population means appear to be unequal. However, you know that there's a chance that your conclusion is wrong. It's possible that the null hypothesis is true, and your observed difference is one of the remote events that can occur. In fact, that's what the observed significance level tells you—how often you would expect to see a difference at least as large as the one you observed when the null hypothesis is true.

When your observed significance level is too large for you to reject the null hypothesis that the means are equal, two explanations are possible. The first explanation is that really there is no difference between the two population means (a conjecture you can't prove), or that there is a small difference, and you can't detect it. For example, if there really is an average difference of five minutes of television viewing between the Internet users and the non-users, it's of little consequence that you can't identify it. Differences of $10 in annual income or one point on a standardized test are rarely important.

The second explanation is more troublesome: there is an important difference between the two groups, but you did not detect it. How can this happen? One reason you did not reject the null hypothesis when it is false may be that the sample size is small and the observed result doesn't appear to be unusual. Remember, when the sample size is small, many outcomes are compatible with the null hypothesis being true. For example, if you flip a coin only three times, you'll never be able to exclude the possibility that it is a fair coin.

? *Why not?* If a coin is fair and you flip it three times, the probability of observing either three heads or three tails is 0.25. That means that one out of four times when you flip a fair coin three times, you will get all heads or all tails. So even if you're flipping a coin that has two heads or two tails, the results you get (all heads or all tails) are compatible with the null hypothesis that the coin is fair. The outcome is not unlikely if the null hypothesis is true. If you flip the coin 10 times, getting all heads or all tails would be very unusual. ■■■

Even if there is a large difference in television viewing between the Internet users and the non-users, you may not be able to detect it if you have five cases in each of the two groups. That's because the observed sample difference may be compatible with many population values, including 0.

Your ability to reject the null hypothesis when it is false also depends on the variability of the observed values. If your observed values have a lot of variability, the range of plausible values for the true population difference will be broad. For the same sample size, as the variability decreases, the range of plausible population values does, too. Remember, when you calculate a t statistic, you divide the observed difference by the standard error of the difference. The standard error of the difference depends on both the sample variances and the sample sizes.

Power

Power is the statistical term used to describe your ability to reject the null hypothesis when it is false. It is a probability that ranges from 0 to 1. The larger the power, the more likely you are to reject the null hypothesis when it is false. Power depends on how large the true difference is, your sample size, the variance of the difference, and the significance level at which you are willing to reject the null hypothesis. Although a detailed discussion of power is beyond the scope of this book, let's consider a simple example because power is so important in data analysis.

? *Why does power depend on all of these factors?* All of these factors are involved in the computation of the t statistic, which is what determines whether you reject the null hypothesis. You'll have a large t statistic if the numerator is large and the denominator is small. The numerator of the t statistic is the difference between the two sample means. So, the larger the population difference, the more likely it is that you will have a large numerator for the t statistic. The denominator of the t statistic depends on the variances of the groups and the number of cases in each group. You'll have a small denominator if the sample variances are small and the sample sizes are large. ■■■

Monitoring Death Rates

You're the CEO of a large hospital chain that is under increasing pressure to monitor quality. Insurance companies and business coalitions want to see if you're doing a good job before they sign contracts with your organization. Although there are many components that contribute to hospital quality, death rates of hospitalized patients are of major concern to all involved. Since death rates depend on type of disease, you have to examine death rates separately for patients with different diseases. Assume that a 10% death rate is the norm for the condition you want to study. In order to compare your hospital's performance to the norm, how many hospitalized cases of the disease should you include in your sample?

? *Is it fair to compare a hospital's observed mortality rate to a national or state norm?* Comparing mortality rates is tricky. However, with the increasing emphasis on cost and quality of medical care, it's being done more and more often. If hospitals have different patient mixes—that is, if sicker or more complicated cases are concentrated in particular types of hospitals—then it's not fair to expect all hospitals to have the same mortality rates. Sophisticated statistical techniques are used to adjust observed mortality rates for differences in patient characteristics. ■■■

To answer the sample size question, look at Table 14.1. Each of the columns of the table corresponds to a possible true death rate for your hospital. That's the value you would get if you looked at all patients with the diagnosis of interest treated at your hospital chain. It's not the same as the death rate you observe in a sample of patients. Each of the rows of the table corresponds to a different sample size. The entry in each of the cells of the table is the power—that is, the probability that you reject the null hypothesis that the true rate for your hospital is 10%, the norm. The criterion used to reject the null hypothesis is a two-tailed observed significance level of 0.05 or less.

With a true 2% death rate and a sample size of 20, you have only a 36% chance of rejecting the false null hypothesis

Table 14.1 Probability of rejecting the null hypothesis† using a two-tailed significance level of 0.05

	Your hospital's true death rate					
Sample size	2%	5%	15%	20%	30%	40%
20	.36	.14	.10	.24	.64	.90
50	.72	.28	.19	.52	.95	*
80	.90	.41	.27	.72	*	*
100	.95	.49	.33	.81	*	*
160	.99	.68	.49	.95	*	*
200	*	.78	.58	.98	*	*
300	*	.92	.75	*	*	*
400	*	.97	.86	*	*	*
500	*	.99	.92	*	*	*
800	*	*	.99	*	*	*

† The null hypothesis is that the death rate is 10%.

* Indicates probabilities > 0.99.

? *What is the null hypothesis that I'm testing here?* The null hypothesis is that your hospital's death rate for a particular condition is the same as the norm—10%. The norm is not a value you estimate. It's a value that was established previously on the basis of large-scale studies. This is the same situation as in Chapter 10 where you tested whether a new treatment for a disease has the same cure rate as the established treatment. You can use the same binomial test to test this hypothesis. ■■■

Look at the column labeled 2%. If your hospital's true mortality rate is 2% and you count the number of deaths in a random sample of 20 cases, there is only a 36% chance that you will correctly reject the null hypothesis that the true mortality rate for your hospital is 10%, using an observed significance level of 0.05 or less for rejecting the null hypothesis. That means that two out of three times, you will fail to reject the null hypothesis when in fact it's false. You'll fail to identify your hospital as an exceptional performer when it really is. If you increase the sample size to 80, there is a 90% chance that you will correctly reject the null hypothesis. With a sample size of 80, you're much more likely to detect your hospital's good performance. As you can see in Table 14.1, for each of the hypothetical hospital rates (the columns), as you increase your sample size, your power increases.

Now look at the row that corresponds to a sample size of 100. You see that the power is 95% when the true hospital value is 2%. That means that 95% of the time when you take a sample of 100 cases and your hospital's true death rate is 2%, you will correctly reject the null hypothesis that your hospital's true rate is 10%. However, if your true value is 5%, meaning that your hospital's mortality is half of the national average, your probability of detecting that difference is only 49%. Similarly, if your true hospital rate is 15%, a 50% increase over the population rate, with a sample size of 100, you stand only a 33% chance of detecting it. (With a sample of 500 patients, you stand more than a 90% chance of detecting it.) From the table, you see that the larger the difference from the population rate, the easier it is to detect.

? *Why aren't the power values the same for 5% and 15% true death rates? They are both 5% different from the hypothetical rate of 10%.* The reason for this is that if you have a population in which only 5% of the cases die, the variability of the sample death rates will be smaller than if you have a population in which 15% of the cases die. If the population probability of dying is 0, whenever you take a sample there's only one possible outcome—everyone's alive. You're not going to see any variability from sample to sample. If the population probability of dying is 5%, you'll see somewhat more variability but not that much more. Regardless of the sample size that you take, your sample will consist of mostly living people. If the probability of dying increases to 15%, the variability of possible sample outcomes increases as well. The population isn't as homogeneous as before. In fact, at 50%, you'll have the largest possible variability, since that's when your population is most diverse. You expect to see all kinds of sample values. ■■■

In summary, when you fail to reject the null hypothesis, you should be cautious about the conclusions you draw. In particular, if your sample size is small, you may be failing to detect even large differences. That's why, when you design a survey or experiment, you should make sure to include enough cases so that you'll have reasonable power to detect differences that are of interest.

Does Significant Mean Important?

In Table 14.1, you see that as you increase your sample size, you are able to detect smaller and smaller discrepancies from the null hypothesis. For example, if you have many thousands of cases, you might be able to detect that your hospital rate is really 10.1%, compared to the norm of 10%. Or you might reject the null hypothesis that there is no difference in television viewing between Internet users and non-users, based on a difference of 10 minutes a day. Just because you are able to reject the null hypothesis doesn't mean you've uncovered an important or large difference. For very large sample sizes, even very small differences may cause you to reject the null hypothesis. Before you conclude that you've found something important, evaluate the observed difference on its own merits. Is a difference of one month of education really a worthwhile finding? Does a difference of one point in a test score really tell you anything?

Summary

How can you test the null hypothesis that two population means are equal, based on the results observed in two independent samples?

- The one-sample *t* test is appropriate only when you want to test hypotheses about one population mean.
- For the independent-samples *t* test, you must have two unrelated samples from normal distributions, or the sample size must be large enough to compensate for non-normality.
- You can't prove that the null hypothesis is true.
- Power is the probability of correctly rejecting a false null hypothesis. If you have small sample sizes, you may be unable to detect even large population differences.

What's Next?

Now that you know how to test hypotheses about two independent means, you're ready to consider a somewhat more complicated problem. How can you tell if *more* than two means are different from each other? That's what Chapter 15 is about.

How to Obtain an Independent-Samples T Test

The SPSS Independent-Samples T Test procedure tests the null hypothesis that the population mean of a variable is the same for two groups of cases. It also displays a confidence interval for the difference between the population means in the two groups.

To obtain an independent-samples *t* test, you must indicate the variable(s) whose means you want to compare, and you must specify the two groups to be compared.

▶ To open the Independent-Samples T Test dialog box (see Figure 14.10), from the menus choose:

Analyze
  Compare Means ▶
    Independent-Samples T Test...

Figure 14.10 Independent-Samples T Test dialog box

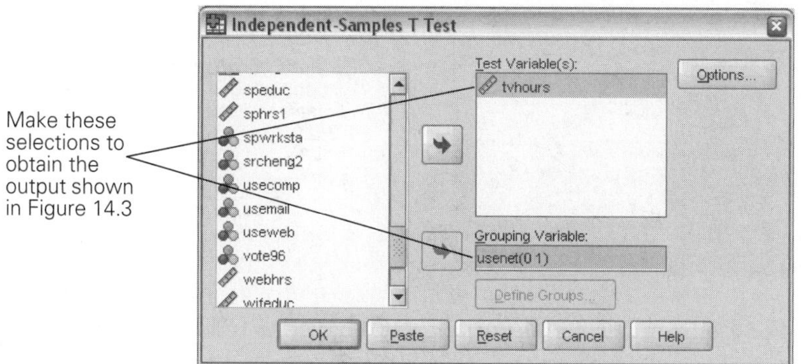

In the Independent-Samples T Test dialog box:

- Select the variable whose mean you want to test and move it into the Test Variable(s) list. You can move more than one variable into the list to test all of them between two groups of cases.
- Select the variable whose values define the two groups and move it into the Grouping Variable box. Click Define Groups and indicate how the groups are defined.
- Click OK.

For each test variable, SPSS calculates a *t* statistic and its observed significance level.

Define Groups: Specifying the Subgroups

After you move a variable into the Grouping Variable box, click Define Groups to open the Define Groups dialog box, as shown in Figure 14.11.

Figure 14.11 Independent-Samples T Test Define Groups dialog box

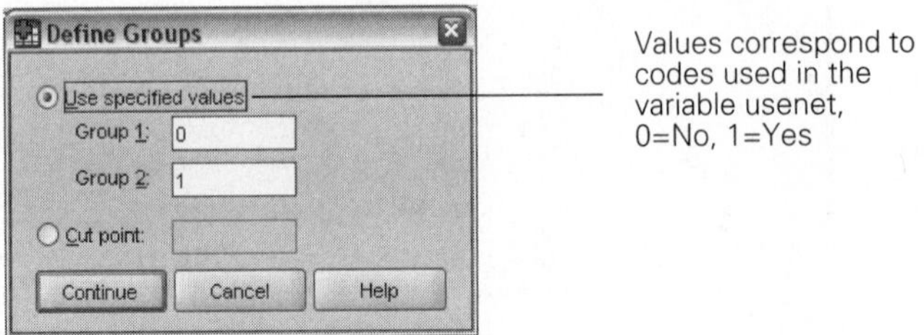

For a numeric grouping variable, you have the following alternatives:

Use specified values. If each group corresponds to a single value of the grouping variable, select this option and enter the values for Group 1 and Group 2. Other values of the grouping variable are ignored.

Cut point. If one group corresponds to small values of the grouping variable and the other group to large values, select this option and enter a value that separates the groups. Cases that exactly equal the cut point are included in the second group. (If you don't want to remember that, enter a cut point that doesn't occur in your data. To compare codes 1 and 2 with codes 3 and 4, enter a cut point of 2.5 and be sure.)

String Grouping Variable

For a string grouping variable, a cut point isn't available. The Define Groups dialog box simply asks for the two values that identify the groups you want to compare.

Options: Confidence Level and Missing Data

In the Independent-Samples T Test dialog box, click Options to open the Independent-Samples T Test Options dialog box, as shown in Figure 14.12. This dialog box allows you to change the confidence level for the confidence interval for the population difference between the means of the two groups and to control the handling of missing values.

Figure 14.12 Independent-Samples T Test Options dialog box

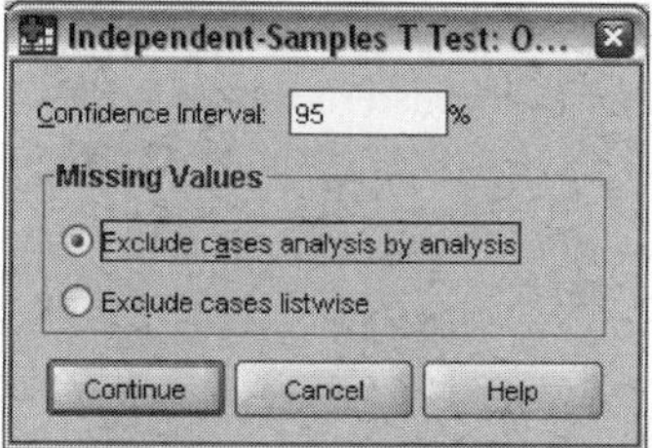

Confidence Interval. Defines the desired confidence interval (usually 95 or 99).

Missing Values. Two alternatives control the treatment of missing data for multiple test variables.

Exclude cases analysis by analysis. All cases that have valid data for the grouping variable and a test variable are used in the statistics for that test variable.

Exclude cases listwise. Only the cases that have valid data for the grouping variable and all specified test variables are used. This ensures that all of the tests are performed using the same cases but doesn't necessarily use all of the available data for each test.

Exercises

Statistical Concepts

1. For the following studies, indicate whether an independent-samples or paired *t* test is appropriate:

a. You want to study regional differences in consumer spending. You randomly select a sample of consumers in the Midwest and the East and track their spending patterns.

b. You want to study differences in the spending habits of teenage boys and girls. You select 100 brother-sister pairs and study their spending behavior.

c. You want to compare error rates for 20 employees before and after they attend a quality improvement workshop.

d. Weight is obtained for each subject before and after Dr. Nogani's new treatment. The hypothesis to be tested is that the treatment has no effect on weight loss.

e. The Jenkins Activity Survey is administered to 20 couples. The hypothesis to be tested is that husbands' and wives' scores do not differ.

f. Subjects are asked their height and then a measurement of height is obtained. The hypothesis to be tested is that self-reported and actual heights do not differ.

g. You want to compare the durability of two types of socks. You have people wear one type of sock on one foot and another type of sock on the other. You see how long it takes for a hole to appear. (Assume they wash them periodically!)

2. You want to compare the average ages of people who buy and who don't buy a product. Suppose you obtain the following results:

		N	Mean	Std. Deviation	Std. Error Mean
AGE	No	100	29.45	15.56	1.56
	Yes	100	38.00	15.49	1.55

			Equal variances assumed	Equal variances not assumed
Levene's Test for Equality of Variances	F		.052	
	Sig.		.820	
t-test for Equality of Means	t		-3.900	-3.900
	df		198	198.000
	Sig. (2-tailed)		.000	.000
	Mean Difference		-8.550	-8.550
	Std. Error Difference		2.196	2.196
	95% Confidence Interval of the Difference	Lower	-12.880	-12.880
		Upper	-4.220	-4.220

a. Write a short paragraph summarizing your findings.

b. Have you proved that in the population the two groups have different mean ages?

c. What is a plausible range for the true difference?

d. What would happen to the observed significance level if the difference and the standard deviations in the two groups remained the same but the sample sizes were tripled?

3. A market research analyst is studying whether men and women find the same types of cars appealing. He asks 150 men and 75 women to indicate which one of the following types of cars they would be most likely to buy: two-door with trunk, two-door with hatchback, convertible, four-door with trunk, four-door with hatchback, or station wagon. Each of the possible responses is assigned a code number. The analyst runs a *t* test and finds that the average value for males and females appears to differ ($p = 0.008$). He doesn't know how to interpret his output, so he comes to you for advice. Explain to him what his results mean.

4. You are interested in whether the size of a company, as measured by the number of employees, differs between companies that offer pension plans and those that don't. You select a random sample of 75 companies and obtain information about the number of employees and availability of a pension plan. You run a two-independent-samples *t* test and find that there is not a statistically significant difference between the two types of companies ($p = 0.237$). A colleague of yours conducts a similar study. She polls 200 companies and finds a similar difference in the number of employees between the two types of companies. However, she claims that she found a significant difference ($p = 0.002$). Is this possible? Explain.

5. You are interested in whether average family income differs for people who find life exciting and for those who don't. You take a sample of people at a local museum on Sunday afternoon and find that there is a $5,000 difference in income between the two groups. You do a *t* test and find the observed significance level to be 0.03.

   Your friend is also studying the same problem. She takes a sample of people in a department store on Saturday afternoon. She finds a $10,000 difference in family income between the two groups. But when she does a *t* test, she finds an observed significance level of 0.2.

   Discuss these studies, their shortcomings, and possible reasons for the contradictory results.

Data Analysis

Use the *gss.sav* data file to answer the following questions:

1. Perform the appropriate analyses to test whether the average number of hours of daily television viewing (variable *tvhours*) is the same for men and women. Write a short summary of your results, including appropriate charts to illustrate your findings. Be sure to look at the distribution of hours of television viewed separately for men and women.

   a. Based on the results you observed, is it reasonable to conclude that in the population, men and women watch the same amount of television?

b. If you found a statistically significant difference between average hours watched by men and women, would you necessarily conclude that men and women do not watch the same amount of television? What other nonstatistical explanations are possible for your findings?

2. Discuss the possible advantages of a paired design to analyze differences between men's and women's television viewing. Discuss the drawbacks as well.

3. Some people claim that they would continue to work if they struck it rich, and others say that they would not (variable *richwork*). Use the two-independent-samples *t* test to identify possible differences between the groups in age, education, television viewing, and so on. Write a paper summarizing your findings.

4. What distinguishes people who believe in life after death from those who do not (variable *postlife*)? Use the available data to identify differences between the two groups. Write a short paper summarizing your results.

5. Consider people who use the Internet and those who don't (*usenet*).

a. What is the average income for people who use the Internet and those who don't? (Use *rincdol.*) Do you have enough evidence to reject the null hypothesis that the average income is the same for the two groups? Which group makes more money?

b. Write a short paper discussing differences between Internet users and non-users.

Use the *marathon.sav* data file to answer the following questions:

6. If you consider the Chicago marathon runners to be a sample from some population of marathon runners, you can test hypotheses about differences in completion times based on gender and age.

a. Describe the age and gender composition of those who completed the Chicago marathon.

b. Test the null hypothesis that men and women have the same average completion times.

c. Even if men and women of a given age ran as fast, given the age distributions in the two groups, would you expect to reject the null hypothesis that average completion times are the same for men and women? Explain.

d. Select only runners in the age group 25–39 (*agecat8* equals 3). Test the null hypothesis that the average time to completion is the same for men and women. Summarize the results.

Use the *manners.sav* data file to answer the following questions:

7. Consider the average age for people who think the world would be a better place if people said please and thank you and those who think it would make no difference (*pleasthx*).

  a. Make histograms of age for the two groups. Is age approximately normally distributed in the two groups?

  b. What assumptions do you need to make in order to test the null hypothesis that the average age is the same in the two groups?

  c. What is the average age for people who think the world would be a better place if people said please and thank you? For those who think it would not make a difference?

  d. Test the null hypothesis that in the population the average age of those selecting the two responses is equal. What can you conclude?

  e. What is the 95% confidence interval for the difference in average ages? Based on the confidence interval, would you reject the null hypothesis that the average difference in age is 10 years? Five years?

Use the *salary.sav* data file to answer the following questions:

8. Use the Select Cases facility to restrict the analysis to clerical workers only (variable *jobcat* equals 1).

  a. Test the assertion that male and female clerical workers have the same average starting salaries. Summarize your findings.

  b. You are 95% confident that the true difference in average beginning salaries for male and female clerical workers is in what range?

  c. How often would you expect to see a difference at least as large in absolute value as the one you observed if, in fact, male and female clerical workers have the same beginning salaries?

  d. Evaluate how well your data meet the assumptions needed for a two-independent-sample t test.

9. The bank claims that male clerical workers are paid more than female clerical workers because they have more formal education. Do the data support this assertion? Explain.

10. Consider office trainees (variable *jobcat* equals 2). The women trainees have hired you to show that the bank discriminates against women office trainees by paying them less. Analyze the data, and prepare a summary of your findings.

11. The bank has now hired you to refute the claim that it discriminates against women office trainees. What evidence can you come up with to support the bank's position? Write a summary of your findings on behalf of the bank.

Use the *electric.sav* data file to answer the following questions:

12. Write a short report discussing the claim that average diastolic blood pressure in 1958 (variable *dbp58*) is the same for men who were alive in 1968 and for men who were not (variable *vital10*). Include appropriate summary statistics and charts.

13. There is one man with a diastolic blood pressure of 160 mm Hg. Rerun the independent-samples t test, excluding him from the analysis. How does your conclusion change? What can you conclude about the effect of outliers on the results of a t test?

14. Test the null hypothesis that average diastolic blood pressure is the same for men who smoke and for men who don't smoke. Write a paragraph summarizing your conclusions. Include appropriate charts. Be sure to consider the effect of the outlier on your results.

Use the *schools.sav* data file to answer the following question:

15. Look at schools that are above and below the median percentage of low income for all Chicago schools (variable *medloinc*). Are there differences between the two groups in the school performance variables? Summarize your results.

Use the *renal.sav* data file to answer the following question:

16. Use the independent-samples t test to identify differences between cardiac surgery patients who developed acute renal failure (variable *type* equals 1) and those who did not (variable *type* equals 0). Identify the variables in the data file for which a t test is appropriate. Summarize your findings.

Use the *buying.sav* data file to answer the following question:

17. Test the null hypothesis that the family buying score (variable *famscore*) is the same when pictures are shown and when they are not (variable *picture*). Test the null hypothesis that the average buying score for the husband (variable *hsumbuy*) is the same with and without pictures. Repeat for the average buying score for wives (variable *wsumbuy*). Summarize your results.

15 One-Way Analysis of Variance

How can you test the null hypothesis that several population means are equal?

- What is analysis of variance?
- What assumptions about the data are needed to use analysis-of-variance techniques?
- How is the *F* ratio computed, and what does it tell you?
- Why do you need multiple comparison procedures?

As a college student, you're used to working long and hard. But you probably harbor expectations that once that degree is in hand, you'll be on Easy Street. Those late nights of studying will magically turn into late nights of fancy dining, glamorous sports events, and long weekends in expensive resorts. Unfortunately, you've already seen that as a college graduate, you can no longer expect a 40-hour work week. College graduates claim to work more than 40 hours a week. Is the long week a curse of just college graduates, or is everyone suffering from the expansion of the work week? That's the question you'll address in this chapter. In order to do so, you'll learn how to test the null hypothesis that several independent population means are equal. The technique you'll use is called **analysis of variance**, usually abbreviated as ANOVA.

▶ This chapter uses the *gssft.sav* data file, which includes only people holding full-time jobs. For instructions on how to obtain the One-Way ANOVA output shown in this chapter, see "How to Obtain a One-Way Analysis of Variance" on p. 324.

Hours in a Work Week

You've already determined in Chapter 12 that the 95% confidence interval for the average number of hours worked per week by people with college or graduate degrees is from 46.04 hours to 47.96 hours. Based on the confidence interval, it's possible that college graduates work almost an extra 8 hours each week over the normal 40 hours. That's not a particularly pleasant thought, especially since "overtime" is not likely to be compensated.

Describing the Data

To see how people with other educational backgrounds fare, look at Figure 15.1. You see that the average work week for all full-time employees is 45.62 hours. (It's the entry for the *Mean* column in the last row, labeled *Total*.) The average work week ranges from a low of 44.95 hours for people with only a high school diploma to a high of 48.19 hours for people with graduate degrees.

From the menus choose:

Analyze
Compare Means ▶
One-Way ANOVA...

Select the variables hrs1 and degree, as shown in Figure 15.10.

In the One-Way ANOVA Options dialog box, select Descriptive, as shown in Figure 15.12.

Figure 15.1 Descriptive statistics for hours worked

HRS1 Number of hours worked last week

	N	Mean	Std. Deviation	Std. Error	95% Confidence Interval for Mean Lower Bound	95% Confidence Interval for Mean Upper Bound	Minimum	Maximum
Less than HS	111	45.03	10.138	.962	43.12	46.93	15	87
High school	808	44.95	10.723	.377	44.21	45.69	6	89
Junior college	131	45.69	11.669	1.020	43.67	47.70	20	89
Bachelor	286	46.37	10.413	.616	45.16	47.58	15	89
Graduate	151	48.19	9.729	.792	46.62	49.75	24	80
Total	1487	45.62	10.647	.276	45.08	46.16	6	89

The average work week ranges from 44.95 hours to 48.19 hours

In the column labeled *Std. Deviation*, you see that the smallest variability in hours worked is for people with graduate degrees, while the largest is for people with junior college degrees. The next row, labeled *Std. Error*, tells you how much the sample means vary in repeated samples of the same size from the same population. For each group, it's the standard deviation divided by the square root of the sample size. The smallest standard error is for high school graduates, since they are the largest group.

? *What's special about 89 hours? Why is that the largest value for three of the five groups?* That's another arbitrary limit set by the General Social Survey during the days of punch cards and sorters. The GSS won't let you report that you work 24/7, although you can watch television that long.

Confidence Intervals for the Group Means

In the next two columns of Figure 15.1, you see for each group the 95% confidence interval for the population value of the average hours worked per week. You are 95% confident that the true average work week for those with less than a high school diploma is between 43.12 and 46.93 hours. For those with a graduate degree, you are 95% confident that the true average work week is between 46.62 and 49.75 hours.

In the Error Bar dialog box in Legacy Dialogs, select the variables hrs1 and degree.

Figure 15.2 Plot of sample means and 95% confidence intervals

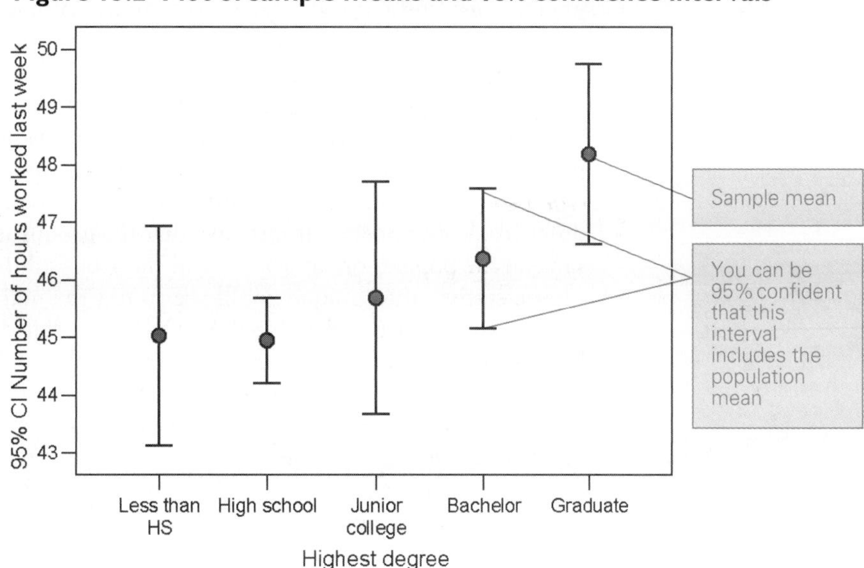

Plots of the means and confidence intervals are shown in Figure 15.2. You see that the 95% confidence interval for high school graduates is the narrowest. That's because there are so many of them in the sample. Many of the confidence intervals in Figure 15.2 overlap. That tells you that some of the values that are plausible for the true work week in one group are also plausible for the true work week in the others. The exception is the confidence interval for those with graduate degrees. It doesn't overlap the confidence interval for those with a high school education.

? *Can you tell from the plot if the 40-hour work week is a reasonable guess for the true hours worked per week?* Sure. Remember, if a value doesn't fall in the 95% confidence interval for the mean, you can reject the hypothesis that it's a plausible population value. You see in Figure 15.1 that the value 40 is not included in any of the confidence intervals. That means you can reject the hypothesis that it's a reasonable value for any of the groups. It appears that the 40-hour work week may be a thing of the past, regardless of your education level. ■■■

Testing the Null Hypothesis

The descriptive statistics and plots suggest that there are differences in the average work week among the five education groups. Now you need to figure out whether the observed differences in the samples may be attributed to just the natural variability among sample means or whether there's reason to believe that some of the five groups have different values in the population for average hours worked.

The null hypothesis says that the population means for all five groups are the same. That is, there is no difference in the average hours worked for people in the five education categories. The alternative hypothesis is that there is a difference. The alternative hypothesis doesn't say which groups differ from one another. It just says that the groups means are not all the same in the population; at least one of the groups differs from the others.

The statistical technique you'll use to test the null hypothesis is called analysis of variance (abbreviated ANOVA). It's called analysis of variance because it examines the variability of the sample values. You look at how much the observations within each group vary as well as how much the group means vary. Based on these two estimates of variability, you can draw conclusions about the population means. If the sample means vary more than you expect based on the variability of the observations within the groups, you can conclude that the population means are not all equal.

SPSS contains several different procedures that perform analysis of variance. In this chapter, you'll use the One-Way ANOVA procedure. It's called one-way analysis of variance because cases are assigned to different groups based on their values for one variable. In this example, you form the groups based on the values of the *degree* variable. The variable used to form groups is called a **factor**. In Chapter 16, you'll learn how to test hypotheses when cases are classified into groups based on their values for two factors.

Assumptions Needed for Analysis of Variance

The Kruskal-Wallis test, described in Chapter 18, requires more limited assumptions about the data.

Analysis of variance requires the following assumptions:

- Independent random samples have been taken from each population. (The groups you are comparing are regarded as distinct populations.)
- The populations are normally distributed.
- The population variances are all equal.

Independence. The independence assumption means that there is no relationship between the observations in the different groups and between the observations in the same group. For example, if you administer four different treatments to each individual, you cannot use the one-way analysis-of-variance procedure to analyze the data. Observations from the same individual appear in each of the groups, so they are not independent. (In this situation, you must use an extension of the paired-samples *t* test. It's called repeated measures analysis of variance, a topic not covered in this book.) Observations within a group are also not independent if conditions are changing with time. For example, if you are explaining a task to subjects and your instructions get better with time, early subjects may not perform as well as later subjects. In this situation, the response of the subject depends on the point in time that he or she entered into the study. Consecutive subjects will be similar to each other.

Normality. The normality assumption in analysis of variance can be checked by making histograms or normal probability plots for each of the groups. In practice, the analysis of variance is not heavily dependent on the normality assumption. As long as the data are not extremely non-normal, you do not have to worry. (If your sample sizes in the groups are small, you should be aware of the impact of unusual observations, which can have a big effect on the mean and standard deviation. You can rerun the analysis without the unusual point to make sure that you reach the same conclusions.)

Equality of Variance. The equality-of-variance assumption can be checked by examining the spread of the observations in the boxplot. You can also compute the Levene test for equality of variance, which is available in the Explore and One-Way procedures. In practice, if the number of cases in each of the groups is similar, the equality-of-variance assumption is not too important.

? *What should I do if I suspect that my data violate the necessary assumptions?* Well, it depends on which assumption is being violated. Analysis of variance is quite insensitive to violations of the normality assumption if the sample sizes are large. In contrast, if the number of cases in each group is quite different, the equality of variance assumption is quite important. The Brown-Forsythe and Welch robust F tests should be used when the assumption of equal variances in all groups appears questionable. These are both available in the SPSS One-Way Analysis of Variance procedure. Also, you can use multiple comparison procedures, which don't assume that variances in all groups is equal. You can also use statistical tests that make fewer assumptions about the distributions of the variables.

The situation is considerably more complicated if you're worried about whether the groups are somehow biased. That is, you're concerned that one or more of your samples differs in some important way from the population of interest. For example, if you want to compare four medical treatments, and the participating physicians have assigned the sickest patients to a particular group, you've got a real problem. You may not be able to draw any correct conclusions from your data. That's why it's very important when comparing several treatments or conditions to make sure that the subjects are randomly assigned to the different groups. *Randomly* doesn't mean *haphazardly*. It means that you must have a well-organized system for random assignment of cases. ■■■

Analyzing the Variability

In analysis of variance, the observed variability in the sample is divided (partitioned, in statistical lingo) into two parts: variability of the observations *within* a group about the group mean, and variability *between* the group means.

? *Why are we talking about variability? Aren't we testing hypotheses about means?* Yes, we're testing hypotheses about population means, but as you've seen in previous chapters, your conclusions about population means are always based on looking at the variability of sample means. You have to determine if your sample mean is outside of the usual range of variability of sample means from the population.

In analysis of variance, you'll look at how much your observed sample means vary. You'll compare this observed variability to the expected variability if the null hypothesis that all population means are the same is true. If the sample means vary more than you'd expect, you have reason to believe that this extra variability is because some of groups don't have the same population mean. (If you have two independent groups, you'll get the same results using ANOVA or the equal-variance t test.) ■■■

Let's look a little more closely now at the two types of variability and how they are used to test the null hypothesis that the population values for average hours worked per week are the same for people in the five education categories. The game plan is as follows: You want to know whether your sample means vary more than you would expect if the null hypothesis is true. First, you'll see how much the observations in a group vary, and then you'll see how much the sample means vary among themselves. If the sample means vary more than you expect, you'll reject the null hypothesis.

Within-Groups Variability

The **within-groups estimate of variability**, as its name suggests, is a variance estimate based on how much the observations within a group vary. The sample variance of each group estimates within-groups variability. One of the assumptions of analysis of variance is that all groups come from populations with the same variance. That makes it possible for you to average the variances in each of the groups to come up with a single number, which is the within-groups variance. (You'll see later how this averaging is done. You can't just add up the sample variances and divide by the number of groups.)

You might wonder why you can't just put all of your observations together and compute the variance. The reason is that you don't know if all of the groups have the same population mean. If they don't, pooling all of the values together will give you the wrong answer. For example, suppose that all people without a high school diploma work exactly 40 hours a week, all people with a high school diploma work exactly 43 hours a week, and all people with a college degree work exactly 45 hours. The variance in each of the groups is 0, since the values within a group don't vary at all. The correct estimate of the within-groups variance is also 0. If you compute the variance for all cases together, it wouldn't be close to 0. The observed variability would be the result of differences in the means of the three groups.

Between-Groups Variability

You have a sample mean for each of the groups in your study. If all of the groups have the same number of cases, you can find the standard deviation of the sample means. What would that tell you? If all the groups come from populations with the same mean and variance, the standard deviation of the sample means tells you how much sample means from the same population vary. The standard deviation of the sample means is an estimate of the standard error of the mean.

From the standard error of the mean, you can estimate the standard deviation of the observations. You do this by multiplying the standard error of the mean by the square root of the number of cases in a group.

? *Where did that come from?* The standard error of the mean is the standard deviation of the observations divided by the square root of the sample size. So, using simple algebra, the standard deviation is the standard error of the mean multiplied by the square root of the sample size. Thus,

$$\text{standard error} = \frac{\text{standard deviation}}{\sqrt{\text{sample size}}}$$

and

$$\text{standard deviation} = \text{standard error} \times \sqrt{\text{sample size}}$$

■■■

If you square the estimate of the standard deviation, you have a quantity that's called the **between-groups estimate of variability**. It's called the between-groups estimate of variability because it's based on how much sample means vary *between* the groups.

Comparing the Two Estimates of Variability

You now have two estimates of how much the observations within a group vary: the within-groups estimate and the between-groups estimate. These two estimates differ in a very important way—the between-groups estimate of variance will be correct only if the null hypothesis is true. If the null hypothesis is false, the between-groups estimate of variance will be too large. The observed variability of the sample means will be the result of two factors: the variability of the observations within a group and the variability of the population means. The within-groups estimate of variability doesn't depend on the null hypothesis being true. It's always a good estimate.

Your decision about the null hypothesis will be based on comparing the between-groups and the within-groups estimates of variability. You'll see how much the number of hours worked varies for individuals in the same education group. This will give you the within-groups estimate of variability. Then you'll see how much the means of the five groups vary. Based on this, you'll calculate the between-groups estimate of variability. If the between-groups estimate is sufficiently larger than the within-groups estimate, you'll reject the null hypothesis that all of the means are equal in the population.

The Analysis-of-Variance Table

The estimates of variability that we've been talking about are usually displayed in what's called an **analysis-of-variance table**. Figure 15.3 is the analysis-of-variance table for the test of the null hypothesis that the population value for average hours worked per week is the same for people in five categories of education. By looking at this table, you'll be able to tell whether you have enough evidence to reject the null hypothesis.

Figure 15.3 Analysis-of-variance table

To obtain this output, select the variables hrs1 and degree in the One-Way ANOVA dialog box, as shown in Figure 15.10.

		Sum of Squares	df	Mean Square	F	Sig.
HRS1 Number of hours worked last week	Between Groups	1557.919	4	389.480	**3.459**	**.008**
	Within Groups	166892.2	1482	112.613		
	Total	168450.1	1486			

Ratio of mean squares

Probability of obtaining F ratio at least this large when null hypothesis is true

The two estimates of variability are shown in the column labeled *Mean Square*. Their ratio is in the column labeled *F*. If the null hypothesis is true, you expect the ratio of the between-groups mean square to the within-groups mean square to be close to 1, since they are both estimates of the population variance. Large values for the *F* ratio indicate that the sample means vary more than you would expect if the null hypothesis were true. Since you are dealing with samples, you know that the ratios will vary in magnitude.

You can tell if your observed *F* ratio of 3.46 is large enough for you to reject the null hypothesis by looking at the observed significance level, which is labeled *Sig*. You see that the probability of obtaining an *F* ratio of 3.46 or larger when the null hypothesis is true is 0.008. Only 8 times in 1000, when the null hypothesis is true, would you expect to see a ratio this large or larger. So you can reject the null hypothesis. It's unlikely that the number of hours worked per week is the same for the five groups in the population.

Now that you know the punch line, let's see where all the numbers are coming from.

Estimating Within-Groups Variability

You need three steps to compute the within-groups estimate of variability:

1. First, you must compute what's called the **within-groups sum of squares**. Take all of the standard deviations in Figure 15.1 and square them to obtain variances. Then multiply each variance by one less than the number of cases in the group. Finally, add up the values for all of the groups. The within-groups sum of squares is:

$$(10.14^2 \times 110) + (10.72^2 \times 807) + (11.67^2 \times 130) + (10.41^2 \times 285) + (9.73^2 \times 150) = 166,892$$

Equation 15.1

You see this number in the second row of Figure 15.3 in the column labeled *Sum of Squares*. (You have to use more decimal places for the standard deviation than shown above to get exactly the answer given.)

2. Next, you must compute the degrees of freedom. That's easy to do. For each group, you compute the number of cases minus 1 and then add up these numbers for all of the groups. In this example, the degrees of freedom are:

$$\text{degrees of freedom} = 110 + 807 + 130 + 285 + 150 = 1482$$

Equation 15.2

This number is shown in the *Within Groups* row of Figure 15.3, in the column labeled *df* (for degrees of freedom).

3. Finally, divide the sum of squares by its degrees of freedom, to get what's called a **mean square**. This is the estimate of the average variability in the groups. It's really nothing more than an average of the variances in each of the groups, adjusted for the fact that the number of observations in the groups differs. Your estimate of the variance for the number of hours worked, based on the variability of the observations within each of the groups, is 112.61.

Estimating Between-Groups Variability

You also need three steps to calculate the between-groups estimate of variability.

1. First, you compute the **between-groups sum of squares**. Subtract the overall mean (the mean of all of the observations) from each group mean. Then square each difference, and multiply the square by the number of observations in its group. Finally, add up all the results. For this example, the between-groups sum of squares is:

$$\begin{aligned}
&111 \times (45.03 - 45.62)^2 \\
&+ 808 \times (44.95 - 45.62)^2 \\
&+ 131 \times (45.69 - 45.62)^2 \\
&+ 286 \times (46.37 - 45.62)^2 \\
&+ 151 \times (48.19 - 45.62)^2 = 1558
\end{aligned}$$

Equation 15.3

2. Next, you must compute the degrees of freedom. The degrees of freedom for the between-groups sum of squares is just the number of groups minus 1. In this example, there are five education groups, so the degrees of freedom for the between-groups sum of squares is 4.

3. Finally, calculate the between-groups mean square by dividing the between-groups sum of squares by its degrees of freedom. The between-groups mean square is 389.

Calculating the F Ratio

You now have the two estimates of the variability in the population: the within-groups mean square and the between-groups mean square. The *F* ratio is simply the ratio of these two estimates. Take the between-groups mean square and divide it by the within-groups mean square:

$$F = \frac{\text{between-groups mean square}}{\text{within-groups mean square}} = \frac{389.48}{112.61} = 3.46$$ **Equation 15.4**

(Remember, the within-groups mean square is based on how much the observations within each of the groups vary. The between-groups mean square is based on how much the group means vary among themselves.) If the null hypothesis is true—that the average hours worked per week is the same for the five groups—the two numbers should be close to each other. If you divide one by the other, the result should be close to 1.

As you see, the ratio of the two estimates is 3.46. Does that mean you automatically reject the null hypothesis? No. You know that your sample ratio will vary, even if the null hypothesis is true. You need to figure out how often you would expect to see a sample value of 3.46 or greater when the null hypothesis is true. That is, you need to determine whether your sample results are unlikely if the null hypothesis is true.

The observed significance level is calculated by comparing your observed *F* ratio to values of the *F* distribution. The observed significance level depends on both the observed *F* ratio and the degrees of freedom for the two mean squares.

? *What's the* F *distribution?* Like the normal and *t* distributions, the ***F* distribution** is defined mathematically. It's used when you want to test hypotheses about population variances. The Central Limit Theorem doesn't work for variances. Their distributions are not normal. The ratio of two sample variances from normal populations has an *F* distribution. The *F* distribution is indexed by two values for the degrees of freedom—one for the numerator and one for the denominator. The degrees of freedom depend on the number of observations used to calculate the two variances.

In Figure 15.3, you see that the observed significance level for this example is 0.008. Since the value is small, you can reject the null hypothesis that the average hours worked per week in the population is the same for the five groups. The observed sample results are not likely to occur when the null hypothesis is true.

Multiple Comparison Procedures

A statistically significant *F* ratio tells you only that it appears unlikely that all population means are equal. It doesn't tell you which groups are different from each other. You can reject the null hypothesis that all population means are equal in a variety of situations. For example, it may be that the average hours worked differs for all of the five groups. Or it may be that only one or two of the groups differ from the rest. Usually, when you've rejected the null hypothesis, you want to pinpoint exactly where the differences are. To do this, you must use multiple comparison procedures.

? *Why do you need yet another statistical technique? Why can't you just compare all possible pairs of means using* t *tests?* The reason for not using many *t* tests is that when you make many comparisons involving the same means, the probability increases that one or more comparisons will turn out to be statistically significant, even when all the population means are equal. This is known as the **multiple comparison problem.**

For example, if you have five groups and compare all pairs of means, you're making 10 comparisons. When the null hypothesis is true, the probability that at least one of the 10 observed significance levels is less than 0.05 is about 0.29. With 10 means (45 comparisons), the probability of finding at least one significant difference is about 0.63. The more comparisons you make, the more likely it is that you'll find one or more pairs to be statistically different, even if all population means really are equal.

Multiple comparison procedures protect you from calling differences *significant* when they really aren't. This is accomplished by adjusting the observed significance level for the number of comparisons that you are making, since each comparison provides another opportunity to reject the null hypothesis. The more comparisons you make, the larger the difference between pairs of means must be for a multiple comparison procedure to call it statistically significant. That's why you should look only at differences between pairs of means that you are interested in. When you use a multiple comparison procedure, you can be more confident that you are finding true differences. ■■■

Many multiple comparison procedures are available. They differ in how they adjust the observed significance level. One of the simplest is the **Bonferroni procedure.** It adjusts the observed significance level by multiplying it by the number of comparisons being made. For example, if you are making five comparisons, the observed significance level for each comparison must be less than 0.05/5, or 0.01, for the difference to be significant at the 0.05 level. The Bonferroni procedure assumes that the variances in all groups are the same. You can test this assumption using the Levene test, shown in Figure 15.4.

In the One-way ANOVA Options dialog box, select Homogeneity of variance test, as shown in Figure 15.12.

Figure 15.4 Levene test of equality of variances

	Levene Statistic	df1	df2	Sig.
HRS1 Number of hours worked last week	.383	4	1482	.821

Based on the Levene test, you can't reject the null hypothesis that the variances in all of the groups are equal, so it's reasonable to use a multiple comparison procedure that assumes equality of group variances. If the assumption of equal variances is violated, you should use multiple comparison procedures that don't assume equal variances.

If you want to compare all five education groups to one another, you can form 10 unique pairs of groups. Statistics for all pairs of group comparisons using the Bonferroni multiple comparison procedure are shown in Figure 15.5.

In the One-Way ANOVA Post Hoc dialog box, select Bonferroni, as shown in Figure 15.11.

Figure 15.5 Bonferroni multiple comparisons for hours worked

Dependent Variable: HRS1 Number of hours worked last week

Bonferroni

(I) DEGREE Highest degree	(J) DEGREE Highest degree	Mean Difference (I-J)	Std. Error	Sig.	95% Confidence Interval	
					Lower Bound	Upper Bound
0 Less than HS	1 High school	.08	1.074	1.000	-2.94	3.10
	2 Junior college	-.66	1.369	1.000	-4.51	3.19
	3 Bachelor	-1.34	1.187	1.000	-4.68	2.00
	4 Graduate	-3.16	1.327	.174	-6.89	.57
1 High school	0 Less than HS	-.08	1.074	1.000	-3.10	2.94
	2 Junior college	-.74	1.000	1.000	-3.55	2.07
	3 Bachelor	-1.42	.730	.521	-3.47	.63
	4 Graduate	**-3.24***	.941	.006	-5.88	-.59
2 Junior college	0 Less than HS	.66	1.369	1.000	-3.19	4.51
	1 High school	.74	1.000	1.000	-2.07	3.55
	3 Bachelor	-.68	1.120	1.000	-3.83	2.47
	4 Graduate	-2.50	1.267	.488	-6.06	1.06
3 Bachelor	0 Less than HS	1.34	1.187	1.000	-2.00	4.68
	1 High school	1.42	.730	.521	-.63	3.47
	2 Junior college	.68	1.120	1.000	-2.47	3.83
	4 Graduate	-1.82	1.067	.887	-4.82	1.18
4 Graduate	0 Less than HS	3.16	1.327	.174	-.57	6.89
	1 High school	**3.24***	.941	.006	.59	5.88
	2 Junior college	2.50	1.267	.488	-1.06	6.06
	3 Bachelor	1.82	1.067	.887	-1.18	4.82

* The mean difference is significant at the .05 level.

There are a lot of numbers, but they're not hard to understand. Each row corresponds to a comparison of two groups. The first row is for the comparison of the less-than-high-school group to the high-school group. The last row is for the comparison of the graduate group to the bachelor's degree group. The difference in hours worked between the two groups is shown in the column labeled *Mean Difference*. Pairs of means that are significantly different from each other are marked with an asterisk. You see that people with graduate degrees work significantly longer than people with only a high school education. No two other groups are significantly different from one another. The table shows all possible pairs of groups twice. There is a row for the comparison of bachelor to graduate and another row for the comparison of graduate to bachelor. These two rows are identical, except for the sign of the mean difference.

The column labeled *Std. Error* (of the difference) is calculated from the within-groups estimate of the standard deviation and the sample sizes in each of the two groups. The observed significance level for the test of the null hypothesis that the two groups come from populations with the same mean is shown in the column labeled *Sig.* Looking down the column of observed significance levels, you see that two of them are less than 0.05. (Note that the mean differences for these pairs are marked with an asterisk.) The 95% confidence interval for the mean difference gives you a range of values that you expect would include the true population difference between the two groups. For example, it's possible that the true difference between the hours worked by people with graduate degrees and people with a high school education is anywhere between 0.59 and 5.88 hours. Note that the confidence intervals for the pairs that are significantly different from one another do not include the value 0. The confidence intervals are also modified to take into account the fact that 10 pairs of means are being compared. They are wider than they would be if only one pair of means were being compared.

Television Viewing, Education, and Internet Use

In the previous chapter, you examined the relationship between hours of television viewing per night and Internet use. You found that Internet users watched significantly fewer hours of television than non-users. However, since Internet users are better educated than non-users, you were concerned about the possible role of differences in education on the observed difference in hours of television viewing. You can use analysis of variance to test the null hypothesis that average hours of television viewing is the same for all education groups. Based on Figure 15.6, the analysis-of-variance table for television hours and education, you can reject the null hypothesis that the average hours of television viewing is the same for all five education groups.

Figure 15.6 Analysis of variance for television viewing

To obtain this output, open the gss.sav data file. From the menus choose:

Analyze
Compare Means ▸
One-Way ANOVA...

Select the variables tvhours and degree.

		Sum of Squares	df	Mean Square	F	Sig.
TVHOURS Hours per day watching TV	Between Groups	400.646	4	100.162	16.242	.000
	Within Groups	5525.314	896	6.167		
	Total	5925.960	900			

To identify where the statistically significant differences are, look at Figure 15.7.

In the One-Way ANOVA dialog box, click Post Hoc. Then select Bonferroni, as shown in Figure 15.11.

Figure 15.7 Bonferroni multiple comparisons for television viewing

Dependent Variable: TVHOURS Hours per day watching TV

Bonferroni

(I) DEGREE Respondent's highest degree	(J) DEGREE Respondent's highest degree	Mean Difference (I-J)	Std. Error	Sig.	95% Confidence Interval	
					Lower Bound	Upper Bound
0 Less than HS	1 High school	1.12*	.227	.000	.48	1.76
	2 Junior college	1.87*	.367	.000	.84	2.90
	3 Bachelor	1.80*	.284	.000	1.00	2.60
	4 Graduate	2.37*	.369	.000	1.33	3.41
1 High school	0 Less than HS	-1.12*	.227	.000	-1.76	-.48
	2 Junior college	.75	.331	.246	-.19	1.68
	3 Bachelor	.68*	.235	.040	.02	1.34
	4 Graduate	1.25*	.333	.002	.31	2.19
2 Junior college	0 Less than HS	-1.87*	.367	.000	-2.90	-.84
	1 High school	-.75	.331	.246	-1.68	.19
	3 Bachelor	-.07	.372	1.000	-1.11	.98
	4 Graduate	.50	.441	1.000	-.74	1.74
3 Bachelor	0 Less than HS	-1.80*	.284	.000	-2.60	-1.00
	1 High school	-.68*	.235	.040	-1.34	-.02
	2 Junior college	.07	.372	1.000	-.98	1.11
	4 Graduate	.57	.374	1.000	-.48	1.62
4 Graduate	0 Less than HS	-2.37*	.369	.000	-3.41	-1.33
	1 High school	-1.25*	.333	.002	-2.19	-.31
	2 Junior college	-.50	.441	1.000	-1.74	.74
	3 Bachelor	-.57	.374	1.000	-1.62	.48

*. The mean difference is significant at the .05 level.

You see that people with less than a high school diploma watch significantly more television than each of the other education groups. You also see that high school graduates watch more television than college graduates and more television than people with graduate degrees. People with junior college, college, and graduate degrees do not differ significantly from one another in average hours of television viewing. This means that you must control for differences in education when studying television viewing by Internet users and non-users.

One of the shortcomings of the analyses in the previous chapter was the failure to take into account the actual amount of Internet use. People were classified into one of two groups based only on whether they used the Internet. You couldn't tell whether people who used the Internet for short periods of time differed in television viewing from people who used the Internet for longer periods of time.

Figure 15.8 is an analysis-of-variance table for hours of television viewing when people are classified into groups based on the amount of time they spend on the Internet. You can reject the null hypothesis that the average hours of television viewing is the same for all groups.

Open the gss.sav file. In the One-Way ANOVA dialog box, select the variables tvhours and netcat. Make sure that the code of 1 for netcat is not set to Missing.

Figure 15.8 Analysis of variance for television hours by Internet use

		Sum of Squares	df	Mean Square	F	Sig.
TVHOURS Hours per day watching TV	Between Groups	271.713	4	67.928	10.719	.000
	Within Groups	5564.174	878	6.337		
	Total	5835.887	882			

Since p<0.0005, you reject the null hypothesis

Looking at Figure 15.9, you see that there are no significant differences in television viewing time for people who use the Internet. The only significant differences are between those who use the Internet and those who don't.

In the One-Way ANOVA dialog box, click Post Hoc. Then select Bonferroni, as shown in Figure 15.11.

Figure 15.9 Bonferroni multiple comparisons for Internet use

Dependent Variable: TVHOURS Hours per day watching TV

Bonferroni

(I) NETCAT Internet usage category	(J) NETCAT Internet usage category	Mean Difference (I-J)	Std. Error	Sig.	95% Confidence Interval	
					Lower Bound	Upper Bound
1 Not Internet user	2 One hour or less	1.22*	.335	.003	.28	2.16
	3 1+ to 4 hours	1.24*	.256	.000	.52	1.96
	4 4+ to 10 hours	.82*	.259	.016	.09	1.55
	5 More than 10	1.13*	.271	.000	.37	1.89
2 One hour or less	1 Not Internet user	-1.22*	.335	.003	-2.16	-.28
	3 1+ to 4 hours	.02	.389	1.000	-1.08	1.11
	4 4+ to 10 hours	-.40	.391	1.000	-1.50	.70
	5 More than 10	-.09	.399	1.000	-1.21	1.03
3 1+ to 4 hours	1 Not Internet user	-1.24*	.256	.000	-1.96	-.52
	2 One hour or less	-.02	.389	1.000	-1.11	1.08
	4 4+ to 10 hours	-.42	.325	1.000	-1.33	.50
	5 More than 10	-.11	.334	1.000	-1.05	.83
4 4+ to 10 hours	1 Not Internet user	-.82*	.259	.016	-1.55	-.09
	2 One hour or less	.40	.391	1.000	-.70	1.50
	3 1+ to 4 hours	.42	.325	1.000	-.50	1.33
	5 More than 10	.31	.337	1.000	-.64	1.26
5 More than 10	1 Not Internet user	-1.13*	.271	.000	-1.89	-.37
	2 One hour or less	.09	.399	1.000	-1.03	1.21
	3 1+ to 4 hours	.11	.334	1.000	-.83	1.05
	4 4+ to 10 hours	-.31	.337	1.000	-1.26	.64

* The mean difference is significant at the .05 level.

? *Is it reasonable to conclude that the average hours of television viewing is the same for all amounts of Internet use?* No. The amount of Internet use reported by the GSS respondents is very limited. Note how few people say that they spend more than 10 hours on the Internet per week. If you want to see whether there is a relationship between the amount of Internet use and the amount of television viewing, you need to study people who report a broader range of hours on the Internet. ■■■

Summary

How can you test the null hypothesis that several population means are equal?

- Analysis of variance is a statistical technique that is used to test hypotheses about two or more population means.
- To use analysis of variance, your groups should be random samples from normal populations with the same variance.
- The F ratio is the ratio of two estimates of the population variance: the between-groups and the within-groups mean squares.
- The analysis-of-variance F test does not pinpoint which means are significantly different from each other. That's why multiple comparison procedures, which protect you against calling too many differences significant, are used to identify groups that appear to be different from each other.

What's Next?

In this chapter, you learned how to test the null hypothesis that when one variable is used to classify the cases into groups, several population means are equal. That accounts for the name *one-way analysis of variance*. In Chapter 16, you will learn how to use analysis-of variance-techniques when cases are classified into groups based on more than one variable. You'll look at the relationship between average hours worked and highest degree for males and females.

How to Obtain a One-Way Analysis of Variance

The SPSS One-Way ANOVA procedure tests the null hypothesis that the population mean of a variable is the same in several groups of cases. (If there are only two groups, it is equivalent to the independent-samples t test.) In addition, the One-Way ANOVA procedure can display multiple comparison statistics to evaluate the differences between all possible pairs of group means.

▶ To open the One-Way ANOVA dialog box (see Figure 15.10), from the menus choose:

Analyze
 Compare Means ▶
  One-Way ANOVA...

Figure 15.10 One-Way ANOVA dialog box

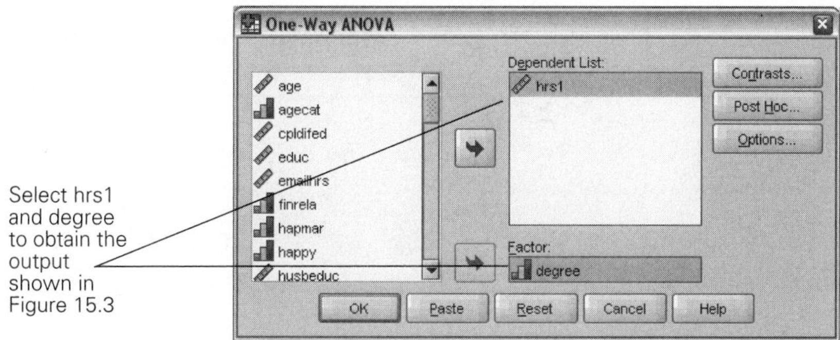

To obtain a one-way analysis of variance, you must indicate the variable(s) whose means you want to compare, and you must specify the groups to be compared. In the One-Way ANOVA dialog box:

▶ Select the variable whose mean you want to test and move it into the Dependent List. You can move more than one variable into the Dependent List to test all of them across the same set of groups.

▶ Select the variable whose values define the groups and move it into the Factor box.

▶ Click OK.

For each test variable, SPSS calculates an *F* statistic and its observed significance level.

Post Hoc Multiple Comparisons: Finding the Difference

If the overall *F* test indicates that subgroup means are significantly different, it is usually important to determine which categories of the factor variable are significantly different from which other categories. To do so, click Post Hoc. This opens the One-Way ANOVA Post Hoc Multiple Comparisons dialog box, as shown in Figure 15.11.

Figure 15.11 Post Hoc Multiple Comparisons dialog box

Select to obtain the Bonferroni test, as shown in Figure 15.5, Figure 15.7, and Figure 15.9

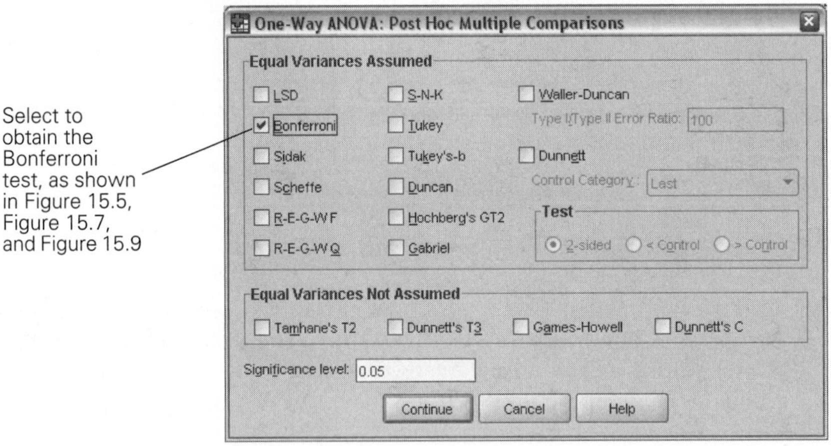

Tests. Many different tests for multiple comparisons are available. For information about a test, point to the test and click the right mouse button.

Options: Statistics and Missing Data

In the One-Way ANOVA dialog box, click Options. In the One-Way ANOVA Options dialog box (see Figure 15.12), you can request additional statistics as well as control the treatment of missing data.

Figure 15.12 One-Way ANOVA Options dialog box

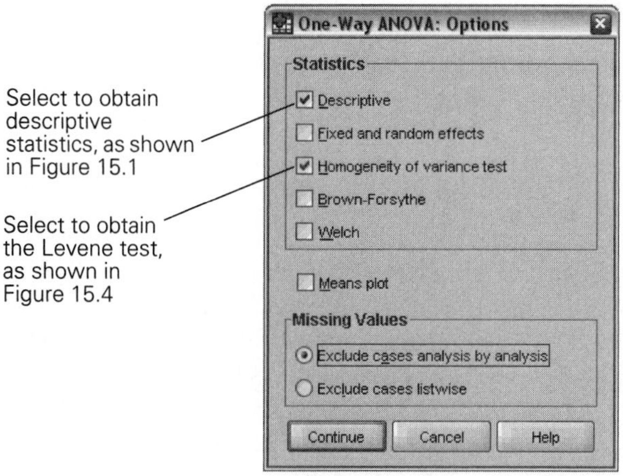

Select to obtain descriptive statistics, as shown in Figure 15.1

Select to obtain the Levene test, as shown in Figure 15.4

Additional statistics available are:

Descriptive. For each group, the number of cases, mean, standard deviation, standard error of the mean, minimum, maximum, and a 95% confidence interval for the population mean.

Homogeneity-of-variance. Calculates the Levene statistic, testing whether the variance of the dependent variable(s) is equal for all of the groups.

Means plot. A plot of group means.

Missing Values. The missing value alternatives control the treatment of missing data for analyses using multiple dependent variables:

Exclude cases analysis by analysis. All cases that have valid data for the grouping variable and a test variable are used in the statistics for that dependent variable.

Exclude cases listwise. Uses only the cases that have valid data for the grouping variable and all specified dependent variables. This ensures that all of the tests are performed using the same cases.

Exercises

Statistical Concepts

1. A market researcher wants to see whether people in four regions of a city buy the same brand of dish detergent. He takes a random sample of people in the different areas and asks them which of 10 brands (coded from 1 to 10) they purchase most often. He enters the data into SPSS and runs the One-Way ANOVA procedure. The observed significance level for his *F* value is 0.00001. What can he conclude based on these results?

2. You want to test the null hypothesis that, in the population, there is no difference in the average age of people who buy three models of a car. You run an analysis of variance and obtain the following output:

		Sum of Squares	df	Mean Square	F	Sig.
AGE	Between Groups	46.643	2	23.322	1.083	.339
	Within Groups	25060.546	1164	21.530		
	Total	25107.189	1166			

a. Is there sufficient evidence to reject the null hypothesis?

b. What conclusions can you draw from the table?

c. If you compute the average ages for the three groups, would you expect them to be similar or quite different?

3. Based on the following table, which groups are significantly different from each other using the Bonferroni test and a significance level of 0.05?

Dependent Variable: STRENGTH

Bonferroni

(I) CONTENT	(J) CONTENT	Mean Difference (I-J)	Std. Error	Sig.	95% Confidence Interval	
					Lower Bound	Upper Bound
5% hardwood	10% hardwood	-4.00	1.886	.280	-9.52	1.52
	15% hardwood	-5.33	1.886	.062	-10.85	.19
	20% hardwood	-9.50*	1.886	.000	-15.02	-3.98
10% hardwood	5% hardwood	4.00	1.886	.280	-1.52	9.52
	15% hardwood	-1.33	1.886	1.000	-6.85	4.19
	20% hardwood	-5.50	1.886	.051	-11.02	2.16E-02
15% hardwood	5% hardwood	5.33	1.886	.062	-.19	10.85
	10% hardwood	1.33	1.886	1.000	-4.19	6.85
	20% hardwood	-4.17	1.886	.234	-9.69	1.35
20% hardwood	5% hardwood	9.50*	1.886	.000	3.98	15.02
	10% hardwood	5.50	1.886	.051	-2.16E-02	11.02
	15% hardwood	4.17	1.886	.234	-1.35	9.69

*. The mean difference is significant at the .05 level.

4. You are interested in comparing four methods of teaching. You randomly assign 20 students to each of the four methods and then administer a standardized test at the end of the study.

a. What null hypothesis are you interested in testing?

b. What statistical procedure might you use to test the hypothesis?

c. What assumptions are necessary for the statistical procedure you have selected?

Data Analysis

Use the *gss.sav* data file to answer the following questions:

1. In the General Social Survey people classified themselves as being very happy, pretty happy, or not too happy (variable *happy*). Consider the relationship between happiness and age.

a. Compute basic descriptive statistics for each of the three happiness groups.

b. Make boxplots of age for the three groups.

c. Does the assumption of equal variances in the groups appear reasonable? The assumption of normality?

d. Perform a one-way analysis of variance on these data. What can you conclude? Which groups are significantly different from one another using the Bonferroni test?

e. From the analysis-of-variance table, estimate the variance of the ages within each happiness group. What is your estimate of the standard deviation within the groups? How does this compare to the actual standard deviations of each group in the table of descriptive statistics?

f. What are the three sample means that you have observed in the table of descriptive statistics? Based on the three sample means, what is your estimate of the variance of the ages within each happiness group?

g. What is the value of the ratio of the two variances?

h. If the null hypothesis is true, how often would you expect to see a ratio of sample variances at least this large?

2. Run a one-way analysis of variance to test the null hypothesis that the average years of education (variable *educ*) is the same for men and women. Run an independent-samples t test to test the same hypothesis. Compare the results. (If you square the equal-variance t value, you'll get the F value from the analysis of variance.)

3. You're interested in seeing whether there is a relationship between happiness and number of hours of television viewed a day (variables *happy* and *tvhours*). Perform the appropriate analyses and summarize your results. Include appropriate descriptive statistics.

4. For the General Social Survey data, formulate four hypotheses that can be tested using analysis-of-variance procedures. (Do not use hypotheses from the chapters or exercises.) Explore two of these hypotheses and write a short paper summarizing your results.

5. Test whether Dole, Clinton, and Perot supporters (variable *pres96*) differ in average age and education. Summarize your findings.

6. Consider respondents' average income (variable *rincdol*).

a. Use the Means procedure to calculate average income for different educational levels (variable *degree*). What is the average income for all respondents? For people with less than a high school degree? For people with a graduate degree?

b. Test the null hypothesis that the average income is the same for all degree groups. What assumptions do you have to make?

c. Using the Bonferroni multiple comparison procedure, determine which groups are significantly different from one another. Summarize your findings.

d. Does staying in school pay off?

7. Examine the relationship between the amount of time on the Internet per week (*netcat*) and age (*age*), education (*educ*), and respondent's income (*rincdol*). Write a short paper. Be sure to indicate which groups are significantly different from one another.

Use the *library.sav* data file to answer the following questions:

8. Consider the relationship between average age and the categories of library use (variable *libusefq*).

   a. From the Graphs menu, choose Interactive and then Histogram. Drag the variable *age* onto the horizontal axis. Drag *libusefq* into the Panel Variables box. Then click OK. This produces separate histograms of age for each of the library frequency categories.

   b. What assumptions do you need in order to use analysis of variance? Based on the histograms in the previous question, do you have reason to believe that your data don't come from populations that meet the assumptions?

   c. Assume that the necessary assumptions for analysis of variance are met. Can you reject the null hypothesis that, in the population, the average age for people in all library use categories is the same?

   d. Use the Bonferroni multiple comparison procedure to determine which groups are significantly different from one another.

   e. Write a report summarizing your findings.

9. The variable *libusfut* contains responses to the question *How often do you plan to use the library in the future?*. This question was asked only of people who said that they use the Internet. Determine the characteristics that predict whether a person plans to use the library in the future and how often. Prepare a report.

Use the *crimjust.sav* data file to answer the following questions:

10. Use the Count function on the Transform menu to create a new variable that is, for each person, the count of the number of times he or she chose the response "prison" (code 1) for the 13 crimes described in *crime1* to *crime13*. Call the new variable *cntprisn*.

   a. Make a frequency table and histogram of *cntprisn*. Is the distribution of the variable symmetric? What is the average number of times a respondent chose a prison sentence?

   b. Assume that the assumptions necessary for using a one-way analysis of variance are met. Test the null hypothesis that in the population, the average number of times prison is selected is equal for the five education categories (*educ*). Do you reject the null hypothesis?

c. Use the Bonferroni procedure to identify which groups are significantly different from one another. Summarize the results.

d. Why is the difference between high school graduates and college graduates (0.92) significant, while the difference between postgraduate and less than high school (–1.29) is not?

Use the *salary.sav* data file to answer the following questions:

11. Use Select Cases on the Data menu to restrict the analysis to clerical workers only (variable *jobcat* equals 1). Test the null hypothesis that the average current salary (variable *salnow*) is the same for the four gender/race groups (variable *sexrace*). Be sure to evaluate the assumptions needed for analysis of variance. If you reject the null hypothesis, indicate which groups are different from each other. Write a short paper summarizing your findings.

12. Repeat the analysis in question 11, but this time test the null hypothesis about the average years of education for clerical employees (variable *edlevel*) in the four gender/race groups. Summarize your findings.

16 Two-Way Analysis of Variance

How can you test hypotheses about population means when you have two factors?

- What is a factor?
- What kinds of hypotheses can be tested when you have two or more factors?
- What is an interaction?
- What assumptions are necessary for using an analysis of variance when you have two or more factors?
- What problems do you encounter if you have an unequal number of cases in the cells?

Despite the many advances made by women, the equitable distribution of household tasks remains problematic. Women, especially those with small children, work a much longer week than their husbands when unpaid work is included. One report claims that women work an average of 76 hours per week, of which 33 hours are unpaid work at home (Walker and Woods). In this chapter, you'll examine the average hours worked by men and women at their full-time, paid jobs, not counting unpaid work at home.

In Chapter 15, you used one-way analysis of variance (ANOVA) to test the null hypothesis that the average hours worked per week is the same for people in five categories of education. You classified the cases into groups based on their values for a single variable—highest degree earned. To study hours of paid employment for men and women, you must classify cases into groups based on the values of two variables: degree and gender. In this chapter, you'll learn how to test hypotheses about the equality of population means when you classify your cases into groups based on two factors. The technique you'll use is an extension of one-way analysis of variance. Predictably, it's called **two-way analysis of variance.**

▶ This chapter uses the *gssft.sav* data file, which contains information only for people who work full time. For instructions on how to obtain the ANOVA output shown in the chapter, see "How to Obtain a GLM Univariate Analysis" on p. 354.

The Design

In Chapter 15, you found differences in the average hours worked for full-time employees with different educational backgrounds. In particular, you saw that people with graduate degrees fared poorly. On average, they worked a 48-hour week. The conclusions you reached about average hours worked were based on the responses of males and females combined. The effects of gender were ignored. Cases were classified into groups based only on the values of the degree variable. Based on the results of the one-way analysis of variance, you could not say anything about differences between men and women in hours worked per week.

If you want to see whether there are differences in average hours worked by full-time employees based both on degree and gender, you must classify cases into groups based on their values for both variables. Since there are five values for the degree variable and two categories of the gender variable, you will have 10 cells, one for each possible combination of degree and gender. Based on the observed means and standard deviations in these 10 cells, you can statistically test whether average hours worked is the same in the population for people with different educational backgrounds, whether it is the same for males and for females, and whether the relationship between average hours worked and degree is the same for males and females.

To obtain this output, from the menus choose:

Analyze
General Linear Model ▸
Univariate...

Select the variables hrs1, degree, and sex, as shown in Figure 16.13.

In the Options dialog box, select Descriptive statistics.

Figure 16.1 Hours worked by degree and gender

Dependent Variable: HRS1 Number of hours worked last week

DEGREE Highest degree	SEX Respondent's Sex	Mean	Std. Deviation	N
0 Less than HS	1 Male	45.48	8.356	66
	2 Female	44.36	12.370	45
	Total	45.03	10.138	111
1 High school	1 Male	47.39	11.656	419
	2 Female	42.31	8.908	389
	Total	44.95	10.723	808
2 Junior college	1 Male	49.45	12.789	64
	2 Female	42.09	9.231	67
	Total	45.69	11.669	131
3 Bachelor	1 Male	48.26	10.848	144
	2 Female	44.45	9.616	142
	Total	46.37	10.413	286
4 Graduate	1 Male	49.53	10.397	95
	2 Female	45.91	8.062	56
	Total	48.19	9.729	151
Total	1 Male	47.82	11.246	788
	2 Female	43.15	9.336	699
	Total	45.62	10.647	1487

Examining the Data

As always, your first step should be to just look at the data. Make stem-and-leaf plots, histograms, or bar charts. Calculate summary statistics for the groups. See if you can get an idea of what's going on in your data. Figure 16.1 shows descriptive statistics for the 10 combinations of gender and degree.

In the Define Clustered Bar Summaries for Groups of Cases legacy dialog box, select Other statistic, and then select hrs1. Select degree for Category Axis and sex for Define Clusters by.

Figure 16.2 Bar chart of hours worked by degree and gender

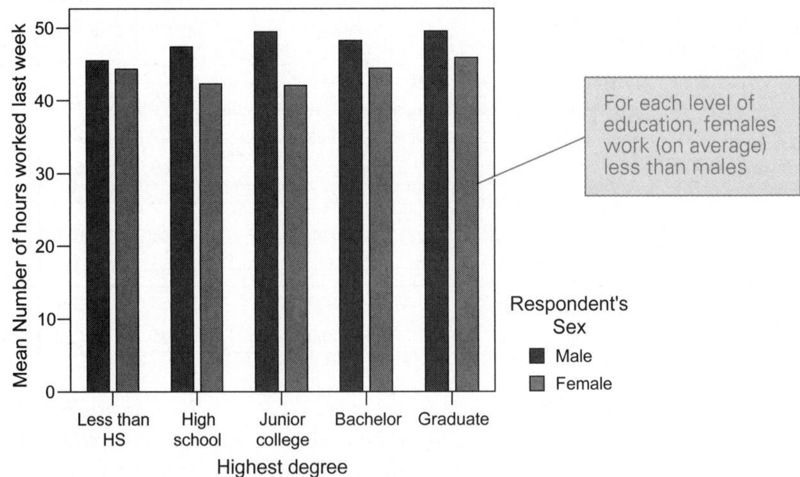

The means are plotted in Figure 16.2, which is a bar chart of the average number of hours worked. There are five clusters of bars, one for each degree category. Within each cluster, there are two bars, one for males and one for females. The height of the bar tells you the average hours worked for each group. The results are certainly interesting. For each level of education, on average, females in the sample work less than males. The largest difference between males and females is for those with junior college degrees. In the sample, men with junior college degrees work an average of seven hours more than women with junior college degrees.

? *Do you really believe that?* There are several nonstatistical explanations that come to mind. First of all, the interviewers simply asked people how long they worked the previous week. That means that men didn't have to *actually* work longer than women, they only had to say they did. It's also possible that women stay on the job for fewer hours than men, because they have so many other obligations to fulfill. ■■■

Figure 16.3 Box plot of hours worked by degree and gender

In the Define Clustered Boxplot Summaries for Groups of Cases legacy dialog box, select hrs1 for Variable, degree for Category Axis, and sex for Define Clusters by.

Activate the chart in the Chart Editor. From the Elements menu, choose Hide Data Labels.

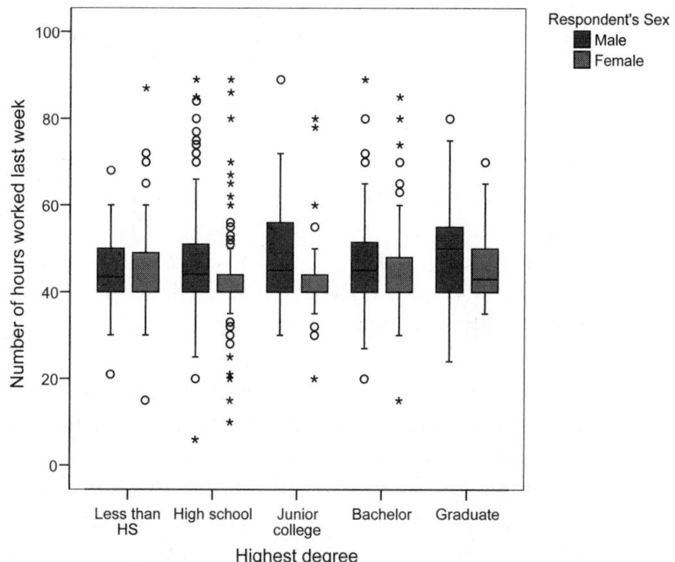

You can also examine differences between the 10 groups of cases by making a box-and-whiskers plot like the one shown in Figure 16.3. This plot conveys much more information than the bar chart. It tells you about the distribution of values in each of the groups. You see that for each of the five degree categories, the median hours worked is higher for men than for women. The length of the boxes for the men is somewhat larger than for the women. That tells you that the interquartile ranges for hours worked are larger for the men than for the women. There are some outlying and extreme points on the plot. Since the groups are reasonably large and all of the extreme values are believable, we won't pursue them further.

Testing Hypotheses

In this example, there are three different questions about the population that need to be answered:

1. Are the average hours worked the same for the five degree categories?
2. Are the average hours worked the same for men and women?
3. Is the relationship between average hours worked and degree the same for men and for women?

Notice that the first two questions both involve only one of the factor variables. The first question asks about degree, ignoring the possible effect of gender. The second asks about the gender effect, ignoring degree status. Both of these questions are hypotheses about what are known as main effects.

? *What's a main effect?* In analysis-of-variance terms, **main effects** are the effects of each of the individual factors, averaged over the other factors. For example, if you are studying the effects of marital status, gender, and degree on average hours worked using a three-way analysis of variance, the effect of each of the three variables, considered individually, is called a main effect. ■■■

The last question involves both of the factors simultaneously. It's called a test of the **interaction** between degree and gender. In statistics, the word interaction has a meaning similar to that in everyday language. When you talk about the interaction of a heavy lunch and a boring class on sleepiness, you usually mean that, considered together, the two factors have more (or less) of an effect than when they are considered individually. In statistics, when you talk about an interaction, you refer to an effect that is larger or smaller than would be predicted based on the main effects of the factors.

Interactions between Factors: An Example

To learn more about the statistical meaning of interaction, consider the following example. You are interested in studying the effect of four different methods of teaching statistics to three types of students: high school students, undergraduate students, and graduate students. You have 12 independent groups of students, one for each combination of teaching method and student type. For each student, you have a score on the standardized statistics final. Figure 16.4 shows a line plot of the average scores for the 12 groups of students.

Figure 16.4 No interaction between factors

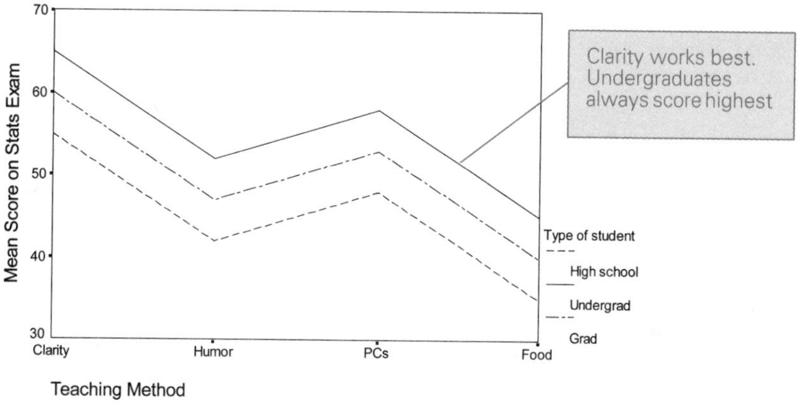

On the horizontal axis, you see the four teaching methods. Each of the lines represents one of the student groups. What can you tell from this plot? For all four teaching methods, undergraduate students score highest, and high school students score lowest. For all three types of students, the first method appears to result in the highest average score and the fourth method, in the lowest average score. Method 3 is a little more effective than method 2.

It's particularly important to notice that the four methods seem to work in the same way for the three student groups. That is, the shape of the three lines is similar for high school students, undergraduate students, and graduate students. The distance between any pair of lines is the same for the entire plot. This means that there is no interaction between teaching method and type of student. When there is no interaction, it makes sense to talk about the main effects—the teaching method main effect and the student type main effect. The first method is the best teaching method. Undergraduate students score highest on the standardized test. It's reasonable to reach each of these conclusions.

When there is no interaction, you can predict the average score for a student with the following equation:

$$\text{predicted score} = \text{mean score} + \text{teaching method effect} + \text{student type effect}$$

Equation 16.1

The equation tells you that the average predicted score depends only on teaching method and the type of student. That's because all four teaching methods work the same for the three types of students.

? *Where did that equation come from?* Assume that the average score for all participants is 50. Teaching method 1 raises the predicted score by 10 points. Teaching method 2 lowers it by 3, method 3 increases it by 3, and method 4 decreases the predicted score by 10 points. Similarly, being a high school student lowers the score by 5 points, being a graduate student leaves it unchanged, and being an undergraduate student adds 5 points to the score. Now, using the formula, you can compute the average scores for students in each of the 12 groups. For example, the predicted average score for undergraduates taught by method 1 is:

$$\text{predicted score} = \text{mean score} + \text{method 1 effect} + \text{undergraduate effect}$$

thus:

$$\text{predicted score} = 50 + 10 + 5 = 65$$

Similarly, the average score for high school students taught by method 2 is $50 - 3 - 5 = 42$. To compute the average score for a group, you just look at the main effects. That is, you just consider the teaching method and the student type. The effects of the teaching methods are the same for all types of students. ■■■

Figure 16.5 is a plot that shows an interaction between the teaching methods and the student types.

Figure 16.5 Average scores with interaction effect

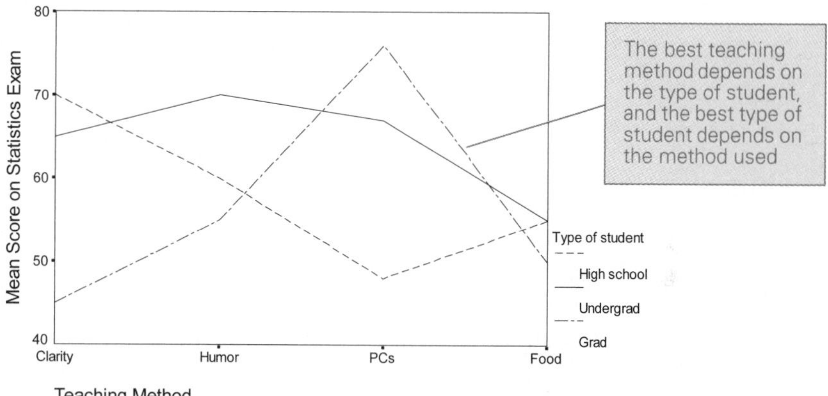

This plot is more difficult to interpret than the previous one. Notice that, unlike the lines in Figure 16.4, the lines for groups cross one another. The difference between any pair of lines is not the same for the four teaching methods. The effect of the teaching methods is not the same for all three student groups. Method 1 performs best for high school students, method 2 performs best for undergraduates, while method 3 is the best for graduate students. In this situation, it doesn't make sense to talk about a best teaching method or a best student type. The best method depends on the type of student, and the best type of student depends on the teaching method used. You can no longer predict the average score for a group from just the effect of the method and the effect of the student type. You also need a term for the combination of each method and student type. That is, you need a student-type by teaching-method interaction term.

Degree and Gender Interaction

Look at Figure 16.6, the line plot for the average hours worked by men and by women. Is there an interaction between gender and education? You see that the two lines don't cross. The shapes of the lines for male and female samples are not too dissimilar. That suggests that there is no interaction. You don't expect pairs of lines drawn from real data to have exactly the same shape, even if there is no interaction between gender and education in the population. Lines based on samples of data would show some interaction. (Remember, you didn't expect sample differences in means to be exactly 0 when the population means were equal.) Your goal is to determine whether the interaction observed in the sample is large enough for you to believe that it also exists in the population.

Figure 16.6 Line plot of observed means

To obtain this plot, in the Plots dialog box for GLM Univariate move degree to the Horizontal Axis box and move sex to the Separate Lines box. Then click Add.

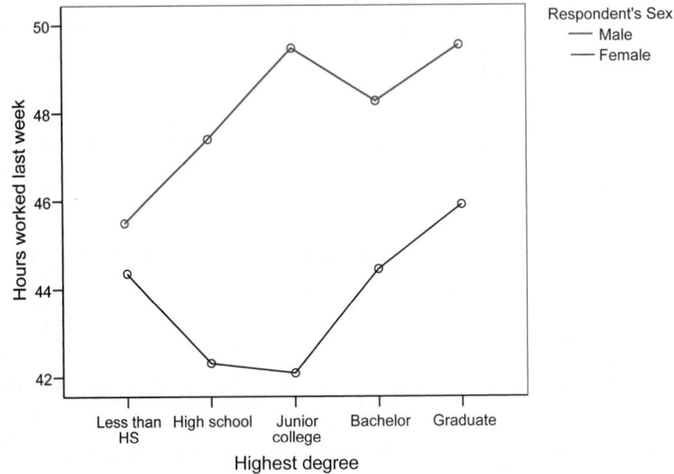

Necessary Assumptions

Now that you know what main effects and interactions are, you're ready to test hypotheses about them using two-way analysis of variance. But first a word about the assumptions you have to make. To use two-way analysis of variance, you'll need to make assumptions similar to those required for one-way analysis of variance. The data for each cell must be an independent random sample from a normal population with a constant variance. (**Constant variance** is another way of saying that the population variance is the same for all cells.) You can check these assumptions the same way you do for one-way analysis of variance. You can check the normality assumption by examining normal probability plots for each of the cells. You can look at the spread of your observations in a box-and-whiskers plot to see if the variability is markedly different in the groups. Later in this chapter, you'll see how to use the GLM Univariate Analysis procedure to check these assumptions.

Analysis-of-Variance Table

The two-way analysis-of-variance table for testing hypotheses about the population values for average hours worked is shown in Figure 16.7. It's very similar to the one-way analysis-of-variance table you saw in Chapter 15. What has changed is the number of hypotheses that you are testing. In one-way analysis of variance, you tested a single hypothesis. Now you can test three hypotheses: one about the main effect of degree, one about the main effect of gender, and one about the degree-by-gender interaction.

Figure 16.7 Two-way analysis-of-variance table

From the menus choose:
Analyze
General Linear Model ▸
Univariate...
Select the variables hrs1, degree, and sex, as shown in Figure 16.13.

Dependent Variable: HRS1 Number of hours worked last week

Source	Type III Sum of Squares	df	Mean Square	F	Sig.
Corrected Model	10069.317[1]	9	1118.813	10.434	.000
Intercept	1829883.119	1	1829883.119	17064.808	.000
DEGREE	1284.700	4	321.175	2.995	.018
SEX	3824.545	1	3824.545	35.666	.000
DEGREE * SEX	697.971	4	174.493	1.627	.165
Error	158380.765	1477	107.231		
Total	3263268.000	1487			
Corrected Total	168450.082	1486			

1. R Squared = .060 (Adjusted R Squared = .054)

You don't reject the null hypothesis of no interaction

In one-way analysis of variance, you will recall, you test whether the sample means vary too much to be from populations with the same mean. Basically, the same thing is happening here. The major difference is that you now have three sets of means to consider: the five degree means, the two gender means, and the 10 cell means.

1. First, you see how much the observations within the 10 cells vary. The *Error Mean Square* in Figure 16.7 tells you that. (In one-way analysis-of-variance output, the same thing is called the *Within Groups Mean Square.*) This estimate of the population variance doesn't depend on the null hypotheses being true.

2. Then, you see how much the sample means of the five degree groups vary. From the variability of these sample means, you get the mean square for degree. Similarly, from the variability of the male and female means, you get the mean square for sex. From the variability of all 10 cell means, you get the mean square for the interaction. Instead

of having just one between-groups mean square as you did in one-way analysis of variance, you now have three different between-groups mean squares: one for degree, one for gender, and one for the interaction. Remember that these estimates of the population variance depend on the appropriate null hypothesis being true. For example, the degree mean square will estimate the population variance only if the null hypothesis that all five degree means are equal is true.

3. The *F* ratios are computed by dividing each of the main effect and interaction mean squares by the error mean square. You look at the observed significance level for each observed *F* ratio to see if you can reject the corresponding null hypothesis.

Testing the Degree-by-Gender Interaction

The first test that you want to look at in Figure 16.7 is for the two-way interaction between degree and sex (the two factors). You test the interaction term first, since you've seen that it can be troublesome to talk about main effects if there is a significant interaction between the factors. The null hypothesis for the interaction term is that the effect of type of degree on average hours worked is the same for males and females in the population.

In Figure 16.7, the observed significance level for the no-interaction hypothesis is 0.16, so you don't reject the null hypothesis that there is no interaction between the two variables. The effect of the type of degree on hours worked seems to be similar for males and females. The absence of interaction tells you that it's reasonable to believe that the difference in average hours worked between males and females is the same for all degree categories. You haven't shown otherwise.

? *What would it mean if I were to find an interaction between degree and gender?* That would mean that the relationship between degree and hours worked was different for males and females. For example, it might be that males with graduate degrees work more than males with any other degrees, while the same is not true for females. If you find an interaction, then you can't comment on just the degree category or the gender category. Instead, you have to talk about males with graduate degrees or females with high school diplomas. You have to consider degree and gender together. ■■■

Since you didn't find an interaction between degree and gender, it makes sense to test hypotheses about the main effects of degree and gender. You can ask whether the population means for the five degree groups are the same and whether the population means for the two gender groups are the same.

? *Why did I have to use a two-way analysis of variance for this problem? I could have computed a one-way analysis of variance for the degree groups and a two-sample* t *test for the gender groups?* The two-way analysis of variance lets you test the interaction between the two variables. You couldn't do that otherwise. As you've seen, testing for interaction is very important. If you find an interaction, it doesn't make much sense to talk about gender or degree differences alone. They must be considered together. ■■■

Testing the Main Effects

Now let's see whether there is a degree main effect. That is, are the population means the same for the five degree groups. In Figure 16.7, you see that the *F* statistic for the degree main effect is 3.00. The observed significance level is 0.018, so you can reject the null hypothesis. That's not surprising. That's what you concluded in Chapter 15.

? *Why do I get different numbers for the degree sum of squares and mean square in the ANOVA table when I do a two-way analysis of variance and when I do a one-way analysis of variance?* If you don't have the same number of cases in each cell and have two or more factors, the computation of the ANOVA table can be quite complicated. That's because you can't neatly separate out the effects of each of the variables. You can get different values, depending on which of several methods you use. Stick with the default method in GLM Univariate Analysis. ■■■

To see whether you can reject the null hypothesis that the average hours worked is the same for males and females, look at the row labeled *SEX* in Figure 16.7. You see that the observed significance level for the sex main effect is very small, less than 0.0005. You can reject the null hypotheses that the average paid work week is the same for males and for females.

Removing the Interaction Effect

Since you didn't find an interaction effect, you can rerun the analysis, eliminating the sex-by-degree interaction term from the model. The results are shown in Figure 16.8.

In the Model dialog box for GLM Univariate, choose the Custom alternative. Move degree and sex into the Model list, as shown in Figure 16.14.

Figure 16.8 ANOVA table without the interaction term

Dependent Variable: HRS1 Number of hours worked last week

Source	Type III Sum of Squares	df	Mean Square	F	Sig.
Corrected Model	9371.346[1]	5	1874.269	17.449	.000
Intercept	1865205.198	1	1865205.198	17364.790	.000
DEGREE	1292.698	4	323.175	3.009	.017
SEX	7813.426	1	7813.426	72.742	.000
Error	159078.736	1481	107.413		
Total	3263268.000	1487			
Corrected Total	168450.082	1486			

1. R Squared = .056 (Adjusted R Squared = .052)

The interaction sum of squares is now included in the Error

The ANOVA table hasn't changed much. The biggest difference is that you no longer see a row for the degree-by-sex interaction. The interaction sum of squares is now part of the error sum of squares. Most of the other sums of squares have changed as well. That's because when you don't have equal numbers of cases for all combinations of values of the factors, it's impossible to uniquely isolate the independent contribution of the effects. The contribution of each effect depends on the other effects in the model. However, when you remove nonsignificant effects, like the interaction term in this example, your results will remain fairly similar. Both the degree and the sex main effects are still statistically significant. The reason for including nonsignificant effects in the error term is to increase the power of the tests by increasing the degrees of freedom for the error mean square.

? *What are those other rows in the ANOVA table?* Let's start from the bottom. The *Corrected Total* row tells you how much the observed values differ from the mean of all the cases combined. To obtain the corrected total sum of squares, subtract the overall mean from each case, square the difference, and sum up all of the differences. Remember, if you divide this sum by the number of cases minus 1, you have the variance of the observations. The corrected total sum of squares tells you how much variability there is in hours worked. Since the variability is measured about the mean, it is said to be "corrected" for the mean. The *Total* sum of squares is calculated in the same way, except that you don't subtract the mean before squaring the values. This is not a particularly informative number.

You already know that in the next row the mean square for the error term is an estimate of how much the observations vary within a group. This estimate doesn't depend on the null hypothesis being true. It's always correct if you've fit a model that contains all main effects and interactions. That's why it serves as the comparison variance for the other estimates of variance, which are dependent on a null hypothesis being true. All of the *F* ratios have the mean square for the error term in the denominator.

You should recognize the next two rows. They contain the variance estimates obtained from the degree groups and from the gender groups. In this example, the degree and gender mean squares are much larger than the error mean square, and the significance level is small leading you to reject the two null hypotheses that there is no degree effect and that there is no gender effect.

Ignore the row labeled *Intercept* and look instead at the row labeled *Corrected Model* in Figure 16.8. The corresponding sum of squares tells you the amount of variability in the dependent variable that can be "explained" or attributed to all of the effects that you've specified. For example, you see that the corrected total sum of squares is 168,450. Of this total, 9371 can be attributed to differences in education and gender. So 5.6% of the total variability in hours worked can be explained by education and gender. This proportion, labeled *R Squared*, is shown at the bottom of the table. It appears that education and gender, although significantly related to hours worked, leave much of the observed variability unexplained. The *F* test in the corrected model row tells you whether your effects, considered together, are statistically significant predictors of the dependent variable. You'll learn more about these concepts in Part 4 of this book. ■■■

Where Are the Differences?

Based on the analysis of variance table, you conclude that there are differences in the average hours worked between men and women and between the five education groups. You have not found an interaction between degree and degree. That is, the effect of education on hours worked doesn't seem to be different for men and women.

Since the gender factor has only two levels, when you reject the null hypothesis of no gender effect, you don't need to pinpoint where the differences are. Men claim to work more hours than women. When you reject the null hypothesis that there is no degree effect, you are still left with an unanswered question: Which of the education groups are different from one another? It's possible that all groups are different from one another, or that only some of the groups differ. In Chapter 15, you used the Bonferroni test to identify the groups that are statistically different from one another. You can use the same techniques when you have more than one factor. You do the analyses separately for each of the main effects.

? *Should I always look at all possible comparisons between groups?* Absolutely not. Whenever possible, you should restrict the number of groups you compare. Remember, the more groups you compare, the less likely you are to detect true differences. ■■■

Multiple Comparison Results

Figure 16.9 contains multiple comparisons based on the Bonferroni method. Notice that there aren't any comparisons for the gender variable. That's because there's only one comparison that can be made, and its significance is based on the *F* statistic for gender. The results in this table don't differ much from those you obtained previously (see Figure 15.5) when gender was the only factor in a one-way analysis of variance. Those with graduate degrees worked significantly more hours than those who only graduated from high school. (Remember, mean differences between pairs of means that are significantly different from one another are designated with an asterisk.)

In the Post Hoc dialog box, move degree and sex into the Post Hoc Tests For list. Select the Bonferroni test.

Figure 16.9 Bonferroni multiple comparisons for full factorial design

Dependent Variable: HRS1 Number of hours worked last week

Bonferroni

(I) Highest degree	(J) Highest degree	Mean Difference (I-J)	Std. Error	Sig.	95% Confidence Interval Lower Bound	Upper Bound
0 Less than HS	1 High school	.08	1.048	1.000	-2.87	3.03
	2 Junior college	-.66	1.336	1.000	-4.42	3.10
	3 Bachelor	-1.34	1.158	1.000	-4.60	1.92
	4 Graduate	-3.16	1.295	.148	-6.80	.48
1 High school	0 Less than HS	-.08	1.048	1.000	-3.03	2.87
	2 Junior college	-.74	.975	1.000	-3.48	2.00
	3 Bachelor	-1.42	.712	.466	-3.42	.58
	4 Graduate	-3.24*	.918	.004	-5.82	-.66
2 Junior college	0 Less than HS	.66	1.336	1.000	-3.10	4.42
	1 High school	.74	.975	1.000	-2.00	3.48
	3 Bachelor	-.68	1.092	1.000	-3.75	2.39
	4 Graduate	-2.50	1.236	.435	-5.97	.98
3 Bachelor	0 Less than HS	1.34	1.158	1.000	-1.92	4.60
	1 High school	1.42	.712	.466	-.58	3.42
	2 Junior college	.68	1.092	1.000	-2.39	3.75
	4 Graduate	-1.82	1.042	.811	-4.75	1.11
4 Graduate	0 Less than HS	3.16	1.295	.148	-.48	6.80
	1 High school	3.24*	.918	.004	.66	5.82
	2 Junior college	2.50	1.236	.435	-.98	5.97
	3 Bachelor	1.82	1.042	.811	-1.11	4.75

Based on observed means.

*. The mean difference is significant at the .05 level.

Checking Assumptions

At the beginning of the chapter, you learned that several assumptions are required for testing hypotheses with analysis of variance. You have to have independent random samples from normal populations with the same variance.

You can check these assumptions by comparing the values you actually observed to those predicted by the ANOVA model. If you fit a model that contains all of the main effects and all of the interactions, the observed and predicted means for each cell are equal. If you omit some of the interaction effects, the observed and predicted cell means don't have to be the same.

To obtain the Levene test, select Homogeneity tests in the GLM Univariate Options dialog box.

Figure 16.10 is a line plot of the predicted means for the 10 gender-and-degree combinations, when you fit an analysis-of-variance model without the interaction effect. Since there is no interaction term, the predicted shape of the lines for males and females is the same. Compare this plot to Figure 16.6. You see that the predicted means aren't the same as the observed means, but they're quite close.

The predicted value is the same for all cases with the same values of the factor variables. For example, all men with graduate degrees are predicted to work 49.9 hours. (Don't worry about where this number came from. Just remember that it is the sum of the overall mean and the estimated gender and education main effects.)

Figure 16.10 Predicted means from the main effects model

In the Model dialog box for GLM Univariate, choose the Custom alternative. Move degree and sex into the Model list.

In the Plots dialog box for GLM Univariate, move degree to the Horizontal Axis box and move sex to the Separate Lines box.

Then click Add.

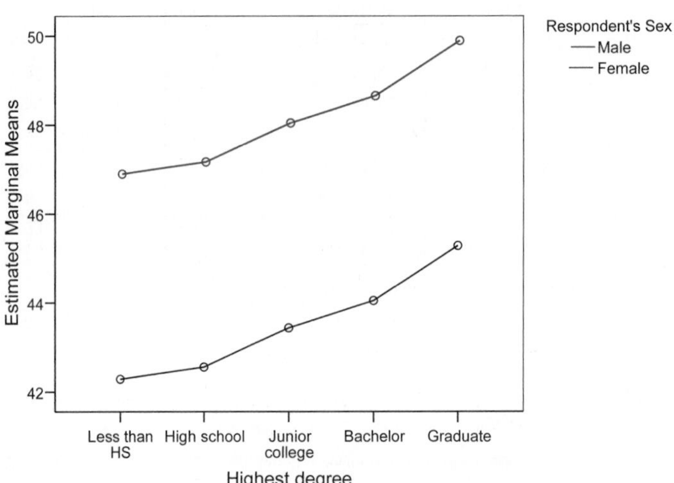

You can look at the difference between the value actually observed for each case and that predicted by the ANOVA model. This difference is called a **residual**. For example, for a man with a graduate degree who actually worked 51 hours, the residual is 1.1 hours. As you'll see in Chapter 22, residuals are an important part of analyses that fit models to data. The residual analyses shown in that chapter are also applicable to ANOVA models. If the assumptions are met, the residuals should be normally distributed and they should have a constant variance.

To check whether the residuals in all cells have the same variance, you can use the Levene test for equality of variance. This is an extension of the same test you use in the independent samples *t* test to check whether the two population variances are equal. If your observed significance level is small, you reject the null hypothesis that the variance is constant for all cells.

? *What should I do if the Levene test is significant?* If you have similar numbers of cases in each of the cells, you don't really have to worry about the equality of variance assumption. Your results shouldn't be affected too much. Analysis of variance is said to be **robust** against unequal variances if the sample sizes are equal. Robustness means that the results from the procedure don't change very much if a particular assumption is violated.

To obtain plots of variances and standard deviations against the mean, select Spread vs. level plot in the Options dialog box.

If your sample sizes are quite different, you can obtain **spread-versus-level** plots. These are plots of the dependent variable's variances and standard deviations against the cell means. You often find that as the mean increases so does the standard deviation or the variance. For example, if you are studying children's heights in different grades, you will find more variability in the higher grades than in the lower grades. If you see a linear relationship on your spread-versus-level plot between the *variance* and the mean (and all of your data values are positive), taking the square root of your data values might make the variances in the cells more comparable. If you see that there is a linear relationship between the *standard deviation* and the mean (and all of your data values are positive), you should consider taking the log of your data values. Once you've transformed your data, you must rerun the entire analysis and determine whether the assumptions are better satisfied. You can also use a multiple comparison procedure that does not assume equality of variances. ■■■

A Look at Television

You rejected the null hypothesis that men and women who hold full-time jobs report working the same number of hours. For each degree category, women on average report working fewer hours each day. Let's see if this purported shortfall in work hours results in more television viewing, instead of more hours spent on home responsibilities.

Figure 16.11 is a line chart of daily hours of television viewing for men and women. You see that for every degree category, except for graduate degree, on average men watch more television than women. (There are only 94 full-time workers with graduate degrees who were asked the television question.) From the analysis-of-variance table in Figure 16.12, you see that there is a significant difference in the average hours of television watched per day for men and women. Men spend more time watching television than do women. There is also a significant degree effect, as you would expect based on the results in Chapter 15. (You don't need any fancy multiple comparison procedure for the gender comparison since there are only two groups.)

In the GLM Univariate dialog box, specify tvhours as the dependent variable. Click Model and select Full factorial.

Click Plots to open the Plots dialog box (Figure 16.15). Select degree for Horizontal Axis, and select sex for Separate Lines.

Click Add, then Continue, and finally, OK.

Figure 16.11 Line plot of television hours

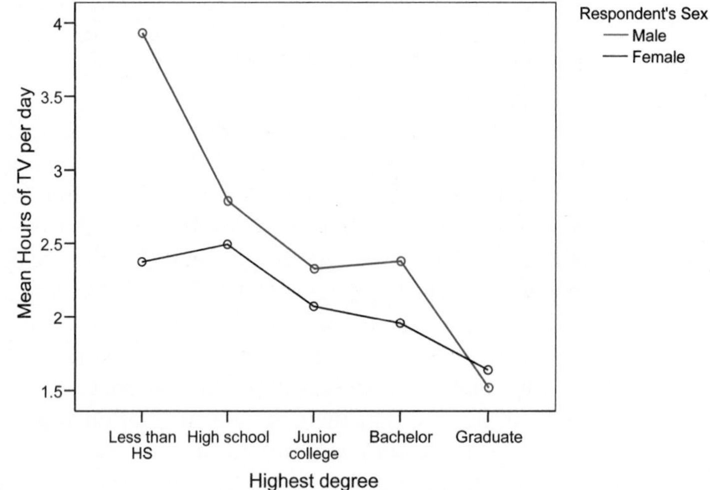

Figure 16.12 Analysis of variance for television hours

Dependent Variable: TVHOURS HOURS PER DAY WATCHING TV

Source	Type III Sum of Squares	df	Mean Square	F	Sig.
Corrected Model	229.972[1]	9	25.552	5.716	.000
Intercept	3149.101	1	3149.101	704.449	.000
SEX	32.886	1	32.886	7.357	.007
DEGREE	142.888	4	35.722	7.991	.000
SEX * DEGREE	32.351	4	8.088	1.809	.125
Error	4278.082	957	4.470		
Total	10351.000	967			
Corrected Total	4508.054	966			

1. R Squared = .051 (Adjusted R Squared = .042)

Extensions

In this example, you had two factor variables: gender and degree. Analysis of variance can be used to test hypotheses about any number of factors. For example, if your cases are classified by gender, degree, and marital status, you can test the equality of means for the two sexes, five degree categories, and five marital status categories. You can also test hypotheses about the interactions of pairs of factors (gender and degree, gender and marital status, degree and marital status) as well as the interaction of all three factors.

Summary

How can you test hypotheses about population means when you have two factors?

- You can use analysis of variance to test hypotheses about means when your cases are grouped on the basis of several variables or factors.
- With analysis of variance, you can test whether the population means are equal for all of the categories of a factor and whether there is an interaction between two factors.
- An interaction is present when the effect of one factor is not the same for all of the categories of the other factor.
- For analysis of variance, you must assume that you have independent samples from normal populations with the same variance.
- If you have different numbers of cases in the cells, several different methods can be used to calculate the analysis-of-variance table. There is no longer a unique solution.

What's Next?

You've learned to test a variety of hypotheses about population means. Now you'll turn your attention to testing hypotheses about data that are best summarized in a crosstabulation. You'll learn how to test whether the two variables that make up the rows and columns of a crosstabulation are independent.

How to Obtain a GLM Univariate Analysis

This section shows how to obtain a factorial analysis of variance using the GLM Univariate Analysis procedure, which compares means of a dependent variable for groups defined by the factor variables. This procedure can be used for analysis-of-variance models with one or more factor variables or covariates and a single dependent variable. The factor variables may be fixed or random; in this chapter, we've looked only at fixed factors.

The default model is the **complete factorial model**. This is a model that includes all main effects and all possible interaction terms. You can modify the model to omit some of these effects or to include factor-by-covariate interactions. You can use multiple-comparison procedures to identify which groups are significantly different from one another. Optionally, you can request descriptive statistics and plots of predicted means, as well as a wide range of diagnostic statistics and plots for evaluating the assumptions.

▶ To open the GLM Univariate dialog box (see Figure 16.13), from the menus choose:

Analyze
 General Linear Model ▶
  Univariate...

Figure 16.13 Univariate dialog box

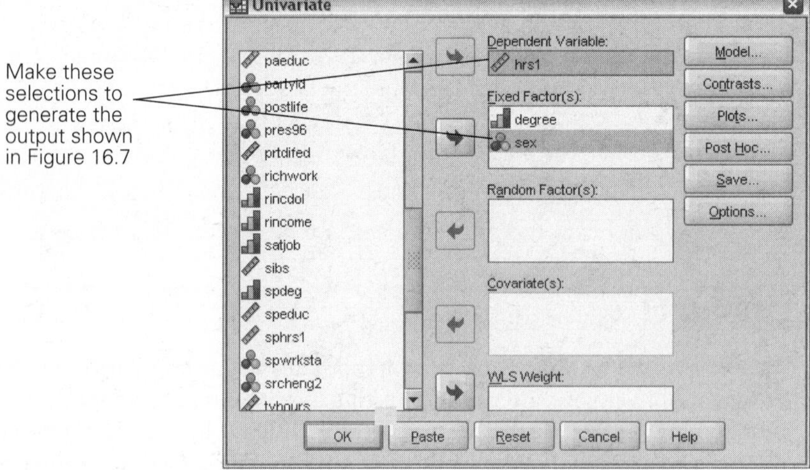

To obtain a factorial analysis of variance, you must specify the dependent variable and the factor variables and covariates. In the GLM Univariate dialog box:

- Select the variable whose mean you want to test and move it into the Dependent Variable box. You can select only one dependent variable.
- Select the variables whose values define the groups and move them into the Fixed Factor(s) list.
- To obtain the default full factorial model that contains all main effects and interactions, click OK.

GLM Univariate: Model

If you don't want to fit the default full factorial model, you must specify a custom model using the GLM Univariate Model dialog box, as shown in Figure 16.14.

Figure 16.14 Univariate Model dialog box

Univariate: Model
Specify Model
Full factorial
Custom
Factors & Covariates:
degree
sex
Model:
degree
sex
Build Term(s)
Type:
Interaction
Sum of squares: Type III
Include intercept in model
Continue
Cancel
Help

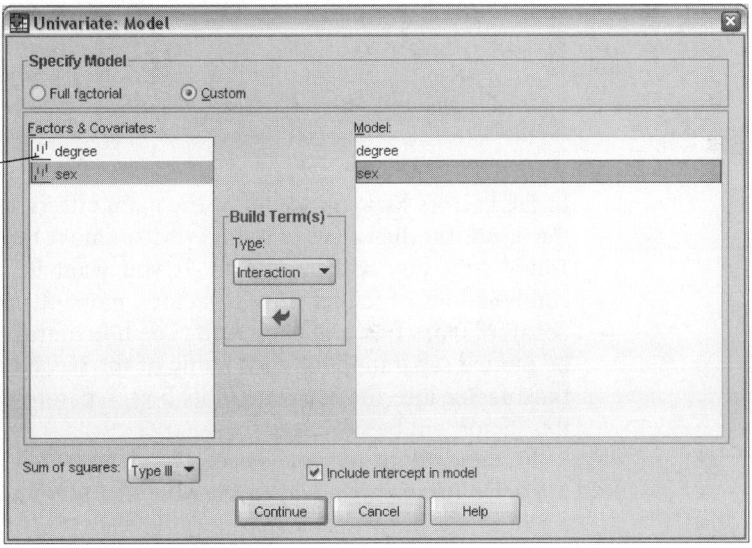

This produces the output shown in Figure 16.8

- For Specify Model, select Custom.
- For main effects, select the variables individually for the Model variable list.

▶ To specify interactions, select two or more variables together and click the Build Term(s) button. To include *all* interaction terms of a particular order for a set of factors, select the factors together and from the Build Term(s) drop-down list, choose the desired order number.

GLM Univariate: Plots

To examine plots of the predicted means from an ANOVA model, open the Plots dialog box, as shown in Figure 16.15.

Figure 16.15 Univariate Profile Plots dialog box

These selections will generate the plot shown in Figure 16.6 if a full factorial model is selected

In the Factors list, you see all of the main effects in your model. To plot the means for the values of a single factor, move that factor into the Horizontal Axis box and click Add. If you want to see the means for all combinations of values of two factors, move the second factor into the Separate Lines box and click Add. The horizontal axis factor means will be plotted separately for each value of the second factor. If you move a third factor into the Separate Plots box, separate plots for each value of this factor will be produced.

In these plots, if you select the default full factorial model, the predicted means are equal to the observed sample means, so you obtain a plot of the *observed* means.

GLM Univariate: Post Hoc

To pinpoint differences between all possible pairs of values of a factor variable, select the factors to be tested. Move them into the Post Hoc Tests For list. Additionally, select one of the multiple comparison procedures shown in Figure 16.16.

Figure 16.16 Univariate Post Hoc Multiple Comparisons for Observed Means dialog box

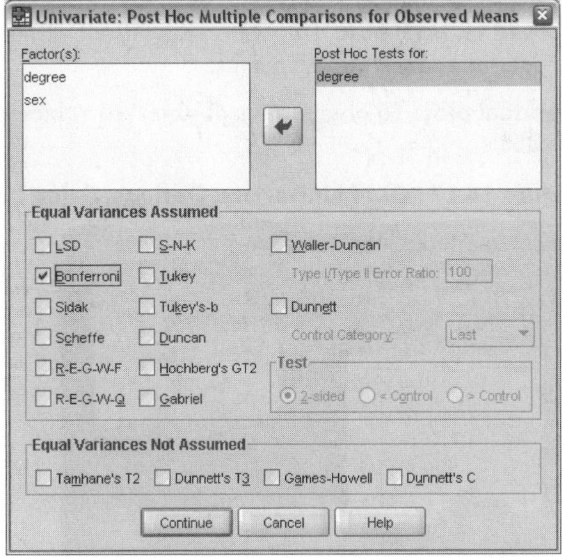

Many different tests are available. For information about a test, point to its name in the dialog box and click the right mouse button.

GLM Univariate: Options

In the GLM Univariate Options dialog box, as shown in Figure 16.17, you can select:

Descriptive statistics. To produce observed means, standard deviations, and counts for all cells.

Homogeneity tests. To calculate the Levene statistic for testing equality of variance for all cells.

Spread vs. level plot. To obtain plots of cell variances and cell standard deviations against the cell means.

Residual plot. To obtain plots of observed values, predicted values, and residuals.

Figure 16.17 GLM Univariate Options dialog box

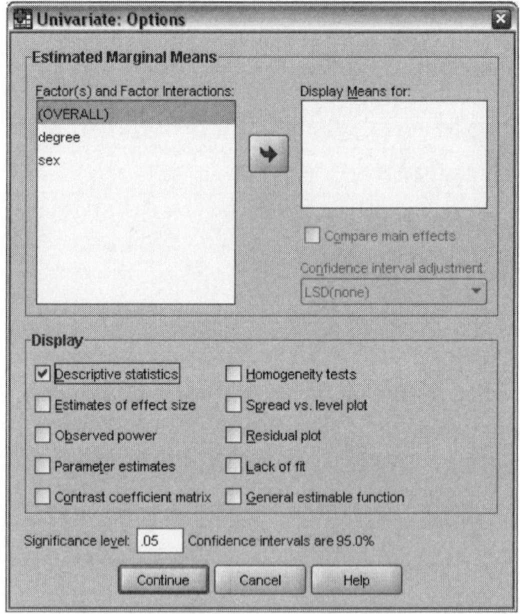

GLM Univariate: Save

In the GLM Univariate Save dialog box, as shown in Figure 16.18, you can save predicted values of the dependent variable, various residuals, and statistics that identify unusual points. For information about any of these, point to its name in the dialog box and click the right mouse button.

Figure 16.18 Univariate Save dialog box

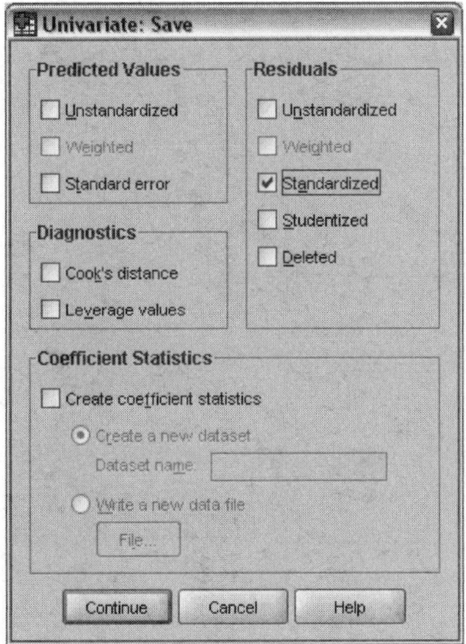

Exercises

Statistical Concepts

1. For each of the following situations, identify the statistical test that you would use to test the hypothesis of interest:

   a. You are interested in examining three temperatures and four combinations of ingredients to see if they all result in the same maximum height of a cake.

   b. You want to know if people in four regions of the country spend the same amount of money on fast food.

c. You want to know if workers on an assembly line are more productive when they are offered an incentive. You measure the productivity of the same workers before and after the incentive program.

d. You want to compare whether the average waiting time in the checkout line is the same for two chains of stores.

2. You are interested in observing changes in spending patterns for clothes during the year. For each person in the study, you obtain quarterly expenditures for clothing and then perform an analysis of variance with the respondent's gender as one factor and season of the year as the second factor. Do you see any problems with this analysis?

3. Based on the following plot of cell means of the average number of days of work missed during the year, do you think there is an interaction between gender and job classification?

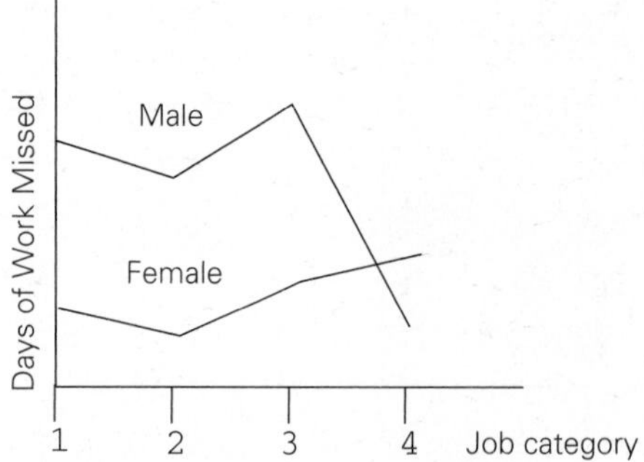

4. Complete the following two-way analysis-of-variance table:

Dependent Variable: HOURS PER DAY WATCHING TV

Source	Type III Sum of Squares	df	Mean Square	F	Sig.
Corrected Model	35.631[1]	2			.028
Intercept	6411.570	1			.000
ANOMIA5	13.179	1			.103
SEX	21.174	1			.039
Error	4718.143	951			
Total	12704.000	954			
Corrected Total	4753.774	953			

1. R Squared = .007 (Adjusted R Squared = .005)

5. You are studying the effects of store location and store hours on sales. You find that there is a significant interaction between location and hours. Is it reasonable to then test for the effect of location and the effect of store hours? Why or why not?

Data Analysis

Use the *gss.sav* data file to answer the following questions:

1. Use two-way analysis of variance to test the following hypotheses. Summarize your findings for each.

    a. Average years of education (variable *educ*) are the same for people employed full time and people employed part time (codes 1 and 2 for variable *wrkstat*).

    b. Average years of education are the same for males and females.

    c. The relationship between education and work status (variable *wrkstat*) is the same for males and females.

2. You are interested in studying the voting behavior of males and females. Use Select Cases to restrict your analysis to respondents with codes 1 (voted) and 2 (did not vote) for *vote96*. Determine if there is a relationship between age and voting, and if it is the same for males and females. Summarize your findings. Is there a significant gender main effect? What does it mean?

3. Repeat the analysis in question 2, but this time examine years of education instead of age.

4. Select two hypotheses of your choice that can be tested with two-way analysis of variance and test them. Write a paper summarizing your findings.

5. Consider the relationship between the number of hours a person works (*hrs1*), the happiness of their marriage (variable *hapmar*), and the person's gender. Use two-way analysis of variance to analyze the data. Indicate what hypothesis you are testing and what your conclusions are about each of them.

6. Money doesn't buy happiness we're told. Test this adage by looking at the relationship between total family income for last year (*incomdol*), gender (*sex*), and happiness of one's marriage (*hapmar*). What do you conclude? Is the relationship between total family income and marital happiness the same for men and women?

7. Consider the relationship between total family income in dollars (*incomdol*) and respondent's highest degree (*degree*) and gender (*sex*).

    a. Plot the average family income for each of the degree categories separately for males and females. Describe the plot.

    b. Test the null hypothesis that the relationship between degree and income is the same for males and females.

    c. Test the null hypothesis that average total family income is the same for all of the degree categories. What do you conclude?

d. Test the null hypothesis that average family income is the same for men and women. What do you conclude?

e. Repeat the analysis using *rincdol,* the respondent's income for the last year. Do your conclusions change?

8. Examine the relationship between the amount of Internet use (*netcat*), gender (*sex*), and education (*educ*), age (*age*), and respondent's income in dollars (*rincdol*). Write a paper summarizing your observations. Include appropriate charts and statistical analyses.

Use the *siqss.sav* data file to answer the following questions:

9. The variable *fsleep* is the number of hours of sleep people reported on the follow-up questionnaire. Examine the relationship between hours of sleep, gender, and Internet use that day (*intfolup*). Write a paragraph summarizing your findings. Include appropriate charts and descriptive statistics, as well as conclusions from a two-way analysis of variance.

10. Consider the relationship between hours of sleep (*fsleep*), educational category (*edcatr*), and gender (*sex*). Test the null hypothesis that on average men and women sleep the same amount of time, that people in the three education categories sleep the same average number of hours, and that there is no interaction between the two variables. What can you conclude?

Use the *crimjust.sav* data file to answer the following questions:

11. Consider the variable *cntprisn*, which you created in the exercises for one-way analysis of variance. (It is the count of the number of times a respondent selected the response *prison* for the 13 crimes described in *crime1* to *crime13*.)

a. Make a plot of the average times prison was chosen for each combination of education (*educ*) and gender. Describe the plot.

b. Using two-way analysis of variance, test the null hypothesis that in the population the average number of times *prison* is selected for the 13 crimes is the same for the five education categories, the two genders, and the education and gender combinations. What do you conclude?

c. Use the Bonferroni test to identify which education groups are significantly different from one another. What do you conclude?

d. Repeat the previous analysis for gender (*sex*) and age (*age*). Summarize your conclusions.

12. Examine the relationship between ratings of the criminal justice system (*cjsrate*) and the average count of prison sentences (*cntprisn*). Summarize your conclusions.

13. Examine the relationship between *cntprisn,* gender (*sex*), and ratings of the criminal justice system (*cjsrate*). Write a summary.

Use the *salary.sav* data file to answer the following questions:

14. You are interested in seeing whether current salaries (variable *salnow*) for clerical employees (variable *jobcat* equals 1) are related to gender and minority status (variables *sex* and *minority*).

   a. Run the appropriate analyses and interpret your results. Be sure to evaluate how well the data meet the required assumptions.

   b. Is there evidence of an interaction between gender and minority status? If there is, what does it mean?

15. Using the *salary.sav* data, select a hypothesis that can be tested using two-way analysis of variance and test it. Write a short explanation of your results.

17 Comparing Observed and Expected Counts

How can you test the null hypothesis that two variables are independent?

- What are observed and expected counts?
- How do you compute the chi-square statistic?
- What assumptions are needed for the chi-square test of independence?
- What is a one-sample chi-square test?
- Why is sample size important?

Manners maketh man (William of Wykeham, fourteenth century; lyricized by Sting, 1988). If you're a student at New College of Oxford University, you recognize the inscription from the grillwork on one of your iron gates. But it's unlikely, unless you've been in school a *really* long time, that a class in manners is still required for your graduation. Perhaps it should be, based on the results of an ABC News Manners Express poll[1]. In this chapter, you'll examine the data from that survey. You'll determine whether people think manners are important and how important they are. You'll also find out if college graduates report holding doors open for others more frequently than nongraduates.

You'll be dealing with tables of counts instead of means, so you can't use the hypothesis testing procedures for means that you've used in the previous chapters. Instead, you'll use the **chi-square test** to test hypotheses about data that are counts.

▶ This chapter uses the *manners.sav* data file. The chi-square test output shown can be obtained by using the SPSS Crosstabs procedure. (For more information about Crosstabs, see Chapter 8.)

1. ABC News. 1999. *ABC News Manners Express Poll*, May 1999 (computer file). ICPSR version. Horsham, Pa.: Chilton Research Services (producer). Ann Arbor, Mich.: Inter-university Consortium for Political and Social Research (distributor).

Freedom or Manners?

The ABC News survey asked many questions, such as whether people you encounter are polite, whether manners are worse now than they were 20 years ago, and whether you swear in public, open doors for others, or engage in other behaviors both polite and not. One of the more unusual questions was, "Which is more important to society: allowing people freedom of expression even if it means tolerating bad manners or enforcing good manners even if it means limiting freedom of expression?" Figure 17.1 is a crosstabulation of answers to this question by gender.

You can obtain a crosstabulation using the Crosstabs procedure, as discussed in Chapter 8.

In the Crosstabs dialog box, select sex and freedman.

Click Cells and select Row percentages.

Figure 17.1 Crosstabulation of freedom and manners by gender

			FREEDMAN Which is more important to society:		
			Allowing people freedom of expression/tolerate bad manners	Enforcing good manners/limit freedom of expression	Total
SEX Sex of respondent	Male	Count	268	195	463
		Row %	57.9%	42.1%	100.0%
	Female	Count	232	240	472
		Row %	49.2%	50.8%	100.0%
Total		Count	500	435	935
		Row %	53.5%	46.5%	100.0%

51% of women said that good manners should be enforced over freedom of expression

You see that 58% of men spoke out in favor of freedom of expression compared to 49% of women. Based on these results, do you think that in the United States population, men and women have different attitudes about the relative importance of freedom of expression and manners? Certainly in this sample, men supported freedom of expression more often than women. But, as usual, the sample results are not what you're interested in. You want to know what you can conclude about the population based on the observed sample results. You want to know whether you have enough evidence to reject the null hypothesis that, in the population, the same percentage of men and women think that freedom of expression is more important than good manners.

Observed and Expected Counts

The basic element of a crosstabulation is the count of the number of cases in each cell of the table. The statistical procedure that you'll use to test the null hypothesis is based on comparing the observed count in each of the cells to the expected count. The expected count is simply the number of cases that you would expect to find in a cell if the null hypothesis is true. Here's how the expected counts are calculated.

Calculating Expected Counts

If the null hypothesis is true, you expect men and women to answer the question in the same way. That is, you expect the *percentage* giving each of the responses to be the same. You don't expect the same *number* of men and women to agree with the statement because you don't necessarily have the same number of men and women in your sample.

In the column marginals in Figure 17.1, you see that in the sample, 53.5% of all respondents thought that freedom of expression was more important and 46.5% thought that manners were more important. If the null hypothesis is true, these are the best estimates for the *percentages* that you would expect for both men and women. To convert the percentages to the actual number of cases in each of the cells, multiply the expected percentages by the numbers of men and women. For example, the expected number of women who think that freedom of expression is more important is

$$53.48\% \times 472 = 252.4 \qquad \textbf{Equation 17.1}$$

Similarly, the expected number of women who think that good manners are more important is

$$46.52\% \times 472 = 219.6 \qquad \textbf{Equation 17.2}$$

For men, the expected values are calculated in the same way, substituting the number of men (463) for the number of women (472) in the above two equations.

? *Is there a simple way that I can remember how to calculate expected values?* Sure. The following rule is equivalent to what you've just done: To calculate the expected number of cases in any cell of a crosstabulation, multiply the number of cases in the cell's row by the number of cases in the cell's column and divide by the total number of cases in the table. Try it. You'll see that it always works. ■■■

To obtain observed and expected counts, click Cells in the Crosstabs dialog box.

See Chapter 8 for more information.

Figure 17.2 Crosstabulation with expected values and residuals

			FREEDMAN Which is more important to society:		
			Allowing people freedom of expression/tolerate bad manners	Enforcing good manners/limit freedom of expression	Total
SEX Sex of respondent	Male	Count	268	195	463
		Expected Count	247.6	215.4	463.0
		Residual	20.4	-20.4	
	Female	Count	232	240	472
		Expected Count	252.4	219.6	472.0
		Residual	-20.4	20.4	
Total		Count	500	435	935
		Expected Count	500.0	435.0	935.0

Residual is the difference between observed and expected counts

You see in Figure 17.2 the observed and expected counts for all four cells. The last entry in each cell is the **residual**, the difference between the observed and expected counts. A positive residual means that you observed more cases in a cell than you would expect if the null hypothesis were true. A negative residual indicates that you observed fewer cases than you would expect if the null hypothesis were true.

The sum of the expected counts for any row or column is the same as the observed count for that row or column. For example, the expected counts for women add up to the observed number of women. Similarly, the expected counts for the number favoring manners add up to the observed number of cases favoring manners. Another way of saying this is that the residuals add up to 0 across any row and down any column.

The Chi-Square Statistic

When you test the null hypothesis that two population means are equal, you compute the t statistic, and then, using the t distribution, calculate how unusual the observed value is if the null hypothesis is true. To test hypotheses about data that are counts, you compute what's called a chi-square statistic and compare its value to the chi-square distribution to see how unlikely the observed value is if the null hypothesis is true.

? *What assumptions are needed to use the chi-square test?* All of your observations must be independent. This implies that an individual can appear only once in a table. You can't let a person choose two favorite car colors and then make a table of color preference by gender. (Each person would appear twice in such a table.) It also means that the categories of a variable can't overlap. (For example, you can't use the age groups less than 30, 25–40, and 35–90.) Also, most of the expected counts must be greater than 5 and none less than 1. ■■■

To compute the Pearson chi-square statistic, do the following:

1. For each cell, calculate the expected count by multiplying the number of cases in the cell's row by the number of cases in the cell's column and dividing the result by the total count.
2. For each cell, find the difference between the observed and expected counts.
3. For each cell, square the difference.
4. For each cell, divide the squared difference by the expected count for the cell.
5. Add up the results of the previous step for all of the cells.

In the current example, the value for the Pearson chi-square statistic is

$$\frac{(268-247.6)^2}{247.6} + \frac{(195-215.4)^2}{215.4} + \frac{(232-252.4)^2}{252.4} + \frac{(240-219.6)^2}{219.6} = 7.16$$

Equation 17.3

If the null hypothesis is true, you expect the observed and expected values to be similar. Of course, even if the null hypothesis is true, the observed and expected values won't be identical, since the results you observe in a sample vary somewhat around the true population value. As before, you have to determine how often to expect a chi-square value at least as large as the one you've calculated if the null hypothesis is true.

To determine whether a chi-square value of 7.16 is unusual, you compare it to the chi-square distribution. Like the *t* distribution, the chi-square distribution depends on a parameter called the degrees of freedom. The degrees of freedom for the chi-square statistic depend not on the number of cases in your sample, as they did for the *t* statistic, but on the number of rows and columns in your crosstabulation. The degrees of freedom for the chi-square statistic are

$$(\text{number of rows in the table} - 1) \times (\text{number of columns in the table} - 1)$$

Equation 17.4

For this example, there is one degree of freedom, since there are two rows and two columns.

? *What's the logic behind the calculation of the degrees of freedom?* For any row or column of a crosstabulation, the residuals sum to 0. This means that you can tell what the expected values must be for the last row and last column of a table without doing any calculations other than summing the expected values in the preceding rows or columns. The number of cells for which you have to calculate expected values is equal to the number of cells when you remove the last row and the last column from your table. The number of cells in a table when one row and one column are removed is the number of rows minus 1 multiplied by the number of columns minus 1, which is the formula for the degrees of freedom.

To obtain the Pearson chi-square test along with a crosstabulation, click Statistics in the Crosstabs dialog box.

See Chapter 8 for more information.

Figure 17.3 Pearson chi-square test

	Value	df	Asymp. Sig. (2-sided)	Exact Sig. (2-sided)	Exact Sig. (1-sided)
Pearson Chi-Square	7.161[2]	1	.007		
Continuity Correction[1]	6.815	1	.009		
Likelihood Ratio	7.171	1	.007		
Fisher's Exact Test				.009	.005
Linear-by-Linear Association	7.154	1	.007		
N of Valid Cases	935				

1. Computed only for a 2x2 table

2. 0 cells (.0%) have expected count less than 5. The minimum expected count is 215.41.

Pearson chi-square (from Equation 17.3)

In Figure 17.3, you see that the observed significance level for the Pearson chi-square value of 7.16 with 1 degree of freedom is 0.007 This means that if the null hypothesis is true, you expect to see a chi-square value with 1 degree of freedom at least as large as 7.16 about 7 times out of 1000. Since the observed significance level is small, you can reject the null hypothesis that men and women answer the question the same way. It appears that women prefer good manners to freedom of expression more often than men.

? *What's all that other stuff in Figure 17.3 along with the Pearson chi-square?* The continuity-corrected chi-square is a modification of the Pearson chi-square for 2×2 tables. Most statisticians agree that the modification is unnecessary, so you can ignore it. The likelihood-ratio chi-square is a statistic very similar to the Pearson chi-square. For large sample sizes, the two statistics are close in value. The linear-by-linear association test is a measure of the linear association between the row and column variables. It's useful only if both the row and column variables are ordered from smallest to largest. Ignore it in other situations.

If you have a table with two rows and two columns, you'll also find something called Fisher's exact test in your output. The advantage of Fisher's exact test is that it is appropriate for 2×2 tables in which the expected value in one or more cells is small. In general, Fisher's exact test is less likely to find true differences. Statistically, a test like this is called **conservative.** ■■■

Given the importance of manners to women, let's see if women claim to behave more politely than men. Figure 17.4 is a crosstabulation of gender and whether one has held the door open for a stranger in "the last few months." Almost 95% of men and 92% of women said they held open a door. Based on the observed significance level for the chi-square statistic, you cannot reject the null hypothesis that men and women are equally likely to report that they open doors for strangers. The observed significance level is 0.069, which is greater than the customary 0.05. Can you conclude that in the population from which the sample was selected, men and women are equally likely to report that they open doors for strangers? No. You haven't shown that they are equally likely to open doors, since you can't prove that the null hypothesis is true. You can conclude only that you don't have enough evidence to reject the null hypothesis as being false.

In the Crosstabs dialog box, select sex as the row variable and opendoor as the column variable.

In the Cell Display dialog box, select Observed and Expected counts, Row percentages, and Unstandardized residuals.

In the Statistics dialog box, select Chi-square.

Figure 17.4 Crosstabulation of door-opening by gender

			OPENDOOR In the last few months have you: Held a door open for someone you didn't know?		
			Yes, something done in the last few months	No, not done in the last few months	Total
SEX Sex of respondent	Male	Count	467	25	492
		Expected Count	459.8	32.2	492.0
		Row %	94.9%	5.1%	100.0%
		Residual	7.2	-7.2	
	Female	Count	477	41	518
		Expected Count	484.2	33.8	518.0
		Row %	92.1%	7.9%	100.0%
		Residual	-7.2	7.2	
Total		Count	944	66	1010
		Expected Count	944.0	66.0	1010.0
		Row %	93.5%	6.5%	100.0%

Chi-Square Tests

	Value	df	Asymp. Sig. (2-sided)	Exact Sig. (2-sided)	Exact Sig. (1-sided)
Pearson Chi-Square	3.318[2]	1	.069		
Continuity Correction[1]	2.870	1	.090		
Likelihood Ratio	3.354	1	.067		
Fisher's Exact Test				.075	.045
Linear-by-Linear Association	3.314	1	.069		
N of Valid Cases	1010				

1. Computed only for a 2x2 table

2. 0 cells (.0%) have expected count less than 5. The minimum expected count is 32.15.

In the previous examples, you tested whether men and women responded in the same way to the question about the importance of manners over freedom of expression and whether they were equally likely to open doors for strangers. The null hypothesis that you were testing can be stated in several equivalent ways. You can say the null hypothesis is that the percentage preferring freedom of expression is the same for both genders or that the percentage opening doors for strangers is the same for men and women. Another way of stating the null hypothesis is that gender and response are independent.

Independence means that knowing the value of one of the variables for a case tells you nothing about the value of the other variable. For example, if marital status and happiness with life are independent, knowing a person's marital status gives you no information about how happy they are with life. Gender and the perceived importance of manners over freedom of expression don't seem to be independent. If you know that a person is a female, you know that she is less likely to favor freedom of expression over good manners than a male.

A Larger Table

The chi-square test can be used to test the hypothesis of independence for a table with any number of rows and columns. The idea is the same as for the two-row and two-column table. As an example, look at the effect of age on answers to the question about whether watching television makes children more respectful of others. Figure 17.5 is a crosstabulation of these two variables.

In the Crosstabs dialog box, select agecat as the row variable and tvrespct as the column variable.

In the Cell Display dialog box, select Observed and Expected counts, Row percentages, and Unstandardized residuals.

Figure 17.5 Crosstabulation of opinion of television by age

			TVRESPCT Thinking specifically about children, do you think most of the television programs they watch encourage them to be more respectful of others, less respectful, or do children's TV programs have no real effect on their respect for others?			
			More respectful of others	Less respectful of others	TV programs have no real effect	Total
AGECAT Age category	18-24	Count	28	35	32	95
		Expected Count	15.6	60.3	19.1	95.0
		Row %	29.5%	36.8%	33.7%	100.0%
		Residual	12.4	-25.3	12.9	
	25-34	Count	37	79	44	160
		Expected Count	26.2	101.6	32.2	160.0
		Row %	23.1%	49.4%	27.5%	100.0%
		Residual	10.8	-22.6	11.8	
	35-44	Count	40	147	35	222
		Expected Count	36.4	140.9	44.7	222.0
		Row %	18.0%	66.2%	15.8%	100.0%
		Residual	3.6	6.1	-9.7	
	45-54	Count	28	119	26	173
		Expected Count	28.4	109.8	34.8	173.0
		Row %	16.2%	68.8%	15.0%	100.0%
		Residual	-.4	9.2	-8.8	
	55-64	Count	6	91	17	114
		Expected Count	18.7	72.4	23.0	114.0
		Row %	5.3%	79.8%	14.9%	100.0%
		Residual	-12.7	18.6	-6.0	
	65 and over	Count	10	106	29	145
		Expected Count	23.8	92.0	29.2	145.0
		Row %	6.9%	73.1%	20.0%	100.0%
		Residual	-13.8	14.0	-.2	
Total		Count	149	577	183	909
		Expected Count	149.0	577.0	183.0	909.0
		Row %	16.4%	63.5%	20.1%	100.0%

From the row percentages, you see that almost 30% of respondents from 18–24 years of age think that watching television makes children more respectful of others. Not surprisingly, only 7% of those 65 and over share this opinion. It appears that as age increases, the percentage of people who believe that watching television makes children more respectful of others decreases.

To test the null hypothesis that age and perception of television's influence are independent, you compute a chi-square statistic for this table the same way you did for a 2×2 table. For example, if the null hypothesis is true, the expected number of people under 25 years old who find that watching television makes children more respectful is 15.6. (This can be calculated by multiplying the overall percentage of people who find that watching television increases respect, 16.4%, by 95, the number of people under 25.)

The Pearson chi-square value for the table is shown in Figure 17.6. You see that the observed significance level is less than 0.0005, which leads you to reject the null hypothesis that age and perception of television's influence are independent. By looking at the residuals in Figure 17.5, you see that people under 25 have large positive residuals for "more respectful of others." That means that the observed number of people in those cells is larger than that predicted by the independence hypothesis. By examining the residuals in a crosstabulation, you can tell where the departures from independence are.

Figure 17.6 Chi-square test for independence

In the Statistics dialog box, select Chi-square.

	Value	df	Asymp. Sig. (2-sided)
Pearson Chi-Square	73.737[1]	10	.000
Likelihood Ratio	77.132	10	.000
Linear-by-Linear Association	2.381	1	.123
N of Valid Cases	909		

1. 0 cells (.0%) have expected count less than 5. The minimum expected count is 15.57.

After the chi-square statistics are printed, SPSS tells you what the smallest expected count is in any cell of the table. In this example, the minimum expected frequency is 15.57. This is important because if too many of the expected values in a table are less than 5, the observed significance level based on the chi-square distribution may not be correct. As a general rule, you should not use the chi-square test if more than 20% of the cells have expected values of less than 5 or if the minimum expected frequency is less than 1.

? *What should I do if one of these conditions is not satisfied?* If your table has more than two rows and two columns, you can see if it makes sense to combine some of the rows or columns. For example, if you have few people in older age groups, you can combine them into a single category. You could create one age group of 55 and older instead of two separate groups. ■■■

Does College Open Doors?

Now let's consider the question of education and opening doors. Figure 17.7 is a crosstabulation of education and opening doors for strangers. Based on the observed significance level (p = 0.008), you can reject the null hypothesis that education and opening doors are independent. As educational level increases, so does reporting of opening doors for strangers. About 88% of people without a high school diploma report opening doors compared to about 96% of those with college degrees. Can you conclude that college makes you well-mannered? Of course not. College may make you more likely to report politeness rather than actually be polite. Or it may be that polite people go to college. Of course, it is possible that memories of carrying all those books and computers may make you more sympathetic to people in need of assistance. (In the exercises, you'll see whether other behaviors, such as rudeness, swearing, and obscene gestures at other drivers, are less likely for college graduates.)

In the Crosstabs dialog box, select degree as the row variable and opendoor as the column variable.

Figure 17.7 Chi-square test for education and opening doors

			OPENDOOR In the last few months have you: Held a door open for someone you didn't know?		
			Yes, something done in the last few months	No, not done in the last few months	Total
DEGREE Highest degree	Less than high school	Count	89	12	101
		Row %	88.1%	11.9%	100.0%
	High school	Count	277	27	304
		Row %	91.1%	8.9%	100.0%
	Some college	Count	249	12	261
		Row %	95.4%	4.6%	100.0%
	College degree	Count	312	14	326
		Row %	95.7%	4.3%	100.0%
Total		Count	927	65	992
		Row %	93.4%	6.6%	100.0%

Chi-Square Tests

	Value	df	Asymp. Sig. (2-sided)
Pearson Chi-Square	11.720[1]	3	.008
Likelihood Ratio	11.147	3	.011
Linear-by-Linear Association	10.352	1	.001
N of Valid Cases	992		

1. 0 cells (.0%) have expected count less than 5. The minimum expected count is 6.62.

A One-Sample Chi-Square Test

So far, you've used the chi-square test to test for independence in a crosstabulation of two variables. You can also use the chi-square test to test null hypotheses about the distribution of values of a single variable. That is, you can see whether the distribution of observed counts in a frequency table is compatible with a set of expected counts. The expected counts are specified by the null hypothesis that you want to test. For example, you can test the hypothesis that people are equally likely to think that manners today are better, the same, or worse than they were 20 years ago.

You can obtain this output using the Chi-Square Test procedure, as described in "Chi-Square Test" on p. 403 in Chapter 18.

In the Chi-Square Test dialog box, select the variable mancomp and All categories equal.

Figure 17.8 Chi-square test for manners comparison

	Observed N	Expected N	Residual
Better today	42	328.7	-286.7
Worse	741	328.7	412.3
About the same	203	328.7	-125.7
Total	986		

Expected counts if null hypothesis is true

Test Statistics

	MANCOMP Compared to 20 or 30 years ago, do you think most people's manners are better today, worse, or about the same?
Chi-Square[1]	815.381
df	2
Asymp. Sig.	.000

1. 0 cells (.0%) have expected frequencies less than 5. The minimum expected cell frequency is 328.7.

Figure 17.8 shows counts of the number of people who think manners are better, the same, or worse than they were 20 years ago. Before you looked at the data, you might have thought that people were equally likely to choose any one of the three possible answers. To test the null hypothesis that the three responses are equally likely in the population, you have to determine the expected counts for each of the categories. That's easy to do. For this hypothesis, the expected count for each category is just the total number of cases divided by 3.

You calculate the chi-square statistic the same way as before. Square each of the residuals (difference between observed and expected), divide by the expected count, and sum up for all of the cells. In Figure 17.8, you see that the chi-square value is a whopping 815.4. The degrees of freedom are 2, one less than the number of categories in the table. Based on the observed significance level, you can handily reject the null hypothesis.

Let's try another test, this time specifying unequal numbers of expected counts for the categories. You want to test the null hypothesis that 10% of people will admit to being rude to someone in public in the last few months. If 1009 people answered the question, you would expect 100.9 to admit to public rudeness and 908.1 not to.

Figure 17.9 Chi-square test for rudeness

	Observed N	Expected N	Residual
Yes, something done in the last few months	181	100.9	80.1
No, not done in the last few months	828	908.1	-80.1
Total	1009		

Test Statistics

	RUDE In the last few months have you: Gotten impatient with someone in public and spoken rudely to them?
Chi-Square[1]	70.653
df	1
Asymp. Sig.	.000

1. 0 cells (.0%) have expected frequencies less than 5. The minimum expected cell frequency is 100.9.

Observed results are highly unlikely if null hypothesis is true

The results of this test are shown in Figure 17.9. From the residuals, you see that more people admitted to being rude than expected if 10% of the population is rude. Again, the chi-square statistic is large and the observed significance level is small, so you reject the null hypothesis that in the population 10% of people admit to rudeness in the last several months.

Power Concerns

You know that when testing hypotheses about means, your ability to reject the null hypothesis when it's false—the **power** of a test—depends not only on the size of the discrepancy from the null hypothesis but also on the sample size. The same is true, of course, for chi-square tests. The value of the chi-square statistic depends on the number of observations in the sample as well as on the difference between the observed and expected values. For example, if you leave the table percentages unchanged but multiply the number of cases in each cell by 10, the chi-square value will be multiplied by 10 as well. This means that if you have small sample sizes, you may not be able to reject the null hypothesis even when it's false. Similarly, for large sample sizes, you may find yourself rejecting the null hypothesis even when the departures from independence are quite small and of limited practical importance.

When one or both of the variables in your crosstabulation is measured on an ordinal scale (for example, good/better/best), the chi-square test is not as powerful as some other statistics for detecting departures from independence. These other statistics make use of the additional information available for ordinal variables to measure both the strength and the direction of the relationship between two variables. If examination of the residuals in such a table leads you to suspect that there are departures from independence, you should use one of the measures described in Chapter 19.

Summary

How can you test the null hypothesis that two variables are independent?

- In a crosstabulation, the observed count is the number of cases in a particular cell.
- An expected count is the number of cases predicted if the two variables are independent.
- The chi-square statistic is based on a comparison of observed and expected counts.
- To use the chi-square test, your observations must be independent, and most of the expected values must be at least 5.
- A one-sample chi-square test is used to test whether a sample comes from a population with specified probabilities for the occurrence of each value.

What's Next?

Many of the statistical tests you've used so far to test hypotheses about population means require that the distribution of values in the population is normal or that the sample size is large enough to compensate for nonnormality. In Chapter 18, you'll learn about statistical tests that require fewer assumptions about the distribution of the data.

Statistical Concepts

1. Consider the following table:

			Age of Respondent				Total
			18 - 29	30 - 45	46 - 59	60 & Over	
Job or Housework	Very satisfied	Count	75	234	123	59	491
	Mod satisfied	Count	119	215	98	40	472
	A little dissatisfied	Count	33	58	29	7	127
	Very dissatisfied	Count	14	21	13	5	53
Total		Count	241	528	263	111	1143

   a. Calculate the number of cases you would expect in each cell if the two variables are independent.

   b. For each cell, calculate the difference between the observed and the expected number of cases.

   c. Calculate the chi-square statistic for the table.

   d. What are the degrees of freedom for the table?

   e. What null hypothesis are you testing with the chi-square statistic you computed?

2. The observed significance level for a chi-square value of 7.83753 with 3 degrees of freedom is 0.0495.

   a. What conclusion would you draw about the relationship between the two variables based on the observed significance level?

   b. How often would you expect to see a chi-square value at least as large as the one you observed if the two variables are independent?

   c. If you reject the null hypothesis that the two variables are independent, can you conclude that one of the variables causes the other?

3. You are studying the relationship between productivity and length of employment. Both variables are coded into three categories. You compute a chi-square test of independence and find the observed significance level to be 0.35. A personnel consultant does a similar study using the same criteria and categories. He calculates a chi-square test of independence and finds an observed significance level of 0.002.

You examine his results and notice that the percentage of cases in each cell of the table is almost identical to the percentages that you observed. You conclude that he doesn't know how to calculate a chi-square value. Give another explanation for why his chi-square could differ from yours.

4. Which pairs of variables do you think are independent, and which are dependent?

   a. Zodiac sign and number of hamburgers consumed per week

   b. Severity of a disease and prognosis

   c. Shoe size and glove size

   d. Color of car and highest degree earned

   e. Husband's highest degree and wife's highest degree

Data Analysis

Use the *gss.sav* data file to answer the following questions:

1. Examine the relationship between education (*degree*) and perception of life (*life*). Can you reject the null hypothesis that education and perception of life are independent? Make a bar chart that graphically summarizes your findings.

2. Can you reject the null hypothesis that amount of Internet use (*netcat*) and perception of life (*life*) are independent? Explain.

3. Test the null hypothesis that men and women are equally likely to believe that there is a life after this one (variable *postlife*). What can you conclude?

   a. Which variable in the table is the dependent variable?

   b. If the null hypothesis is true, what's your best guess for the percentage of people who believe in life after death? The percentage who don't believe in life after death?

   c. Calculate the expected values for the four cells. Compare them to those you get from SPSS. If you know the expected value for one cell, can you calculate the expected values for the other cells? How and why?

4. Test whether belief in life after death and highest degree earned (variable *degree*) are independent. What do you conclude?

5. See if the relationship between belief in life after death and highest degree earned is the same for males and for females. What can you conclude?

6. Test the null hypothesis that men and women were equally likely to vote for Perot, Dole, and Clinton (variable *pres96*). (You'll have to exclude people whose

response was *Other*. You can do this by making the code 4 [*Other*] a missing value or by selecting cases for which *pres96* does not equal 4.) Summarize your findings.

7. Use the Crosstabs procedure to look for other differences among supporters of the three presidential candidates.

8. The variable *satjob* measures how satisfied a person is with his or her work. Examine the variables in the *gss.sav* file that might be related to *satjob*. Write a short paper detailing your findings. Don't restrict your analyses to variables for which crosstabulation is appropriate. Use all of the techniques you've learned so far.

9. Look at the relationship between zodiac sign (*zodiac*) and a person's perception of life (variable *life*). What problems do you see with using the chi-square statistic for the *zodiac* by *life* crosstabulation? Fix the problem and rerun the table. What can you conclude?

10. Test the null hypothesis that all zodiac signs are equally likely. What do you conclude?

11. Since all zodiac signs don't have exactly the same number of days, you should compute expected counts using the actual number of days in a zodiac sign. From your favorite astrology column, figure out how many days there are in each sign and then test the null hypothesis that the observed and expected counts don't differ.

12. Test the null hypothesis that people are equally likely to find life dull, routine, and exciting (*life*).

Use the *crimjust.sav* data file to answer the following questions:

13. Many different factors, such as age (*age*), education (*educ*), and gender (*sex*), may be related to one's perception of the criminal justice system (*cjsrate*) and crime (*vlntcmp5*, *crmcomp5*). Write a report that outlines what you have uncovered based on your analysis of the Vermont data. Be sure to indicate which associations are not likely due to chance.

14. If you or someone in your household is a victim of crime, you are exposed to many facets of the criminal justice system. Examine the impact of a having a victim of crime in your household on ratings of judges (*judgrate*), police (*policrat*), prosecutors (*proscrat*), and juries (*juryrate*). Write a paper summarizing your findings. (Or revise the paper you have already written in a previous chapter to highlight those differences that are not likely to be due to chance.)

Use the *manners.sav* data file to answer the following question:

15. Revise the paper that you wrote describing the relationship between manners, their importance and practice, and gender, age, education, and income to highlight those differences that are not likely to be due to chance.

Use the *salary.sav* data file to answer the following questions:

16. Test the hypothesis that highest degree earned (variable *degree*) and gender/race group (variable *sexrace*) are independent. (Variable *degree* was computed in the exercises for Chapter 8.) Summarize your conclusions. Based on the residuals, which cells contain fewer than the expected number of cases and which cells contain more than the expected number of cases?

17. Test the null hypothesis that job classification (variable *jobcat*) and gender/race group are independent.

a. What problem do you see with the chi-square test for the *jobcat* by *sexrace* table?

b. What can you do to the table so the chi-square test is appropriate?

c. Based on the new table, what can you conclude about the null hypothesis?

Use the *renal.sav* data file to answer the following question:

18. In Chapter 14, you used the independent-samples t test to identify differences between cardiac surgery patients who developed renal failure (variable *type* equals 1) and those who did not (variable *type* equals 0). Now use the Crosstabs procedure to look for differences between the two groups. Identify the variables for which a chi-square test is appropriate. Summarize your findings.

18 Nonparametric Tests

What are nonparametric tests, and when are they useful?

- When do you use the sign test?
- What is the Wilcoxon signed-rank test, and what is its advantage over the sign test?
- What is a nonparametric alternative to the independent-samples *t* test?
- When do you use the Kruskal-Wallis test?
- How can you test hypotheses when each subject has multiple measurements?

We can compare our understanding of the effects of the Internet on society to the blind men in J.G. Saxe's poem, "who grasp at an elephant and debate whether it's a wall, a spear, a snake, a tree, a fan, or a rope." The conclusion, "Each was partly in the right and all were in the wrong," may describe both the blind men and current ideas regarding the effects of the Internet.

The New York Times headline posed "How Lonely Is the Life That Is Lived Online?" in presenting research conducted by Norman H. Nie of the Stanford Institute for the Quantitative Study of Society. Nie and his colleagues collected information on how 6,000 Americans spend their time each day. Their conclusion: the Internet isolates people from family, friends, and even business associates (Nie et al., 2003). In this chapter, you will analyze a subset of the data from the Nie study. You'll compare self-reports of time on the Internet to those obtained from a carefully administered time diary. (See Chapter 3 for more information on the Nie study.) You'll test if people claim to receive and send the same number of e-mail messages per day. You'll determine whether Internet users spend the same amount of time with their family and friends as do non-users.

The statistical tests that you will use in this chapter are called **nonparametric tests** because they make limited assumptions about the underlying distributions of the data. Most of the statistical procedures you have used so far require fairly detailed assumptions about the populations from which the samples are selected. For example, to use analysis of variance, you have to assume that each group is an independent random sample from a normal population and that the group variances are equal. You know that many of the procedures work reasonably well even when the assumptions are not completely met.

However, when you analyze data, especially from small samples, you'll encounter situations in which there are serious departures from the necessary assumptions. In such situations, you need procedures that require less stringent assumptions about the data. Collectively, these procedures are called distribution-free or nonparametric tests. They do not require that the data come from a particular distribution, although some do require assumptions about the shapes of the underlying distributions. The disadvantage of nonparametric tests is that they are less likely to find true differences when assumptions for parametric procedures are met. Also, the hypotheses tested by nonparametric tests are different sometimes. For example, you test hypotheses about medians instead of means.

▶ This chapter uses the ***siqss.sav*** data files. For information about how to obtain the nonparametric test output shown in this chapter, see "How to Obtain Nonparametric Tests" on p. 403.

Nonparametric Tests for Paired Data

Among the clear problems in Internet research are determining how much time a person spends on the Internet, how much of it is job related, and how much of it is personal. Attempts to relate Internet use to other measures are doomed unless reasonably accurate measurements of Internet usage time are obtained. Unfortunately, Internet time, like television viewing time, is not well defined. If you're talking on the phone to a friend while surfing the Web, are you engaged in isolating Internet time or are you spending time with friends?

In the Nie study, two values for time on the Internet are available: an estimate based on the time diary and an estimate from follow-up questions. Both values are for "yesterday." Before testing in more detail whether these two measures differ, let's consider a basic question: do the two measures agree on whether a person even used the Internet "yesterday"?

? *Did Nie and his colleagues analyze the data using nonparametric tests?* No, they used sophisticated statistical methods to analyze data from 6,000 respondents. They controlled for a large number of variables to make sure that there weren't differences other than Internet use between the two groups. In this chapter, we're looking at some very simple hypotheses that can be tested using a subset of the Nie data. The analyses here are deliberately simple. This is an introductory book! ■■■

Figure 18.1 is a crosstabulation of Internet use based on the time diary and on the follow-up interview. The results are troubling. You see that 314 people who claimed on the follow-up interview to have used the Internet were not identified by the time diary as having used the Internet. Additionally, three people who reported Internet use in their time diaries did not report it in the follow-up questionnaire.

? *How is this possible?* The time-diary information was obtained for six hours, one randomly selected from each of the following blocks: midnight to 5 A.M.; 6 A.M. to 9 A.M.; 10 A.M. to 1 P.M.; 2 P.M. to 5 P.M.; 6 P.M. to 8 P.M.; and 9 P.M. to 11 P.M. The time spent on a particular activity was assumed to be the same for all hours in the same time block. If that weren't the case, the estimates from the time diary and the follow-up questionnaire might be quite different. The time diary asked for detailed information for only six hours, since it would have been burdensome for people to provide the information for all 24 hours. If you find differences between the time diary and follow-up values, they may be due to either a faulty report of totals or problems with the assumption that activities are constant within a time period. ■■■

It's not very often that two sets of measurements agree perfectly. There will always be people who give contradictory information, even when asked exactly the same question twice. For people who have contradictory identifications by the time diary and the follow-up methods, it's important to determine whether disagreement in both directions are equally likely. That is, are "yesterday's" Internet users and non-users, as defined by either method, equally likely to have different classifications by the other method?

From the Analyze menu choose:

Nonparametric Tests ▶ 2 Related Samples...

Move intdiary and intfolup into the Test Pairs(s) list. Select McNemar.

Figure 18.1 Crosstabulation of Internet use by diary and follow-up

INTDIARY Internet use based on diary	INTFOLUP Internet use based on followup	
	No	Yes
No	361	314
Yes	3	96

Consider Figure 18.1 again. The cases on the main diagonal of the table—those for whom the two classifications are identical—don't provide you with any information about whether the two types of cases are equally likely to have different classifications. The other two cells do. You see that a total of 317 people were classified differently by the time diary and the follow-up questionnaire. If the number of (+, –) and (–, +) pairs (where the first entry is for the time diary and the second entry for the self-report) are equal in the population, you would expect 158.5 people to fall into each of the two cells. (That's simply the 317 people with different classifications, divided by 2.) You know that sample results from a population vary, so you don't expect to find exactly 158.5 people in each cell. Instead, you expect to find a count that is compatible with the null hypothesis.

You can use McNemar's test to test the null hypothesis that people are equally likely to fall into the two contradictory classification categories. McNemar's test is computed like the usual chi-square test, but only the two cells in which the classifications don't match are used. In Figure 18.2, you see that the chi-square value is 303 and the observed significance level is less than 0.0005. You reject the null hypothesis. It's apparent that the follow-up questionnaire identifies many more people as Internet users than does the diary method.

McNemar's test is often used in before/after designs. For example, a voter indicates a preference for one of two candidates. Then he or she watches a campaign speech or receives some other propaganda, after which the voter is again asked for a preference. What the campaign staff wants to know is whether the intervention is effective. If people who change their opinion are equally likely to switch to both candidates, the intervention is ineffective. If the switch is predominantly to the candidate footing the bill, the intervention is deemed effective.

Figure 18.2 McNemar's test

	INTDIARY Internet use based on diary & INTFOLUP Internet use based on followup
N	774
Chi-Square[1]	303.155
Asymp. Sig.	.000

1. Continuity Corrected

? *Why don't you use the usual chi-square test for a crosstabulation?* The chi-square test tests a different hypothesis from the one we're interested in here. It tests whether the two methods for determining Internet use "yesterday" are independent. We're interested only in people who are classified differently. We want to know whether they are evenly split between the two cells. ■■■

Sign Test

You know that the time-diary method failed to identify as Internet users quite a few people who said on a follow-up questionnaire that they had used the Internet "yesterday." The explanation may be that the time diary sampled only six hours of the day. Or it may be that people don't remember whether they used or did not use the Internet on the previous day. It's not clear whether the time diary or the questionnaire is closer to the truth.

Whether someone used the Internet at all on a particular day is part of the larger question of how much time was actually spent on the Internet. After all, if you're studying the impact of the Internet on social activities, there's probably a big difference between someone who spends 10 minutes a day checking e-mail and someone who surfs for six hours a day. Failure to detect small amounts of time spent online is probably far less important than failure to detect large amounts of time online.

Let's look now at the actual time spent online as determined by the two methods. Only people who are identified by both methods as having used the Internet yesterday will be included. The results will be easier to interpret when we eliminate people whose correct classification is uncertain. Figure 18.3 is a stem-and-leaf plot of the differences between the diary time and the follow-up time. For each case, you simply subtract the time (in minutes) reported on the follow-up questionnaire from the time estimate based on the time diary. Times seem to cluster about 0, but there are some large differences.

Select cases with values of 1 for both intfolup and intdiary. In the Explore dialog box, select diftime and Stem-and-leaf.

Figure 18.3 Stem-and-leaf plot of differences

```
Diary minus followup time on Internet Stem-and-Leaf Plot

 Frequency    Stem &  Leaf

     4.00 Extremes     (=<-300)
     4.00       -2 .  1146
    10.00       -1 .  0222224488
    21.00       -0 .  111222233334566666667
    36.00        0 .  000000000012222333344666666666666999
    16.00        1 .  0022222222346688
     5.00        2 .  14444

 Stem width:     100.00
 Each leaf:        1 case(s)
```

You can use the paired t test to test the null hypothesis that the average reported time online is the same for the diary and follow-up methods. Since the t test is based on means, however, you have to worry about the possible impact of the outliers on the results. You know that the arithmetic mean is quite sensitive to very large or small values that are far removed from the rest of the data points. The median is not. It may be more appropriate to use the sign test to test the null hypothesis that the median difference between the two members of a pair is 0. To use the sign test, you don't have to make any assumptions about the shapes of the distributions from which the data are obtained. The only requirement is that the different pairs of observations be selected independently and that the values can be ordered from smallest to largest. That's because the test is based on seeing which of a pair of values is larger.

Computing the sign test is easy. You count the number of cases in each of three categories—time diary and follow-up times are the same (these are called **tied cases**), diary times exceed follow-up times, and follow-up times exceed diary times. If the null hypothesis is true, you would expect to see similar numbers of cases in the last two categories. That is, the number of cases in which diary times exceed follow-up times should be roughly the same as the number of cases in which follow-up times exceed diary times. Just as in McNemar's test, the number of tied cases doesn't provide any useful information about differences.

From the Analyze menu choose:

Nonparametric Tests ▶ 2 Related Samples...

Select netimfol and netimdry.
Select Sign test.

Figure 18.4 Sign test comparing diary and follow-up times

		N
NETIMFOL Followup: time spent on Internet (min) - NETIMDRY Diary: expanded time spent on Internet (min)	Negative Differences[1]	47
	Positive Differences[2]	39
	Ties[3]	10
	Total	96

1. NETIMFOL Followup: time spent on Internet (min) < NETIMDRY Diary: expanded time spent on Internet (min)
2. NETIMFOL Followup: time spent on Internet (min) > NETIMDRY Diary: expanded time spent on Internet (min)
3. NETIMDRY Diary: expanded time spent on Internet (min) = NETIMFOL Followup: time spent on Internet (min)

Follow-up estimates are greater than time-diary estimates for 39 cases

	NETIMFOL Followup: time spent on Internet (min) - NETIMDRY Diary: expanded time spent on Internet (min)
Z	-.755
Asymp. Sig. (2-tailed)	.450

In Figure 18.4, you see that of the 96 people who reported Internet use yesterday using both the time diary and the follow-up questionnaire, 10 had exactly the same times for both. Follow-up times were less than diary times for 47 people (negative differences), while the reverse was true for 39 people (positive differences). The number of positive differences is somewhat less than the number of negative differences; however, you know that in samples, you wouldn't expect them to be exactly equal. The observed significance level in Figure 18.4 tells you that there's little reason to doubt that the median difference is 0 in the population. More than 40% of the time, you would expect to see a sample difference at least as large as the one you observed when the null hypothesis is true.

? *What's the difference between the sign test and McNemar's test?* Not much. For dichotomous variables, the sign test is called McNemar's test. (A **dichotomous variable** has only two values, typically coded 0 and 1.) For two dichotomous variables, if you square the z value in the sign test output, you'll get the chi-square value for McNemar's test. The observed significance level for the two statistics is identical. ■■■

Wilcoxon Test

When you calculate the sign test, all you look at is which of the two numbers for a pair is larger. You ignore the magnitude of the difference. For example, if the difference between the diary and the follow-up measure is two minutes, you treat the difference the same way as if it were 250 minutes. The sign test ignores a lot of useful information about the data.

The **Wilcoxon matched-pairs signed-rank test** is a nonparametric test that uses the information about the size of the difference between the two members of a pair. That's why it's more likely to detect true differences when they exist. However, the Wilcoxon test requires that the differences be a sample from a symmetric distribution. That's a less stringent assumption than requiring normality, since there are many other distributions besides the normal distribution that are symmetric. From Figure 18.3, it appears that the symmetry assumptions may not be unreasonable for these data. Again, since you have a sample, you don't expect the sample distribution to be exactly symmetric.

Make the selections as shown in Figure 18.18.

Figure 18.5 Wilcoxon test for Internet times

		N	Mean Rank	Sum of Ranks
NETIMFOL Followup: time spent on Internet (min) - NETIMDRY Diary: expanded time spent on Internet (min)	Negative Ranks	47[1]	43.59	2048.50
	Positive Ranks	39[2]	43.40	1692.50
	Ties	10[3]		
	Total	96		

1. NETIMFOL Followup: time spent on Internet (min) < NETIMDRY Diary: expanded time spent on Internet (min)
2. NETIMFOL Followup: time spent on Internet (min) > NETIMDRY Diary: expanded time spent on Internet (min)
3. NETIMDRY Diary: expanded time spent on Internet (min) = NETIMFOL Followup: time spent on Internet (min)

	NETIMFOL Followup: time spent on Internet (min) - NETIMDRY Diary: expanded time spent on Internet (min)
Z	-.768[1]
Asymp. Sig. (2-tailed)	.443

1. Based on positive ranks.

To calculate the Wilcoxon test, first you find the difference between the two values for each pair. Next, for all cases in which the difference is not 0, you rank the differences from smallest to largest, ignoring the sign of the differences. That is, the smallest difference in absolute value is assigned a rank of 1, the second smallest difference is assigned a rank of 2, and so on. In the case of ties (equal differences), you assign the average rank to the tied cases.

? *You do what to the tied cases?* Consider the following eight differences: 0, 2, –3, 3, –3, 4, 7, and –10. They are arranged from smallest to largest, ignoring the sign. You ignore all 0 differences. Then you assign a rank of 1 to the smallest number, which is 2. To get the ranks for the next three cases, which are tied in value, you give them the next available ranks. In this case, they are 2, 3, and 4. Then you find the average of the ranks you've given to the tied cases—the average of 2, 3, and 4, which is 3—and use that as the rank for all of them. The ranks for the seven non-zero differences are then 1, 3, 3, 3, 5, 6, and 7. ■■■

Once you have the ranks, you calculate the average of the ranks separately for the positive and negative differences. If the null hypothesis is true, you expect the mean rank to be similar for the two groups. Since you replace the observed differences with ranks, the effect of people with very large differences between the diary and follow-up is less severe than if you were to calculate a *t* test using the observed data values.

In Figure 18.5, you see that the mean rank for cases in which diary is greater than follow-up is 43.59, while for the cases for which follow-up is greater than diary, the mean rank is 43.40. From the two-tailed significance level, you see that the difference in mean ranks is not large enough for you to reject the null hypothesis that the population mean difference between diary and follow-up times is 0. So, even based on this more powerful test, you're unable to reject the null hypothesis.

? *If nonparametric tests require so few assumptions about the data, why not just use them all of the time?* The disadvantage to nonparametric tests is that they are usually not as good at finding differences when there are differences in the population and the assumptions for the parametric tests are met. Another way of saying this is that nonparametric tests are not as powerful as tests that assume an underlying normal distribution, the so-called **parametric tests**. That's because nonparametric tests ignore some of the available information. For example, you've just seen that data values are replaced by ranks in the Wilcoxon test.

In general, if the assumptions of a parametric test are plausible, you should use the more powerful parametric test. You've seen that many of these tests can handle reasonable violations of the assumptions. That is, they are **robust**. (That's a compliment for a statistical procedure!) Nonparametric procedures are most useful for small samples when there are serious departures from the required assumptions. They are also useful when outliers are present, since the outlying cases won't influence the results as much as they would if you used a test based on an easily influenced statistic, such as the mean. ■■■

Who's Sending E-mail?

In the follow-up questions, people were asked for the number of e-mails they sent and received yesterday and whether they were personal or work related. You can use the Wilcoxon test to test the null hypothesis that the average number of e-mails sent was equal to the average number of e-mails received. From the descriptive statistics in Figure 18.6, you see that on average, people who use the Internet claim to have sent four personal e-mails and received nine. From the maximum values, you see that some people reported very large numbers for mail sent and received. This would be troublesome if you were going to use a *t* test, since the large values would affect the mean and variance. In the Wilcoxon test, ranks are used instead of absolute differences, so the effect of strange data points is greatly diminished. Based on the observed significance level in Figure 18.7, you reject the null hypothesis for personal e-mail. (Is it spam?) When you repeat the analysis for the number of work e-mails, you get similar results. People claim to receive more mail than they send.

? *What should I do if I'm not sure if I should be using a nonparametric test or a parametric test?* When in doubt, do them both! If you reach the same conclusions based on both types of tests, there's nothing to worry about. If the results from the nonparametric test are not significant, while those from the parametric test are, try to figure out why. Do you have one or two data values that are much smaller or larger than the rest? If so, they may be affecting the mean and having a large impact on your conclusions. Examine them carefully to make sure they're okay. If the problem is with the non-normal distribution of data values, see if you can transform the data to better conform with the parametric assumptions. If your transformation is successful, you can use one of the more powerful parametric procedures for your analysis. ■■■

Select all cases. In the Two-Related-Samples dialog box, select emgetprs and emsendpr. Select Wilcoxon.

Figure 18.6 Descriptive statistics for personal e-mails sent and received

	N	Mean	Std. Deviation	Minimum	Maximum
EMGETPRS Emails received - personal	322	9.34	27.887	0	426
EMSENDPR Emails sent - personal	313	3.94	7.685	0	100

		N	Mean Rank	Sum of Ranks
EMSENDPR Emails sent - personal - EMGETPRS Emails received - personal	Negative Ranks	174[1]	124.24	21618.00
	Positive Ranks	51[2]	74.65	3807.00
	Ties	80[3]		
	Total	305		

1. EMSENDPR Emails sent - personal < EMGETPRS Emails received - personal

2. EMSENDPR Emails sent - personal > EMGETPRS Emails received - personal

3. EMGETPRS Emails received - personal = EMSENDPR Emails sent - personal

Figure 18.7 Wilcoxon test for personal e-mails sent and received

	EMSENDPR Emails sent - personal - EMGETPRS Emails received - personal
Z	-9.139[1]
Asymp. Sig. (2-tailed)	.000

1. Based on positive ranks.

Mann-Whitney Test

Each day has a fixed length. If you spend time on the Internet, does that time replace time spent with family and friends, or does the Internet save you time and let you spend more time with family and friends? Figure 18.8 is a pair of histograms of active time spent with family for people who used and did not use the Internet yesterday, as determined by the diary values. You see that the distribution of times is markedly non-normal. Since the two-independent-sample t test depends on the assumption of normality, a nonparametric alternative to the t test may be more appropriate.

The **Mann-Whitney test** is the most commonly used alternative to the independent-samples t test. If you want to use this test to test the null hypothesis that the population means are the same for the two groups, the shape of the distributions must be the same in both groups. This implies that the population variances for the two groups must be the same. It doesn't matter what the shape is, but it has to be the same in the two groups. If you simply want to test whether one population has larger values than the other, you don't have to worry about the shapes of the distributions being the same. Strictly speaking, the variables should come from continuous distributions so that there are no ties. However, the test performs reasonably well when there are ties.

The actual computation of the Mann-Whitney test is simple. You rank the combined data values for the two groups. Then you find the average rank in each group.

Figure 18.8 Histograms of active time spent with family

From the Graphs menu choose:
Legacy Dialogs
Histogram...

Select etafam as the variable and intdiary for Panel by Columns.

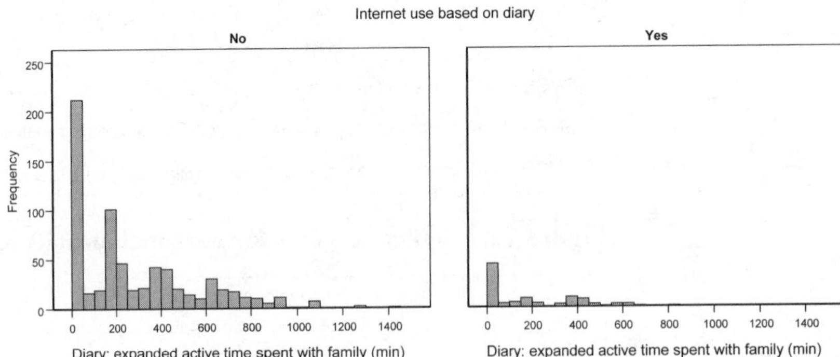

In Figure 18.9, you see the average ranks for active time spent with family, active time spent with friends, and time spent with no active interactions with anyone ranked separately for those who used the Internet and those who didn't. For both active time spent with family and active time spent with friends, the average rank is smaller for Internet users. Since the rank of 1 is assigned to the smallest value, that means that Internet users had smaller times for both of the variables. Internet users spent more time with no active interactions with others, since the average rank for Internet users is larger than that for non-users.

Make the selections shown in Figure 18.16.

Figure 18.9 Average ranks of time spent

	Internet use based on diary	N	Mean Rank	Sum of Ranks
Diary: expanded active time spent with family (min)	No	675	396.27	267483.50
	Yes	99	327.69	32441.50
	Total	774		
Diary: expanded active time spent with friends (min)	No	675	388.94	262537.50
	Yes	99	377.65	37387.50
	Total	774		
Diary: expanded time spent with no active interactions with anyone (min)	No	675	373.78	252301.00
	Yes	99	481.05	47624.00
	Total	774		

In Figure 18.10, you see that you can reject the null hypothesis that Internet users and non-users spent the same amount of active time with family and without active interactions with anyone. You can't reject the null hypothesis that Internet users and non-users spent the same amount of active time with friends. You must look at the average ranks for the two groups to determine which has the smaller values. The sign of the *z* statistic in Figure 18.10 doesn't tell you that. Note that it's negative in all three situations.

Figure 18.10 Test statistics for time spent by Internet use

	ETAFAM Diary: expanded active time spent with family (min)	ETAFRI Diary: expanded active time spent with friends (min)	ETNOACT Diary: expanded time spent with no active interactions with anyone (min)
Mann-Whitney U	27491.500	32437.500	24151.000
Wilcoxon W	32441.500	37387.500	252301.000
Z	-2.902	-.609	-4.463
Asymp. Sig. (2-tailed)	.004	.543	.000

? *What if you used the follow-up classification for Internet users?* You get the same results. Internet users spend less time with others. ■■■

Kruskal-Wallis Test

You found that Internet users spend less active time with others than do non-users. However, since time on the Internet varies greatly, the simple classification "Internet user" doesn't tell you very much. You'd like to see if there is a relationship between the amount of time on the Internet and the amount of time spent with others. The **Kruskal-Wallis test** is a nonparametric alternative to one-way analysis of variance. It is computed exactly like the Mann-Whitney test, except that there are more groups. If you want to test the null hypothesis that all population means are equal, the data must be independent samples from populations with the same shape. Again, this is a less stringent assumption than having to assume that the data are from normal populations. The assumption of equal variances, however, remains. If you only want to test whether values from one population are larger than values from another, you don't need the assumption that the shapes are the same.

Figure 18.11 is output from the Kruskal-Wallis test when you form Internet use groups based on the diary information. Only people who reported Internet use yesterday are included. For active time spent with family and active time spent with friends, you cannot reject the null hypothesis that values for the three groups of users are equal. You can, however, reject the null hypothesis that values for time alone are similar in the three groups.

Make the selections shown in Figure 18.17.

Figure 18.11 Time spent by Internet users

	time online diary	N	Mean Rank
Diary: expanded active time spent with family (min)	Use 2 hours or less	37	55.95
	2 to 4 hours	45	45.38
	more than 4 hours	17	49.29
	Total	99	
Diary: expanded active time spent with friends (min)	Use 2 hours or less	37	49.34
	2 to 4 hours	45	50.94
	more than 4 hours	17	48.94
	Total	99	
Diary: expanded time spent with no active interactions with anyone (min)	Use 2 hours or less	37	39.28
	2 to 4 hours	45	54.77
	more than 4 hours	17	60.71
	Total	99	

	ETAFAM Diary: expanded active time spent with family (min)	ETAFRI Diary: expanded active time spent with friends (min)	ETNOACT Diary: expanded time spent with no active interactions with anyone (min)
Chi-Square	2.978	.157	8.773
df	2	2	2
Asymp. Sig.	.226	.924	.012

Friedman Test

The Nie study sampled six hours of a person's day. If you assume that people (not college students) sleep eight hours a day, that would be a little more than a one-third of all waking hours. Does that adequately represent a person's day? The only way to answer that question is to compare estimates from partial-day data to estimates from a 24-hour day. We don't have that information. However, you can look at how long people spent on various activities during each of the six-hour periods. You can test the hypothesis that the minutes of Internet use are constant over the six hours. Since you have six measurements for each person, the samples are not independent. (Measurements taken on the same subject like this are called **repeated measures.**) For repeated measures, you can't use the Kruskal-Wallis test that requires independent samples. Instead, you will use the Friedman test to test hypotheses when several measurements are obtained from the same person. To calculate the Friedman test, for each person you rank the values for each of the related variables and then compare the average ranks.

Let's consider whether the median minutes online are equal for the six hours for which measurements are available. In Figure 18.12, you see that the average minutes online are largest for the fifth block, 6 P.M. to 8 P.M., and smallest for the first, midnight to 5 A.M. (Remember, these aren't college students.) The average ranks for the six hours are shown in Figure 18.13. Since the first hour has the smallest value and the fifth hour has the largest, it isn't surprising that the average rank is smallest for the first hour and largest for the fifth. Based on the chi-square value in Figure 18.13, you can reject the null hypothesis that time online is constant during the six time periods. It's not surprising that Internet activity, as well as other activities, vary by time of the day. That's why it's important to look at several time periods, as the Nie study did.

Make the selections shown in Figure 18.19.

Figure 18.12 Minutes online for each of six hours

	N	Mean	Std. Deviation	Minimum	Maximum
TONLINE1 time spent online, hour 1	99	3.5354	13.57617	.00	60.00
TONLINE2 time spent online, hour 2	99	9.9495	19.38460	.00	60.00
TONLINE3 time spent online, hour 3	99	7.2727	16.16674	.00	60.00
TONLINE4 time spent online, hour 4	99	7.1212	16.87425	.00	60.00
TONLINE5 time spent online, hour 5	99	11.0606	21.34833	.00	60.00
TONLINE6 time spent online, hour 6	99	10.6061	21.40800	.00	60.00

Figure 18.13 Friedman test

	Mean Rank
TONLINE1 time spent online, hour 1	3.11
TONLINE2 time spent online, hour 2	3.65
TONLINE3 time spent online, hour 3	3.55
TONLINE4 time spent online, hour 4	3.45
TONLINE5 time spent online, hour 5	3.70
TONLINE6 time spent online, hour 6	3.55

N	99
Chi-Square	12.895
df	5
Asymp. Sig.	.024

? *What can we conclude from all of this?* Internet research is a complex problem. It's difficult to even accurately determine how long and how often people use the Internet. Any differences you find in social behaviors between users and non-users have to be "controlled" for age, education, and other characteristics that are associated with Internet use. If college graduates use the Internet and also tend to be more isolated, you have to be careful not to conclude that it's the Internet use that's causing isolation. Measures of time spent on social behaviors are also fraught with problems. Think about a typical day in your life and try to classify all of your activities into categories. That's a pretty daunting task. ■■■

Summary

What are nonparametric tests, and when are they useful?

- The advantage of nonparametric tests is that they require fewer assumptions than other tests.
- The disadvantage is that they are less powerful than other tests, meaning they are not as good at finding differences when they exist in the population when parametric assumptions are met.
- The sign test is a less powerful alternative to the paired t test.
- The Wilcoxon matched-pairs signed-rank test is used to test the same null hypothesis as the sign test. It is usually more powerful than the sign test.
- The Mann-Whitney test is used to test the hypothesis that two independent groups come from populations with the same distribution. It is an alternative to the independent-samples t test.
- The Kruskal-Wallis test is a nonparametric alternative to one-way analysis of variance.
- The Friedman test is used to test hypotheses when several observations are made on the same subject.

How to Obtain Nonparametric Tests

This section shows how to obtain nonparametric tests from SPSS. The Nonparametric Tests submenu, accessed from the Analyze menu, offers several different procedures for tests that make limited assumptions about your data.

Chi-Square Test

The chi-square test tests the distribution of a categorical variable against the hypothesis that each category has a specified proportion of cases in the population. To open the Chi-Square Test dialog box (see Figure 18.14), from the menus choose:

Analyze
 Nonparametric Tests ▸
  Chi-Square...

Figure 18.14 Chi-Square Test dialog box

Open the manners.sav data file and select the variable mancomp and All categories equal to obtain the test shown in Figure 17.8

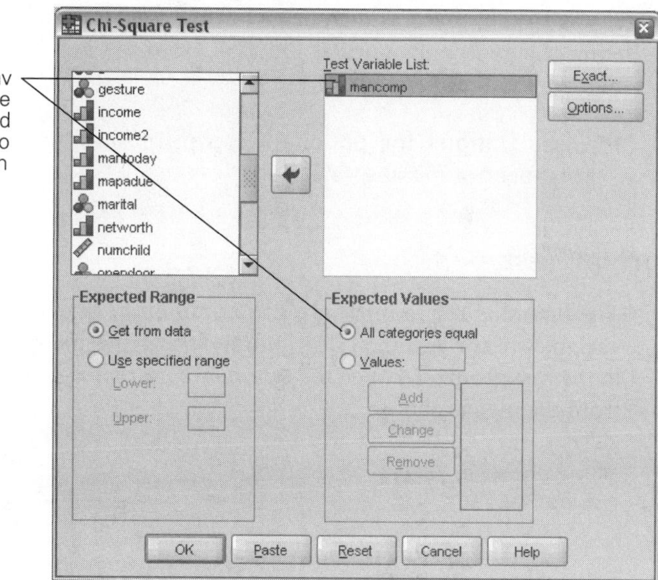

▶ Select one or more categorical variables and move them into the Test Variable List. Then specify the hypothesis that you want to test.

▶ Select one of the alternatives for Expected Range. The expected range must be a sequence of consecutive integers. (If it isn't, use Automatic Recode on the Transform menu first.) Get from data does not impose an assumption about values that are expected to occur. **Use specified range** tells SPSS to assume that the test variables have values in the range between the values you specify as lower and upper.

▶ Select one of the alternatives for Expected Values. **All categories equal** tests the hypothesis that the values of the test variables are equally distributed across the categories in the expected range. **Values** lets you enter a list of values that are proportional to your expectations. For each category in the expected range, type in a value and click **Add** to add it to the bottom of the list. If you click a value in the list, you can remove it or change it to a newly specified value by clicking the appropriate button.

You must enter one expected value for each category in the expected range. The values must appear in the list in order: the first expected value in the list corresponds to the first (lowest) category in the expected range, and so on.

The expected values must all be positive, but their scale doesn't matter. If you expect twice as many cases in the first of two categories, you can enter the expected values as 2 and 1, or as 66 and 33, or as any pair of numbers in that proportion.

▶ Click **Options** for optional descriptive statistics or to control the treatment of missing values.

Binomial Test

The binomial test compares the distribution of one or more dichotomous variables to the binomial distribution with a specified probability of being in the first group. To open the Binomial Test dialog box (see Figure 18.15), from the menus choose:

Analyze
  Nonparametric Tests ▶
    Binomial...

Figure 18.15 Binomial Test dialog box

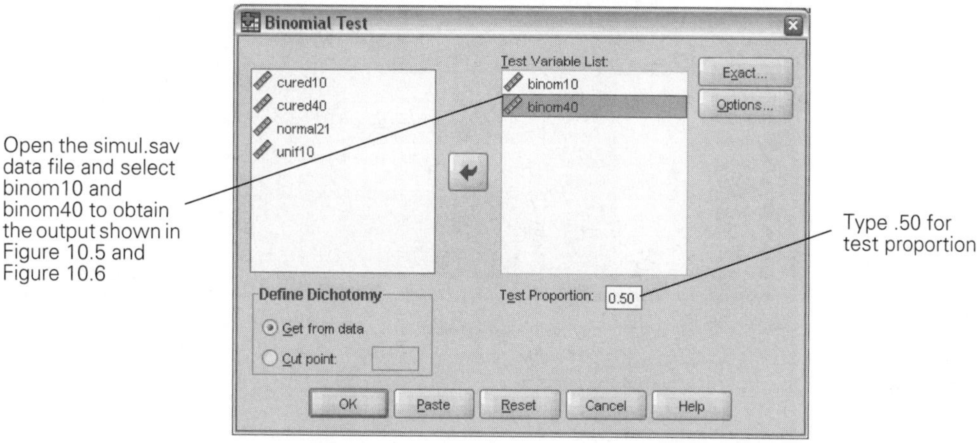

Open the simul.sav data file and select binom10 and binom40 to obtain the output shown in Figure 10.5 and Figure 10.6

Type .50 for test proportion

▶ Select one or more variables and move them into the Test Variable List.

If they are dichotomous (if they have precisely two categories), use the **Get from data** alternative in the Define Dichotomy group. Otherwise, select **Cut point** and enter a value. Data values less than the cut point form the first group, while data values equal to or greater than the cut point form the second group.

▶ Specify the test proportion (the proportion that should be in the first group if the null hypothesis is true) and click **OK**.

▶ Click **Options** for optional descriptive statistics or to control the treatment of missing values.

Two-Independent-Samples Tests

The two-independent-samples tests compare the distribution of one or more numeric variables between two groups. To open the Two-Independent-Samples Tests dialog box (see Figure 18.16), from the menus choose:

Analyze
  Nonparametric Tests ▸
    2 Independent Samples...

Figure 18.16 Two-Independent-Samples Tests dialog box

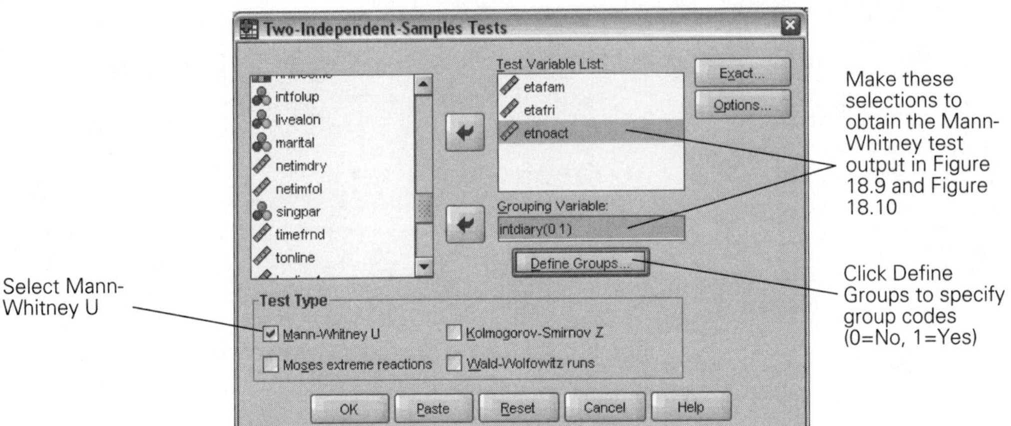

- Select one or more numeric variables and move them into the Test Variable List.
- Select a numeric grouping variable with a code corresponding to each of the two groups and move it into the Grouping Variable box. Click Define Groups and indicate the group codes.
- Select one or more of the tests in the Test Type group and click OK.
- Click Options for optional descriptive statistics or to control the treatment of missing values.

Test Type. The available tests are:

Mann-Whitney U. Based on ranking the observations in the two groups.

Moses extreme reactions. Tests whether the range (excluding the lowest 5% and the highest 5%) of an ordinal variable is the same in the two groups.

Kolmogorov-Smirnov Z. Based on the maximum differences between the observed cumulative distributions in the two groups.

Wald-Wolfowitz runs. Based on the number of runs within each group when the cases are placed in rank order.

Define Groups (Two Independent Samples)

The Define Groups dialog box for the two-independent-samples tests asks you to enter an integer for group 1 and an integer for group 2. Cases with other values of the grouping variable are ignored.

Several-Independent-Samples Tests

The several-independent-samples tests compare the distribution of one or more numeric variables between several groups. To open the Tests for Several Independent Samples dialog box (see Figure 18.17), from the menus choose:

Analyze
  Nonparametric Tests ▸
    K Independent Samples...

Figure 18.17 Tests for Several Independent Samples dialog box

Make the selections shown to obtain the Kruskal-Wallis output in Figure 18.11.

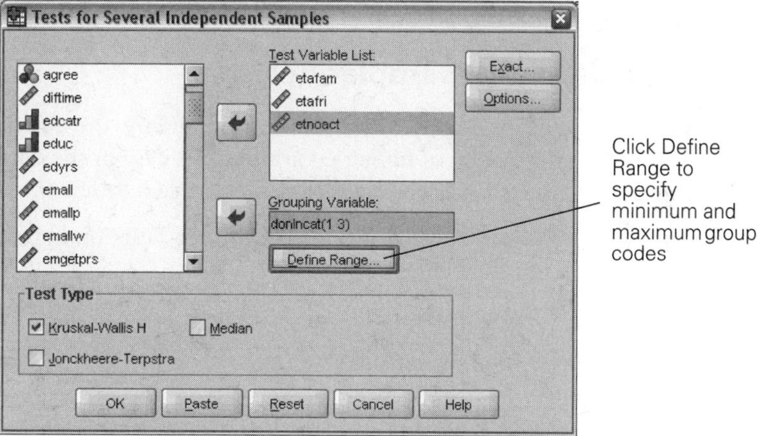

Click Define Range to specify minimum and maximum group codes

- Select one or more numeric variables, measured at the ordinal level, and move them into the Test Variable List.
- Select a numeric grouping variable with a code corresponding to each of the several groups and move it into the Grouping Variable box. Click **Define Range** and indicate the range of group codes.
- Select one or more of the tests in the Test Type group and click **OK**.
- Click **Options** for optional descriptive statistics or to control the treatment of missing values.

Test Type. The available tests are:

Kruskal-Wallis H. Tests whether the distribution of ordinal variables is the same in all groups by comparing the sum of ranks in the groups.

Jonckheere-Terpstra. An alternative to the Kruskal-Wallis test. Under certain conditions, it is more powerful.

Median. Tests whether the groups are sampled from a population in which the median of the test variable is the same.

Define Range (K Independent Samples)

The Define Range dialog box for the several-independent-samples tests asks you to enter an integer for the minimum value of the grouping variable and another integer for its maximum value. Non-integer values of the grouping variable encountered in the data are truncated to integers, and cases with truncated values outside of the range specified here are ignored.

Two-Related-Samples Tests

The two-related-samples tests compare the distribution across one or more pairs of numeric variables. (A case in the data file must include the values for both related samples, in two different variables.)

To open the Two-Related-Samples Tests dialog box (see Figure 18.18), from the menus choose:

Analyze
 Nonparametric Tests ▶
  2 Related Samples...

Figure 18.18 Two-Related-Samples Tests dialog box

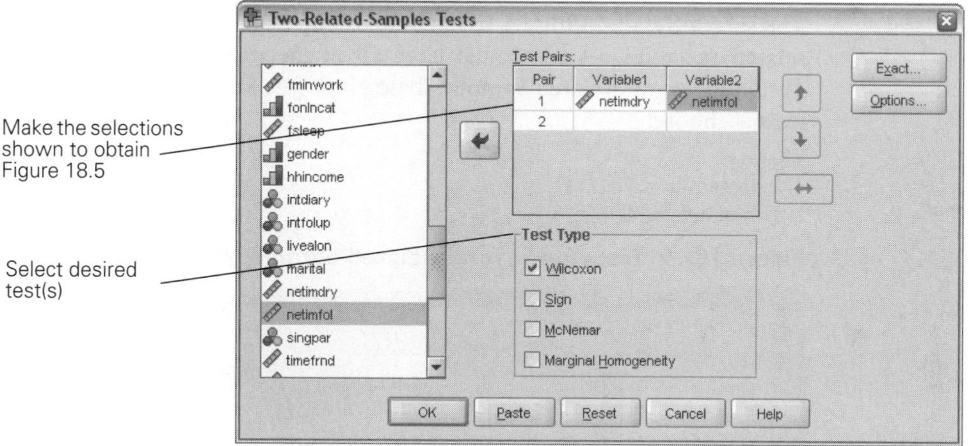

- Move the two variables you want to compare into the Test Pairs list.
- Select any other pairs of variables in the same way.
- Select one or more of the tests in the Test Type group and click **OK**.
- Click **Options** for optional descriptive statistics or to control the treatment of missing values.

Test Type. The available tests are:

Wilcoxon. Based on the ranks of the absolute values of the difference between the two variables, compared between cases in which that difference is positive and cases in which it is negative.

Sign. Based on a comparison of positive and negative differences between the two variables, using either the binomial distribution or (for more than 25 cases) a normal approximation.

McNemar. Tests whether the two possible combinations of unlike values for the variables are equally likely. Requires that the test variables be dichotomous, with the same two categories for each pair.

Marginal Homogeneity. An extension of the McNemar test for more than two categories.

Several-Related-Samples Tests

The several-related-samples tests compare the distribution across several related variables. (A case must have all of the variables.) To open the Tests for Several Related Samples dialog box (see Figure 18.19), from the menus choose:

Analyze
  Nonparametric Tests ▶
    K Related Samples

Figure 18.19 Tests for Several Related Samples dialog box

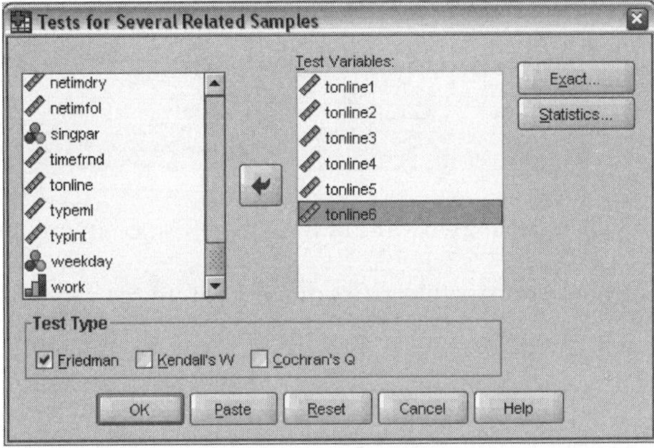

▶ Click each of the variables that you want to compare. Move them into the Test Variables List.

▶ Select the test that you want and click OK.

▶ Click Statistics for optional descriptive statistics.

Test Type. The available tests are:

Friedman. Tests the null hypothesis that *K*-related variables come from the same population.

Kendall's W. Normalization of the Friedman statistic. A measure of agreement that ranges from 0 to 1.

Cochran's Q. Tests the null hypothesis that *K*-related binary variables have the same mean.

Options: Descriptive Statistics and Missing Values

In any of the Nonparametric Tests dialog boxes, open the Options dialog box (see Figure 18.20) by clicking **Options**.

Figure 18.20 Chi-Square Test Options dialog box

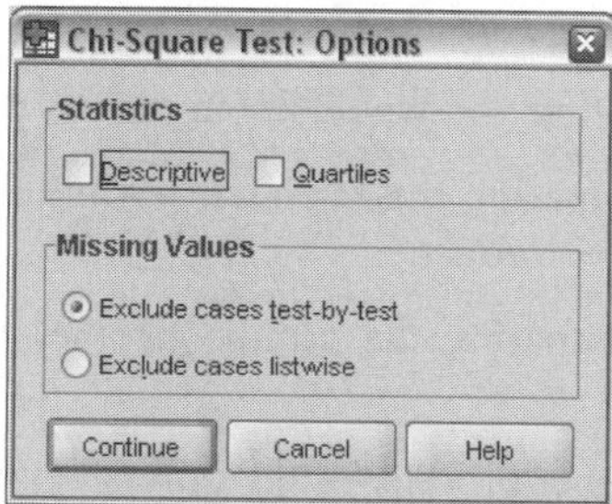

Statistics. The Statistics options allow you to display statistics for the test variables:

> **Descriptive.** These statistics include the mean, minimum, maximum, standard deviation, and the number of nonmissing cases.
>
> **Quartiles.** Displays the 25th, 50th, and 75th percentiles.

Missing Values. The Missing Values alternatives control the exclusion of cases with missing data when more than one set of variables is tested. These alternatives are not available for the tests of several related samples.

> **Exclude cases test-by-test.** Calculates each test on the basis of all cases that have valid data for the variables used in that test.
>
> **Exclude cases listwise.** Calculates all tests on the basis of the cases that have valid data for all the variables used in any test. This ensures that all of the tests are performed using the same cases.

Exercises

Statistical Concepts

1. Indicate which nonparametric test you would use in each of the following situations:

a. You are interested in comparing the satisfaction rankings given by male and female purchasers of a new product.

b. You are interested in comparing the family incomes of purchasers of four different types of products.

c. You are interested in whether product rating differs before and after use.

2. State a hypothesis that can be tested by each of the following procedures:

a. Sign test

b. Wilcoxon signed-rank test

c. McNemar test

d. Kruskal-Wallis test

3. For each of the following situations, indicate which statistical test(s) you would use:

a. You want to know if patients lose weight during radiation therapy, so you conduct a study in which you weigh each patient before and after radiation therapy.

b. You want to know if women marry men who earn more than they do, so you select 100 working couples and obtain salaries for both spouses.

c. You want to know whether men and women are equally likely to like opera.

d. You want to know whether Eskimos, Alaskans, and Canadians have the same average heart rate.

e. You want to know if there is a difference in high school GPA for students who complete college and those who enroll but do not complete it.

f. You want to know if four treatments for curing acne are equally effective. Each of 50 adolescents receives one of the treatments for four months and is then classified as *improved, same,* or *worsened.*

g. You want to know whether men and women in four regions of the country have the same average cholesterol levels.

h. You want to know if the average IQ of schizophrenics is 100.

Data Analysis

Use the *gss.sav* data file to answer the following questions:

1. Use a nonparametric test to see if the median difference in years of education between husbands and wives is 0 (variables *husbeduc* and *wifeduc*). What do you conclude? Compare your results to those from a paired-samples *t* test. If the assumptions are met for the paired *t* test, which test should you use?
2. Compute a nonparametric test to see whether the distribution of hours of television watched per day (variable *tvhours*) is the same for people in the five degree categories (variable *degree*). What do you conclude? How do your conclusions compare to those from a one-way analysis of variance?
3. Select four hypotheses based on the *gss.sav* data that you can test using nonparametric procedures. Perform the appropriate analyses. Check the requisite assumptions.
4. Use the sign test to test whether the median difference between years of education for a person's mother (variable *maeduc*) and father (variable *paeduc*) is 0. Can you reject the null hypothesis?
5. Make a histogram of the difference between father's and mother's education (variable *prtdifed*). Use the Wilcoxon test to test whether the median difference is 0. Which test is more powerful, the sign test or the Wilcoxon test?

Use *gssft.sav* to answer the following questions:

6. The variables *husbft* and *wifeft* tell you whether a husband and wife are employed full time. Use the sign test to test whether husbands and wives are equally likely to be employed full time. What do you conclude?
7. Use the McNemar test to test the same hypothesis as in the previous question. What is the relationship between the two tests? Square the z value from the sign test. What does this correspond to in McNemar's test?
8. Use a nonparametric test to see if there is a difference in hours worked for males and females (variable *hrs1*). What do you conclude?
9. Restrict your analysis to husbands and wives, both of whom are employed full time. Perform a nonparametric test to see if husbands and wives work the same number of hours a week (variables *husbhr* and *wifehr*). What factors are you controlling for in this analysis that you weren't in the previous chapter? What can you conclude?

Use the *salary.sav* data file to answer the following questions:

10. Use a nonparametric test to see if there is a relationship between gender (variable *sex*) and beginning salary (variable *salbeg*) for clerical employees (*jobcat* equals 1). Compare your findings to those you obtain from a parametric test of the same hypothesis. Summarize the results.

11. Repeat the analysis in question 8 for years on the job (variable *time*). What can you conclude about possible differences in time on the job between males and females? Perform a parametric test of the same hypothesis. Which test do you think is more appropriate in this situation? Why?

12. Use a nonparametric test to see whether current salaries (variable *salnow*) for clerical employees differ for the four gender/race groups (variable *sexrace*). Compare your results from those from a parametric analysis. Summarize your conclusions.

13. Select a hypothesis of interest that can be tested with a nonparametric procedure. Test the hypothesis and write a brief summary of your results.

Use the *schools.sav* data file to answer the following question:

14. Using nonparametric tests, rerun some of the analyses you performed on the *schools.sav* data in previous chapters. For example, use the sign test and the Wilcoxon signed-rank test to look at changes in school scores between 1993 and 1994. Do any of your conclusions change?

15. Use the *electric.sav* data file to answer the following questions:

a. Use a nonparametric test to test the hypothesis that there was no difference in cigarette smoking at the beginning of the study (variable *cgt58*) between those who were alive 10 years later and those who were not (variable *vital10*).

b. Make a histogram of the distribution of cigarettes per day for the two groups. Does the distribution appear to be normal? Is the shape of the distribution the same in the two groups?

c. What hypothesis can you test if the shape of the distribution is not the same in the two groups?

16. Use a nonparametric test to test the whether cigarette consumption (variable *cgt58*) is related to education (variable *educcat*). What can you conclude?

19 Measuring Association

How can you measure the strength of the relationship between two categorical variables?

- What are measures of association, and why are they useful?
- Is there a single best measure of association?
- Why is the chi-square statistic not a good measure of association?
- What is proportional reduction in error?
- When a measure of association equals 0, does that always mean the two variables are unrelated?
- For variables measured on an ordinal scale, how can the additional information about order be incorporated into a measure of association?
- How do you measure the agreement between two raters?

Citizens, with the possible exception of criminals who evade justice through some vagary of the law, are quick to find fault with the criminal justice system. Prisons are expensive to maintain and overcrowded, prosecutors are overburdened, and judges are erratic. In response to such concerns, the state of Vermont embarked in 1991 on a massive reorganization of its crime and corrections program. For nonviolent offenders, a sentence called "reparative probation," under the supervision of community boards, was introduced. For more serious offenders, special programs were designed within the prison system. In March of 1999, a telephone survey of 601 randomly selected adults was conducted to assess public attitudes about the reforms.[1]

1. John Doble and Judith Greene. 1999. *Attitudes toward crime and punishment in Vermont: Public opinion about an experiment with restorative justice*, 1999 (computer file). Englewood Cliffs, N.J.: Doble Research Associates, Inc. (producer), 2000. Ann Arbor, Mich.: Inter-university Consortium for Political and Social Research (distributor), 2001.

In this chapter, you'll use the survey to determine how Vermont residents felt about the criminal justice system in their state and how they rated the performance of judges, prosecutors, juries, and prisons. In particular, you'll study the strength and nature of the relationships between these variables.

Many different statistical techniques are used to study the relationships among variables. In Chapter 20 through Chapter 24, you'll learn about linear regression models, which are used to predict the values of a dependent variable from a set of independent variables. Regression analysis is used when your dependent variable has many possible values and is measured on an interval or ratio scale. In this chapter, you'll look at techniques that are useful for measuring the strength and nature of relationships between two categorical variables. Categorical variables have a limited number of possible values, and their joint distribution can be examined with a crosstabulation.

▶ This chapter uses the *crimjust.sav* data file. For instructions on how to obtain crosstabulations and associated statistics, see Chapter 8.

Components of the Justice System

You can see from the frequency table in Figure 19.1 that only 3.2% of the 601 people surveyed rated the criminal justice system in Vermont as *Excellent*. Over 50% of the sample characterized the criminal justice system as *Only fair* or *Poor*. Since the criminal justice system has many components—judges, prosecutors, juries, police, and prisons, among others—respondents were asked to rate the performance of each separately. One of the questions that arises is how important each of the parts is to the total evaluation of the criminal justice system. For example, if prosecutors are viewed as *Good*, is the entire system more likely to be viewed as *Good* than if judges are viewed as *Good*?

In the Frequencies dialog box, select the variable cjsrate.

Figure 19.1 Frequency table of opinions of criminal justice system

		Frequency	Percent	Valid Percent	Cumulative Percent
Valid	Excellent	19	3.2	3.3	3.3
	Good	257	42.8	44.2	47.4
	Only fair	241	40.1	41.4	88.8
	Poor	65	10.8	11.2	100.0
	Total	582	96.8	100.0	
Missing	Not sure/don't know	19	3.2		
Total		601	100.0		

Figure 19.2 is a crosstabulation of the ratings of the criminal justice system and the ratings of judges. You see that 27% of the people who rated judges as *Excellent* rated the criminal justice system as *Excellent* as well. Sixty percent of the people who rated the judges as *Poor* rated the overall system as *Poor* as well.

Figure 19.2 Crosstabulation of ratings of criminal justice system and judges

In the Crosstabs dialog box, select judgrate as the row variable and cjsrate as the column variable. In the Cell Display dialog box, select Observed counts and Row percentages.

			CJSRATE RATE JOB DONE: CJ SYSTEM OVERALL				
			Excellent	Good	Only fair	Poor	Total
JUDGRATE RATE JOB DONE: VT'S JUDGES OVERALL	Excellent	Count	10	20	7		37
		Row %	27.0%	54.1%	18.9%		100.0%
	Good	Count	6	172	88	10	276
		Row %	2.2%	62.3%	31.9%	3.6%	100.0%
	Only fair	Count	3	54	117	21	195
		Row %	1.5%	27.7%	60.0%	10.8%	100.0%
	Poor	Count		4	17	31	52
		Row %		7.7%	32.7%	59.6%	100.0%
Total		Count	19	250	229	62	560
		Row %	3.4%	44.6%	40.9%	11.1%	100.0%

Figure 19.3 shows the relationship of the ratings of the overall system to the ratings of prosecutors. Again, you see a relationship between the two variables. Almost 69% of the people who rated prosecutors as *Good* rated the criminal justice system as *Good* as well; 68% of those who rated prosecutors as *Only fair* rated the criminal justice system in the same way.

In the Crosstabs dialog box, select proscrat as the row variable and cjsrate as the column variable. In the Cell Display dialog box, select Observed counts and Row percentages.

Figure 19.3 Crosstabulation of ratings of criminal justice system and prosecutors

			CJSRATE RATE JOB DONE: CJ SYSTEM OVERALL				
			Excellent	Good	Only fair	Poor	Total
PROSCRAT RATE JOB DONE: VT'S PROSECUTORS	Excellent	Count	8	11	3	1	23
		Row %	34.8%	47.8%	13.0%	4.3%	100.0%
	Good	Count	7	192	71	9	279
		Row %	2.5%	68.8%	25.4%	3.2%	100.0%
	Only fair	Count	1	40	141	24	206
		Row %	.5%	19.4%	68.4%	11.7%	100.0%
	Poor	Count		4	10	27	41
		Row %		9.8%	24.4%	65.9%	100.0%
Total		Count	16	247	225	61	549
		Row %	2.9%	45.0%	41.0%	11.1%	100.0%

Are ratings of judges or prosecutors more closely related to the ratings of the overall system? That's not an easy question to answer, since the word *related* can have many different meanings. For this problem, a "perfect" relationship is easy to define. A perfect relationship is one in which all people gave the same ratings to the overall system and a particular component. It's the "imperfect" relationships that can be quantified in many different ways. Numerous and varied *measures of association* are used to describe the relationships between categorical variables. A measure of association is just a number whose magnitude tells you how strongly two variables are related. In general, measures of association range in absolute value from 0 to 1. If the two variables are measured on an ordinal scale, a positive sign tells you that the values of the two variables increase together, while a negative sign tells you that as the values of one variable increase, the values of the other variable decrease. The larger the absolute value of the measure, the stronger the relationship between the two variables.

There are many different measures of association, since there are many different ways to define *association*. The measures differ in how they can be interpreted and in how they define perfect and intermediate levels of association. They also differ in the level of measurement required for the variables. For example, if two variables are measured on an ordinal scale, it makes sense to talk about their values increasing or decreasing together. Such a statement is meaningless for variables measured on a nominal scale.

When you have two variables measured on a nominal scale, you're limited in what you can say about their relationship. It doesn't make sense to say that marital status increases as religious affiliation increases, or that automobile color decreases with increasing state of residence. If the categories of the variables don't have a meaningful order, it doesn't make sense to say that their values increase together or that the values of one variable increase as the values of the other variable decrease. In this situation, you can't talk about the direction of the association—all you can do is measure its strength.

? *Why not just calculate the percentage of people who gave exactly the same ratings to the overall system and to each component?* That's certainly a reasonable starting point for this problem, since the rating systems are the same for the two variables. Almost 60% of respondents gave exactly the same ratings to the overall system and to judges. For prosecutors, 67% gave the same ratings. If one of the percentages is 100%, you know that the variables are perfectly related. Cohen's kappa, a measure of agreement between ratings of two variables or between two raters, is discussed in "Measuring Agreement" on p. 431. Two variables can be strongly related without having exactly the same values, so it's important to consider measures that define association in other ways. ■■■

Proportional Reduction in Error

One way you can quantify the strength of the relationship between the ratings of the criminal justice system and the ratings of judges and prosecutors is to determine how well you can predict one from the other. That is, from the rating for judges, predict the rating for the overall system. Similarly, from the rating for prosecutors, predict the rating for the overall system. Compare this to your prediction for the overall rating of the system if you had no other information. Measures of association that look at how much better you can predict the values of a dependent variable when you know the values of an independent variable are called **proportion reduction in error (PRE) measures.** PRE measures compare the errors in two different situations: one in which you don't use the independent variable for prediction and one in which you do.

- In the first situation, you predict how a person feels about the criminal justice system, knowing only that in the sample, 3.4% of people rate it as *Excellent*, 44.6% rate it as *Good*, 40.9% rate it as *Only fair*, and 11.1% rate it as *Poor*.
- In the second situation, you have an additional piece of information available—the ratings of the judges.

? *Why don't these percentages match the values in Figure 19.1?* Only cases that rated both the judges and the overall system are included in the percentages used for the PRE calculations. ■■■

Calculating Lambda

If you have no information about a person's rating of the judges, what would you predict as their overall rating of the criminal justice system? If you're interested in making as few errors as possible, you should predict the criminal justice rating that occurs most often in the sample. *Good* was the most frequent category, so that's your best guess when you don't know anything else.

Now count the number of cases in Figure 19.2 that you'd misclassify if you predicted *Good* for everyone. Your prediction would be wrong for the 19 people who said *Excellent*, the 229 people who said *Only fair*, and the 62 people who said *Poor*. The total number of misclassified people is the sum, 310. This is the error for the first situation, when you know only how often the different values of the dependent variable occur in the sample.

Let's take a look at the second situation. The rule is straightforward—for each category of the *independent* variable, predict the category of the *dependent* variable that occurs most frequently. If you know that someone rates judges as *Excellent*, predict *Good* for the criminal justice system because that's the most frequent choice of people who rate judges as *Excellent*. Using this rule, you'd incorrectly classify 17 people who rated judges as *Excellent*—10 who rated the system as *Excellent* and 7 who rated the system as *Only fair*.

Similarly, for someone who rates judges as *Good*, predict *Good* for the criminal justice system, since that's the most frequent choice of people who rank judges as *Good*. You are wrong for the 104 people who rated judges, but not the system, as *Good*. Predict *Only fair* for people who rated judges *Only fair*. You're wrong for 78 such people. Predict *Poor* for those who rated judges as *Poor*. You incorrectly classify 21 such people.

Now compare the errors. In the first situation, you incorrectly classified 310 people. In the second situation, you incorrectly classified 220 (17 + 104 + 78 + 21) people. The **lambda statistic** tells you the proportion by which you reduce your error in predicting the dependent variable when you use the independent variable. That's why it's called a proportional reduction in error measure. Lambda (λ) is calculated as

$$\lambda = \frac{\text{Misclassified in situation 1} - \text{Misclassified in situation 2}}{\text{Misclassified in situation 1}}$$

Equation 19.1

For Figure 19.2,

$$\lambda = \frac{310 - 220}{310} = \frac{90}{310} = 0.29$$

Equation 19.2

Using the ratings for judges, you reduced your error in predicting the ratings of the criminal justice system by 29%.

The largest value that lambda can be is 1. Figure 19.4 shows a table in which lambda is 1. For each category of the independent variable (*city*), there is one cell with all of the cases. If you guess that value for all cases, you make no errors. The introduction of the independent variable lets you predict perfectly, and it results in a 100% reduction in error rate.

To obtain lambda along with the crosstabulation, select Lambda in the Crosstabs Statistics dialog box.

Figure 19.4 Crosstabulation in which lambda equals 1

Count

		CITY				
		Chicago	New England	Miami, New York	Other	Total
LIFE	Exciting	22				22
	Pretty routine		22	44		66
	Dull				484	484
Total		22	22	44	484	572

			Value	Asymp. Std. Error[1]	Approx. T[2]	Approx. Sig.
Nominal by Nominal	Lambda	Symmetric	.875	.023	9.848	.000
		LIFE Dependent	**1.000**	.000	10.198	.000
		CITY Dependent	.750	.046	8.638	.000
	Goodman and Kruskal tau	LIFE Dependent	1.000	.000		.000[3]
		CITY Dependent	.814	.016		.000[3]

1. Not assuming the null hypothesis.
2. Using the asymptotic standard error assuming the null hypothesis.
3. Based on chi-square approximation

100% reduction in error—if you know the city, you can perfectly predict view of life

A value of 0 for lambda means that the independent variable is of no help in predicting the dependent variable. When two variables are statistically independent, lambda is 0; but a lambda of 0 does not necessarily imply statistical independence. As with all measures of association, lambda measures association in a very specific way—reduction in error when values of one variable are used to predict values of the other. If this particular type of association is absent, lambda is 0. Even when lambda is 0, other measures of association may find association of a different kind. No measure of association is sensitive to every type of association imaginable.

Goodman and Kruskal's tau is a modification of lambda. Instead of predicting the group with the largest number of cases when making the prediction, you use a proportional prediction rule. That is, you guess each category with a probability equal to the proportion of cases in that category. For example, if one-third of your sample rates the criminal justice system as *Good*, you would guess *Good* one-third of the time.

Two Different Lambdas

Lambda is not a symmetric measure. Its value depends on which variable you predict from which. Suppose that instead of predicting the overall rating, you tried to predict the reverse—judges' ratings from the ratings of the criminal justice system. You'd get a different value for lambda. You'd get a value of 0.19.

Both of these values for lambda are shown in Figure 19.5. The first is in the line labeled with *CJSRATE Dependent*, and the second is in the line labeled with *JUDGRATE Dependent*.

To obtain lambda along with the crosstabulation, select Lambda in the Crosstabs Statistics dialog box.

Figure 19.5 Lambda statistics

			Value	Asymp. Std. Error[1]	Approx. T[2]	Approx. Sig.
Nominal by Nominal	Lambda	Symmetric	.242	.041	5.341	.000
		JUDGRATE RATE JOB DONE: VT'S JUDGES OVERALL Dependent	.190	.051	3.370	.001
		CJSRATE RATE JOB DONE: CJ SYSTEM OVERALL Dependent	.290	.039	6.503	.000
	Goodman and Kruskal tau	JUDGRATE RATE JOB DONE: VT'S JUDGES OVERALL Dependent	.136	.022		.000[3]
		CJSRATE RATE JOB DONE: CJ SYSTEM OVERALL Dependent	.143	.022		.000[3]

1. Not assuming the null hypothesis.
2. Using the asymptotic standard error assuming the null hypothesis.
3. Based on chi-square approximation

Knowing judges' ratings helps you to predict ratings of the criminal justice system

? *How can I test a hypothesis that a measure of association is 0 in the population?* This doesn't involve anything new. You just have to calculate the probability that you'd obtain a value at least as large (in absolute value) as the one you observed if the value is 0 in the population. SPSS prints the approximate significance levels for some of the measures of association we've discussed in this chapter. You can't get the *t* statistic by dividing the measure by the standard error displayed, since the standard error displayed is not calculated assuming that the null hypothesis is true. All of the measures of association reported in this chapter are statistically significant. That is, you can reject the null hypothesis that their value is 0.

Symmetric Lambda

In the previous example, you considered the rating of the overall criminal justice system to be the dependent variable. But this need not be the case. It's certainly possible that judges' ratings are dependent on the perception of the criminal justice system as a whole. If you have no reason to consider one of the variables dependent and the other independent, you can compute a **symmetric lambda** coefficient. For example, if you are studying the relationship between judges' ratings and police officers' ratings, you have no reason to consider one as dependent and the other as independent. In this situation, you predict the first variable from the second and then the second variable from the first, and then calculate a combined value.

In this example, if judges' ratings are predicted without any other information, you misclassify 284 people (all people who didn't choose *Good*). Knowing the overall ratings, you still misclassify 230 people. When you calculated lambda with overall ratings as the dependent variable, you found that using judges' ratings decreased the number misclassified from 310 to 220.

Symmetric lambda is the sum of the two misclassification differences, divided by the total number misclassified without additional information. In other words, you just add up the numerators for the two lambdas, then add up the denominators, and then divide:

$$\text{Symmetric } \lambda = \frac{(284 - 230) + (310 - 220)}{284 + 310} = 0.242 \qquad \textbf{Equation 19.3}$$

This number is the symmetric lambda in Figure 19.5.

? *Is it really possible for variables to be related and still have a lambda of 0? That doesn't sound right.* Actually, this can happen easily. For example, consider Figure 19.6. The two variables are clearly related. Chicagoans are unlikely to be unsatisfied, while Los Angelenos are unlikely to be satisfied. However, *Indifferent* is the response that occurs most often in each city. You predict that value whether or not you know the city. Since knowing the independent variable doesn't help prediction at all, lambda equals 0. Remember: a measure of association is sensitive only to a particular kind of association. ■■■

Figure 19.6 Variables related but lambda equals 0

The best guess for the dependent variable is always Indifferent, regardless of the city

			CITY City			Total
			Chicago	New York	L.A.	
SATISFAC Satisf	Satisfied	Count	19	10	1	30
		Column %	47.5%	25.0%	2.5%	25.0%
	Indifferent	Count	20	20	20	60
		Column %	50.0%	50.0%	50.0%	50.0%
	Unsatisfied	Count	1	10	19	30
		Column %	2.5%	25.0%	47.5%	25.0%
Total		Count	40	40	40	120
		Column %	100.0%	100.0%	100.0%	100.0%

			Value	Asymp. Std. Error[1]	Approx. T[2]	Approx. Sig.
Nominal by Nominal	Lambda	Symmetric	.129	.027	4.328	.000
		SATISFAC Satisf Dependent	.000	.000	.[3]	.[3]
		CITY City Dependent	.225	.049	4.328	.000

1. Not assuming the null hypothesis.

2. Using the asymptotic standard error assuming the null hypothesis.

3. Cannot be computed because the asymptotic standard error equals zero.

Measures of Association for Ordinal Variables

Lambda can be used as a measure of association for variables measured on any scale. All it requires is that the variables have a limited number of distinct values. You can't compute a lambda for salary and age unless you recode the variables to have a smaller number of values. The ratings you used in the previous example are both ordinal variables. However, you didn't use the order information in computing lambda. You can interchange the order of the rows and columns in any way you want (for example, you can put *Good* before *Excellent*, or *Poor* before *Good*) and not change the value of lambda.

There are measures of association that make use of the additional information available for ordinal variables. They tell you not only about the strength of the association but also about the direction. If judges' ratings increase as overall ratings increase, you can say that the two variables have a **positive relationship**. If, on the other hand, the values of one variable increase while those of the other decrease, you can say that the variables have a **negative relationship**. You can't make statements like these about nominal variables, since there's no order to the categories of the variables. Values can't increase or decrease unless they have an order.

Concordant and Discordant Pairs

Many ordinal measures of association are based on comparing pairs of cases. For example, look at Table 19.1, which contains a listing of the values of *cjsrate* (the criminal justice system rating) and *judgrate* (the rating for judges) for three cases.

Table 19.1 Values of variables cjsrate and judgrate

	cjsrate	judgrate
Case 1	1	2
Case 2	2	3
Case 3	3	2

Consider the pair of cases, case 1 and case 2. Both case 2 values are larger than the corresponding values for case 1. That is, the value for *cjsrate* is larger for case 2 than for case 1, and the value for *judgrate* is larger for case 2 than for case 1. Such a pair of cases is called **concordant**. A pair of cases is concordant if the value of each variable is larger (or each is smaller) for one case than for the other case.

A pair of cases is **discordant** if the value of one variable for a case is larger than the value for the other case but the direction is reversed for the second variable. For example, case 2 and case 3 are a discordant pair, since the value of *cjsrate* is larger for case 3 than for case 2, but the value of *judgrate* is larger for case 2 than for case 3.

When two cases have identical values on one or both variables, they are said to be **tied**.

There are five possible outcomes when you compare two cases. They can be concordant, discordant, tied on the first variable, tied on the second variable, or tied on both variables. When data are arranged in a crosstabulation, it's easy to compute the number of concordant, discordant, and tied pairs just by looking at the table and adding up cell frequencies.

If most of the pairs are concordant, the association between the two variables is said to be positive. As values of one variable increase (or decrease), so do the values of the other variable. If most of the pairs are discordant, the association is negative. As values of one variable increase, those of the other variable tend to decrease. If concordant and discordant pairs are equally likely, there is no association.

Measures Based on Concordant and Discordant Pairs

The ordinal measures of association that we'll consider are all based on the difference between the number of concordant pairs (P) and the number of discordant pairs (Q), calculated for all *distinct* pairs of observations. Since we want our measures of association to fall within a known range for all tables, we must standardize the difference, $P - Q$, so that it falls between –1 and 1, where –1 indicates a perfect negative relationship, +1 indicates a perfect positive relationship, and 0 indicates no relationship. The various measures differ in the way they attempt to standardize $P - Q$.

Goodman and Kruskal's Gamma

One way of standardizing the difference between the number of concordant and discordant pairs is to use **Goodman and Kruskal's gamma.** You calculate the difference between the number of concordant and discordant pairs $(P - Q)$ and then divide this difference by the sum of the number of concordant and discordant pairs $(P + Q)$.

In Figure 19.7, you see that the value of gamma for the *cjsrate* by *judgrate* crosstabulation is 0.68.

To obtain gamma along with the crosstabulation, select Gamma in the Crosstabs Statistics dialog box.

Figure 19.7 Goodman and Kruskal's gamma

		Value	Asymp. Std. Error[1]	Approx. T[2]	Approx. Sig.
Ordinal by Ordinal	Gamma	.680	.040	12.778	.000
N of Valid Cases		560			

1. Not assuming the null hypothesis.

2. Using the asymptotic standard error assuming the null hypothesis.

What does this mean? A positive gamma tells you that there are more "like" (concordant) pairs of cases than "unlike" pairs. There is a positive relationship between the two sets of ratings. As judges' ratings increase, so do ratings of the overall system. A negative gamma would mean that as judges' ratings increased, overall ratings decreased.

The absolute value of gamma has a proportional reduction in error interpretation. That is, you use two different rules to make a prediction and see how much you reduce the error by using one rule rather than the other. You are trying to predict whether a pair of cases is concordant or discordant. In the first situation, you classify pairs as like or unlike based on the flip of a coin. In the second situation, you base your decision rule on whether there are more concordant or discordant pairs. If most of the pairs are concordant, you predict "like" for all pairs. If most of the pairs are discordant, you predict "unlike." The absolute value of gamma is the proportional reduction in error when the second rule is used instead of the first.

For example, if only slightly more than half of the pairs of cases are concordant, guessing randomly and classifying all cases as "like" leads to a similar number of misclassified cases—approximately one-half. The value of gamma is close to 0. If all of the pairs are concordant, guessing "like" will result in the correct classification of all pairs. Guessing randomly will classify only half of the pairs correctly. In this situation, the value of gamma is 1.

If two variables are independent, the value of gamma is 0. However, a gamma of 0, like a lambda of 0, does not necessarily mean independence. (If the table is a 2×2 table, however, a gamma of 0 *does* mean that the variables are independent.)

Kendall's Tau-b

Gamma ignores all pairs of cases that involve ties. A measure that attempts to normalize $P - Q$ by considering ties on each variable in a pair separately (but not ties on both variables) is **tau-b**. It's computed as

$$\tau_b = \frac{P - Q}{\sqrt{(P + Q + T_X) \times (P + Q + T_Y)}}$$ **Equation 19.4**

where T_X is the number of ties involving only the first variable and T_Y is the number of ties involving only the second variable. Tau-*b* can have the value of +1 and –1 only for tables that have the same number of rows and columns.

To obtain these statistics, select Kendall's tau-b, Kendall's tau-c, and Somers' d in the Crosstabs Statistics dialog box.

Figure 19.8 Kendall's tau-b and tau-c

		Value	Asymp. Std. Error[1]	Approx. T[2]	Approx. Sig.
Ordinal by Ordinal	Kendall's tau-b	.462	.033	12.778	.000
	Kendall's tau-c	.383	.030	12.778	.000
N of Valid Cases		560			

1. Not assuming the null hypothesis.

2. Using the asymptotic standard error assuming the null hypothesis.

Figure 19.9 Somers' d

			Value	Asymp. Std. Error[1]	Approx. T[2]	Approx. Sig.
Ordinal by Ordinal	Somers' d	Symmetric	.462	.033	12.778	.000
		JUDGRATE RATE JOB DONE: VT'S JUDGES OVERALL Dependent	.464	.034	12.778	.000
		CJSRATE RATE JOB DONE: CJ SYSTEM OVERALL Dependent	.461	.033	12.778	.000

1. Not assuming the null hypothesis.

2. Using the asymptotic standard error assuming the null hypothesis.

In Figure 19.8, the value of tau-*b* is 0.462. Since the denominator is complicated, there's no simple explanation in terms of proportional reduction of error. However, tau-*b* is a commonly used measure of association.

Kendall's Tau-c

A measure that can attain, or nearly attain, the values of +1 and –1 for a table of any size is **tau-c**. It's computed as

$$\tau_c = \frac{2m(P-Q)}{N^2(m-1)}$$ **Equation 19.5**

where m is the smaller of the number of rows and columns and N is the number of cases. For this example, in Figure 19.8 you see that tau-*c* is 0.383. Unfortunately, there is no simple proportional reduction of error interpretation of tau-*c* either.

Somers' d

Gamma, tau-*b*, and tau-*c* are all symmetric measures. It doesn't matter which variable is considered dependent. The value of the statistic is the same. Somers proposed an extension of gamma when one of the variables is considered dependent. It differs from gamma only in that the denominator is the sum of all pairs of cases that are not tied on the independent variable. (In gamma, *all* cases involving ties are excluded from the denominator.) When overall rating of the criminal justice system is considered the dependent variable, Somers' *d*, in Figure 19.9 is 0.461.

Evaluating the Components

Now that you know how to interpret various measures of association, look at Table 19.2. Each column identifies one of the components of the criminal justice system. The rows contain measures of association between each of the components and the overall rating. For measures that are not symmetric, the overall criminal justice system rating is considered the dependent measure. You see that for a particular component, the measures vary quite a bit in absolute magnitude. For example, lambda for *Jury* is 0.19, while gamma is 0.59. If you choose a measure and look at the relative magnitude of a measure for each component, a fairly clear pattern emerges. Prosecutors' ratings are most closely related to overall ratings of the criminal justice system. Then come judges' ratings and prisons' ratings. The rank order of juries and police depends on the measure of association selected. Good prosecutors get convictions, and that's what people associate most strongly with a good criminal justice system.

You have to create your own summary table; SPSS won't do it for you.

Table 19.2 Measures of association with overall criminal justice system rating

Statistic	Jury	Police	Jail	Judge	Prosecutor
lambda	0.19	0.20	0.22	0.29	0.41
Goodman tau	0.08	0.09	0.11	0.14	0.23
Somers' *d*	0.39	0.36	0.41	0.46	0.55
Kendall's tau-*b*	0.34	0.34	0.41	0.46	0.54
Kendall's tau-*c*	0.24	0.27	0.34	0.38	0.44
gamma	0.59	0.53	0.62	0.68	0.77
kappa	0.22	0.25	0.30	0.34	0.46

? *How can I decide what measure of association to use?* No single measure of association is best for all situations. To choose the best one for a particular situation, you must consider the type of data and the way you want to define association. If a certain measure has a low value for a table, this doesn't necessarily mean that the two variables are unrelated. It can also mean that they're not related in the way that the measure can detect. But you shouldn't calculate a lot of measures and then report only the largest. Select the appropriate measures in advance. If you look at enough different measures, you increase your chance of finding significant associations in the sample that do not exist in the population. ■■■

Measuring Agreement

In the Vermont survey, the same four-point rating scale is used for both the overall criminal justice system and the various components. The measures of association that you computed so far didn't require an identical measurement scale for the two variables. For example, if you rated the criminal justice system on a five-point scale and the components on a three-point scale, the measures of association would be computed and interpreted in the same way.

When you do use the same scale for both variables, you can compute additional measures of association that make use of this additional information. Cohen's kappa coefficient provides an index of agreement between ratings of two variables or ratings of the same variable by two raters when the same scale is used for both. For example, you can calculate Cohen's kappa to assess the agreement between two physicians who examine the same patients and determine the severity of their disease. Or you can use it to quantify the agreement between ratings of two television programs assigned by the same viewers.

To see how Cohen's kappa is interpreted and calculated, look at Figure 19.10, the crosstabulation of judges' and prosecutors' ratings. If ratings for prosecutors and judges were in perfect agreement, all of the cases would be on the diagonal of the table. As you can see, that's not the case here.

In the Crosstabs dialog box, select the variables judgrate and proscrat.

In the Crosstabs Statistics dialog box, select Kappa.

In the Crosstabs Cell Display dialog box, select Total.

Figure 19.10 Cohen's kappa

			PROSCRAT RATE JOB DONE: VT'S PROSECUTORS				
			Excellent	Good	Only fair	Poor	Total
JUDGRATE RATE JOB DONE: VT'S JUDGES OVERALL	Excellent	Count	13	17	4	1	35
		% of Total	2.4%	3.1%	.7%	.2%	6.4%
	Good	Count	5	192	63	6	266
		% of Total	.9%	35.4%	11.6%	1.1%	49.0%
	Only fair	Count	3	59	117	12	191
		% of Total	.6%	10.9%	21.5%	2.2%	35.2%
	Poor	Count	2	8	20	21	51
		% of Total	.4%	1.5%	3.7%	3.9%	9.4%
Total		Count	23	276	204	40	543
		% of Total	4.2%	50.8%	37.6%	7.4%	100.0%

If judges' and prosecutors' ratings agree perfectly, all cases would be on the diagonal

		Value	Asymp. Std. Error[1]	Approx. T[2]	Approx. Sig.
Measure of Agreement	Kappa	.395	.033	12.634	.000
N of Valid Cases		543			

1. Not assuming the null hypothesis.

2. Using the asymptotic standard error assuming the null hypothesis.

A simple measure of agreement that comes to mind is the percentage of time that the ratings agree. You see that 63% of the cases fall on the diagonal. The problem with using this as a measure of agreement is that it doesn't take into account how much agreement there should be by chance. That is, even if there is no relationship between judges' and prosecutors' ratings, sometimes they will agree by chance.

Cohen's kappa corrects the observed percentage agreement for chance. It also normalizes the resulting value so that the coefficient always ranges from –1 to +1. A value of 1 indicates perfect agreement, while a value of –1 indicates perfect disagreement. A value of 0 indicates that the similarity between the two raters is the same as you would expect by chance. In Figure 19.10, you see that kappa is 0.40, indicating that there is moderate agreement between the two ratings. In Table 19.2, you can see the kappa statistic for ratings of the overall criminal justice system and each of the components. The conclusions you've reached before don't change based on the kappa statistic. Prosecutors' and judges' ratings agree most closely with ratings of the overall system.

Correlation-Based Measures

When your variables are measured on a scale in which order is meaningful, you can calculate correlation coefficients that measure the strength of the linear association between two variables. Two variables are linearly related if in a scatterplot the points cluster around a straight line. If all of the points fall exactly on a line with a positive slope, the correlation coefficient is 1. If they fall exactly on a line with a negative slope, the correlation coefficient is –1. The absolute value of the correlation coefficient tells you how tightly the points cluster around the line. (A correlation coefficient of 0 doesn't necessarily mean that the two variables are not related. They may be related in a nonlinear way. This is discussed in more detail in Chapter 20.)

Two commonly encountered correlation coefficients are the Pearson correlation coefficient and the Spearman correlation coefficient. Both of these coefficients range in value from –1 to +1. The Pearson correlation coefficient is calculated using the actual data values. The Spearman correlation coefficient, a nonparametric alternative to the Pearson correlation coefficient, replaces the actual data values with ranks.

Figure 19.11 Correlation coefficients for prosecutors' and judges' ratings

Select Correlations in the Crosstabs Statistics dialog box.

		Value	Asymp. Std. Error[1]	Approx. T[2]	Approx. Sig.
Interval by Interval	Pearson's R	.497	.042	13.330	.000[3]
Ordinal by Ordinal	Spearman Correlation	.492	.039	13.128	.000[3]
N of Valid Cases		543			

1. Not assuming the null hypothesis.

2. Using the asymptotic standard error assuming the null hypothesis.

3. Based on normal approximation.

In Figure 19.11, you see that both the Pearson and Spearman correlation coefficients for prosecutors' ratings and judges' ratings are around 0.50. When the two variables in the crosstabulation have a very small number of values, the correlation coefficients are not as informative as they are for larger tables. (The assumptions for testing the null hypothesis that the Pearson correlation coefficient is 0 are discussed in Chapter 21.)

Measures Based on the Chi-Square Statistic

In this chapter, you measured how closely two variables displayed in a crosstabulation are related. Often, the hypothesis of interest is whether the two variables are independent. Since the chi-square test of independence is frequently used when analyzing crosstabulations, there are a variety of measures of association that are based on the chi-square statistic. Unfortunately, the actual value of the chi-square statistic and its associated observed significance level provide little information about the strength and type of association between two variables. All you can conclude from the observed significance level is that two variables are not independent. Nothing more.

You know that the value of the chi-square statistic depends on the sample size, the number of rows and columns in the table, and the extent of the departure from independence. If you multiply all cell frequencies in Figure 19.2 by 10, you also increase the value of the chi-square by a factor of 10. Are the two variables more strongly related just because the cell frequencies have been multiplied by 10? Of course not. The relationship has stayed the same; it's simply the sample size that has increased.

Since the value of the chi-square statistic depends on the sample size, you can't compare chi-square values for the same table from studies with different sample sizes. Another drawback to the chi-square statistic is that many different types of relationships between two variables result in the same chi-square value. The chi-square value doesn't tell you anything about how two variables are related.

Chi-square based measures of association modify the chi-square statistic so that it isn't influenced by sample size and it falls in the range from 0 to 1, with 0 corresponding to no association and 1 to perfect association. Without such adjustments, you can't compare chi-square values from tables with different sample sizes and different dimensions.

In the Statistics dialog box, select Chi-square.

Figure 19.12 Chi-square-based measures

	Value	df	Asymp. Sig. (2-sided)
Pearson Chi-Square	285.825[1]	9	.000
Likelihood Ratio	198.442	9	.000
Linear-by-Linear Association	133.997	1	.000
N of Valid Cases	543		

1. 4 cells (25.0%) have expected count less than 5. The minimum expected count is 1.48.

In the Statistics dialog box, select Contingency coefficient and Phi and Cramer's V.

		Value	Approx. Sig.
Nominal by Nominal	Phi	.726	.000
	Cramer's V	.419	.000
	Contingency Coefficient	.587	.000
N of Valid Cases		543	

1. Not assuming the null hypothesis.

2. Using the asymptotic standard error assuming the null hypothesis.

Phi Coefficient

The phi coefficient is one of the simplest modifications of the chi-square statistic. To calculate a phi coefficient, just divide the chi-square value by the sample size and then take the square root. From Figure 19.12, for the prosecutors' and judges' ratings

$$\phi = \sqrt{\frac{\chi^2}{N}} = \sqrt{\frac{285.8}{543}} = 0.73$$

Equation 19.6

The maximum value of phi depends on the size of the table. If a table has more than two rows or two columns, the phi coefficient can be greater than 1, which is an undesirable feature. For tables with two rows and two columns, SPSS reports a negative value if the sign of the correlation coefficient is negative.

? *What's the test for linear association in Figure 19.12?* It's the Pearson correlation coefficient squared multiplied by the number of cases minus 1. The hypothesis tested by the statistic is that there is no linear relationship between the two variables. It can be used only for variables measured on an ordered scale. ■■■

Coefficient of Contingency

Unlike the phi coefficient, the coefficient of contingency is always less than 1. To calculate it for the data in Figure 19.12, use the following formula:

$$C = \sqrt{\frac{\chi^2}{\chi^2 + N}} = \sqrt{\frac{285.8}{285.8 + 543}} = 0.59$$ **Equation 19.7**

Although the value of C is always between 0 and 1, it can never get as high as 1, even for a table showing what seems to be a perfect relationship. The largest value it can have depends on the number of rows and columns in the table. For example, if you have a 4×4 table, the largest possible value of C is 0.87.

Cramér's V

Cramér's *V* is a chi-square-based measure of association that *can* attain the value of 1 for tables of any dimension. You calculate it for Figure 19.12 using the formula:

$$V = \sqrt{\frac{\chi^2}{N(k-1)}} = \sqrt{\frac{285.8}{543(3)}} = 0.42$$ **Equation 19.8**

where k is the smaller of the number of rows and columns. If the number of rows or columns is 2, Cramér's *V* is identical in value to phi. The values for phi, Cramér's *V*, and the contingency coefficient are shown in Figure 19.12.

Table 19.3 shows the chi-square-based measures for the criminal justice and components' ratings. Again, you see that the actual values of the measures differ. However, if you choose a measure and look across the columns, you see that the relative ordering of the components remains the same.

Table 19.3 Chi-square-based measures with overall criminal justice system rating

Statistic	Jury	Police	Jail	Judge	Prosecutor
chi-square	150.4	185.8	242.3	271.1	350.8
phi	0.52	0.57	0.68	0.70	0.80
Cramér's *V*	0.30	0.33	0.39	0.40	0.46
contingency coefficient	0.46	0.49	0.56	0.57	0.62
number of cases	563	575	525	560	549

? *If the chi-square statistic is so bad as a measure of association, why does it order the components as well as any other measure?* All of the crosstabulations for which the chi-squared statistics were calculated have similar numbers of cases. If twice as many people rated juries as prosecutors, the chi-square statistic for juries would be 351. That would give juries the largest chi-square value, even though juries have the weakest relationship to the overall ratings. ■■■

Chi-square-based measures are difficult to interpret. Although they can be used to compare the strength of association in different tables, the strength of association being compared isn't easily related to an intuitive concept of association.

Summary

How can you measure the strength of the relationship between two categorical variables?

- Measures of association are used to measure the strength of the relationship between two categorical variables.
- Measures of association differ in the way they define association.
- You should select a measure of association based on the characteristics of the data and how you want to define association.
- The chi-square statistic is not a good measure of association. Its value doesn't tell you anything about the strength of the relationship between two variables.
- Measures of proportional reduction in error (PRE) compare the error you make when you predict values of one variable based on values of another with the error when you predict them without information about the other variable.
- Two variables may be related and yet have a value of 0 for a particular measure of association.
- Cohen's kappa measures agreement between two raters.
- Special measures of association are available for ordinal variables. They are based on counting the number of concordant pairs (as one variable increases, so does the other) and the number of discordant pairs (as one variable increases, the other decreases).

What's Next?

In this chapter, you computed coefficients that measure the strength of the association between two variables in a crosstabulation. In the remainder of this book, you'll learn how to measure the strength of the linear relationship between a dependent variable and a set of independent variables. You'll build a linear regression model to predict the value of the dependent variable from the independent variables.

Exercises

Statistical Concepts

1. The following table shows the relationship between depth of hypnosis and success in treatment of migraine headaches (Cedercreutz, 1978). Calculate the appropriate lambda statistic. How can you interpret this value?

			MIGRAINE			
			Cured	Better	No Change	Total
HYPNOSIS	Deep	Count	13	5		18
		% Total				18.0%
	Medium	Count	10	26	17	53
		% Total				53.0%
	Light	Count		1	28	29
		% Total				29.0%
Total		Count	23	32	45	100
		% Total	23.0%	32.0%	45.0%	100.0%

2. Discuss the difference you would see in the gamma statistic if you coded job satisfaction from low to high and condition of health from good to poor, as compared to the value you would get if both job satisfaction and condition of health are coded in the same direction.

Data Analysis

Use the *gss.sav* data file to answer the following questions:

1. Consider the relationship between how happy a person is in general (*happy*) and the happiness of his or her marriage (*hapmar*).

a. Compute the chi-square statistic.

b. Do the two variables appear to be independent? Explain.

c. What does the value of the chi-square statistic tell you about the strength of the relationship between the two variables?

d. What would happen to the value of the chi-square statistic if you tripled the number of cases in each cell of the crosstabulation?

e. Do you think the chi-square test is appropriate for this table? Or are there too many cells with small expected values? Is the smallest expected value *too* small?

2. Compute phi, Cramér's *V*, and the contingency coefficient for the table in question 1. What can you say about the strength of the association between the two variables based on the values of the coefficients?

3. Compute the lambda coefficients for the table in question 1.

a. If you want to predict general happiness without any other information, what would you predict? How many people would you incorrectly classify using this rule?

b. For someone whose marriage is very happy, what would you predict for general happiness? How many people do you misclassify using this rule?

c. What would you predict for general happiness for someone whose marriage is pretty happy? How many people do you misclassify using this rule?

d. What would you predict for general happiness for someone whose marriage is not too happy? How many people do you misclassify using this rule?

e. Based on your answers for questions 3a–3d, compute the lambda statistic for predicting a person's general happiness from the happiness of the marriage.

4. Repeat question 3 but this time predict happiness with marriage from general happiness.

5. Is it obvious whether general happiness or happiness with marriage is the dependent variable? Compute symmetric lambda based on the results you obtained from questions 3 and 4. Do your answers agree with those from SPSS? If not, assume that SPSS is correct and find your mistakes.

6. What does the gamma statistic tell you? How would you interpret gamma if it were negative?

7. For full-time employees only (variable *wrkstat* equals 1), make a crosstabulation of *satjob* and respondent's income in quartiles (you must calculate this variable yourself—call it *income4*). For each cell, obtain the appropriate percentages, observed and expected cell counts, and residuals. Also obtain the chi-square statistic, gamma, and lambda.

a. Can you conclude that income and job satisfaction are independent? Why or why not?

b. What percentage of people with earnings in the top quartile are very satisfied with their jobs? What percentage of people with earnings in the bottom quartile are very satisfied with their jobs?

c. Look at the value for gamma. Explain the meaning of the sign.

d. Which variable in the table would you consider to be the dependent variable? What is the value of lambda when you try to predict the dependent variable from the independent variable?

e. Based on the gamma statistic, is it reasonable to conclude that the association between *happy* and *hapmar* is stronger than the association between *satjob* and *income4*? What can you conclude based on lambda?

8. Now look at the relationship between job satisfaction and income separately for men and women.

   a. Does it make sense to compare the two chi-square values? Why or why not?

   b. Can you conclude that for men, job satisfaction and income are independent? How about for women?

   c. Compare the strength of the association between the two variables for men and for women. What can you conclude?

9. Use appropriate measures of association to identify variables that are associated with a person's perception of life (*life*). Write a paper detailing how strongly various characteristics are related to perception of life (*life*).

Use the *crimjust.sav* data file to answer the following questions:

10. Based on the values of the variable *cntprisn* that you created, compute a new variable, *type*, that has the value 1 if *cntprisn* is less than or equal to 4 (few prison sentences); the value 2 if between 5 and 9 prison sentences are suggested; and the value 3 if 10 or more prison sentences are selected. Make a crosstabulation of *cntprisn* and *type* to make sure that you computed it correctly.

11. Describe the relationship between *type* and *cjsrate* and *type* and *policrat*. Are *type* and *cjsrate* independent? Are *policrat* and *type* independent? What measure of association would you choose to summarize the relationships? Why?

12. Consider the variables *vlntis* (how often a violent offender is sent to prison) and *vlntshld* (how often a violent offender should be sent to prison).

    a. Do most people think that violent offenders are properly handled by the criminal justice system? Do they think that offenders go to prison too seldom or too often?

    b. What do the counts on the diagonal tell you? The counts below the diagonal? The counts above the diagonal?

    c. Summarize the relationship between the two variables. Explain your choice of summary statistic.

20 Linear Regression and Correlation

How can you choose the line that best summarizes the linear relationship between two variables?

- What is the least-squares line?
- What does the slope tell you? The intercept?
- How can you tell how well a line fits the data?
- How do you calculate predicted values and residuals?

How much can sales be expected to increase if the advertising budget is doubled? How much does the cholesterol level rise as fat intake increases? What would you expect the selling price of a company to be, based on its net revenues? To answer questions like these for variables measured on at least an ordinal scale, you must determine if there is a relationship between the two variables, and if so, you must describe that relationship.

In Chapter 9, you saw that plotting the values of two variables is an essential first step in examining the nature of their relationship. From a plot, you can tell whether there is some type of pattern between the values of the two variables or whether the points appear to be randomly scattered. If you see a pattern, you can try to summarize the overall relationship by fitting a mathematical model to the data. For example, if the points on the plot cluster around a straight line, you can summarize the relationship by finding the equation for the line. Similarly, if some type of curve appears to fit the data points, you can determine the equation for the curve.

In this and the remaining chapters in Part 4, you will learn about linear regression models. You will begin with fitting the simplest of mathematical models, a straight line.

▶ This chapter uses the *cntry15.sav* data file. For instructions on how to obtain the linear regression output shown in this chapter, see "How to Obtain a Linear Regression" on p. 460.

Life Expectancy and Birthrate

Figure 20.1 is a scatterplot of life expectancy for females and birthrate per 1000 population for 15 countries. Two of the points, Mongolia and France, are identified by country name. Mongolia has a birthrate of 34 births per 1000 and a life expectancy for females of 68 years. France has a birthrate of 13 and a life expectancy for females of 82 years. Looking at the plot, you see that the points are not randomly scattered over the grid. Instead, there appears to be a pattern. Countries with high birthrates have lower life expectancies than countries with low birthrates. Another way of saying this is that as birthrate increases, life expectancy decreases.

You can create scatterplots using the Graphs menu, as discussed in Chapter 9. Choose Legacy Dialogs, then Scatter/Dot, and then select Simple Scatter. Select the variables lifeexpf and birthrat in the Simple Scatterplot dialog box. Use the Point ID facility in the Chart Editor to identify individual points.

Figure 20.1 Life expectancy and birthrate

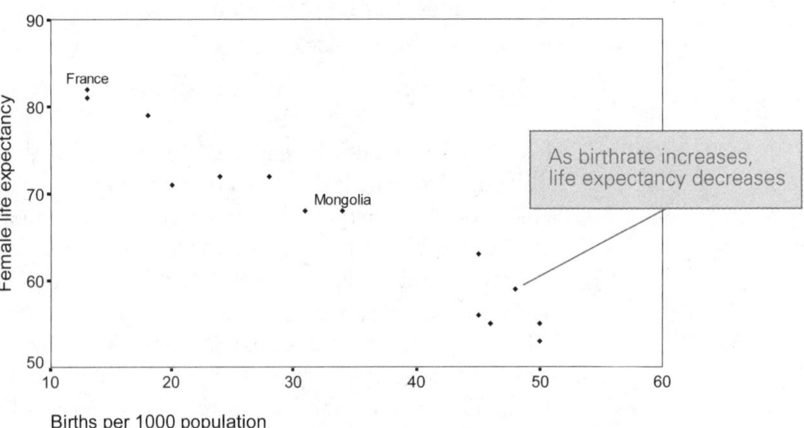

If you had to describe the pattern of points in Figure 20.1, you might say that they cluster around a straight line; that is, the relationship between the two variables is linear. Since life expectancy decreases as birthrate increases, you can also characterize the relationship as "negative." (If life expectancy increased with increasing birthrate, the relationship would be termed "positive.") Of course, the observed points don't fall exactly on a straight line. If they did, the plot would look like Figure 20.2.

Figure 20.2 Perfect linear relationships

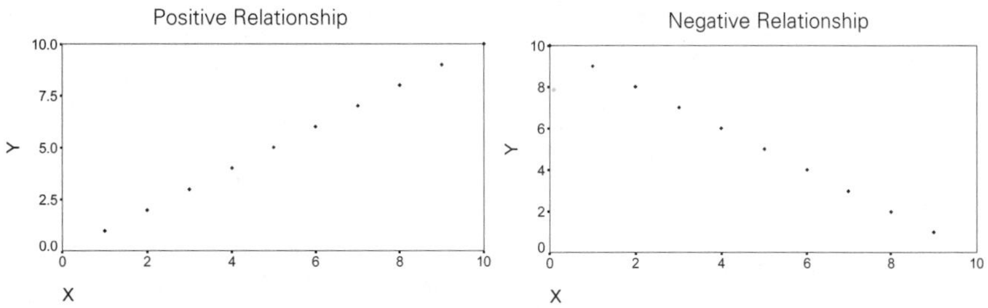

Remember to correct the value for Bhutan, as discussed in Chapter 9.

? *Why aren't we using all of the countries that we plotted in Chapter 9?* Simplicity. Sometimes it's easier to understand what's going on when you have a small number of data points that you can easily manipulate. As an exercise, you can rerun all of the analyses in this chapter on the larger (corrected) data file. You'll also use the larger data file in Chapter 23 to build a multiple regression model to predict life expectancy for females from several independent variables. ■■■

Choosing the "Best" Line

If all of the points fall exactly on a straight line, you don't have to worry about determining which line summarizes the data points best. All you have to do is connect the observed points and you have the line that best fits the data. When the points don't fall exactly on a straight line, choosing a line is more complicated.

You can see in Figure 20.3 that there are many different lines that can be drawn through the observed data values for life expectancy for females and birthrate.

Figure 20.3 Three possible lines

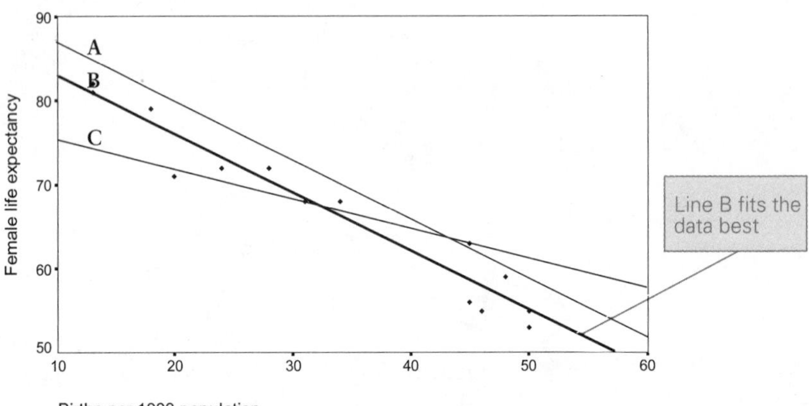

Common sense tells you that some of the lines in Figure 20.3 summarize the data better than others. For example, line A has many more points below it than above it. It doesn't really pass closely through the majority of the data points. Line C has similar numbers of points above and below it, but the distance from the points to the line is unnecessarily large. Line B fits the data best. It passes close to most of the observed data points.

The Least-Squares Line

Line B is not just any line that appears to summarize the data values well. Line B is unique. It is the **least-squares regression line,** which means that of all possible lines that can be drawn on the plot, line B has the smallest sum of squared vertical distances between the points and the line. To understand what this means, look at Figure 20.4, which shows the observed data points and line B, the least-squares regression line.

Figure 20.4 Least-squares regression line

You can add a regression line to a scatterplot after you create it. See "Editing a Scatterplot" on p. 192 in Chapter 9.

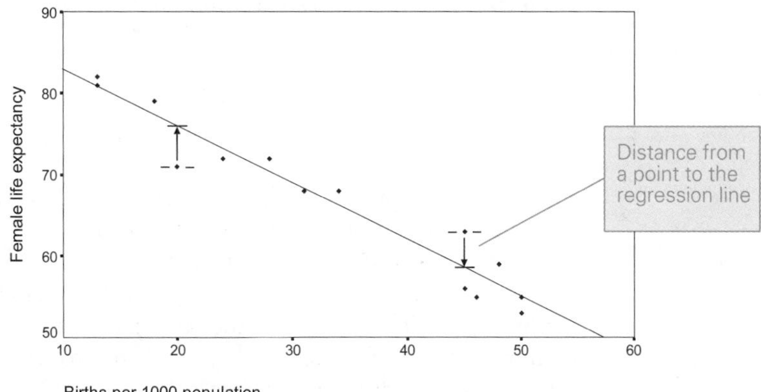

For each of the data points, you can calculate the distance between the point and the regression line by drawing a vertical line from the point to the regression line. (That's the dotted line for some of the points in Figure 20.4.) If you calculate the vertical distances between each of the points and the line, square each of them, and then add them up for all of the points, you have the **sum of squared distances** between the points and the regression line. (You use squared distances, since you don't want positive and negative differences between the points and the line to cancel out.) The least-squares regression line is the line that has the smallest sum of squared vertical distances between the points and the line. Any other line you draw through the points will have a larger sum of squared distances.

The Equation for a Straight Line

Before you learn how to calculate the least-squares regression line, let's consider the equation for a straight line. If y is the variable plotted on the vertical axis and x is the variable plotted on the horizontal axis, the equation for a straight line is

$$y = a + bx \qquad \textbf{Equation 20.1}$$

The value a is called the **intercept**. It is the predicted value for y when x is 0. The value b is called the **slope**. It is the change in y when x changes by one unit. Y is often called the dependent variable, since you can try to predict its values based on the values of x, the independent variable. In this example, the dependent variable is life expectancy for females. You want to predict its values based on birthrate, the independent variable.

Consider the line

life expectancy = 90 – (0.70 × birthrate) **Equation 20.2**

Figure 20.5 is a plot of the line. The slope is –0.70. This tells you that for an increase of 1 in birthrate, there is a decrease in life expectancy of 0.70 years. (For the birthrate variable, an increase of 1 means an increase of 1 live birth per 1000 population per year.) For example, a country with a birthrate of 10 is predicted to have a life expectancy of 83 years. A country with a birthrate of 11 is predicted to have a life expectancy of 82.3 years. The difference between 83 and 82.3 is –0.70 years, which is the value for the slope.

Figure 20.5 Regression line

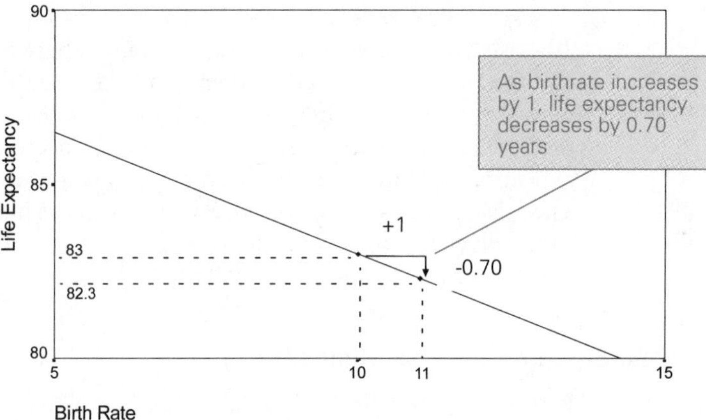

If the value for the slope is positive, you know that as the values of one variable increase, so do the values of the other variable. If the slope is negative, you know that as the values of one variable increase, the values of the other variable decrease. If the slope is large, the line is steep, indicating that a small change in the independent variable results in a large change in the dependent variable. If the slope is small, there is a more gradual increase or decrease. If the slope is 0, the changes in x have no effect on y. In this case, the least-squares regression line is horizontal, as shown in Figure 20.6.

Figure 20.6 No linear relationship (slope = 0)

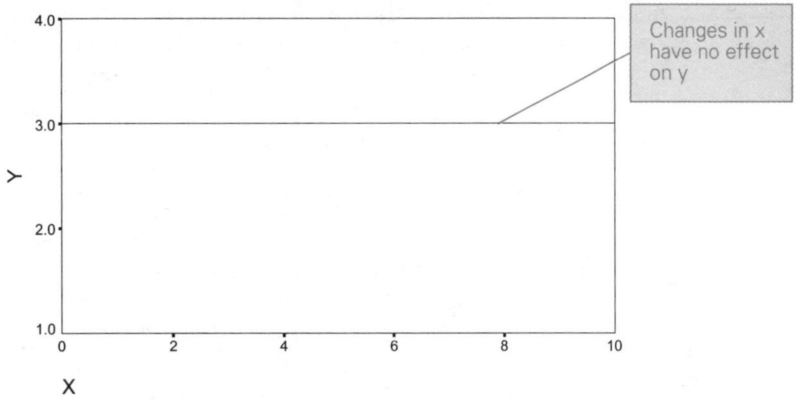

Changes in x have no effect on y

The value of the intercept in Figure 20.5 is 90. That's the point where the line hits the vertical axis; that is, where $x = 0$.

? *Why does the line hit the vertical axis at about 86 in Figure 20.5, not at 90?* The reason is that the vertical axis is not drawn at the point where the birthrate is 0. In Figure 20.5, the scale starts at five births per 1000 population. The predicted life expectancy for five births is 86.5 years. That's the point at which the line crosses the vertical axis. Look at Figure 20.7. The horizontal axis starts at a birthrate of 0, so the intercept is 90, as you would expect.

Figure 20.7 Regression line with horizontal axis starting at 0

You can change the scale of the x or y axis in a scatterplot after you create it. Activate the chart in the Chart Editor and then double-click on the axis to open a Properties dialog box with options for that axis.

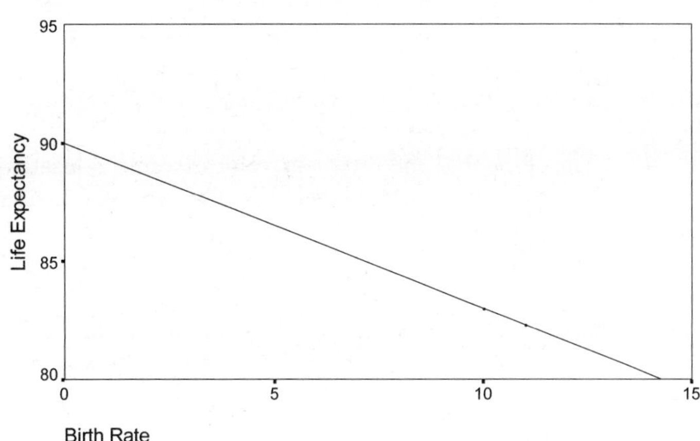

You should be careful in interpreting the intercept. Unless a value of 0 makes sense for the independent variable, the intercept may not have any substantive meaning. In this example, the intercept, 90, is the value for life expectancy when the birthrate is 0. Since there aren't any countries with birthrates of 0, the life expectancy for a hypothetical nonreproducing nation is meaningless.

? *Will I get the same answer for the slope and the intercept if I switch which variable is the dependent variable and which is the independent variable?* No, you'll get different answers. The slope and the intercept are not symmetric measures—it matters which variable is the dependent variable and which is the independent variable. ■■■

Calculating the Least-Squares Line

It's easy to calculate the slope and the intercept when all of the points fall exactly on a straight line. It's somewhat more complicated to calculate them for a least-squares regression line when the points don't fall exactly on the line. It's easiest to let SPSS do the calculations for you. Look at the output shown in Figure 20.8.

Figure 20.8 Linear regression of life expectancy and birthrate

To obtain this regression output, from the menus choose:

Analyze
Regression ▸
Linear...

Select the variables lifeexpf and birthrat, as shown in Figure 20.15.

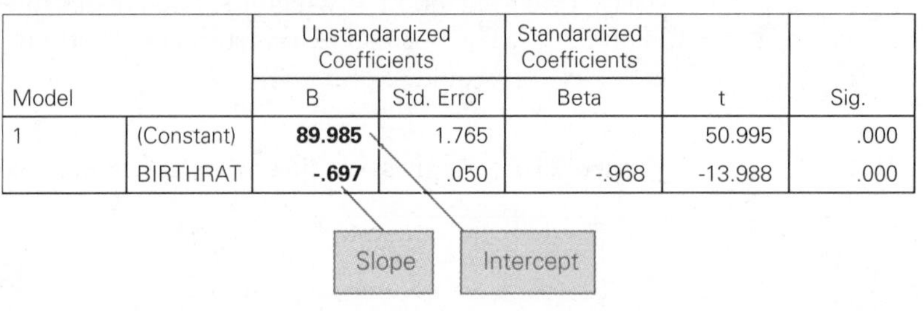

Model		Unstandardized Coefficients		Standardized Coefficients	t	Sig.
		B	Std. Error	Beta		
1	(Constant)	**89.985**	1.765		50.995	.000
	BIRTHRAT	**-.697**	.050	-.968	-13.988	.000

The slope and the intercept values from the least-squares regression are shown in the column labeled *B*. The regression equation is

predicted life expectancy = 89.99 – (0.697 × birthrate) **Equation 20.3**

(You'll learn the meaning of the other numbers in Chapter 21.)

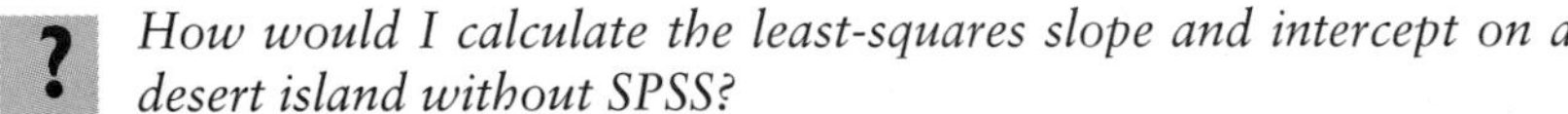

How would I calculate the least-squares slope and intercept on a desert island without SPSS?

You can calculate the slope using

$$b = \sum_{i=1}^{N} \frac{(x_i - \bar{x})(y_i - \bar{y})}{(N-1)s_x^2}$$

For each case, subtract the mean of the independent variable, $\bar{x}$, from the case's value for the independent variable, x_i. Then subtract the mean of the dependent variable, $\bar{y}$, from the case's value for the dependent variable, y_i. Multiply these two differences for each case and then sum them for all of the cases. For the last step, divide this sum by the product of the number of cases minus 1 (N–1) and the variance of the independent variable (s_x^2). Since the least-squares regression line passes through the point ($\bar{x}$, $\bar{y}$), you can calculate the intercept as

$$a = \bar{y} - b\bar{x}$$

Calculating Predicted Values and Residuals

You can use the least-squares regression line to predict the life expectancy for any country whose birthrate you know. For example, the birthrate for Spain is 11 births per 1000 population. You predict the life expectancy of females, based on the regression equation, to be

$$\text{life expectancy} = 89.99 - (0.697 \times 11) = 82.32 \text{ years} \qquad \textbf{Equation 20.4}$$

(You'll get the same predicted value if you find the point on the least-squares line in the plot that corresponds to a birthrate of 11, and read off the corresponding value for life expectancy.) The actual life expectancy for Spanish women is 82 years, so you see that your prediction is excellent. The difference between the observed and predicted value is only –0.32 years.

You see in Figure 20.9 the birthrates and observed and predicted female life expectancies for the 15 countries included in the regression. The last column contains the residual, which is the difference between the observed and predicted value of the dependent variable. You see that the smallest residual in absolute value is 0.08 for the Netherlands. The largest in absolute value is –5.04 for Thailand. A positive residual means that the observed value is greater than the predicted value, while a negative residual means that the observed value is smaller than the predicted value.

The residual for a case is nothing more than the vertical distance from the point to the line. The sign tells you whether the observed point is above or below the least-squares regression line. Another way of saying that the least-squares line has the smallest sum of squared vertical distances from the points to the line is to say that the least-squares line has the smallest sum of squared residuals. As you'll see in Chapter 22 and Chapter 24, the analysis of residuals plays an important role in regression analysis.

To save predicted values and residuals along with the regression, see "Linear Regression Save: Creating New Variables" on p. 465.

To list cases as shown here, from the menus choose:

Analyze
Reports ▸
Case Summaries...

Figure 20.9 Observed and predicted values, and residuals

		BIRTHRAT	LIFEEXPF	Unstandardized Predicted Value	Unstandardized Residual
COUNTRY	Algeria	31	68	68.37	-.37
	Burkina Faso	50	53	55.12	-2.12
	Cuba	18	79	77.43	1.57
	Ecuador	28	72	70.46	1.54
	France	13	82	80.92	1.08
	Mongolia	34	68	66.28	1.72
	Namibia	45	63	58.61	4.39
	Netherlands	13	81	80.92	.08
	North Korea	24	72	73.25	-1.25
	Somalia	46	55	57.91	-2.91
	Tanzania	50	55	55.12	-.12
	Thailand	20	71	76.04	-5.04
	Turkey	28	72	70.46	1.54
	Zaire	45	56	58.61	-2.61
	Zambia	48	59	56.51	2.49

Determining How Well the Line Fits

The least-squares regression line is the line that fits the data best in the sense of having the smallest sum of squared residuals. However, this does not necessarily mean that it fits the data *well*. Before you use the regression line for making predictions or describing the relationship between the two variables, you must determine how well the line fits the data. If the line fits poorly, any conclusions based on it will be unreliable.

? *Can't I tell how well the line fits by looking at how big the slope is?* No. The value of the slope depends not only on how closely two variables are related but also on the units in which they are measured. If you're predicting income in dollars from years of education, the slope will (hopefully) be large. If you're predicting college GPA from SAT scores, the slope will be small, not necessarily because the relationship between the two variables is weak, but because college GPA's range somewhere from 1 to 5, while SAT scores range from 200 to 800. You'll have to multiply SAT scores (which are measured in hundreds) by a small slope to end up with predicted GPA values. ■■■

The Correlation Coefficient

To describe how well the model fits the data, you want an *absolute* measure that doesn't depend on the units of measurement and is easily interpretable. The statistic most frequently used for this purpose is the **Pearson correlation coefficient** (r). It ranges in value from –1 to +1. If all of the points fall exactly on a line with a positive slope, the correlation coefficient has a value of +1. If all of the points fall exactly on a line with negative slope, the correlation coefficient is –1. The absolute value of the correlation coefficient tells you how closely the points cluster around a straight line. Both large positive values (near +1) and large negative values (near –1) indicate a strong linear relationship between the two variables—the points are close to the line.

? *If two variables have a large correlation coefficient, does that mean that one of them causes the other?* Not at all. You can't assume that because two variables are correlated, one of them causes the other. If you find a large correlation coefficient between the ounces of coffee consumed in a day and salary, you can't conclude that drinking more coffee will increase your salary. In certain regions of the world, there is a reasonably large correlation coefficient between the number of storks in an area and the birthrate. No doubt you are aware of the causal link there. ■■■

Unlike the slope and the intercept, the correlation coefficient is a symmetric measure. That means you get the same value regardless of which of the two variables is the dependent variable. That makes sense, since life expectancy is linearly related to birthrate as strongly as birthrate is to life expectancy.

You can calculate the correlation coefficient from the slope using the formula

$$r = b \times \left(\frac{s_x}{s_y}\right)$$ **Equation 20.5**

where s_x and s_y are the standard deviations of the independent and dependent variables. You see from the formula that if the dependent and independent variables are standardized to have standard deviations of 1, the correlation coefficient and the slope are equal.

Look at Figure 20.10, which shows two pairs of variables that have the same least-squares values for the slope and the intercept. They differ in how well the straight-line model fits the data. In Plot A, the data points cluster fairly tightly about the line, while in Plot B, the data points are more widely scattered about the line.

Figure 20.10 Two plots with the same slope and intercept

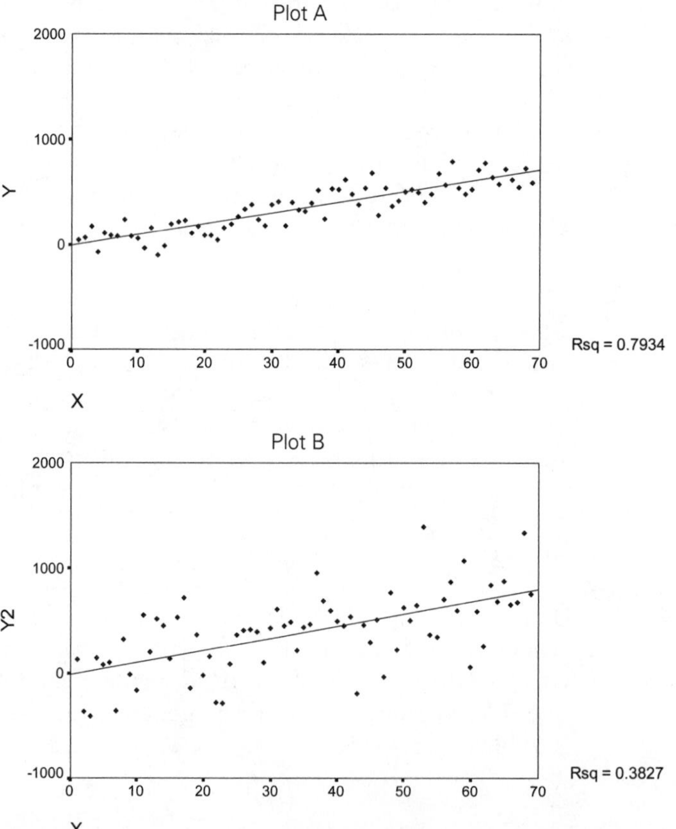

The correlation coefficient for Plot A is about 0.9, while for Plot B it is about 0.6. Since the correlation coefficient for the first plot is larger in absolute value than that for the second, you can tell without looking at the plot that the points cluster more tightly about the line in Plot A than in Plot B.

If there is no *linear* relationship between the two variables, the correlation coefficient is close to 0. A correlation coefficient of 0 does not mean that there isn't any type of relationship between two variables. It is possible for two variables to have a correlation coefficient close to 0 and yet be strongly related in a nonlinear way.

Figure 20.11 Strong nonlinear relationship

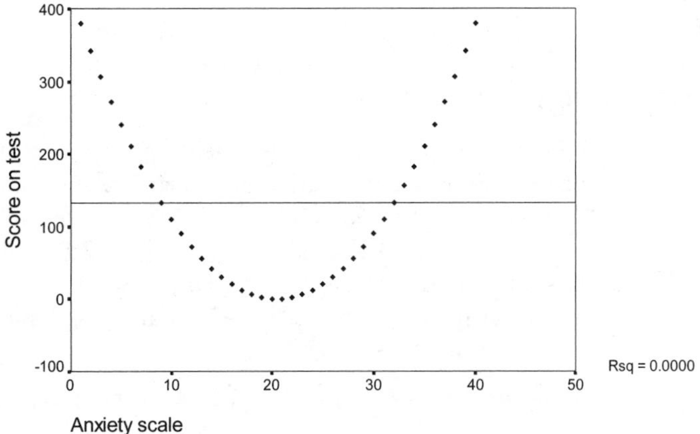

Figure 20.11 is a plot of two variables that have a correlation coefficient of 0 but have a strong *nonlinear* relationship. Test scores are strongly related to anxiety levels but not linearly. You should always plot the values of the two variables before you compute a regression line or a Pearson correlation coefficient. Plotting allows you to detect nonlinear relationships for which the regression line and correlation coefficient are not good summaries.

? *What if I just want to look at correlation coefficients between pairs of variables? Do I have to run the SPSS Regression procedure?* No. There's a special procedure called Correlations, which allows you to calculate correlation coefficients between two or more pairs of variables. You can calculate Pearson correlation coefficients or either of two nonparametric alternatives, called Spearman's and Kendall's correlation coefficients. In Chapter 21, you'll see the circumstances under which you might want to calculate nonparametric correlation coefficients. However, don't calculate correlation coefficients between many pairs of variables and then focus on the large values. The more coefficients you examine, the more likely some will be large due solely to chance. ■■■

Explaining Variability

In Figure 20.4, you see that for the life expectancy and birthrate example, the data points cluster fairly tightly around the straight line. The absolute value of the correlation coefficient between life expectancy and birthrate is 0.97. That's the value labeled *R* in Figure 20.12. The actual correlation coefficient between the two variables is –0.97, since the slope of the regression line is negative. It's unusual to find such a large correlation coefficient between two variables.

Figure 20.12 Linear regression of female life expectancy and birthrate, and analysis-of-variance table

To obtain this analysis, select the variables lifeexpf and birthrat in the Linear Regression dialog box.

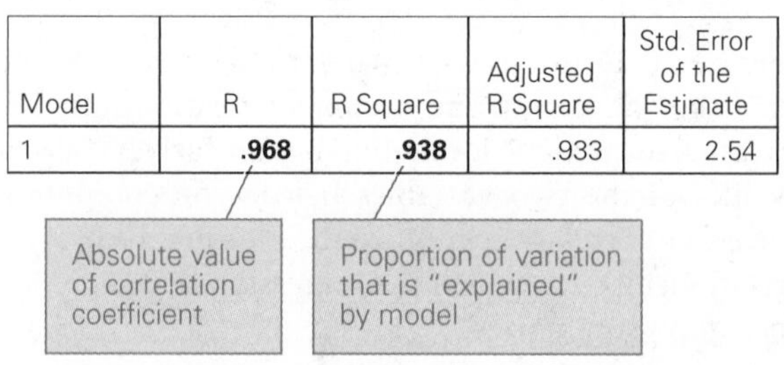

Model	R	R Square	Adjusted R Square	Std. Error of the Estimate
1	**.968**	**.938**	.933	2.54

Model		Sum of Squares	df	Mean Square	F	Sig.
1	Regression	1259.263	1	1259.263	195.653	.000
	Residual	83.671	13	6.436		
	Total	1342.933	14			

If you square the value of the correlation coefficient, you obtain another useful statistic. The square of the correlation coefficient tells you what proportion of the variability of the dependent variable is "explained" by the regression model. In this example, close to 94% of the variability in observed female life expectancies is explained by birthrate.

What does that mean? You know that all countries do not have the same life expectancy. There is substantial variability. One of the explanations for the observed variability is differences in the degree of development of the countries. Birthrate is one indicator of this. If there is a perfect relationship between life expectancy and birthrate, you can attribute all of the observed differences in life expectancy to differences in birthrate. In other words, birthrate would explain all of the observed variability in life expectancy. In this situation, the correlation coefficient and its square are both 1. When all of the data points don't fall exactly on the regression line, you can calculate how much of the observed variability in female life expectancy can be attributed to differences in birthrates.

? *How do you do that?* If you square the residuals for all of the cases and add them up, you have a measure of how much variability in life expectancy is *not* explained by birthrates. In this example, the sum of the squared residuals is 83.67. (This is the *Sum of Squares* for *Residual* in Figure 20.12.) You obtain the total variability of the dependent variable by calculating its variance and multiplying by the number of cases minus 1. That gives you 1,342.93. (This is the sum of the *Regression* and *Residual Sum of Squares* in Figure 20.12.) To calculate the proportion of variability that is *not* explained by the regression, divide 83.67 by 1,342.93. This gives you 0.0623. The proportion of variability explained by the regression is 1 minus the proportion not explained, or 0.9377.

A quick way to arrive at the proportion of explained variability is to square the correlation coefficient between the dependent and independent variable. From the value of R^2 (labeled *R Square* in Figure 20.12), you can see that almost 94% of the variability in life expectancies can be explained by differences in birthrates. The remaining 6% is not explained.

The entry in Figure 20.12 labeled *Adjusted R Square* is an estimate of how well your model would fit another dataset from the same population. Since the slope and the intercept are based on the values in your dataset, the model fits your data somewhat better than it would another sample of cases. The value of adjusted R^2 is always smaller than the value of R^2. In this example, the value of adjusted R^2 is 0.933, slightly less than the R^2 value of 0.938.

Some Warnings

- *Don't use the linear regression equation to make predictions when values of the independent variable are outside of the observed range.* For example, in this dataset, the smallest observed birthrate is 13 for France and the Netherlands. The largest is 50 for Burkina Faso and Tanzania. Although a straight line appears to be a reasonable summary for countries with birthrates between 13 and 50, you have no reason to expect that the same relationship will hold for birthrates much smaller or much larger than those observed. It's certainly possible that life expectancy will decline in a nonlinear fashion for birthrates larger than 50. That's why you should restrict your predictions to countries that have birthrates in the same range as your data. Otherwise, it's possible for your predictions to be way off target.
- *Don't calculate a regression equation unless the relationship between the two variables appears to be linear over the entire observed range of the independent variable.* For example, look at Figure 20.13, a hypothetical plot of birthrates and life expectancies for 20 countries. Note that the countries fall into two distinct clusters. Within each cluster, there is no linear relationship between the two variables. However, if you combine the data and calculate a regression line (shown), you may erroneously conclude that there is a linear relationship. The "relationship" is due solely to the fact that the clusters have different average values for the two variables.

Figure 20.13 Not a linear relationship

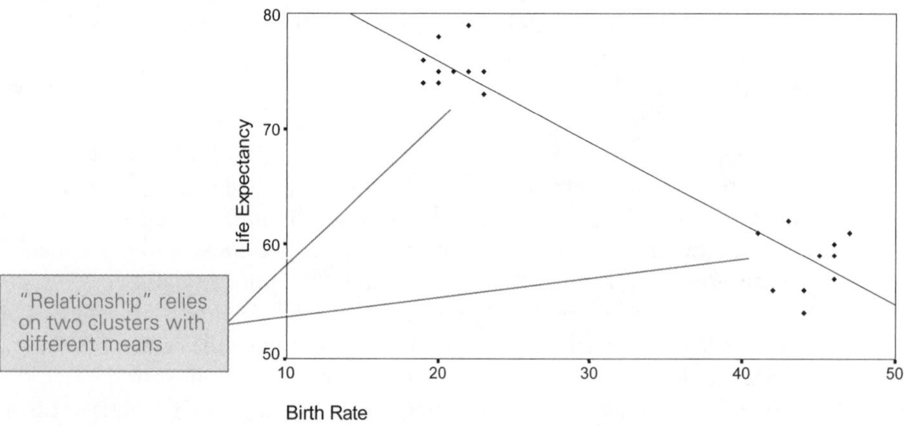

- *Beware of a relationship that depends heavily on a single point.* Consider Figure 20.14. Without the circled point, there is no linear relationship between the two variables. If the indicated point is included in the analysis, the least-squares regression line no longer has a slope of 0. The entire observed relationship is due to a single influential point. In Chapter 22, you'll learn how to identify such influential points.

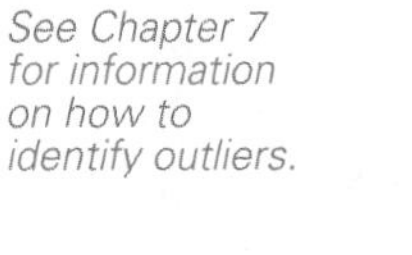

See Chapter 7
for information
on how to
identify outliers.

Figure 20.14 Two unrelated variables

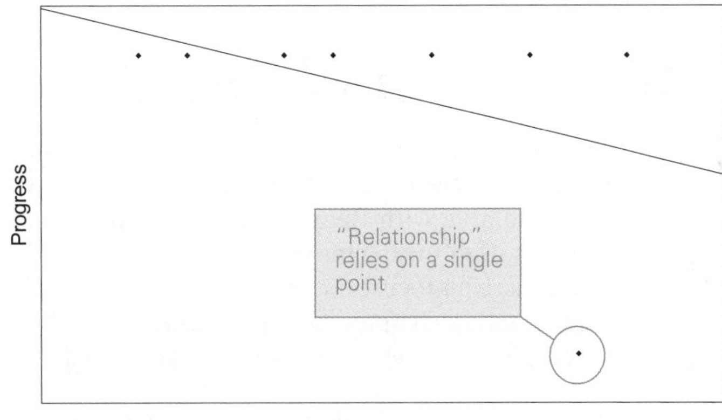

Summary

How can you choose the line that best summarizes the linear relationship between two variables?

- The equation for a straight line is $y = a + bx$. The intercept a is the value of y when x is 0. The slope b tells you how much y increases or decreases for a one-unit change in x.
- The least-squares line has, of all possible lines, the smallest sum of squared vertical distances from the point to the line.
- The absolute value of the Pearson correlation coefficient between two variables tells you how closely the points cluster around a straight line.
- You calculate the predicted value by substituting the observed value of the independent variable into the least-squares regression equation. A residual is the difference between the observed and predicted values.

What's Next?

In this chapter, you simply fit a regression line to a set of data. You did not test any hypotheses about the population from which the sample is selected. In Chapter 21 and Chapter 22, you'll see what assumptions are needed to test hypotheses about the population regression line based on the observed sample. You'll also see how to use residuals from the regression to look for violations of these assumptions.

How to Obtain a Linear Regression

This section describes the basics of obtaining a linear regression analysis using the SPSS Linear Regression procedure. Optional statistics and plots available with Regression are also discussed, including saving the residuals used for diagnostics.

This section focuses on bivariate linear regression, as discussed in this chapter and in Chapter 21 and Chapter 22. Options specific to obtaining a multiple linear regression are discussed in Chapter 23.

▶ To open the Linear Regression dialog box (see Figure 20.15), from the menus choose:

Analyze
  Regression ▶
    Linear...

Figure 20.15 Linear Regression dialog box

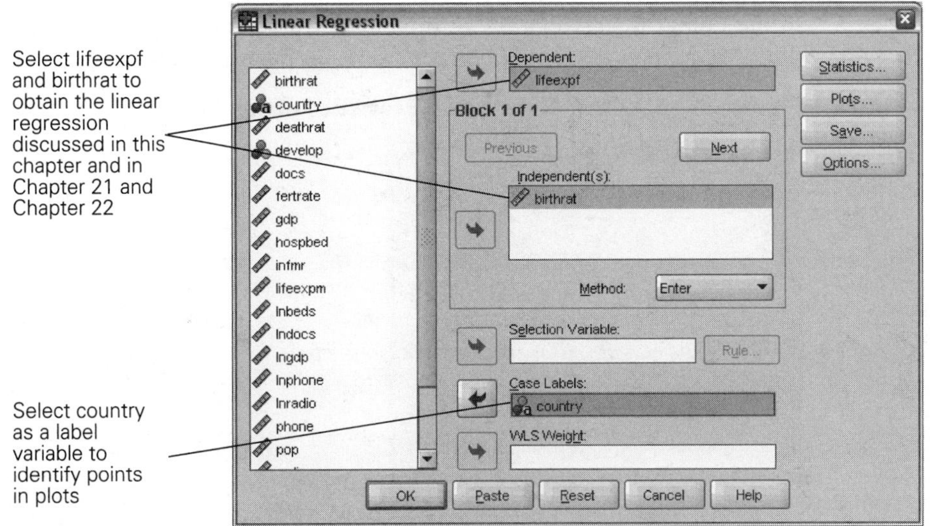

▶ Select the dependent variable and move it into the Dependent list.

▶ Select the independent variable and move it into the Independent(s) list and click **OK**.

This performs a basic regression analysis and displays an analysis-of-variance table for the regression, along with the constant, the regression coefficient, and the other statistics for the regression equation. You can also obtain a multiple linear regression by selecting more than one variable for the Independent(s) list. (Multiple linear regression is discussed in Chapter 23.)

Statistics: Further Information on the Model

In the Linear Regression dialog box, click **Statistics** to open the Linear Regression Statistics dialog box, as shown in Figure 20.16.

Figure 20.16 Linear Regression Statistics dialog box

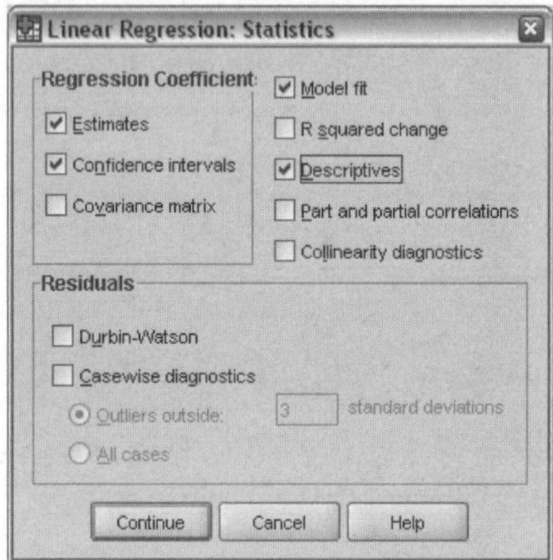

Regression Coefficients. The Regression Coefficients options are:

Estimates. For each independent variable in the model, unstandardized and standardized coefficients, standard error, *t* value, and the two-tailed significance level for the *t* value. For independent variables not in the model, the standardized coefficient if it were in the model, *t* value, significance of *t*, partial correlation with the dependent variable controlling for independent variables in the equation, and minimum tolerance. These statistics are displayed by default. Deselect this option to suppress them.

Confidence intervals. For each unstandardized regression coefficient, a 95% confidence interval.

Covariance matrix. For multiple regression models (with two or more independent variables), a square matrix with covariances of the unstandardized regression coefficients below the main diagonal, correlations above the diagonal, and variances on the diagonal.

Additional available statistics include:

Model fit. The *R* statistic, R^2, adjusted R^2, and the standard error of the estimate, plus an ANOVA table. These statistics are displayed by default. Deselect this option to suppress them.

R squared change. The change in R^2 when an independent variable is added or removed from a regression equation.

Descriptives. Means and standard deviations for all selected variables and a correlation matrix.

Part and partial correlations. Types of correlation coefficients that measure the correlation between a dependent and independent variable, taking into account the effects of the other independent variables.

Collinearity diagnostics. For multiple regression (see Chapter 23), several statistics and tests for collinearity among the independent variables.

Durbin-Watson. Summary statistics for standardized and unstandardized residuals and predicted values, plus the Durbin-Watson statistic measuring serial correlation among the residuals.

Casewise diagnostics. Listing of values of selected variables for cases that have standardized residuals that are in absolute value larger than the specified cutoff.

Residual Plots: Basic Residual Analysis

In the Linear Regression dialog box, click Plots to display the Linear Regression Plots dialog box, as shown in Figure 20.17.

Figure 20.17 Linear Regression Plots dialog box

Select to obtain partial residual plots for multiple regression models, as discussed in Chapter 24

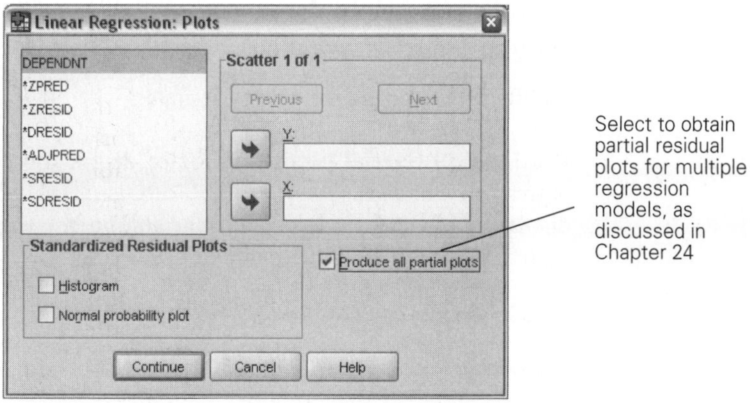

You can request scatterplots with any combination of the dependent variable and any of the residuals listed (these are described in detail below).

▶ To obtain a scatterplot of any two of these, select one of them and move it into the Y (vertical axis) box; then select the other and move it into the X (horizontal axis) box.

For additional plots, click Next and select another pair. To review or modify previously specified plots, click Previous until you reach the plot of interest. You can then select the Y or X variable, move it out of its box, and (if you want) move another variable into its place.

▶ To plot predicted values or residuals against an independent variable, use the Save dialog box to create a new variable containing the predicted values or residuals. (See "Linear Regression Save: Creating New Variables" on p. 465.) Then choose Scatter from the Graphs menu and specify the scatterplot.

Available variables for these plots include:

***ZPRED.** The standardized predicted values of the dependent variable.

***ZRESID.** The standardized residuals.

***DRESID.** Deleted residuals, the residuals for a case when it is excluded from the regression computations.

***ADJPRED.** Adjusted predicted values, the predicted value for a case when it is excluded from the regression computations.

***SRESID.** Studentized residuals.

***SDRESID.** Studentized deleted residuals.

Produce all partial plots. These are diagnostic plots used in multiple regression analysis. They will be discussed in Chapter 24. A plot is displayed for each independent variable in the equation, provided that there are at least two independent variables.

Standardized Residual Plots. The available plots are:

Histogram. A histogram of the standardized residuals, to help you check whether they are normally distributed.

Normal probability plot. A P-P plot of the distribution of standardized residuals against a standard normal distribution.

Linear Regression Save: Creating New Variables

In the Linear Regression dialog box, click **Save** to display the Linear Regression Save New Variables dialog box, as shown in Figure 20.18. Each of the options in this dialog box creates a new variable in your working data file. (This is similar to saving standard scores using the Descriptives procedure, as discussed in "Standard Scores" on p. 92 in Chapter 5.) These variables are used primarily for diagnostics, as discussed in Chapter 22 and Chapter 24.

Figure 20.18 Linear Regression Save New Variables dialog box

Select to save predicted values and residuals, as shown in Figure 20.9

Discussed in Chapter 21. See Figure 21.3.

Distances are discussed in Chapter 24

Standardized and Studentized residuals are discussed in Chapter 22

Discussed in Chapter 24

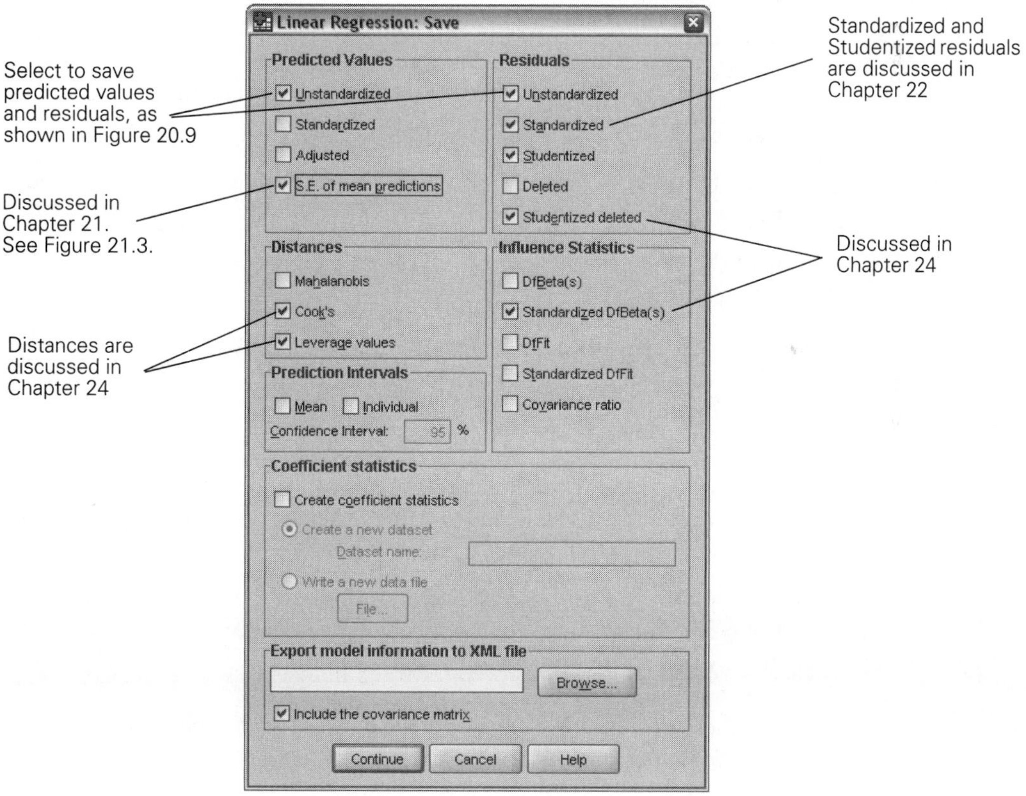

For each option you select, one or more variables will be added to the working data file in the Data Editor. To help you keep track, the names of the newly created variables are displayed at the end of the Notes table in the Viewer.

Summary statistics are calculated and displayed for any new variables requested from the Predicted Values, Distances, and Residuals groups.

Predicted Values. The Predicted Values group includes:

Unstandardized. The value predicted by the regression model for the dependent variable.

Standardized. The predicted value for the dependent variable standardized to have a mean of 0 and a standard deviation of 1.

Adjusted. The predicted value for a case if that case is excluded from the calculation of the regression coefficients.

S. E. of mean predictions. An estimate of the standard error of the mean predicted value.

Distances. The Distances group includes:

Mahalanobis. A measure of the distance of a case from the average values of all of the independent variables.

Cook's. A measure of how much the residuals of all cases would change if the current case were omitted from the calculations.

Leverage values. A measure of how greatly the current case influences the fit of the regression model.

Prediction Intervals. The Prediction Intervals group includes:

Mean. Two new variables containing lower and upper bounds for the prediction interval of the mean value of the dependent variable, for all cases with the given values of the independent variable(s).

Individual. Two new variables containing lower and upper bounds for the prediction interval for the dependent variable for a case with the given values of the independent variables.

Below these options, you can specify the level for the confidence interval by entering a percentage value greater than 0 and less than 100.

Residuals. The Residuals group includes:

Unstandardized. The value of the dependent variable minus its predicted value.

Standardized. The residual divided by an estimate of its standard error.

Studentized. The residual divided by an estimate of its standard error that varies from case to case, depending on the distance of the case's values of the independent variable(s) from the mean values.

Deleted. The residual if the current case were excluded from the calculation of the regression coefficients.

Studentized deleted. The deleted residual divided by an estimate of its standard error.

Influence Statistics. Finally, the Influence Statistics group includes:

DfBeta(s). A new variable for each term in the regression model, including the constant, containing the change in the coefficient for that term if the current case were omitted from the calculations. DfBetas are discussed in Chapter 24.

Standardized DfBeta(s). A new variable for each term in the regression model, including the constant, containing the DfBeta value divided by an estimate of its standard error. Standardized DfBetas are discussed in Chapter 24.

DfFit. The change in the predicted value of the dependent variable if the current case is omitted from the calculations.

Standardized DfFit. The DfFit value divided by an estimate of its standard error.

Covariance ratio. The determinant of the covariance matrix with the current case excluded from the calculations, divided by the determinant of that matrix with the current case included.

Linear Regression Options

In the Linear Regression dialog box, click Options to request regression through the origin or to control the treatment of missing data (see Figure 20.19).

Figure 20.19 Linear Regression Options dialog box

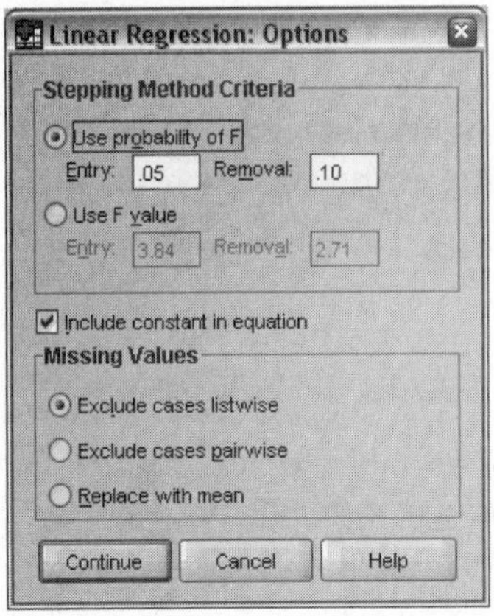

Available options include:

Stepping Method Criteria. These criteria do not apply to bivariate regression. They are described in Chapter 23.

Include constant in equation. Leave this option selected for an ordinary regression equation. Deselect it if you want to constrain the constant term to equal 0. (This leads to regression through the origin, a specialized topic not discussed in this book.)

Missing Values. The available treatments of missing data control the way SPSS obtains the correlation matrix from which the regression statistics are calculated:

Exclude cases listwise. Computes each correlation coefficient using only the cases that have valid data for all variables in the matrix, so that the coefficients in the matrix are all based on the same cases.

Exclude cases pairwise. Computes each correlation coefficient using all of the cases with valid data for the variables involved. In some circumstances, a matrix of such coefficients can be inconsistent (an inconsistent correlation matrix is one that could not possibly be calculated from a single set of cases).

Replace with mean. Missing values for a variable are replaced with the variable's mean value before the correlation matrix is computed.

Exercises

Statistical Concepts

1. Plot the following two points:

Point 1: Age of home = 20, appraised value = \$40,000

Point 2: Age of home = 30, appraised value = \$50,000

a. Write the equation of the straight line that passes through them.

b. What is the value for the intercept?

c. What is the value for the slope?

d. If the line you've drawn predicts appraised value exactly, what would be the appraised value for a 25-year-old home?

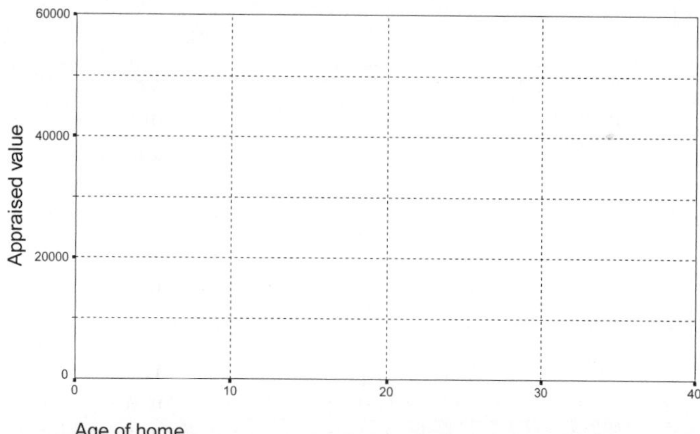

2. A mail-order house is interested in studying the relationship between income and type of product purchased. They take a random sample of orders and then call people to determine family income. They then calculate the correlation coefficient between income and product code. The value is 0.76. Based on this study, what can you conclude about the relationship between income and type of product purchased? Explain.

3. Here are the equations for two regression lines:

Profit = −1000 + 0.25(sales)

Profit = 2000 + 2.0(advertising budget)

Can you tell from the regression lines if the correlation coefficient is larger for profit and sales or for advertising budget and sales? Why or why not?

4. You are a production manager interested in the relationship between defect rate and volume at one of your plants. You've taken a random sample of 160 shifts and recorded the percentage of defective items and the volume. Here is the plot of the two variables and the regression statistics:

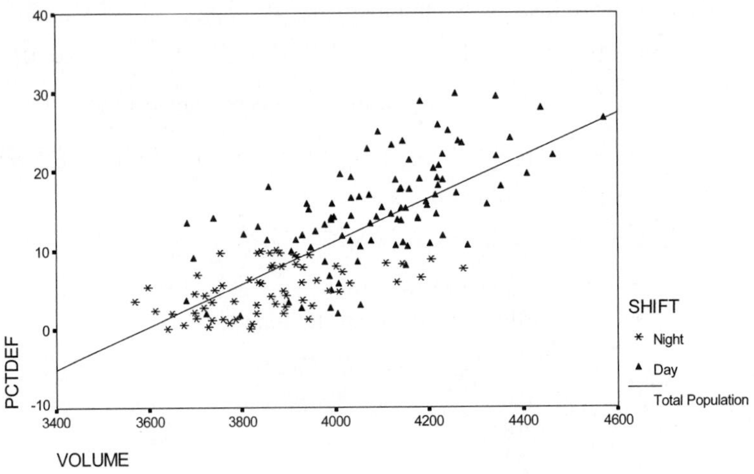

Model	R	R Square	Adjusted R Square	Std. Error of the Estimate
1	.740	.548	.545	4.92

Model		Sum of Squares	df	Mean Square	F	Sig.
1	Regression	4647.124	1	4647.124	191.717	.000
	Residual	3829.839	158	24.239		
	Total	8476.963	159			

Model		Unstandardized Coefficients		Standardized Coefficients		
		B	Std. Error	Beta	t	Sig.
1	(Constant)	-97.073	7.819		50.995	.000
	VOLUME	.027	.002	.740	13.846	.000

a. Is there a relationship between defect rate and volume? If so, is it positive or negative?

b. Which variable is the independent variable and which is the dependent variable?

c. Write out the regression equation and sketch it on the plot.

d. What percentage of the variability in the defect rate can be explained by differences in volume?

e. What defect rate would you predict for a shift with a volume of 4000 units?

f. What defect rate would you predict for a shift with a volume of 9000 units?

g. Would you expect all shifts that produced 4000 items to have the same defect rate?

h. What would you estimate the standard deviation of the distribution of the defect rate to be for a volume of 4000 units?

i. If a particular shift produced 4000 items and had a defect rate of 10%, based on the regression model what would be the residual for the shift?

5. Plot the following points and sketch the least-squares line:

Number of ads	Sales revenues
10	20,000
18	28,000
24	35,000
32	44,000
35	48,000
37	50,000
42	55,000

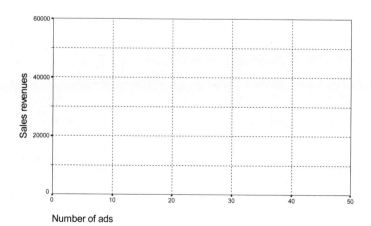

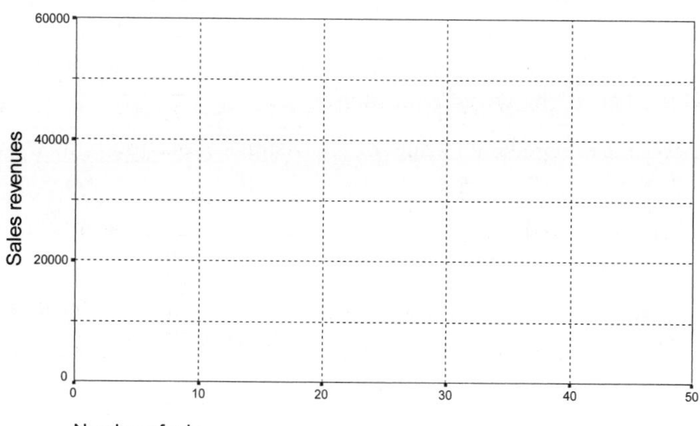

6. For which of the following situations would you consider using linear regression?
   - a. To predict car model preference from age.
   - b. To predict car mileage from its weight in pounds.
   - c. To compare sales performance for three regional offices.
   - d. To examine the relationship between profitability and growth rate.
   - e. To examine the relationship between type of product and customer satisfaction.

Data Analysis

Use the *gss.sav* data file to answer the following questions:

1. Using the Graphs menu, make a scatterplot of husband's education against wife's education (variables *husbeduc* and *wifeduc*). Edit the chart to draw the regression line and print R^2.
   - a. Does there appear to be a linear relationship between the two variables?
   - b. Would you characterize the relationship as positive or negative?
   - c. From the plot, estimate the slope and the intercept. (It's easier to estimate the intercept if you edit the x axis to start at 0.)
   - d. What's the value for the correlation coefficient?
   - e. Describe the points that are far from the regression line.
2. From the Regression procedure, obtain the least-squares estimates for the slope and the intercept.
   - a. Write the regression equation to predict a husband's education from his wife's education. What proportion of the variability in husbands' education can be "explained" by wives' education?
   - b. What is the predicted value for a husband's education if his wife's education is 13 years?
   - c. If the observed value for a husband's years of education is 14 and his wife's 13 years, what's the residual?
   - d. Write the regression equation to predict a husband's education if husbands' and wives' years of education are always the same.
   - e. Write the regression equation to predict a husband's education if there is no linear relationship between husbands' and wives' education.

3. Run the Regression procedure to obtain the least-squares line that predicts a wife's education from her husband's education.

   a. Write the regression equation. Is it the same as that for predicting a husband's education from his wife's education?

   b. What proportion of the variability in wives' education can be explained by husbands' education? Compare this to the proportion of variability in husbands' education that can be explained by wives' education.

4. Run the Descriptives procedure to standardize the values of *husbeduc* and *wifeduc*.

   a. Write the linear regression equation to predict *zhusbedu* from *zwifeduc*. (Instead of displaying 0 for the intercept, SPSS displays a very small number. For example, SPSS might display 15E–19, which means 15 times 10^{-19}. The correct value to use in your equation is 0.)

   b. What's the relationship between the slope and the multiple *R*?

   c. Write the linear regression equation to predict *zwifeduc* from *zhusbedu*. How does that compare to the equation that predicts *zhusbedu* from *zwifeduc*?

   d. Summarize what you learned about the relationship of the slope and correlation coefficient when both the dependent and independent variables are in standardized form.

5. Now let's look at the relationship between a husband's education and his wife's education separately for couples in which the wife is employed full time (variable *wifeft* equals 1) and in which the wife is not employed full time (*wifeft* equals 0).

   a. Use the Graphs menu to make the appropriate scatterplot using *wifeft* as the Set Markers By variable.

   b. Edit the chart so that separate regression lines are drawn for the two types of couples. Print separate R^2 values as well.

   c. Does there seem to be a linear relationship between a husband's and wife's education for both groups of couples? What is the correlation coefficient between a husband's and wife's education for couples in which the wife is employed full time? For couples in which the wife is not employed full time?

6. Select full-time workers only (variable *wrkstat* equals 1). Use the Graphs menu to obtain a plot of hours worked last week (*hrs1*) with education (*educ*). Set the markers by the value of *sex*. Edit the chart to draw regression lines and display R^2 for the subgroups (males and females) and for all cases combined. What can you conclude from the plot?

7. Use the Bivariate Correlations procedure to find all pairs of correlation coefficients between mother's education, father's education, respondent's education, and spouse's education (variables *maeduc*, *paeduc*, *educ*, and *speduc*).

a. What is the correlation coefficient between a person's years of education and the spouse's years of education? Why isn't this the same correlation coefficient as that between a husband's and wife's education?

b. What proportion of the variability in the father's education can be explained by the mother's education?

c. Discuss the statement "The father's education is a much more powerful predictor of the educational attainment of children than is the mother's education."

Use the ***salary.sav*** data file to answer the following questions:

8. Use the Graphs menu to make a scatterplot of current salary on the y axis against beginning salary (variables *salnow* and *salbeg*). Identify the points based on gender.

a. Edit the chart to include the regression line and R^2 for the entire sample. Does the relationship appear to be linear? What proportion of variability in current salaries is "explained" by beginning salary? From the plot, estimate the equation for the regression line to predict current salary from beginning salary.

b. Edit the chart to show separate regression lines for males and females. Does the relationship between current salary and beginning salary appear to be different for males and females?

9. Use the Regression procedure to estimate the least-squares regression line to predict current salary from beginning salary. Write the equation. For a person with a beginning salary of \$12,000, what would you predict for the current salary? If the actual current salary is \$14,000, what is the residual?

10. Rerun the analysis in question 9 to predict beginning salary from current salary. Do you get the same results? Why not?

11. Use the Descriptives procedure to standardize current salary and beginning salary. Rerun the regression model in question 9 using the standardized variables. What's the relationship between the slope and the correlation coefficient?

12. Use the Bivariate Correlations procedure to calculate Pearson correlation coefficients between beginning salary, current salary, job seniority, work experience, and age (variables *salbeg*, *salnow*, *time*, *work*, and *age*). Which pair of variables is most strongly linearly related?

Use the ***country.sav*** data file to answer the following questions:

13. Plot female life expectancy against birthrate (variables *lifeexpf* and *birthrat*) for all countries in the data file. Edit the plot to include the regression line and R^2. What proportion of the variability in female life expectancy is "explained" by birthrate? How does this value compare to that from the data file with 15 countries used in this chapter?

14. What is the linear regression equation to predict female life expectancy from birthrate when all of the cases are used? Did the coefficients change much? Based on the regression model, predict female life expectancy for a country with a birthrate of 20 per 1000 population.

15. Plot male life expectancy (variable *lifeexpm*) against female life expectancy. Edit the chart to draw the regression line and give R^2. Using point selection mode, identify any unusual countries.

a. From the plot, estimate the intercept and slope of the regression equation. Compare them to the intercept and slope from the Regression procedure.

b. If male and female life expectancy were the same, write the regression equation to predict male life expectancy from female life expectancy.

c. If male life expectancy was exactly 75% of female life expectancy, write the corresponding regression equation.

d. What proportion of the variability in male life expectancy is "explained" by female life expectancy?

e. What's the correlation coefficient between male life expectancy and female life expectancy? Is the relationship positive or negative?

f. Based on the regression equation, what's the male life expectancy for a country with a female life expectancy of 60? For a country with a female life expectancy of 40? For a country with a female life expectancy of 35? What reservations do you have about predicting male life expectancy when female life expectancy is 35?

16. Plot female life expectancy against the number of phones per 100 people (variable *phone*).

a. Is the relationship between the two variables linear? Calculate the correlation coefficient between the two variables. Is it a reasonable summary of the relationship? Explain.

b. Plot female life expectancy against the natural log of the number of phones (variable *lnphone*). How would you characterize the relationship? Edit the chart to include the regression line and R^2. What proportion of the observed variability in female life expectancy is "explained" by the log of the number of phones?

c. Run the Regression procedure to estimate the slope and intercept of the regression line that predicts female life expectancy from the log of the number of phones.

Use the *schools.sav* data file to answer the following questions:

17. Plot ACT scores in 1993 (variable *act93*) and the percentage of students taking the ACT in 1993 (variable *pctact93*), using variable *medloinc* (whether a school is above or below the median in percentage of low income students) to set the markers. Edit the chart to include separate regression lines and R^2 for the two groups. Summarize your findings.

18. Look at the relationship between ACT scores (variable *act93*), percentage of low-income students (*loinc93*), percentage of limited-English-proficiency students (*lep93*), and graduation rates (*grad93*). Describe the relationships you see. Based on the plots, which variable do you think is the best predictor of ACT scores?

21 Testing Regression Hypotheses

How can you test hypotheses about the population regression line based on the results you obtain in a sample?

- What is the population regression line?
- What assumptions do you have to make about the data to test hypotheses about the population regression line?
- How do you test the null hypothesis that the slope or the correlation coefficient is 0 in the population?
- What is the difference between the confidence interval for the mean prediction and the prediction interval for an individual case?

In Chapter 20, you summarized the relationship between two variables by fitting a regression line to the sample data. Often, however, you want to do more than that. You want to draw conclusions about the relationship of the two variables in the *population* from which the sample was selected. For example, you want to draw conclusions about the relationship between female life expectancy and birthrate not only for the 15 countries in your sample, but for all nations. In this chapter, you'll learn how to test hypotheses about the population regression line.

The Population Regression Line

Whenever you test a hypothesis, you draw a conclusion about one or more populations based on results you've observed in a sample. For example, in previous chapters you tested a variety of hypotheses about population means based on sample results. You test regression hypotheses in the same way. You draw conclusions about the relationship of two variables in the population based on results you observed in a sample. If you observed the entire population, your regression line would be the "true" or population regression line. However, since you have only a sample from the population, you don't know the true slope and intercept. Instead, you estimate their values based on sample results.

▶ This chapter continues the linear regression analysis discussed in Chapter 20. For instructions on how to obtain the output shown, see "How to Obtain a Linear Regression" on p. 460 in that chapter.

Assumptions Needed for Testing Hypotheses

When you fit a regression line only to summarize the observed relationship between two variables, you must keep two questions in mind:

- Are the variables measured on at least an ordinal scale? It makes no sense to calculate a regression line between variables such as color preference and state of residence. If there is no meaningful order to the values of the variables, the slope and intercept are meaningless statistics.
- Does the relationship between the two variables appear to be linear? You shouldn't use a straight line regression if the data points fall on a curve. (Transforming a nonlinear relationship to a linear one is discussed in Chapter 22.)

If you want to test hypotheses about the population regression line, your data must satisfy additional assumptions. If these assumptions are met, the distributions of all possible sample values of the slope and intercept are normal. You can then easily determine whether observed values of the slope and intercept are "unusual."

- *All of the observations must be independent.* Inclusion of one case in your sample must not influence the inclusion of another case. If you obtain several pairs of values from the same case, you are violating the assumption of independence. For example, when studying the relationship between systolic blood pressure and age, if you measure the same person's blood pressure at three different ages, the observations are not independent.
- *For each value of the independent variable, the distribution of the values of the dependent variable must be normal.* You know that all countries with a birthrate of 13 don't have the same life expectancy. Instead, there is a distribution of life expectancies for each value of birthrate. This distribution must be normal.
- *The variance of the distribution of the dependent variable must be the same for all values of the independent variable.* The variance of the distribution of female life expectancy must be the same for all birthrates.

- *The relationship between the dependent and the independent variable must be linear in the population.* In other words, the means of the distributions of the dependent variable must fall on a straight line.

Figure 21.1 Linear regression assumptions

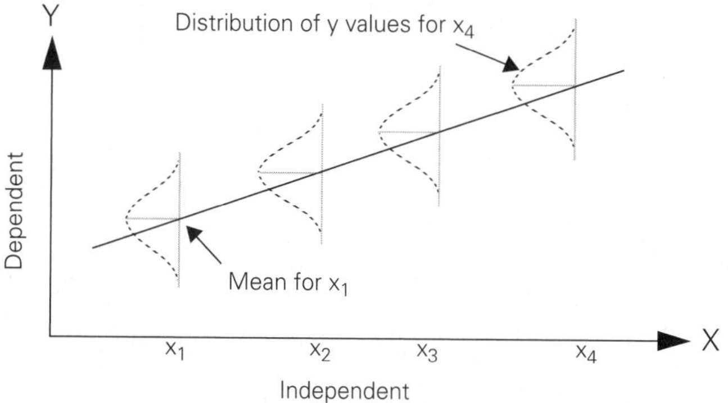

Figure 21.1 illustrates these assumptions. You see that for each value of the independent variable (x), there is a normal distribution of values of the dependent variable (y). The variances of all of the distributions are the same. The means of all of the distributions fall on the population regression line. In the next chapter, you'll learn how to examine your data to see if they violate any of the required assumptions. You'll also see what you can do if the required assumptions are not met.

Testing Hypotheses

To see how to test hypotheses about the population regression line, consider again the life expectancy and birthrate data. Figure 21.2 contains the slope and intercept values calculated in the previous chapter. The slope and intercept values define the least-squares regression line that passes through your data points. These values can also be thought of in another way: they are your best guess or estimate of the unknown population values for the slope and the intercept. Based on these estimates, you can test hypotheses about the population values for the slope and intercept.

Figure 21.2 Regression coefficients and model summary

This is the same analysis described in Chapter 20. Select the variables lifeexpf and birthrat in the Linear Regression dialog box, as shown in Figure 20.15.

In the Linear Regression Statistics dialog box, select Confidence intervals, as shown in Figure 20.16.

In the Linear Regression Save New Variables dialog box, select S.E. of mean predictions, as shown in Figure 20.18.

		Unstandardized Coefficients		Standardized Coefficients			95% Confidence Interval for B	
Model		B	Std. Error	Beta	t	Sig.	Lower Bound	Upper Bound
1	(Constant)	89.985	1.765		50.995	.000	86.173	93.797
	BIRTHRAT	-.697	.050	-.968	-13.988	.000	-.805	-.590

Model	R	R Square	Adjusted R Square	Std. Error of the Estimate
1	.968	.938	.933	2.54

The values that you obtained for the slope and intercept are based on one sample from the population. If you take a different sample from the same population, you will get different values for the slope and intercept. The distributions of all possible values of the slope and intercept are normal if the regression assumptions are met. The standard deviations of these distributions are called the standard error of the slope and the standard error of the intercept. They are estimated from the data. In Figure 21.2, the standard errors of the slope and intercept are shown in the column labeled *Std. Error*. (The estimate of the variance of the dependent variable for each value of the independent variable is the entry labeled *Std. Error of the Estimate* in Figure 21.2.)

Testing that the Slope Is Zero

When you calculate a linear regression, you want to test whether there is a linear relationship between the two variables in the population. This is equivalent to testing the null hypothesis that the population slope is 0. You must calculate the probability of obtaining a sample slope at least as large in absolute value as the one you observed, if the null hypothesis is true. As always, you reject the null hypothesis if this probability, the observed significance level, is small.

In this example, the sample slope is –0.70 and its standard error is 0.05, so the value for the *t* statistic is –0.70/0.05, which is –14. (You use the *t* distribution, since the distribution of the sample slopes is normal and the standard error is not known but is estimated from the sample.) The sample slope is 14 standard error units below the hypothesized value of 0. Since this is a most unlikely event (the observed significance level is less than 0.0005), you can reject the null hypothesis. There appears to be a linear relationship between 1992 female life expectancy and birthrate.

? *Does this mean that the straight line is the correct model for these data?* Not necessarily. All it says is that a straight line is better than a model that does not include the independent variable. For example, if your data points fall on a curve but you fit a straight line to them, you may well reject the null hypothesis that the slope is 0. The values of the two variables may increase or decrease together, though not in a completely linear fashion. Always plot your data values to see if a straight line model is appropriate. ■■■

The test that the slope is 0 is the same as the test that the population correlation coefficient is 0. So, if you reject the null hypothesis that the slope is 0, you can also reject the null hypothesis that the population correlation coefficient is 0. This isn't surprising, since as you saw in Chapter 20, the correlation coefficient is the slope when both the independent and dependent variables are standardized to have a mean of 0 and a standard deviation of 1.

? *If one of the variables doesn't have a normal distribution for each value of the other variable, can I still test hypotheses about the population correlation coefficient?* In this situation, you may want to consider a nonparametric alternative to the Pearson correlation coefficient. For example, you can compute Spearman's correlation coefficient. It's the Pearson correlation coefficient when the data values for each variable are replaced by ranks. It measures the linear relationship between the two sets of ranks. You don't need the assumption of normality to use it. ■■■

You can also test the null hypothesis that the population value of the intercept is 0 using the same procedure as outlined for the slope. This test is of limited interest, however, since it simply tells you whether the regression line passes through the **origin**, which is the point where the values of both variables are 0. To find out whether there is a linear relationship between two variables, you test that the slope is 0.

Confidence Intervals for the Slope and Intercept

The sample values for the slope and intercept are your best guesses for the population values. However, as is always the case when you're estimating population values from a sample, it's most unlikely that the sample values are exactly on target. You can get an idea of a range of possible population values by calculating confidence intervals for the population slope and the intercept. All values that fall within these intervals are plausible population values. In Figure 21.2, you see that the 95% confidence interval for the population slope ranges from –0.805 to –0.590.

? *Where does that come from?* The 95% confidence interval for the slope is calculated just like the 95% confidence interval for the mean. Instead of the mean and the standard error of the mean, you substitute the slope and the standard error of the slope. For example, the lower limit of the 95% confidence interval for the slope is $-0.697 - 2.16 \times 0.05$. (2.16 is the t value with 13 degrees of freedom, which has 2.5% of the area to the right of it, and 0.05 is the standard error of the slope.) Similarly, the upper limit is $-0.697 + 2.16 \times 0.05$. ■■■

Notice that the 95% confidence interval does not include the value 0. The 95% confidence interval will include 0 only if the observed significance level for the test that the slope is 0 is greater than 0.05. You can't reject the null hypothesis that the population slope is any of the values within the confidence interval. (Recall that the same relationship was true for testing hypotheses about means.)

Predicting Life Expectancy

In Chapter 20, you used the regression equation to predict values for life expectancy based on birthrates. The equation for female life expectancy is

$$\text{predicted life expectancy} = 89.99 - 0.697 \times \text{birthrate}$$ **Equation 21.1**

For a country with a birthrate of 30 per 1000 population,

$$\text{predicted life expectancy} = 89.99 - 0.697 \times 30 = 69.08 \text{ years}$$

Equation 21.2

Do you have absolute confidence in this prediction? Of course not. First of all, you know that if you take another sample of countries, you would get slightly different estimates of the slope and intercept. You also know that all countries with birthrates of 30 don't have the same life expectancy. All kinds of values are possible. To make your prediction more useful, you must estimate the variability associated with it. If you know that for a birthrate of 30, the range of plausible values for life expectancy is from 40 to 100, you know that your prediction is pretty useless. On the other hand, if the range of values is from 65 to 73, your estimate is more useful.

Predicting Means and Individual Observations

Before you can estimate the variability of your predicted value, you must decide which of two types of predictions you are interested in making. Do you want to predict the average life expectancy for females for *all* countries with a birthrate of 30? Or do you want to predict the life expectancy for a particular country such as Morocco, which has a birthrate of 30? In both cases, the predicted value is the same; what differs is the variability.

In previous chapters, you saw that means vary less than individual observations. So, you shouldn't be surprised that you can predict the average life expectancy for all countries with the same birthrate with less variability than you can predict the life expectancy for an individual country. When you predict the mean, you just worry about how much the predicted mean differs from the population mean for that birthrate. When you predict an individual case, you also have to worry about how much the individual case differs from the predicted mean for that birthrate.

Consider the unlikely situation when your sample and population regression lines are identical. For each birthrate, you can predict the mean life expectancy values perfectly, but you still can't predict values for individual countries perfectly. That's because the values for individual countries don't fall exactly on the regression line—only the average values for countries with the same birthrate do.

Whether you are predicting the value for the mean or for an individual case, the variability of the prediction also depends on the values of the independent variable. Predictions are most stable for values of the independent variable close to the sample mean. That's because the regression line always passes through the point corresponding to the mean of the dependent variable and the mean of the independent variable. Different samples from the same population don't change the predicted value as much for points close to the mean as they do for points farther away. As the distance from the mean increases, so does the variability associated with the prediction. For the 15 countries in your sample, the average birthrate is 32.87. So, you would expect the variability of the predicted value to be smallest for birthrates close to 33.

Standard Error of the Predicted Mean

Figure 21.3 is a plot of the standard error of the predicted mean life expectancy for different values of birthrate.

Figure 21.3 Plot of standard error of predicted mean

In the Simple Scatterplot legacy dialog box, select the variables birthrat and sep_1.

The variable sep_1 is created by the Regression procedure. See "Linear Regression Save: Creating New Variables" on p. 465 in Chapter 20.

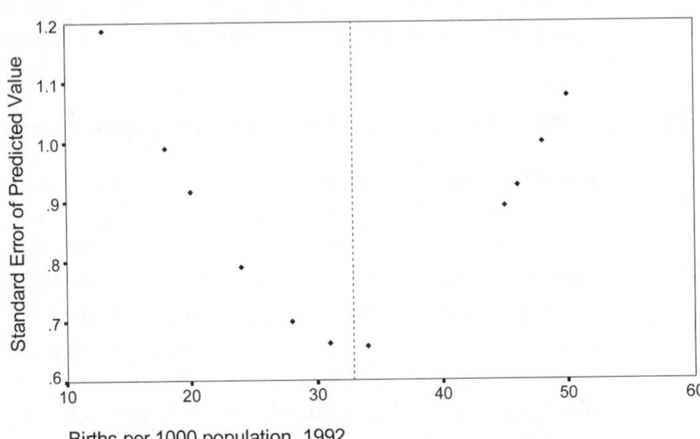

The vertical line at 32.9 is the average birthrate for all cases. You see that the standard error is smallest at the mean. The farther birthrates are from the sample mean, the larger the standard error of the predicted means.

? *How do you compute the standard error of the predicted mean?* You can calculate the standard error of the predicted mean female life expectancy for all countries with a birthrate of 30 (denoted as X_0) using the following formula:

$$S_{\hat{Y}} = S\sqrt{\frac{1}{N} + \frac{(X_0 - \bar{X})^2}{(N-1)S_X^2}} = 2.54\sqrt{\frac{1}{15} + \frac{(30 - 32.87)^2}{14(13.60)^2}} = 0.67$$

In the formula, S is the standard error of the estimate, N is the number of cases in the sample, $\bar{X}$ is the mean, and S_X^2 is the variance of the independent variable. The standard error of the estimate is your best guess for the standard deviation of the dependent variable for any value of the independent variable. It's simply the standard deviation of the residuals, using the number of cases minus 2 in the denominator instead of the number of cases minus 1. You can find it in Figure 21.2.

Confidence Intervals for the Predicted Means

Once you have the standard error of the predicted mean value, you can calculate a confidence interval for the population mean for any value of the independent variable.

In Figure 21.4, you see a plot of the data, the regression line, and the 95% confidence interval for the mean predictions. As you expect, the confidence interval is narrowest for values close to 32.9, the sample mean. It grows wider as the distance from the average birthrate increases.

Figure 21.4 Data, regression line, and 95% confidence intervals

To obtain this scatterplot, select the variables lifeexpf and birthrat in the Simple Scatterplot dialog box.

In the Chart Editor, from the Elements menu choose Fit Line at Total. In the Properties dialog box, select Confidence Intervals for Mean. Click Apply.

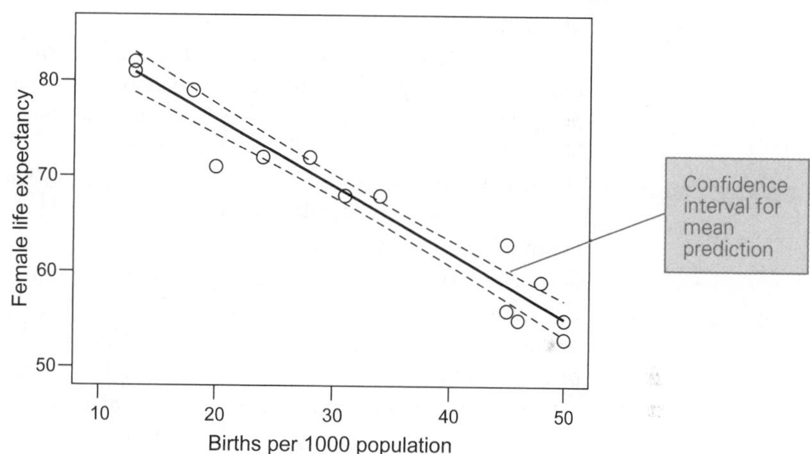

Consider a birthrate of 30. The predicted value for the average female life expectancy for countries with a birthrate of 30 is 69.08 years. The standard error of the mean prediction is 0.67. The 95% confidence interval for the mean prediction is from 67.62 years to 70.51 years. You calculate the 95% confidence interval for the mean prediction the same way you've calculated other confidence intervals. The lower limit is $69.08 - 2.16 \times 0.67$; the upper limit is $69.08 + 2.16 \times 0.67$. (2.16 is the t value with 13 degrees of freedom, which has 2.5% of the area to the right of it.)

Although you don't know if this particular confidence interval includes the population value for the average female life expectancy in 1992 for countries with a birthrate of 30, you do know that 95 times out of 100, the 95% confidence interval will include the true population mean.

Prediction Intervals for Individual Cases

Now consider what's involved in predicting the female life expectancy for Morocco, a country with a birthrate of 30. Your predicted value is the same as before: 69.08 years. What changes is the standard error of the prediction. The standard error of the individual prediction is

$$S_{ind} = \sqrt{S^2 + S_{\hat{Y}}^2} = \sqrt{2.54^2 + 0.67^2} = 2.63$$ **Equation 21.3**

where S is the standard error of the estimate and $S_{\hat{Y}}$ is the standard error of the mean prediction.

The standard error of the individual prediction depends both on how much the mean prediction varies and how much the values of the dependent variable vary for a particular value of the independent variable. When you predict an individual observation, you have two sources of error: the regression line differs from the actual population line and an individual observation differs from the mean. The standard error of the mean prediction takes care of the first source of variability, and the standard error of the estimate takes care of the second. (For large sample sizes, the standard error of an individual prediction will be equal to the standard error of the estimate. That's why it's called the standard error of the estimate.)

For an individual prediction, you can calculate an interval closely akin to a confidence interval. It's called a **prediction interval.** This is a range of values that you expect to include the actual value for a particular case. (Confidence intervals are for population values like the mean, slope, or intercept, not for values of individual cases.) A 95% prediction interval is calculated similarly to a 95% confidence interval for the mean prediction. The standard error of the individual prediction is used in place of the standard error of the mean prediction.

The 95% prediction interval for Morocco is from 63.40 years to 74.73 years. The true female life expectancy for Morocco is 67 years, so the true value falls within the prediction interval. As you would expect, the 95% prediction interval for Morocco is wider than the 95% confidence interval for the predicted mean value for all countries with a birthrate of 30 (67.62 years to 70.51 years). That's because the standard error of the individual prediction is always larger than the standard error of the mean prediction.

Figure 21.5 Prediction intervals and confidence intervals

To add prediction intervals, activate the previous chart in the Chart Editor. From the Elements menu, choose Fit line at Total. In the properties dialog box, select Confidence Intervals Individual and then click Apply.

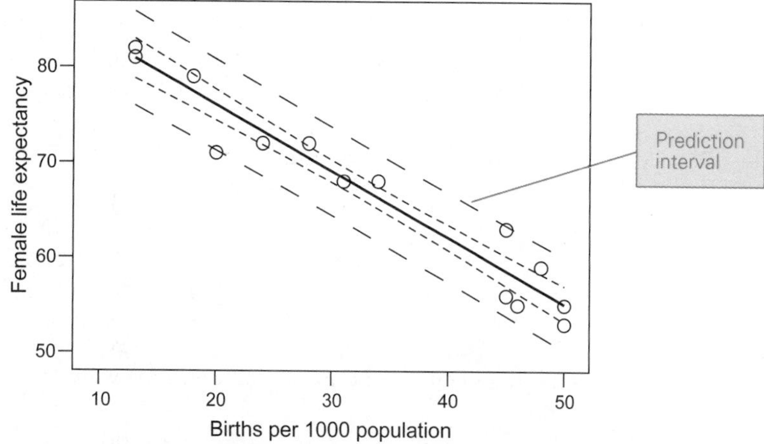

Figure 21.5 illustrates much of what we've been talking about in this chapter. You see the 15 data points and the least-squares regression line. The two bands closest to the regression line show the 95% confidence intervals for the predicted mean. The distance between the two bands is narrowest at 32.87, the mean birthrate for all cases. You see that quite a few of the data points don't fall within these bands. That's because the interval is for predicting mean values, not values for individual cases. The prediction intervals, shown by the two outermost bands, are much wider than the confidence intervals. In this example, all of the observed data points fall within the prediction intervals, though that need not be the case.

Summary

How can you test hypotheses about the population regression line based on the results you obtain in a sample?

- The population regression line is the line that describes the relationship between the dependent variable and the independent variable in the population.
- To draw conclusions about the population regression line, you must assume that for each value of the independent variable, the distribution of values of the dependent variable is normal with the same variance. The means of these distributions must all fall on the population regression line.
- The confidence interval for the mean prediction provides you with a range of values that you expect will include the population mean value. The prediction interval for an individual case provides you with a range of values that you expect will contain the observed value for a particular case.

What's Next?

In this chapter, you learned about the assumptions necessary for testing hypotheses about a population regression line. You also learned how to test hypotheses and compute confidence intervals for commonly used regression statistics and predicted values. In Chapter 22, you'll see how to examine your data to see if they violate these assumptions.

How to Obtain a Bivariate Correlation

This section shows how to obtain Pearson correlation coefficients and two different nonparametric correlation coefficients from SPSS using the Bivariate Correlations procedure. The Pearson correlation coefficient is appropriate for variables measured at the interval level, while the Kendall and Spearman coefficients assume only an ordinal level of measurement.

In addition, the Correlations command can:

- Display full or condensed information about statistical significance.
- Display univariate statistics, cross-product deviations, and covariances.
- Compute one- or two-tailed observed significance levels.

▶ To open the Bivariate Correlations dialog box (see Figure 21.6), from the menus choose:

Analyze
 Correlate ▶
  Bivariate...

Figure 21.6 Bivariate Correlations dialog box

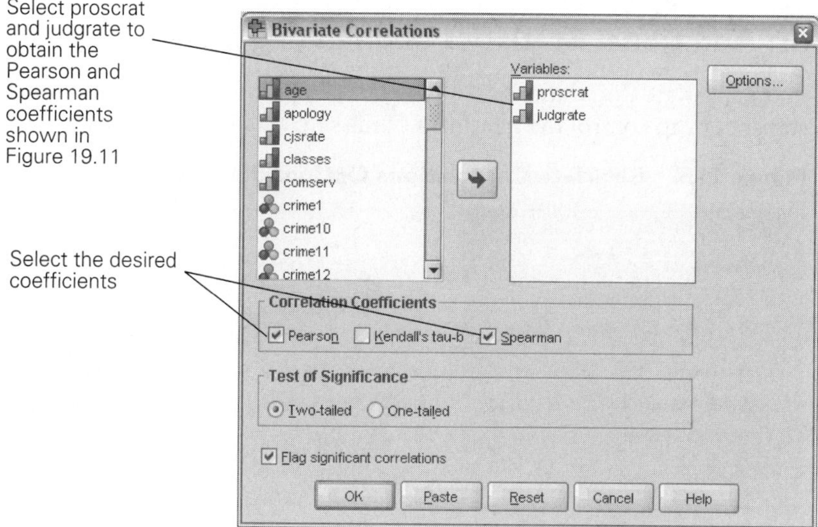

▶ In the Bivariate Correlations dialog box, select two or more numeric variables and move them into the Variables list.

This produces a matrix of correlation coefficients for all pairs of the selected variables.

Correlation Coefficients. At least one type of correlation must be selected. You can choose one or more of the following:

Pearson. Describes the strength of the linear association between variables measured at the interval level.

Kendall's tau-b. Describes the strength of the association between variables measured at the ordinal level.

Spearman. A rank-order correlation coefficient that also measures association at the ordinal level. This is simply the Pearson correlation when the data values are replaced by ranks.

Test of Significance. Select one of the alternatives to obtain two-tailed or one-tailed tests of statistical significance.

Flag significant correlations. If you select Flag significance correlations, correlations significant at the 0.05 level are marked with an asterisk, and those significant at the 0.01 level with two asterisks.

Options: Additional Statistics and Missing Data

In the Bivariate Correlations dialog box, click Options to obtain additional statistics or to control the treatment of missing data (see Figure 21.7).

Figure 21.7 Bivariate Correlations Options dialog box

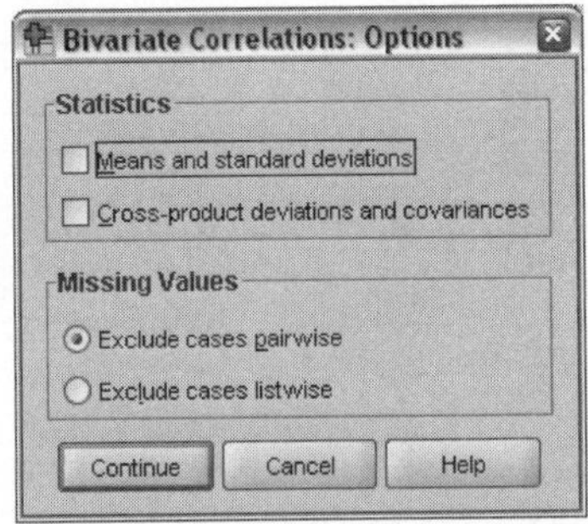

Statistics. The optional statistics are:

Means and standard deviations. Displayed for the variables in the correlation matrix.

Cross-product deviations and covariances. Displayed for each pair of variables.

Missing Values. The available treatments of missing data are:

Exclude cases pairwise. This computes each correlation coefficient using all of the cases with valid data for the two variables involved.

Exclude cases listwise. This computes each correlation coefficient using only the cases that have valid data for all variables in the matrix, so that the coefficients in the matrix are all based on the same cases.

How to Obtain a Partial Correlation

This section shows how to obtain partial correlation coefficients among a group of variables from SPSS, while controlling for one or more additional variables. SPSS first calculates a matrix of zero-order Pearson correlation coefficients among all variables, including the control variables. It then uses these zero-order coefficients to compute the matrix of partial correlation coefficients among variables in the Variables list, controlling for all of the specified control variables.

▶ To open the Partial Correlations dialog box (see Figure 21.8), from the menus choose:

Analyze
  Correlate ▶
    Partial...

Figure 21.8 Partial Correlations dialog box

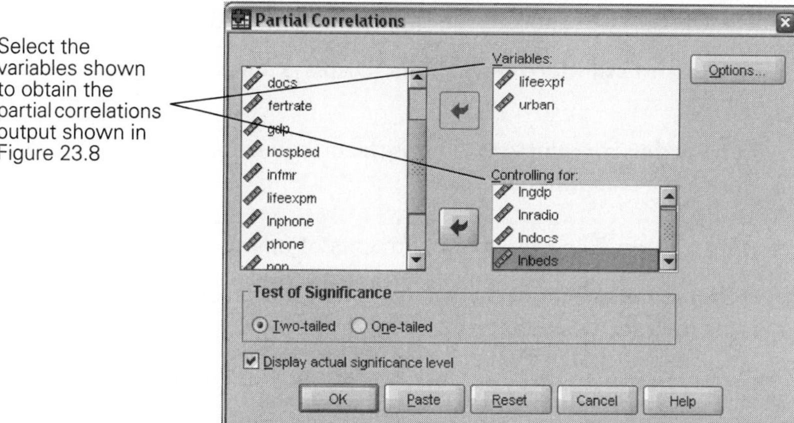

▶ Select two or more numeric variables and move them into the Variables list.

▶ Move one or more numeric control variables into the Controlling For list.

SPSS calculates and displays the observed significance level. If you deselect the Display actual significance level option, significance levels are not displayed, but correlations significant at the 0.05 level are marked with one asterisk, and those significant at the 0.01 level with two asterisks.

Test of Significance. Select one of the alternatives to obtain two-tailed or one-tailed tests of statistical significance.

Options: Additional Statistics and Missing Data

In the Partial Correlations dialog box, click the Options button to obtain additional statistics or to control the treatment of missing data (see Figure 21.9).

Figure 21.9 Partial Correlations Options dialog box

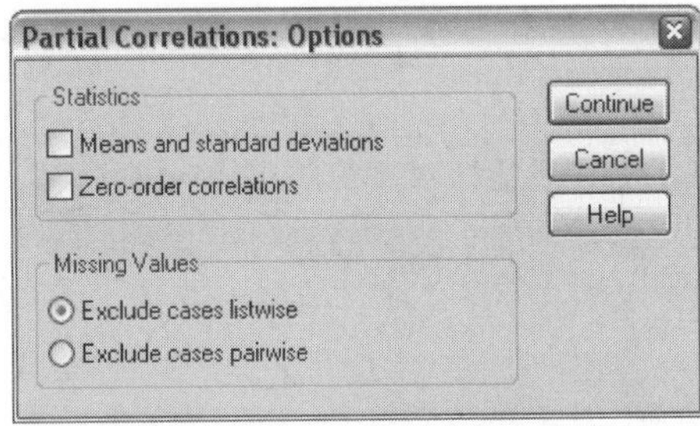

Statistics. You can choose from the following options:

Means and standard deviations. Displays statistics for the variables in both lists.

Zero-order correlations. This is the matrix from which partial correlations are computed.

Missing Values. The available treatments of missing data are:

Exclude cases listwise. This computes each correlation coefficient using only the cases that have valid data for all variables in both variable lists, so that all partial correlation coefficients in the matrix are based on the same set of cases.

Exclude cases pairwise. This computes each zero-order correlation using all of the cases with valid data for the two variables involved. This uses as much of the data as possible, but means that the partial correlation coefficients may not be based on the same set of cases.

Exercises

Statistical Concepts

1. As a personnel manager, you're interested in studying the relationship between salary and years on the job for sales associates. You're also interested in whether males and females are similarly reimbursed. You have obtained salary and experience data from a sample of 1,000 associates.

   a. Explain how you would go about analyzing the data.

   b. If the years of experience were less than five for all your associates, what difficulties do you see with drawing conclusions about salaries for employees with more than five years of experience?

2. You are reading an article that presents the following equation for the relationship of weekly dollars spent for restaurant meals and total family income:

   restaurant dollars = 2.1 + 0.12(weekly income)

   The observed significance level for the slope is reported as 0.03.

   a. From the value of the slope, can you tell how well the model fits the data?

   b. Interpret the observed significance level reported for the equation.

   c. If the value of the slope in the regression model is tripled, does that mean that the model fits the data better? Does it mean that the observed significance level will necessarily be smaller?

   d. Is it possible to reject the null hypothesis that the slope is 0 and not reject the null hypothesis that there is no linear relationship between the two variables?

3. You obtain the following regression statistics for the relationship between defect rate and volume at one of your plants. You have a random sample of results from 160 shifts at the plant.

Model	R	R Square	Adjusted R Square	Std. Error of the Estimate
1	.740	.548	.545	4.92

Model		Sum of Squares	df	Mean Square	F	Sig.
1	Regression	4647.124	1	4647.124	191.717	.000
	Residual	3829.839	158	24.239		
	Total	8476.963	159			

Model		Unstandardized Coefficients		Standardized Coefficients	t	Sig.
		B	Std. Error	Beta		
1	(Constant)	-97.073	7.819		50.995	.000
	VOLUME	.027	.002	.740	13.846	.000

a. What are the null and alternative hypotheses?

b. What is the population of interest? What is the sample?

c. On the basis of the output, what can you conclude about the null hypothesis?

d. Can you reject the null hypothesis that the slope is 0?

e. Can you reject the null hypothesis that there is no linear relationship between the dependent and independent variables?

f. Can you reject the null hypothesis that the population correlation coefficient is 0?

g. What would you predict the defect rate to be on a day when the volume is 4200 units? What would you predict the average defect rate to be for all days with production volumes of 4200?

h. In what way do the two estimates of the defect rate in question 3g differ? (Calculations not required.)

4. Here are two equations for predicting sales volume:

predicted sales = 120 + 6(advertising budget)

predicted sales = 50 + 20(sales force)

Which of the two equations is the better predictor of sales? Explain your answer.

5. A marketing analyst writes his own regression analysis program. The program produces the following regression line and 95% confidence interval for the mean prediction. Do you think the new program is producing correct results? Why or why not?

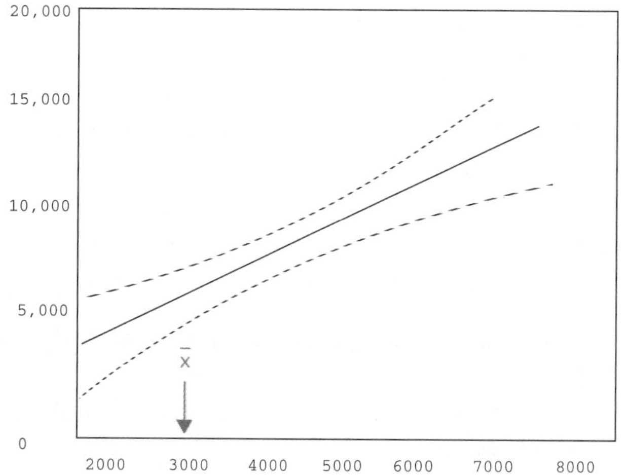

6. Fill in the missing entries in the following output:

Model		Unstandardized Coefficients		Standardized Coefficients	t	Sig.
		B	Std. Error	Beta		
1	(Constant)		2.728		1.576	.213
	X	1.700	.823	.766		.131

Would you reject the null hypothesis that there is no linear relationship between X and the dependent variable? On what do you base your decision?

Data Analysis

Use the *gss.sav* data file to answer the following questions:

1. Run a regression equation to predict father's education from mother's education (variables *paeduc* and *maeduc*). Include 95% confidence intervals for the slope and intercept. Save the standard error of the mean prediction.

   a. Write the linear regression equation to predict father's education from mother's education.

b. Based on the results of the linear regression, can you reject the null hypothesis that there is no linear relationship between father's and mother's education?

c. What proportion of the variability in mother's education is explained by father's education?

d. How can you tell from the slope if the correlation coefficient between the two variables is positive or negative?

e. What can you conclude about the population correlation coefficient based on what you know about the slope? Can you reject the null hypothesis that the population correlation coefficient is 0?

f. Based on the 95% confidence interval for the slope, can you reject the null hypothesis that the population value for the slope is 1? Explain.

2. Based on the regression equation developed in question 1, answer the following:

a. What do you predict for father's education for a person who has a mother with 12 years of education?

b. What do you predict for average father's education for all people who have a mother with 12 years of education?

c. Using the Graphs menu, plot the standard error of the mean prediction (*sep_1*) against the independent variable. For what value of the independent variable is the standard error of the mean prediction smallest? Why?

d. Using the Graphs menu, plot father's education against mother's education. Edit the chart so that it shows the points, the regression line, and the 95% confidence interval for the mean prediction. Explain what the confidence interval for the mean prediction tells you.

e. Most of the data points do not fall within the 95% confidence interval for the mean prediction. Why is that?

f. Edit the chart to remove the 95% confidence intervals for the mean prediction. Instead, plot the 95% prediction interval. What is a 95% prediction interval? Why is it so much wider than the 95% confidence interval? Explain why most of your data points fall within this interval.

3. Develop a regression model to predict hours worked (variable *hrs1*) for full-time employees (*wrkstat* equals 1) based on their years of education (variable *educ*).

a. Write the regression equation.

b. Test the null hypothesis that the slope is 0.

c. Determine the proportion of the variability in the dependent variable that is explained by the independent variable.

d. Graph your results. Include 95% confidence intervals for the mean prediction.

4. Develop a regression equation that predicts the number of hours of television viewing (variable *tvhours*) from some other characteristic, such as education, income in dollars, and so forth. Summarize your findings. Be sure to make appropriate plots.

Use the *salary.sav* data file to answer the following questions:

5. Estimate the regression model that predicts current salary (variable *salnow*) from beginning salary (variable *salbeg*). Save the standard error of the predicted mean.

a. Can you reject the null hypothesis that there is no linear relationship between the two variables?

b. Based on the 95% confidence interval for the slope, can you reject the null hypothesis that the population value for the slope is 2? Explain.

6. Estimate the standard deviation of current salary for each value of beginning salary.

7. Plot the standard error of the predicted mean against the values of beginning salary. At what point is the standard error the smallest? Why?

8. Using the Graphs menu, plot the points, the regression line, and the 95% confidence interval for the mean prediction. Compare this to a plot with the 95% prediction intervals for an individual observation. Explain why the 95% prediction intervals are wider than the 95% confidence intervals for the mean.

Use the *country.sav* data file to answer the following questions:

9. Estimate the regression equation that predicts male life expectancy (variable *lifeexpm*) from female life expectancy (variable *lifeexpf*). Obtain 95% confidence intervals for the slope and the intercept. Save the standard error of the mean prediction.

a. Can you reject the null hypothesis that there is no linear relationship between female life expectancy and male life expectancy?

b. What is the 95% confidence interval for the population slope? Based on it, can you reject the null hypothesis that the population value for the slope is 0?

c. What is the correlation coefficient between female life expectancy and male life expectancy? Is it positive or negative?

d. When female life expectancy increases by one year, how much does male life expectancy increase?

10. Plot the standard error of the mean prediction against the values of female life expectancy. For what value of female life expectancy is the standard error of the mean prediction smallest? What does the value correspond to? What can you conclude from the plot?

11. Plot male life expectancy against female life expectancy. Include the regression line and the 95% confidence interval for the mean prediction.

   a. Explain why it doesn't bother you that many of the observed data points are not within the confidence interval. Why is the 95% confidence interval for the mean prediction so narrow?

   b. What average value for male life expectancy would you predict for all countries with a female life expectancy of 65?

   c. What male life expectancy would you predict for Botswana, a country with a female life expectancy of 65?

   d. Do you think your prediction for Botswana or for all countries with an observed female life expectancy of 65 is closer to the truth?

   e. Edit the chart to include the 95% prediction intervals. Explain why these are wider than the corresponding 95% confidence intervals for the mean prediction.

12. Repeat question 9 using two variables of your choice from the *country.sav* file.

Use the file *bodyfat.sav* to answer the following questions:

13. Plot percent body fat (*pctfat1*) against the height (*height*) and weight (*weight*). Obtain correlation coefficients between all pairs of variables.

   a. Does there appear to be a linear relationship between percent body fat and height? Explain your conclusion.

   b. Does there appear to be a linear relationship between percent body fat and weight?

   c. Can you reject the null hypothesis that the population values for the correlation coefficients in a and b are 0?

14. Calculate the regression equation to predict body fat from weight. Repeat for height.

   a. Test the null hypothesis that each of the population slopes is 0. What is the relationship of the test that the population slope is 0 to the corresponding test that the population correlation coefficient is 0?

   b. What proportion of the observed variability in percent body fat can be explained by differences in weight? In height?

22 Analyzing Residuals

How can you tell if the assumptions necessary for hypothesis testing in regression are violated?

- How can you use residuals to check the assumptions of independence, linearity, normality, and constant variance?
- What can you do if one or more of the assumptions are violated?
- What are influential points, and why are they important?

When you begin studying the relationship between two variables, you usually don't know whether your data violate the assumptions needed for regression analysis. You don't know whether there is a linear relationship between the two variables, much less whether the distribution of the dependent variable is normal and has the same variance for all values of the independent variable. An important part of regression analysis is checking whether the required assumptions of linearity, normality, constant variance, and independence of observations are met.

When you are fitting models to data, residuals play a very important role. By examining the distribution of the residuals and their relationships to other variables, you can detect departures from the regression assumptions. In this chapter, you'll examine the residuals from the linear regression model you developed in Chapter 20.

▶ This chapter primarily analyzes the residuals from the linear regression described in Chapter 20, which uses the *cntry15.sav* data file. To obtain the output shown, you must first run the analysis described in that chapter and save residuals as described in "Linear Regression Save: Creating New Variables" on p. 465.

Residuals

As you saw in Chapter 20, a residual is what's left over after the model is fit. It's the difference between the observed value of the dependent variable and the value predicted by the regression line. If the assumptions required for a regression analysis are met, the residuals should have the following characteristics:

- They should be approximately normally distributed.
- Their variance should be the same for all values of the independent variable.
- They should show no pattern when plotted against the predicted values.
- Successive residuals should be approximately independent.

To create these variables, run the regression described in Chapter 20 and save new variables, as shown in Figure 20.18.

To obtain the casewise listing shown at the right, from the menus choose:

Analyze
Reports ▸
Case Summaries...

Figure 22.1 Residuals from linear regression

Country	Births per 1000 population	Female life expectancy	Unstandardized Predicted Value	Unstandardized Residual	Standardized Residual	Studentized Residual
Algeria	31	68	68.37	-.37	-.15	-.15
Burkina Faso	50	53	55.12	-2.12	-.84	-.92
Cuba	18	79	77.43	1.57	.62	.67
Equador	28	72	70.46	1.54	.61	.63
France	13	82	80.92	1.08	.43	.48
Mongolia	34	68	66.28	1.72	.68	.70
Namibia	45	63	58.61	4.39	1.73	1.85
Netherlands	13	81	80.92	.08	.03	.04
North Korea	24	72	73.25	-1.25	-.49	-.52
Somalia	46	55	57.91	-2.91	-1.15	-1.23
Tanzania	50	55	55.12	-.12	-.05	-.05
Thailand	20	71	76.04	-5.04	-1.99	-2.13
Turkey	28	72	70.46	1.54	.61	.63
Zaire	45	56	58.61	-2.61	-1.03	-1.10
Zambia	48	59	56.51	2.49	.98	1.07

In this chapter, you'll learn how to check your residuals to see if they violate any of the criteria just mentioned. But first, you'll learn about several types of residuals that you can use when searching for problems. These residuals are all modifications of the usual residual. Their advantage is that they make it easier for you to spot problems.

Standardized Residuals

Figure 22.1 contains observed and predicted female life expectancies, as well as residuals, for the countries you used to develop your regression model. You see that Ecuador has an observed female life expectancy of 72 years, while the regression model predicts 70.46. The residual for Ecuador is therefore +1.54 years. Since the residual is positive, it means that the observed life expectancy for Ecuador is larger than that predicted by your model. A negative residual for Ecuador would tell you that the observed life expectancy was smaller than the predicted value.

Without looking at the residuals for the other cases, can you tell if Ecuador's residual is large or small compared with the other countries? The answer is no. You really can't judge the relative size of a residual by looking at its value alone (just like you can't tell whether you did brilliantly or poorly on a test without knowing how the rest of the class did).

It's easier to determine the relative magnitudes of residuals if you standardize them so that they have a mean of 0 and a standard deviation of 1. (Remember from Chapter 5 that a standard score—*z* score—tells you how many standard deviations above or below the mean an observation falls.) To calculate the standardized residual, you divide the observed residual by the estimated standard deviation of the residuals. (This is the entry labeled *Std. Error* in Figure 20.8. For this example, it's 2.54.) There's no need to subtract the mean, since the mean of the residuals is 0 when an intercept term is in the model. Standardized residuals have a standard deviation of slightly less than 1, since the standard error of the estimate is a little larger than the sample standard deviation of the residuals.

If the distribution of residuals is approximately normal, you know that about 95% of the standardized residuals should be between –2 and +2. Ninety-nine percent of the standardized residuals should be between –2.58 and +2.58. Cases with standardized residuals outside this range are unusual. If there are many cases with standardized residuals larger than 2, that may indicate that the model fits the data poorly. Of course, even if the model fits well, you expect to see about 5% of the cases with standardized residuals greater than 2 in absolute value.

For Ecuador, the standardized residual is 1.54/2.54, which is 0.61. That's a little more than one-half of a standard deviation above the mean. It's not at all unusual. In Figure 22.1, you see that only Thailand has a standardized residual that is close to 2 in absolute value.

Studentized Residuals

When you compute a standardized residual, all of the observed residuals are divided by the same number. However, in Chapter 20, you saw that the variability of the predicted value is not constant for all points but depends on the value of the independent variable. Cases with values of the independent variable close to the sample mean have smaller variability for the predicted value than cases with values far removed from the mean. The **Studentized residual** takes into account the differences in variability from point to point. You calculate it by dividing the observed residual by an estimate of the standard deviation of the residual at that point. The Studentized residual makes it easier to see violations of the regression assumptions, so it's preferred to the standardized residuals.

If the regression assumptions are met, you can use the t distribution, with the degrees of freedom equal to the number of cases minus the number of coefficients (including the intercept), to calculate the probability of observing a Studentized residual at least as large in absolute value as the one observed. For example, the probability of observing a Studentized residual as large in absolute value as that for Thailand (–2.13) is approximately 0.053. So, the residual for Thailand is somewhat unusual but not extreme enough to cause real concern.

When you single out the largest residuals in a sample to see if they are unlikely, you should multiply the probabilities from the t distribution by the number of residuals in your sample. This adjusts for the fact that you are looking at many residuals. It prevents you from calling too many residuals unlikely. If the corrected probability is less than 0.05, you can be confident that the residual is unlikely. If you multiply the observed probability for Thailand by 15, Thailand's residual is not at all unusual.

See "Linear Regression Save: Creating New Variables" on p. 465 for information on how to save residuals.

See Appendix B for instructions on computing the CDF function.

? *How can I get these probabilities from SPSS?* First, you have to save the Studentized residuals from the regression. You do this by selecting Studentized residuals in the Linear Regression Save New Variables dialog box. They are automatically assigned a name such as *sre_1*. Then you use the Compute Variable dialog box to compute a new variable that contains the probability for each case (see "Example: Cumulative Distribution Function" on p. 609 in Appendix B for an example of how to compute a cumulative distribution function). The formula is:

```
2*(1–cdf.t(abs(sre_1),13))
```

This will give you the desired probabilities using a t distribution with 13 degrees of freedom. To use the formula for another dataset, you must replace the number 13 with the appropriate degrees of freedom. Use the degrees of freedom for the *Residual Mean Square* in the ANOVA table. ■■■

Checking for Normality

If the regression assumptions are met, the distribution of the ordinary residuals and the standardized residuals should be approximately normal. For samples larger than 30 cases or so, the distribution of Studentized residuals should be normal as well. (For smaller samples, the distribution of the Studentized residuals should be a t distribution, which has more observations in the tails than the normal distribution.)

The first step for assessing normality is to make a stem-and-leaf plot or a histogram of the residuals. (The distributions of the residuals and standardized residuals will look the same, since you're dividing all of the residuals by a constant to get standardized residuals.) From these displays, you can judge the shape of the distribution of the residuals as well as identify outlying values. Figure 22.2 is a stem-and-leaf plot of the standardized residuals.

You can obtain stem-and-leaf plots using the Explore procedure, described in Chapter 7. Select the variable zre_1 in the Explore dialog box.

Figure 22.2 Stem-and-leaf plot of standardized residuals

```
ZRE_1     Standardized Residual

Frequency    Stem &  Leaf

    3.00       -1 .  019
    4.00       -0 .  0148
    7.00        0 .  0466669
    1.00        1 .  7

Stem width:   1.00000
Each leaf:       1 case(s)
```

Nothing about the stem-and-leaf plot suggests that the data couldn't be a sample from a normal population. There are no extreme outlying values, and the distribution has a single peak that is more or less in the middle. When you have a small number of cases, it can be difficult to judge if the sample comes from a normal population. Small samples from a normal population don't necessarily look "normal." However, if the distribution is very asymmetrical, or if you have many outliers, the normality assumption is suspect.

You can obtain normal probability plots using the Explore procedure. (See Chapter 7.) In the Explore Plots dialog box, select Normality plots with tests.

Figure 22.3 Q-Q plot of standardized residuals

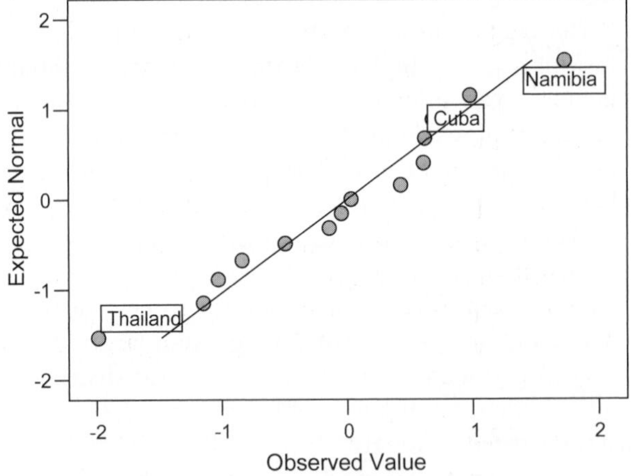

As you recall from Chapter 13, you can use special plots and statistical tests to see if a sample comes from a normal population. Figure 22.3 is a Q-Q plot of the standardized residuals. If the data are a sample from a normal distribution, you expect the points to fall more or less on a straight line. You see that the two largest residuals in absolute value (Thailand and Namibia) are stragglers from the line. This means that they're somewhat larger in absolute value than you would expect.

Figure 22.4 Detrended Q-Q plot of residuals

You can obtain detrended normal probability plots using the Explore procedure. (See Chapter 7.)

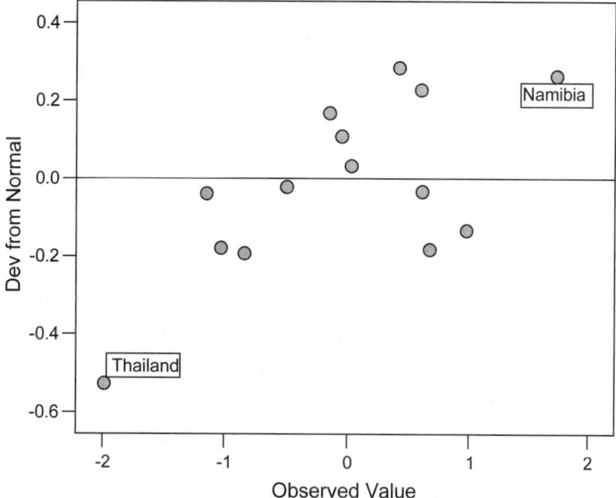

Figure 22.4 is a detrended normality plot. For each of the points in the Q-Q plot, it shows the distance from the observed point to the line. If your data are a sample from a normal population, the points in the detrended normal plot should fall randomly in a band around 0. Again, you see that the residual for Thailand sticks out.

If you look at the results of the statistical tests of normality in Figure 22.5, you see that there's not enough evidence to reject the assumption of normality. However, when the sample is small, the tests are not very powerful. That is, you often won't reject the hypothesis of normality even when it is incorrect. If your sample size is large, the tests of normality may lead you to reject the normality assumption based on small departures that won't affect the regression analysis. As long as the normality assumption is not badly violated, the results of regression analysis will not be seriously affected.

Figure 22.5 Tests of normality of residuals

You can obtain normality tests using the Explore procedure. (See Chapter 7.)

	Kolmogorov-Smirnov[1]			Shapiro-Wilk		
	Statistic	df	Sig.	Statistic	df	Sig.
Standardized Residual	.137	15	.200*	.971	15	.866

*. This is a lower bound of the true significance.

1. Lilliefors Significance Correction

? *What should I do if there are large departures from normality?* Several different regression problems can produce non-normal residuals. It's possible that the population distribution of the dependent variable is not normal. But it's also possible that the residuals appear to be non-normal because the regression model doesn't fit the data, or because the variance of the dependent variable is not constant over the values of the independent variable. Make sure that you have fixed any other known problems in your regression before you worry about the lack of normality of the residuals.

Once you've ruled out other problems as the cause for non-normality of the residuals, you can try to transform the values of the dependent variable. For example, if the distribution of residuals is not symmetric but has a tail toward larger values, you can try taking the log of the dependent variable, provided all of the values are positive. If the tail is toward smaller values, you can try squaring the values of the dependent variable. ■■■

Checking for Constant Variance

To check whether the variance of the dependent variable is the same for all values of the independent variable, you can plot the Studentized residuals against the predicted values. If the variance is constant, you won't see any pattern in the data points. That's the case in Figure 22.6. The residuals appear to be randomly scattered around a horizontal line through 0.

Figure 22.6 Studentized residuals versus predicted values

In the Scatterplot legacy dialog box, select the variables sre_1 and pre_1

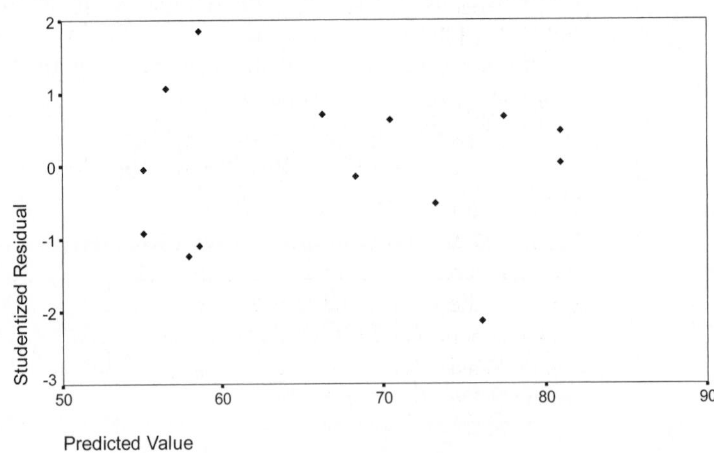

In Figure 22.7, you see a funnel shape. The variability of the residuals is increasing with increasing predicted values. That means the variance of the residuals is smaller for small values of the predicted dependent variable than for larger values.

Figure 22.7 Variance increasing with predicted values of the dependent variable

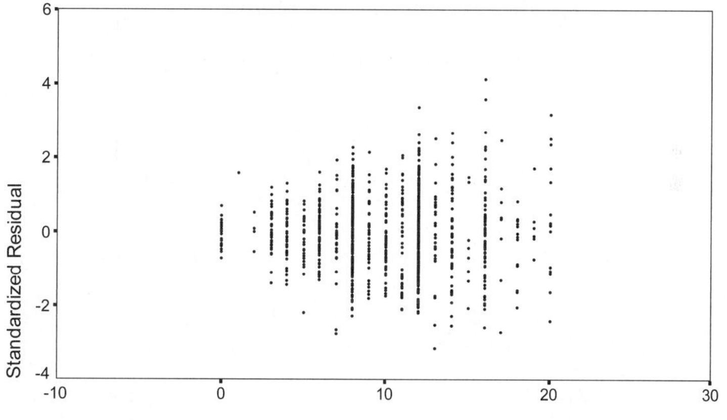

That's not unusual, since the variability of the dependent variable often increases with the value of the independent variable. For example, if you are looking at the relationship between salary and years of education, you would expect to see more variability in MBA salaries than in those of high school graduates.

? *What should I do if the variance of the dependent variable appears not to be constant?* If the variance of the dependent variable isn't constant, you can try transforming the values of the dependent variable and then going back and rerunning the regression using the transformed variable in place of the original variable. If the variance of the dependent variable increases linearly with values of the independent variable, and all values of the dependent variable are positive, try taking the square root of the dependent variable. If the standard deviation increases linearly with increasing values of the independent variable, try taking logs of the dependent variable.

Checking Linearity

The first step of any regression analysis is plotting the dependent variable against the independent variable. You should fit a linear regression model only if the points cluster around a straight line. You can also evaluate the linearity assumption by plotting the Studentized residuals against the predicted values. If the relationship between the dependent variable and the independent variable is not linear, you will see a curve in the plot.

Figure 22.8 Female life expectancy and phones per 100

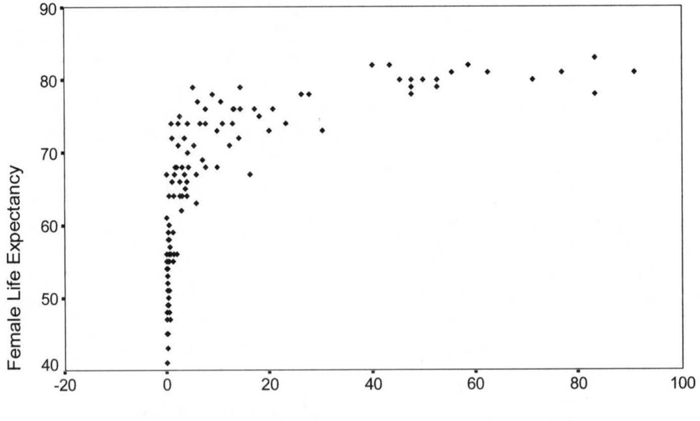

Figure 22.8 is a plot of female life expectancy against the number of phones per 100 people (from the larger data file of statistics on different countries in the world). The relationship between the two variables is not linear. If you run a linear regression with these two variables and then plot the Studentized residuals against the predicted values, you will see a strong nonlinear relationship between the residuals and the predicted values (see Figure 22.9).

Figure 22.9 Studentized residuals versus predicted values

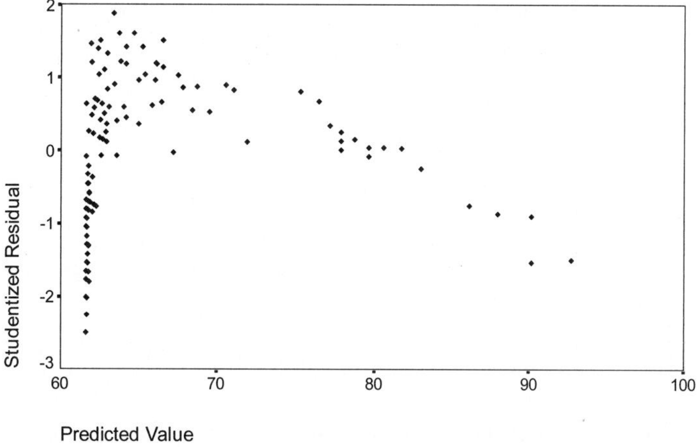

When the relationship between two variables isn't linear, you can sometimes transform the variables to make the relationship linear. For example, if you take the natural log of the number of phones per 100 people, you get the plot shown in Figure 22.10. The relationship between female life expectancy and the natural log of the number of phones per 100 people is now more or less linear.

Figure 22.10 Female life expectancy and natural log of phones per 100

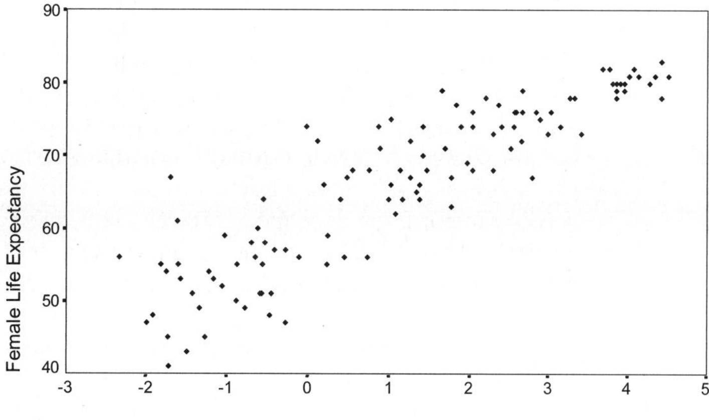

? *How do I decide what transformation to use?* Sometimes you might know the mathematical formula that relates two variables. In this case, you can use mathematics to figure out what transformation you need. This situation happens more often in the physical or biological sciences than in the social sciences. For example, if the equation that relates the dependent variable and the independent variable is

$$Y = AX^B$$

you can take the natural log of both sides of the equation and obtain the equation

$$\log(Y) = \log(A) + B \times \log(X)$$

There is now a linear relationship between the log of the dependent variable (Y) and the log of the independent variable (X). (Remember that $y = a + bx$ is the equation for a straight line.) If the relationship isn't known, you can choose a transformation by looking at the plot of the data. Often, a relationship appears to be nearly linear for part of the data but is curved for the rest. The log transformation is useful for "straightening out" a relationship when the values of the dependent variable increase more quickly than a linear model predicts. The square transformation may be helpful when the data points fall on a downward curve with the values of the dependent variable decreasing more quickly than a linear model predicts. ■■■

When you try to make a relationship linear, you can transform either the independent variable, the dependent variable, or both. If you transform only the independent variable, the distribution of the dependent variable doesn't change. If it was normally distributed with a constant variance for each value of the independent variable, it remains that way. However, if you transform the dependent variable, you change its distribution. For example, if you take logs of the dependent variable, the log of the dependent variable—not the original dependent variable—must be normally distributed with a constant variance. In other words, the regression assumptions must hold for the transformed variables you actually use in the regression equation.

? *Isn't transforming the data more or less cheating, or at least distorting the true picture?* No. All that transforming a variable does is change the scale on which it's measured. Instead of saying that there is a linear relationship between work experience and salary, for example, you say that there is a linear relationship between work experience and the log of salary. It's much easier to build models for relationships that are linear than for those that are not. That's why transforming variables is often a convenient strategy. ■■■

Checking Independence

Another assumption needed for regression hypothesis testing is that all of the observations are independent. This means that the value of one observation is in no way related to the value of another observation. Nonindependence can be a serious problem when the data are gathered in a sequence. For example, if you're looking at the length of time required to perform a new surgery and if surgeons become more proficient as the number of operations they do increases, earlier patients may have longer surgical times than later patients. In such a situation, successive patients will be more similar than you would expect if the observations are independent.

You can check the independence assumption by plotting the Studentized residuals against the sequence variable. (Strictly speaking, the residuals are not independent, since they sum to 0. However, if there are many more points than coefficients, this dependency can be ignored.)

Figure 22.11 Studentized residuals versus order of observations

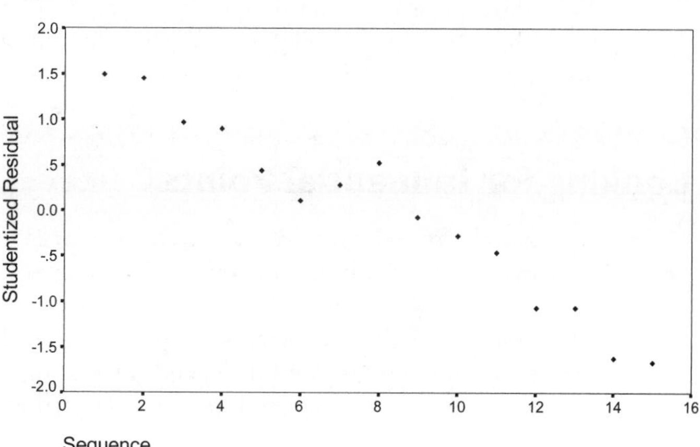

Figure 22.11 is a plot of Studentized residuals against the order in which the observations are taken for the hypothetical study of the length of surgery. You see that the value of the residual is related to the order in which the data are obtained. Early observations have large positive residuals, while later ones have large negative residuals. This is a pattern you might see if you're performing surgery and getting more proficient as you do more operations.

You can use the **Durbin-Watson test** to see if adjacent observations are correlated. This statistic ranges in value from 0 to 4. If there is no correlation between successive residuals, the Durbin-Watson statistic should be close to 2. Values close to 0 indicate that successive residuals are positively correlated, while values close to 4 indicate strong negative correlation. The value of the Durbin-Watson statistic for the plot in Figure 22.11 is 0.13, suggesting that adjacent residuals are positively correlated. To test whether the observed Durbin-Watson statistic is significantly different from 2, you need special tables that are available in books about time series analysis. As a rule of thumb, if your observed value is between 1.5 and 2.5, you need not worry.

A Final Comment on Assumptions

You should always examine your data for violations of the regression assumptions because significance levels, confidence intervals, and other regression tests are sensitive to certain types of violations. These tests cannot be interpreted in the usual fashion if there are serious departures from the regression assumptions. If you carefully examine the residuals, you'll have an idea of whether problems exist. If you observe problems, you can try to remedy them using transformations.

Looking for Influential Points

It's possible for one or more cases to have a large impact on your regression model, especially if the sample size is small. That is, the slope or intercept values change a lot when certain cases are excluded from the computation of the coefficients. This is undesirable since you want a regression model that doesn't depend heavily on the values for a small number of points. You want all points to contribute more or less equally to the model.

For example, look at Figure 22.12, which is the life expectancy and birthrate data with an additional point in the lower left corner. You know that the slope of the regression line without the point is –0.697. With the point, the slope is –0.470, almost a 32% change. You can have SPSS calculate the change in slope when each point is eliminated in turn from the calculation of the regression statistics.

Figure 22.12 Life expectancy and birthrate with outlier

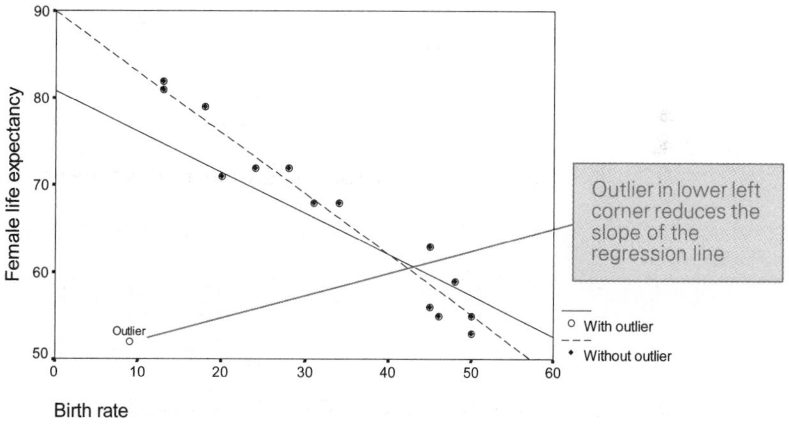

In Figure 22.13, you see the actual change in the slope associated with the removal of each case from the analysis in the column labeled *DFBETA BIRTHRAT*. There's only one point for which the change is not close to 0. That's the outlying point added to the data. An easier way to look for points that have a large impact on the slope is to plot the change in slope against an arbitrary case sequence number, as shown in Figure 22.14. By clicking on the outlying point, you can immediately see what country it belongs to.

To figure out whether a change in slope is large, you have to compare the change to the actual value for the slope. For example, a change of 0.2 is large for a slope of 0.6 but small for a slope of 200. To avoid having to compare the change in slope to the slope, you can look at the standardized change in slope shown in Figure 22.13. You should be suspicious of standardized changes in absolute value larger than $2/\sqrt{N}$, where N is the number of cases in the sample. You can see that the standardized change observed for the outlier (5) is much larger than $(2/\sqrt{16}) = 1/2$.

Figure 22.13 Change in slope

		DFBETA BIRTHRAT	Standardized DFBETA BIRTHRAT
COUNTRY	Algeria	-.00022	-.00149
	Burkina Faso	-.03151	-.21882
	Cuba	-.03204	-.22602
	Ecuador	-.00496	-.03438
	France	-.05131	-.36513
	Mongolia	.00279	.01925
	Namibia	.01630	.11247
	Netherlands	-.04422	-.31153
	North Korea	-.00618	-.04248
	Outlier	**.22713**	**5.00376**
	Somalia	-.02289	-.15873
	Tanzania	-.01708	-.11749
	Thailand	.00192	.01318
	Turkey	-.00496	-.03438
	Zaire	-.01847	-.12772
	Zambia	.00434	.02972

Change in slope

Standardized change in slope

Figure 22.14 Sequence plot of change in slope

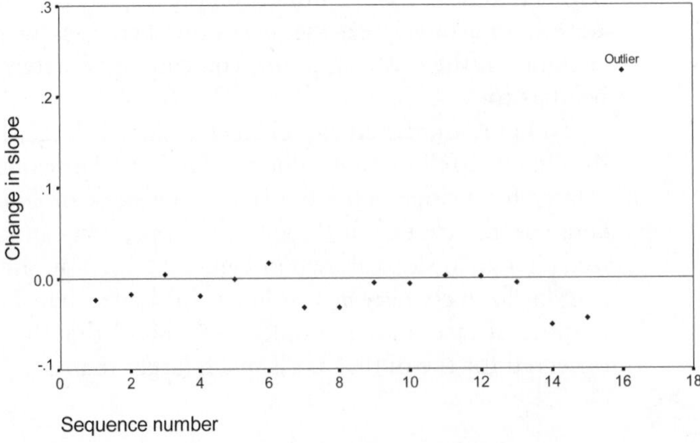

? *What should I do if I find a point that influences the results a lot?* First of all, make sure that it's not the result of errors in data collection or entry. If there are errors, correct them. If the data values are correct, try to determine if the case is unusual in ways other than altering your regression line. For example, if you are looking at the relationship between life expectancy and birthrate and find a country with very high values for both variables, look for an explanation. It may be that the predominant religious affiliation or some other characteristic of the country is different from the rest. In that case, you may want to develop separate regression models for countries with and without the characteristic in question. If you can't identify anything unusual about the point, you are stuck with it. You may want to present the results of your analyses with and without the point. What you shouldn't do is arbitrarily discard points that don't fit the model in some way. ■■■

Studentized Deleted Residuals

To see the impact of a case on the computation of the slope, you calculated the regression with and without the case. The same idea can be extended to other regression statistics. For example, you can calculate what's called the **Studentized deleted residual**, also known as the **jackknifed residual.** The Studentized deleted residual is the Studentized residual for a case when the case is excluded from the computation of the regression statistics. When there are departures from the regression assumptions, you can see them easier using Studentized deleted residuals than you can using Studentized residuals. You won't go wrong if you always use the Studentized deleted residuals for analyzing residuals. (If the regression assumptions are met and you have roughly equal numbers of observations at each value of the independent variable, standardized, Studentized, and Studentized deleted residuals will show the same patterns on residual plots.)

Summary

How can you tell if the assumptions necessary for hypothesis testing in regression are violated?

- You can look for violations of the regression assumptions by examining the residuals. If the assumptions are correct, the distribution of the residuals should be approximately normal with constant variance.
- You can transform your data values to better conform to the necessary regression assumptions.
- It's important to identify influential points, since you don't want your regression results to depend too heavily on any single point.

What's Next?

In this chapter, you saw how to use residuals and related statistics to check the assumptions needed for the linear regression model. In Chapter 23, you'll learn how to build a regression model that contains more than one independent variable.

Statistical Concepts

1. What violations of assumptions, if any, are suggested by the following plots?

a.

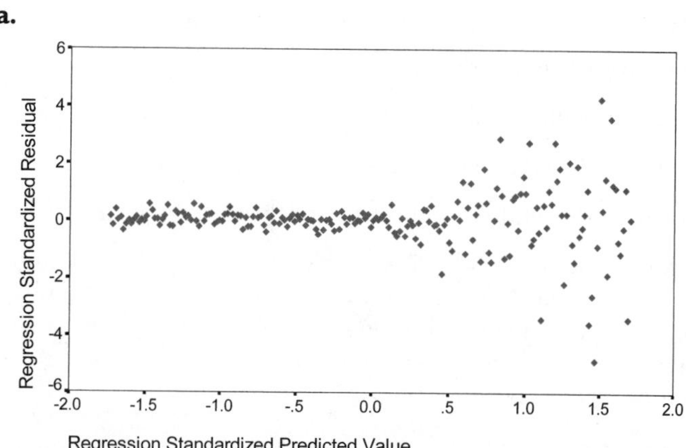
Regression Standardized Residual
6
4
2
0
-2
-4
-6
-2.0 -1.5 -1.0 -.5 0.0 .5 1.0 1.5 2.0

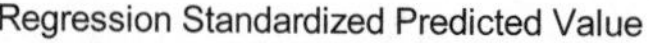
Regression Standardized Predicted Value

b.

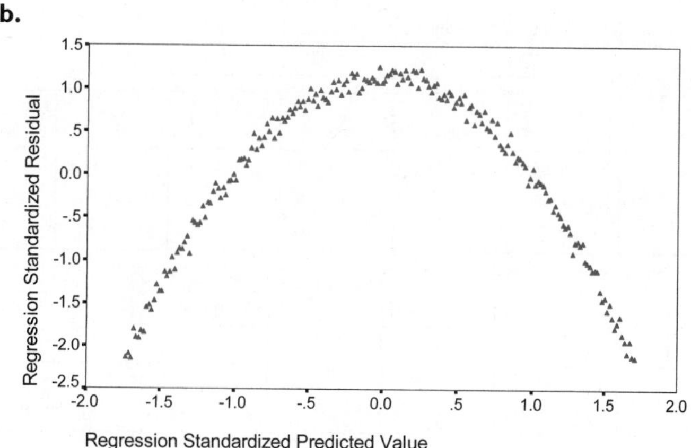
Regression Standardized Residual
1.5
1.0
.5
0.0
-.5
-1.0
-1.5
-2.0
-2.5
-2.0 -1.5 -1.0 -.5 0.0 .5 1.0 1.5 2.0
Regression Standardized Predicted Value

c.

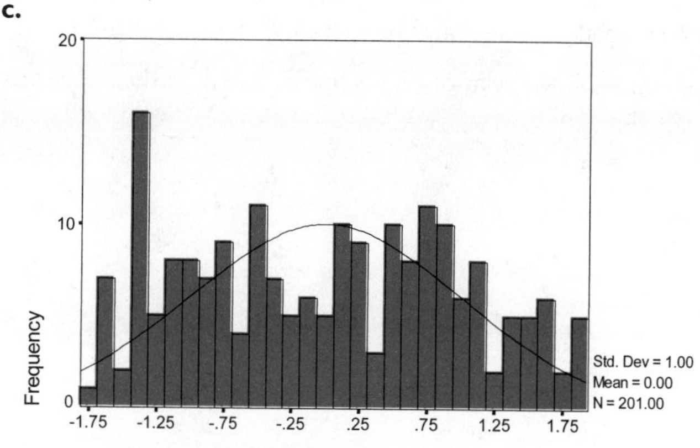
Frequency
20
10
0
-1.75 -1.25 -.75 -.25 .25 .75 1.25 1.75
Std. Dev = 1.00
Mean = 0.00
N = 201.00
Regression Standardized Residual

2. What can you learn from the following?

a. A histogram of the residuals

b. A sequence plot of the residuals

c. A plot of the independent variable and the dependent variable

d. A plot of the residuals against the independent variable

3. You are the decision maker for a major savings and loan. You must estimate what percentage of your CD deposits are likely to be cashed prematurely during the year. You've computed a regression equation that predicts the percentage of premature CD withdrawals from banks based on unemployment rates for the area. You obtain the following regression statistics:

Model	R	R Square	Adjusted R Square	Std. Error of the Estimate
1	.888	.788	.777	1.06

Model		Sum of Squares	df	Mean Square	F	Sig.
1	Regression	79.388	1	79.388	70.675	.000
	Residual	21.343	19	1.123		
	Total		20			

Model		Unstandardized Coefficients		Standardized Coefficients	t	Sig.
		B	Std. Error	Beta		
1	(Constant)	-2.246	.757		-2.966	.008
	UNEMPLOY	1.117	.133	.888	8.407	.000

a. Based on the above statistics, what can you conclude about the relationship between unemployment and premature CD withdrawal?

b. Can you tell if the relationship is linear, based on the above statistics?

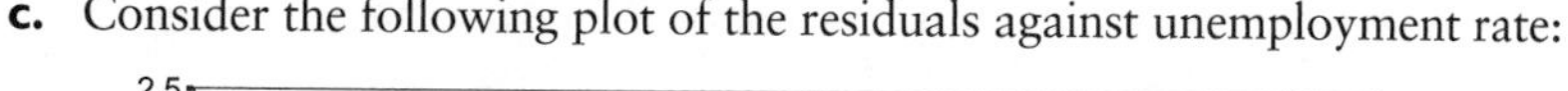

c. Consider the following plot of the residuals against unemployment rate:

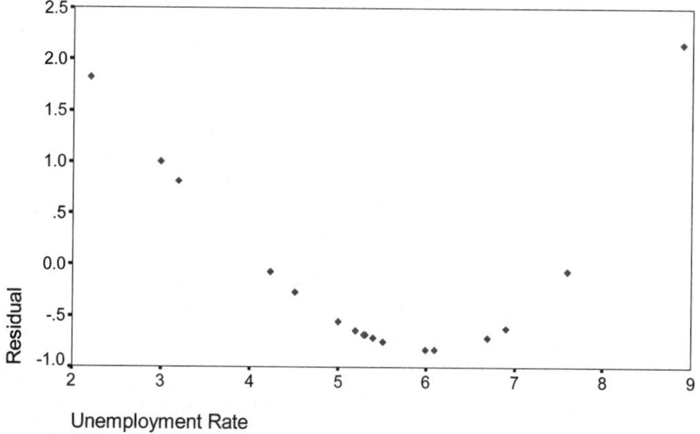

What do you think about the adequacy of the linear model now?

d. Based on the plot of the residuals, sketch the plot of CD withdrawal and unemployment rates.

e. Why do you think the correlation coefficient between the two variables is large?

Data Analysis

Use the *gss.sav* data file to answer the following questions:

1. Build a linear regression model to predict a person's education (variable *educ*) from the father's (variable *paeduc*). Save the Studentized residual, the predicted value, and the change in the coefficients (*dfbeta*).

a. Plot the two variables and draw the regression line.

b. Write the regression equation.

c. Test the null hypothesis that there is no linear relationship between the two variables.

d. What proportion of the variability in a person's education is "explained" by their father's education?

2. For the model you developed in question 1, examine the assumptions you need for linear regression.

   a. Make a histogram and Q-Q plot of the Studentized residual. What do these suggest about the violation of the normality assumption?

   b. Use the Explore procedure to make a stem-and-leaf plot of the Studentized residual and to list the five largest and five smallest values for the Studentized residual. Describe the values for education and father's education for the cases that are outliers.

   c. Make a scatterplot of the Studentized residual against the predicted education. What should you look for in this plot? What assumption are you checking? Do you suspect that the assumption is not met?

   d. Plot the change in slope associated with removing a point against case ID. From the plot, what is the largest change in slope that would occur if you eliminated a point? Use point selection mode to identify the cases with large values for the change in slope. What are their values for the dependent and independent variables?

3. Develop a regression equation to predict husband's education (variable *husbeduc*) from wife's education (variable *wifeduc*). Using the steps outlined in question 2, check the regression assumptions. Summarize your conclusions about how well the data conform to the assumptions necessary for linear regression.

4. Select two variables of your choice from the *gss.sav* file, develop a linear regression model, and test the necessary assumptions. Write a short summary of your results.

Use the *salary.sav* data file to answer the following questions:

5. Develop a regression equation to predict current salary (variable *salnow*) from beginning salary (variable *salbeg*). Save the predicted values, Studentized residuals, and changes in the coefficients.

   a. Make a histogram and a Q-Q plot of the Studentized residuals. Does their distribution appear to be approximately normal? If not, in what way is the distribution not normal?

   b. Plot the Studentized residuals against the predicted values. What kind of pattern are you looking for in this plot? Do you see anything that suggests possible problems? Explain.

   c. Plot the changes in the value of the slope against case ID number. Are there any employees who have a big impact on the estimate of the regression coefficient?

6. Using the Compute facility, create new variables that are the natural log of current salary and of beginning salary. Plot the natural log of current salary against the natural log of beginning salary. How does this plot differ from the one you made of the original values?

7. Repeat question 5 using the transformed values of the salaries. Do you think the regression assumptions are better met for the transformed variables?

Use the *country.sav* data file to answer the following questions:

8. Develop a regression equation to predict female life expectancy (variable *lifeexpf*) from birthrate (variable *birthrat*) for all countries in the data file. Save the predicted values, Studentized residuals, and changes in the values of the slope and the intercept.

   a. Make stem-and-leaf plots or histograms of the Studentized residuals. Make a Q-Q plot as well. Do you think the residuals are normally distributed? If not, how does their distribution differ from a normal distribution?

   b. Plot the Studentized residuals against predicted female life expectancies. What are you looking for in this plot? Is the variability of the residuals pretty much constant over the entire range of values of predicted female life expectancy?

   c. Plot the change in slope against ID number. Identify any points that have a large effect on the value of the slope. Which country will change the slope most if it is excluded from the regression computations? What are its values for female life expectancy and birthrate?

9. Repeat question 8, predicting male life expectancy (variable *lifeexpm*) from female life expectancy.

10. Predict female life expectancy from the number of phones per 100 people (variable *phone*). Save the Studentized residuals and predicted values. Make a stem-and-leaf plot of the Studentized residuals. Test the null hypothesis that the residuals come from a normal population. Plot the residuals against the predicted values. Do you think the data satisfy the regression assumptions?

11. Repeat question 10 using the natural log of the number of phones (variable *lnphones*) instead of the actual number of phones. Examine all of the diagnostic plots. What do you conclude now about violations of the regression assumptions?

Use the *buying.sav* data file to answer the following questions:

12. Build a regression model to predict a husband's buying score (variable *hsumbuy*) from his wife's buying score (*wsumbuy*). Check the regression assumptions. Summarize your results.

13. Build a regression model to predict wife's buying score from her husband's prediction of her score (*hpredsum*). Do you think you have a useful model? Explain.

Use the ***schools.sav*** data file to answer the following questions:

14. Build a linear regression model to predict ACT scores. Use the variable you think is the best predictor as the independent variable. Check all of the necessary assumptions. Write a short report explaining your regression results.

15. Build a linear regression model to predict graduation rates. Again, select the independent variable that you think is the best predictor. Check the regression assumptions.

16. Calculate the correlation coefficients between *act93*, *loinc93*, *grad93*, and *pctact93*. Which variables have the largest correlation coefficient in absolute value? For which pairs of variables is the correlation coefficient a good summary measure of their relationship?

Use the ***bodyfat.sav*** data file to answer the following questions:

17. For the regression equation that predicts percent body fat (*pctfat1*) from weight, save the Studentized residuals and predicted values.

   a. Plot the Studentized residuals against the predicted values. If the regression assumptions are met, how should this plot look? Is there anything unusual about the plot you have made?

   b. Plot the Studentized residuals against the case numbers (*caseno*). What do you look for in this plot? Do you see any problems?

   c. Make a histogram and normal probability plot of the Studentized residuals. Do the residuals appear to come from a normal distribution?

   d. Is there anything unusual about the case that appears as an outlier in the plots?

18. Choose the body circumference measure that you think is most strongly linearly related to percent body fat.

   a. Estimate the linear regression equation to predict percent body fat from the selected circumference measure.

   b. Check the regression assumptions using appropriate diagnostic plots. Summarize your results.

23 Building Multiple Regression Models

How do you build a regression model with more than one independent variable?

- What are partial regression coefficients?
- How can you test the null hypothesis that all of the population partial regression coefficients are 0?
- What can you tell from the partial regression coefficient about the relationship between the dependent variable and an independent variable?
- What are variable selection methods and why are they useful?

In Chapter 20 through Chapter 22, you used linear regression analysis to examine the relationship between a dependent variable and one independent variable. In most real-life situations, however, predicting the values of a dependent variable requires more than a single independent variable. If you want to predict how long a patient stays in the hospital after surgery, you need to consider a myriad of possible predictors: patient's age, severity of illness as measured by numerous laboratory and physical findings, and type of surgery. To predict salary, you must consider a variety of characteristics of the employee and the job: work experience, education, seniority, and type of position. In this chapter, you will learn how to build a linear regression model that has more than one independent variable. The technique you'll use is called **multiple linear regression analysis**.

▶ This chapter uses the *country.sav* data file described in Chapter 9. To duplicate the results shown, you must correct the value for Bhutan in the data file, as described in that chapter. For instructions on how to obtain the multiple regression analysis shown, see "How to Obtain a Multiple Linear Regression" on p. 547.

▶ To make the output more compact, before running the analysis in this chapter, from the menus you should choose:

Edit
 Options...

Click the **Output Labels** tab. In the Pivot Table Labeling group, select **Names** under Variables in Labels Shown As.

Predicting Life Expectancy

In the previous chapters, you predicted female life expectancy from birthrate for a sample of 15 countries. You found that for your sample, birthrate is an excellent predictor, since it explains almost 94% of the observed variability in female life expectancy. Birthrate predicts life expectancy well because both variables are strongly related to a country's economic development and prosperity. In this chapter, you'll try to predict female life expectancy from a combination of variables that measure specific economic and health care delivery characteristics of a country.

Table 23.1 describes the independent variables you have available. For most of the 122 countries, the data are for 1992.

Table 23.1 Predictors of life expectancy

Variable name	Description
urban	Percentage of the population living in urban areas
docs	Number of doctors per 10,000 people
beds	Number of hospital beds per 10,000 people
gdp	Per capita gross domestic product in dollars
radios	Radios per 100 people

The Model

You can write the multiple linear regression equation that predicts female life expectancy from all of the variables in Table 23.1 as

Equation 23.1

$$\text{Predicted life expectancy} = \text{constant} + B_1\text{urban} + B_2\text{docs} + B_3\text{beds} + B_4\text{gdp} + B_5\text{radios}$$

Instead of just an intercept and slope, the multiple linear regression equation contains a constant (analogous to the intercept) and five coefficients (B_1 through B_5)—one for each of the five independent variables. These coefficients are called **partial regression coefficients.** If your data are a sample from a population about which you want to draw conclusions, then the sample partial regression coefficients are estimates of the unknown population coefficients. (The population partial regression coefficients are usually designated with the Greek letter β.)

You can again use the method of least squares to estimate the values of the coefficients. That is, you select the coefficients that result in the smallest sum of squared differences between the observed and predicted values of the dependent variable. Any other coefficients have a larger sum of squared residuals.

? *Can I include a nominal variable, such as type of government or region of the world, in a regression model?* Yes, but not without some effort. If you have a variable that has only two possible values—for example, democracy/not democracy or developed/not developed—you can code the two responses using 0 and 1. Then you can treat the variable like any other variable in the regression model. The partial regression coefficient for the variable tells you how much the predicted value for the dependent variable changes when the code is 1. For example, a coefficient of 0.5 for democracy tells you that life expectancy increases by half of a year for democracies compared to nondemocracies. If you have a nominal variable with more than two categories, you have to create a new set of variables to represent the categories. How that's done is beyond the scope of this book. ■■■

Assumptions for Multiple Regression

To test hypotheses about the population regression line when you have a single independent variable, your data must be a random sample from a population in which the following assumptions are met:

- The observations are independent.
- The relationship between the two variables is linear.
- For each value of the independent variable, there is a normal distribution of values of the dependent variable.
- The distributions have the same variance.

You need only a slight modification of these assumptions for multiple regression. You must assume that the relationship between the dependent and the independent variables is linear and that for each *combination* of values of the independent variables, the distribution of the dependent variable is normal with a constant variance.

Examining the Variables

Before you estimate the coefficients, you must make sure that the independent variables are linearly related to the dependent variable. If they're not, you may have to transform the data as described in the previous chapter. For example, you may have to take logs or square roots of one or both of a pair of variables to make the relationship linear.

Figure 23.1 is a scatterplot matrix of the independent variables and the dependent variable. You're particularly interested in looking at the last row of the matrix. It shows you the relationships between female life expectancy and the other independent variables.

Figure 23.1 Scatterplot matrix

To obtain this matrix, in the Scatterplot Matrix dialog box, select the variables urban, docs, hospbed, gdp, radio, and lifeexpf.

See Chapter 9 for instructions on how to obtain matrix scatterplots.

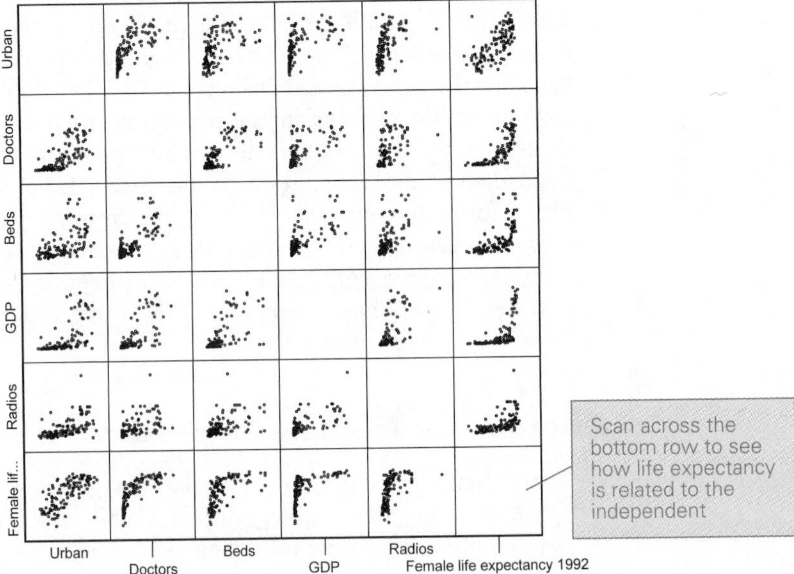

The relationship between female life expectancy and the percentage of the population living in urban areas appears to be more or less linear. The other four independent variables appear to be related to female life expectancy, but the relationship is not linear. Before you can use these variables in the multiple linear regression equation, you have to transform the data so that the relationships are more or less linear. Let's see what happens when you take the natural log of the values of the four independent variables.

To obtain this matrix, in the Scatterplot Matrix dialog box, select the variables urban, lndocs, lnbeds, lngdp, lnradio, and lifeexpf.

See Chapter 9 for instructions on how to obtain matrix scatterplots.

Figure 23.2 Scatterplot matrix after transformations

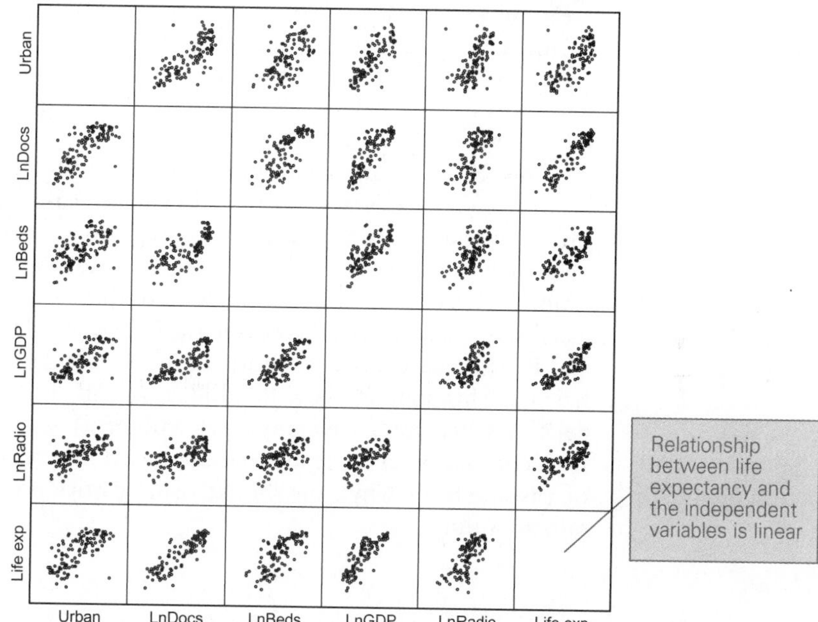

Figure 23.2 is the scatterplot matrix after you've changed the scale on which doctors, hospital beds, per capita gross domestic product, and radios are measured. The transformations were successful. All of the independent variables have a linear relationship with female life expectancy. Now it makes sense to compute the multiple linear regression equation using the values of the transformed variables in place of the original variables. (For simplicity, in the rest of the chapter we'll usually refer to the variables using their original names. So, instead of always saying the log of the number of doctors, we'll say the doctors variable, which means the transformed variable representing the log of the number of doctors per 10,000 population.)

? *What happens if some of the cases have missing values for the dependent variable or the independent variables?* By default, SPSS excludes all cases that have missing values for any of the variables in the regression from the computation of the regression statistics. This is known as **listwise deletion.** (You'll find this term in some of the procedure dialog boxes.) The alternative to listwise deletion is called **pairwise deletion.** If you request pairwise deletion, each correlation coefficient between two variables is calculated using all cases that have values for the two variables. (Calculating the correlation matrix between all pairs of variables is the first step of the regression computations.) For example, if a case has values only for variables A and B, those values are used when calculating the correlation coefficient between variables A and B. The case contributes nothing to the calculation of the other coefficients. This sounds like a good idea, but it's not. If you use pairwise deletion when you have many cases with missing values, you can end up with correlation coefficients that are based on entirely different groups of cases. That's why you should usually stick to listwise deletion of missing values.

Looking at How Well the Model Fits

Figure 23.3 shows the results of the multiple linear regression analysis.

To obtain the regression analysis shown in Figure 23.3, Figure 23.4, and Figure 23.5, from the menus choose:

Analyze
Regression ▶
Linear...

Select the variables shown in Figure 23.16.

Figure 23.3 Model summary statistics

Model	Variables Entered	Variables Removed	Method
1	URBAN, LNBEDS, LNRADIO, LNDOCS, LNGDP	.	Enter

Model	R	R Square	Adjusted R Square	Std. Error of the Estimate
1	.909	.827	.819	4.75

Figure 23.4 ANOVA table

Model		Sum of Squares	df	Mean Square	F	Sig.
1	Regression	11834.950	5	2366.990	104.840	.000
	Residual	2483.490	110	22.577		
	Total	14318.440	115			

Figure 23.5 Regression coefficients

Model		Unstandardized Coefficients		Standardized Coefficients	t	Sig.
		B	Std. Error	Beta		
1	(Constant)	40.779	3.201		12.739	.000
	LNBEDS	1.174	.750	.097	1.565	.120
	LNDOCS	3.965	.568	.555	6.982	.000
	LNGDP	1.626	.619	.225	2.629	.010
	LNRADIO	1.541	.703	.130	2.191	.031
	URBAN	-.007	.033	-.015	-.201	.841

One of the first things you want to do when you run a regression model is to look at is how well the model fits. The entry labeled *R Square* in Figure 23.3 tells you that 82.7% of the observed variability in life expectancy is "explained" by the five independent variables. That's quite a lot, although not as good as when birthrate is used alone. *R* is the correlation coefficient between the observed value of the dependent variable and the predicted value based on the regression model. A value of 1 tells you that the dependent variable can be perfectly predicted from the independent variables. A value close to 0 tells you that the independent variables are not linearly related to the dependent variable. The observed value of 0.91 is quite large, indicating that the linear regression model predicts well.

The analysis-of-variance table in Figure 23.4 is used to test several equivalent null hypotheses: that there is no linear relationship in the population between the dependent variable and the independent variables, that all of the population partial regression coefficients are 0, and that the population value for multiple R^2 is 0.

The test of the null hypothesis is based on the ratio of the regression mean square to the residual mean square. In Figure 23.4, the ratio of the two mean squares, labeled *F*, is 104.84. Since the observed significance level is less than 0.0005, you can reject the null hypothesis that there is no linear relationship between female life expectancy and the five independent variables. At least one of the population regression coefficients is not 0. (This test is sometimes known as the **overall regression F test.**)

? *Where does an analysis-of-variance table come from in regression analysis?* The analysis-of-variance table is similar to the one you calculated in Chapter 15. The total observed variability in the dependent variable is subdivided into two components: that explained by the linear regression (labeled *Regression*) and that not explained by the linear regression (labeled *Residual*).

If there is no linear relationship between the dependent variable and the independent variables, then both the regression mean square and the residual mean square (the variance of the residuals) are estimates of the variance of the dependent variable for each combination of values of the independent variables. If there is a linear relationship, the estimate based on the regression mean square will be too large. How much it is too large depends on the magnitude of the regression coefficients.

You can calculate multiple R^2 from the sums of squares in the analysis-of-variance table. First, add up the regression and residual sums of squares. That's the total sum of squares. Then divide the regression sum of squares by the total sum of squares. This gives you 0.827, which is R^2, the proportion of the variability in the dependent variable explained by the linear regression.

Examining the Coefficients

The coefficients for the independent variables are listed in the column labeled B in Figure 23.5. Using these coefficients, you can write the estimated regression equation as

$$\hat{Y} = 40.78 - 0.007 \times \text{urban} + 3.96 \times \text{lndocs} + 1.17 \times \text{lnbeds} + 1.63 \times \text{lngdp} + 1.54 \times \text{lnradio}$$ **Equation 23.2**

where $\hat{Y}$ is the predicted female life expectancy. In the multiple regression equation, the partial regression coefficient for a variable tells you how much the value of the dependent variable changes when the value of that independent variable increases by 1 and the values of the other independent variables do not change. A positive coefficient means that the predicted value of the dependent variable increases when the value of the independent variable increases. A negative coefficient means that the predicted value of the dependent variable decreases when the value of the independent variable increases.

The coefficient for the doctors variables tells you that predicted female life expectancy increases by 3.96 years for a change of 1 in the value of the doctors variable. Since the doctors variable is measured on a natural log scale, the change refers to an increase of 1 on the log scale. So, instead of telling you how much life expectancy increases when the number of doctors per 10,000 increases by 1, it's telling you how much life expectancy increases when the *logarithm* of the number of doctors per 10,000 increases by 1. (For example, if the number of doctors increases from 2.72 to 7.39 per 10,000 that's a change of 1 on a log scale, since the natural log of 2.72 is 1, and the natural log of 7.39 is 2.)

You test the null hypothesis that the population partial regression coefficient for a variable is 0 using the t statistic and its observed significance level. As in the case of bivariate regression (regression with a single independent variable), you calculate the t statistic by dividing the estimated coefficient by its standard error. You can calculate confidence intervals for the population partial regression coefficients the same way as you did for the slope and intercept.

From Figure 23.5, you can reject the null hypothesis that the coefficients for doctors, radios, and per capita GDP are 0. You can't reject the null hypothesis that the coefficients for urbanization and hospital beds are 0. (If you request 95% confidence intervals for these two coefficients, you'll see that they include the value 0.) You'll learn later in this chapter that this finding doesn't mean that urbanization and hospital beds are not good predictors of female life expectancy when considered alone or in combination with other variables. They just don't contribute significantly to the model being considered.

All of the variables except urbanization have positive coefficients, which means that life expectancy increases with increasing values of the variables. That makes sense. You expect that more doctors, more hospital beds, more radios, and larger per capita GDP's would be associated with longer life expectancies. All of these variables are indicators of the prosperity of a country. Doctors and hospital beds, unlike radios, might even contribute directly to the health of the population. Urbanization has a small negative coefficient, which is unexpected since life expectancy generally increases with urbanization. Since you can't reject the hypothesis that its coefficient is 0 in the population, the fact that the sign is wrong isn't of much concern. You'll look at this variable more closely later.

Interpreting the Partial Regression Coefficients

When there is only one independent variable in a regression model, the interpretation of its regression coefficient is straightforward. If you reject the null hypothesis that the population value for the coefficient is 0, you can conclude that it's quite likely that there is a linear component to the relationship between the dependent variable and the independent variable. In multiple linear regression, the interpretation is considerably more complicated because any conclusion about a particular independent variable depends on the relationship of that variable both to the dependent variable and to the other independent variables in the model.

For example, in Figure 23.5, the partial regression coefficients for urbanization and hospital beds are not statistically different from 0. Can you conclude that these two variables are not linearly related to female life expectancy? Not necessarily. Look at Figure 23.6, which is a matrix of Pearson correlation coefficients for all of the variables in the model. The first part of the table contains the observed correlation coefficients for each pair of variables. The second part of the table contains the one-tailed observed significance levels (the probability of observing a correlation coefficient at least that large and of the same sign when the population correlation coefficient is 0) for all of the coefficients.

You obtain these correlation coefficients as part of the same regression analysis shown in Figure 23.3. In the Linear Regression Statistics dialog box, select Descriptives.

Figure 23.6 Correlation matrix

		LIFEEXPF	LNBEDS	LNDOCS	LNGDP	LNRADIO	URBAN
Pearson Correlation	LIFEEXPF	1.000	.730	.880	.836	.693	.750
	LNBEDS	.730	1.000	.711	.741	.616	.619
	LNDOCS	.880	.711	1.000	.824	.633	.794
	LNGDP	.836	.741	.824	1.000	.716	.789
	LNRADIO	.693	.616	.633	.716	1.000	.668
	URBAN	.750	.619	.794	.789	.668	1.000
Sig. (1-tailed)	LIFEEXPF	.	.000	.000	.000	.000	.000
	LNBEDS	.000	.	.000	.000	.000	.000
	LNDOCS	.000	.000	.	.000	.000	.000
	LNGDP	.000	.000	.000	.	.000	.000
	LNRADIO	.000	.000	.000	.000	.	.000
	URBAN	.000	.000	.000	.000	.000	.

The correlation coefficient between female life expectancy and urbanization is 0.750. Based on the observed significance level, you can reject the null hypothesis that there is no linear relationship between the two variables. Similarly, you see that the correlation coefficient between female life expectancy and hospital beds is 0.730. Again, you can reject the null hypothesis that there is no linear relationship between the two variables. So what's going on? The variables are individually related to life expectancy, but in the multiple regression equation, their coefficients are not significantly different from 0.

The explanation is actually quite simple. Look at Figure 23.6 again. You see that the independent variables are highly correlated with each other. The correlation coefficient between doctors and urbanization is 0.794. The correlation coefficient between hospital beds and doctors is 0.711. That means that if you have a regression model that includes doctors as an independent variable, urbanization and hospital beds don't contribute much unique information. Much of the information they convey is already being supplied by the other independent variables.

? *Why bother looking at scatterplots when I can compute a whole matrix of correlation coefficients so easily?* Correlation matrices are useful for looking at the strength of the linear relationship between pairs of variables. However, don't use them as a substitute for scatterplot matrices, which show you what the relationship between two variables *really* looks like. From a correlation coefficient, you can't tell if a straight line is a good summary of the data or if it's just better than no line. You also can't tell if there are unusual points in your data or if there are distinct clusters of points that make it look like there is a relationship between the variables when there really isn't one.

You should also be careful when looking at a whole matrix of observed significance levels for correlation coefficients. If you have a lot of coefficients, you expect some of them to be statistically significant even if the variables are not linearly related. Out of 100 coefficients, you expect 5 to be significant by chance alone. A simple correction for the fact that you're examining many coefficients is to multiply the observed significance level by the number of coefficients you're looking at. ■■■

Changing the Model

Consider what happens to the multiple regression equation when you remove the doctors variable from the model. Figure 23.7 shows the coefficients for the model with only four independent variables. You see that when the doctors variable is removed, the coefficients of the other variables change. The coefficient for the hospital beds variable is more than twice as large as before and is now significantly different from 0. The coefficient for urbanization changes from –0.007 to +0.085. It's also significantly different from 0. The coefficient for radios is no longer significantly different from 0.

When the independent variables are correlated with each other, the coefficient for a particular variable depends on the other variables included in the model. That's why you must be very careful about the conclusions you draw about individual variables based on a multiple regression model. You would be wrong if you conclude based on Figure 23.3 that urbanization and hospital beds are not linearly related to life expectancy. They certainly are!

To obtain this output, repeat the analysis shown in Figure 23.3 without including lndocs as one of the independent variables.

Figure 23.7 Coefficients with doctors removed from the model

Model		Unstandardized Coefficients		Standardized Coefficients	t	Sig.
		B	Std. Error	Beta		
1	(Constant)	26.860	2.995		8.968	.000
	LNBEDS	2.504	.868	.208	2.886	.005
	LNGDP	3.235	.687	.447	4.712	.000
	LNRADIO	1.390	.841	.117	1.653	.101
	URBAN	.085	.036	.190	2.388	.019

Partial Correlation Coefficients

When you want to measure the strength of the linear relationship between the dependent variable and an independent variable, while "controlling" or keeping constant the effects of other independent variables, you can compute what's called the partial correlation coefficient. The **partial correlation coefficient** is the correlation between two variables when the linear effects of other variables are removed. For example, you can estimate the correlation between female life expectancy and urbanization, controlling for hospital beds, doctors, GDP, and radios. This is done by calculating a linear regression model that predicts each of the two variables from the other independent variables. For example, both urbanization and female life expectancy are separately predicted from beds, GDP, radios, and doctors. The partial correlation coefficient is then the Pearson correlation coefficient between the two sets of residuals.

In SPSS, you can use the Partial Correlations procedure to calculate partial correlation coefficients for any pair of variables, controlling for the linear effects of other variables. For example, Figure 23.8 shows the partial correlation coefficient between urbanization and life expectancy when the linear effects of GDP, radios, doctors, and hospital beds are eliminated.

To obtain these coefficients, from the menus choose:

Analyze
Correlate ▶
Partial...

Select the variables shown in Figure 21.8.

Figure 23.8 Partial correlation between urbanization and life expectancy

				Female life expectancy 1992	Percent urban, 1992
Control Variables	lngdp& lnradio &lndocs & lnbeds	Female life expectancy	Correlation	1.000	-.019
			Significance (2-tailed)	.	.841
			df	0	110
		Percent urban	Correlation	-.019	1.000
			Significance (2-tailed)	.841	.
			df	110	0

The partial correlation coefficient is –0.019. Since the observed significance level is large, you can't reject the null hypothesis that the population value for the partial correlation coefficient is 0. If you look at the observed significance level (0.841), you'll see that it's exactly the same as the observed significance level for the partial regression coefficient for urbanization when the other four independent variables are in the model. That's not a coincidence. The two tests are equivalent. That is, when you test whether a partial regression coefficient is 0, you're also testing whether the partial correlation coefficient between the dependent variable and the independent variable, controlling for the effects of the other independent variables in the model, is 0.

Tolerance and Multicollinearity

The strength of the linear relationships among the independent variables is measured by a statistic called the **tolerance.** For each independent variable, the tolerance is the proportion of variability of that variable that is *not* explained by its linear relationships with the other independent variables in the model. Since tolerance is a proportion, its values range from 0 to 1. A value close to 1 indicates that an independent variable has little of its variability explained by the other independent variables. A value close to 0 indicates that a variable is almost a linear combination of the other independent variables. Such data are called **multicollinear.**

? *How can I tell if multicollinearity is a problem for my data?* Look at the tolerances for each of the independent variables in the model. (These are printed if you select **Collinearity diagnostics** in the Linear Regression Statistics dialog box.) If any of the tolerances are small—say, less than 0.1—multicollinearity may be a problem. If your variables are multicollinear, you may find that although you can reject the null hypothesis that all the population coefficients are 0 based on the overall *F* statistic, none of the individual coefficients in the model is significantly different from 0 based on the *t* statistic. Or you may encounter coefficients with the wrong sign.

For example, you might find that the coefficients for doctors and hospital beds are negative, when you know they should be positively related to life expectancy. In such a situation, you should identify the variables that are almost linear combinations of each other and remove some of them from the model. If the independent variables are very highly related, you may not even be able to estimate a regression model that contains all of them. In this event, SPSS will issue a warning and omit the offending variables from the model. ■■■

Beta Coefficients

A common mistake in regression analysis is equating the magnitude of the partial regression coefficients to the relative importance of the variables. For example, in Figure 23.7 the coefficient for urbanization is quite small compared to that for radios. What does that tell you? Not much. The magnitude of the partial regression coefficient depends, among other things, on the units in which the variable is measured. In this example, urbanization is measured as a percentage, while radios are expressed as the natural log of the number of radios per 100 people. Consider what happens if you express urbanization not as a percentage but as a proportion—for example, the value 50% is changed to 0.50. The coefficients are shown in Figure 23.9. As you can see, the coefficient for urbanization is 100 times as large as that in Figure 23.5. The coefficients for the other independent variables are the same.

Figure 23.9 Coefficients with urbanization divided by 100

Model		Unstandardized Coefficients		Standardized Coefficients	t	Sig.
		B	Std. Error	Beta		
1	(Constant)	40.779	3.201		12.739	.000
	LNBEDS	1.174	.750	.097	1.565	.120
	LNDOCS	3.965	.568	.555	6.982	.000
	LNGDP	1.626	.619	.225	2.629	.010
	LNRADIO	1.541	.703	.130	2.191	.031
	URBPROP	**-.656**	3.270	-.015	-.201	.841

The coefficient for urbanization is 100 times as large as that in Figure 23.5

One way to make partial regression coefficients somewhat more comparable is to calculate what are called *beta weights*. These are the partial regression coefficients when all independent variables are expressed in standardized (z score) form. In the output, these coefficients are in the column labeled *Beta*. You can see by comparing Figure 23.5 with Figure 23.9 that dividing the urbanization variable by 100 does not affect the beta coefficient. However, the values of the beta coefficients still depend on the other independent variables in your model, so they don't reflect in any absolute sense the importance of the individual independent variables.

Building a Regression Model

The five variables you used to predict life expectancy were selected from among a large number of variables that describe a country. No doubt, additional information about the countries could be useful in building a model to predict life expectancy. It's also possible that some of the selected variables are not particularly good predictors. This is the usual situation in model building.

Your goal is to build a simple model that predicts well. If you can predict life expectancy well using two variables instead of five, the simpler model is better. If you include irrelevant variables in a model, you increase the standard errors of the coefficients without improving prediction. If you exclude variables that are important predictors, your model is biased. That is, it doesn't represent the true underlying model.

The following sections describe techniques that are used for identifying a subset of independent variables that results in a "good" regression model. None of the methods results in a model that is "best" in any absolute sense. However, these methods may help you identify a group of variables that predicts the dependent variable reasonably well.

Methods for Selecting Variables

You can build many different models from the same set of independent variables. For example, if you have 5 independent variables, you can build 32 different models: 1 model with only the constant, 5 models with only one independent variable, 10 models with 2 independent variables, 10 models with 3 independent variables, 5 models with 4 independent variables, and 1 model with all 5 independent variables. As the number of independent variables increases, so does the number of possible models.

Although for a small number of independent variables it is possible for you to evaluate all possible models, several methods that do not require as much computation are commonly used. These methods sequentially add or remove variables from a model. The decision to enter or remove a variable is based on how much it changes multiple R^2. Whenever you add an independent variable to a regression model, R^2 increases or remains the same. It never decreases when a variable is added. Similarly, when you remove a variable from a regression model, R^2 decreases or remains the same.

- **Forward selection.** Forward selection starts with a model that contains only the constant term. At each step, you add the variable that results in the largest increase in multiple R^2, provided that the change in R^2 is large enough for you to reject the null hypothesis that the true change is 0, using a preset significance level. (The default criterion is an observed significance level of 0.05 or less. If you increase the significance level, you make it easier for variables to enter the model.) You stop entering variables into the model when there are no more variables that result in a significant increase in R^2.
- **Backward elimination.** In backward elimination, you start with a regression model that contains all of the independent variables. At each step, you remove the variable that changes R^2 least, provided that the change is small enough so that you can't reject the null hypothesis that the true change is 0, using a preset significance level. (The default criterion is an observed significance level of 0.1 or larger.) You stop removing variables when removal of any variable in the model results in a significant change in R^2.
- **Stepwise variable selection.** The most commonly used method for model building is stepwise variable selection. It resembles forward selection except that after you enter a variable into the model, you

remove any variables already in the model that are no longer significant predictors. This means that variables whose importance diminishes as additional predictors are added are removed. Stepwise variable selection is a combination of forward selection and backward elimination. You have two criteria: one for entering a variable and one for removing a variable. You select the first two variables for entry the same way as in forward selection. Then you examine the variables in the model to see if either of them meet the removal criteria. If so, you remove it from the model. At each step, you enter a new variable using the same rules as in forward selection; then you examine the variables already in the model for removal, using the same rules as in backward elimination. You stop when no more variables meet the entry criterion. (The significance level for entering a variable should be smaller than the significance level for removing a variable. Otherwise, if the computer didn't stop you, you might end up entering and removing the same variable over and over.)

Whenever you use one of the variable selection methods to build a model, the observed significance levels for the coefficients are not really correct, since you're looking at many variables and choosing the one with the smallest significance level for entry or the one with the largest significance level for removal. Unfortunately, the true significance level is difficult to compute, since it depends not only on the number of cases and variables but also on the correlations of the independent variables. So, although we'll be referring to observed significance levels for the coefficients as variables are entered and removed from the model, in a strict statistical sense they are not really correct.

? *Why do you look at the statistical significance of the change in multiple R^2 instead of looking at the observed significance levels for the regression coefficients?* In multiple linear regression, there are many statistics that are equivalent. Choosing the variable that results in the largest increase in R^2 is the same as choosing the variable that has the smallest observed significance level for the test of its partial regression coefficient. It's also the same as choosing the variable with the largest absolute value for the t statistic, the variable with the largest absolute value for its partial correlation coefficient, and the variable that results in the smallest residual sum of squares. ■■■

Forward Selection

Let's see what happens when you build a regression model using the forward selection method. You'll start out with a model that has only the constant in it, and at each step you'll enter the variable that results in the largest increase in R^2, provided that the observed significance level for the increase is less than 0.05 (the default). That is, the increase has to be large enough so that you can reject the null hypothesis that the population value for the increase in R^2 is 0. Testing that the change in R^2 is 0 when a variable enters the model is the same as testing that the partial regression coefficient for the variable is 0, so you can look at the observed significance level for the regression coefficient when a variable enters and see if it is less than 0.05.

Figure 23.10 Variables entered

Model	Variables Entered	Variables Removed	Method
1	LNDOCS	.	Forward (Criterion: Probability-of-F-to-enter <= .050)
2	LNGDP	.	Forward (Criterion: Probability-of-F-to-enter <= .050)
3	LNRADIO	.	Forward (Criterion: Probability-of-F-to-enter <= .050)

Figure 23.10 summarizes what happens when you use forward selection to predict female life expectancy from the five independent variables. The first row, labeled *Model 1*, tells you that the natural log of doctors per 10,000 people is the first variable to enter the model. *Model 1* is the regression model with only the constant and the doctors variable. *Model 2* results from the addition of *lngdp* to *Model 1*. Similarly, *Model 3* results from adding *lnradio* to *Model 2*.

Figure 23.11 Model summary

Model	R	R Square	Adjusted R Square	Std. Error of the Estimate
1	.880[1]	.775	.773	5.32
2	.902[2]	.813	.809	4.87
3	.907[3]	.823	.818	4.76

1. Predictors: (Constant), LNDOCS
2. Predictors: (Constant), LNDOCS, LNGDP
3. Predictors: (Constant), LNDOCS, LNGDP, LNRADIO

Summary statistics for the three models are also displayed in Figure 23.11. In the row for *Model 1*, you see that a model with only the constant and the doctors variable explains almost 78% of the variability in female life expectancy. When *lngdp* is added to the model, *R Square* increases to 81.3%, so the change in R^2 when *lngdp* is added is 81.3% – 77.5%, or 3.8%. When *lnradio* is added to the model containing *lndocs* and *lngdp*, R^2 increases by 1%, from 81.3% to 82.3%. Although the increase is not large, it is large enough for you to reject the null hypothesis that the true increase is 0. The urban and beds variables don't appear in any of the three models. That's because R^2 would not have changed enough if they were added to the models.

Figure 23.12 Regression coefficients

Model		Unstandardized Coefficients		Standardized Coefficients		
		B	Std. Error	Beta	t	Sig.
1	(Constant)	57.232	.688		83.233	.000
	LNDOCS	6.290	.318	.880	19.792	.000
2	(Constant)	42.138	3.206		13.143	.000
	LNDOCS	4.261	.513	.596	8.307	.000
	LNGDP	2.493	.519	.345	4.802	.000
3	(Constant)	41.697	3.140		13.278	.000
	LNDOCS	4.123	.505	.577	8.168	.000
	LNGDP	1.871	.566	.259	3.306	.001
	LNRADIO	1.684	.679	.142	2.482	.015

The regression coefficients for all three of the models are shown in Figure 23.12. The observed significance level for all of the coefficients is less than 0.05 because that's the criterion you used to select variables for the model. If the partial regression coefficient for a variable has an observed significance level larger than 0.05, the variable can't enter the model. In this table, you can easily see how the coefficient for a variable changes when other independent variables are added. For example, the coefficient for *lndocs* is 6.29 when it's the only variable in the regression model. When *lngdp* and *lnradio* are added (*Model 3*), the coefficient changes to 4.123. The contribution of a variable to the regression model depends on the other independent variables in the regression model.

Statistics for Excluded Variables

To help you understand what goes on at each step in building the model, look at Figure 23.13. For each model, this table shows you statistics for variables that are *not* in the model. For *Model 1*, the variables ***lnbeds***, ***lngdp***, ***lnradio***, and ***urban*** are not in the model. The values for the standardized regression coefficients, if each of the variables were entered next into the model, are shown in the column labeled *Beta In*. For example, if the ***lnbeds*** variable entered into a model containing only the doctors variable, its beta coefficient would be 0.211. The *t* statistic for testing that the coefficient is 0, and its observed significance level, are shown in the columns labeled *t* and *Sig*. For all of the variables except ***urban***, you can reject the null hypothesis that the regression coefficient is 0.

Figure 23.13 Excluded variables

Model		Beta In	t	Sig.	Partial Correlation	Collinearity Statistics Tolerance
1	LNBEDS	.211	3.492	.001	.312	.495
	LNGDP	.345	4.802	.000	.412	.322
	LNRADIO	.226	4.217	.000	.369	.599
	URBAN	.140	1.943	.054	.180	.370
2	LNBEDS	.120	1.937	.055	.180	.419
	LNRADIO	.142	2.482	.015	.228	.482
	URBAN	.017	.226	.822	.021	.313
3	LNBEDS	.098	1.589	.115	.149	.408
	URBAN	-.022	-.306	.760	-.029	.298

The partial correlation coefficient for ***lnbeds*** is 0.312. That's the correlation coefficient between ***lnbeds*** and ***lifeexpf***, controlling for ***lndocs***, the independent variable already in the model. The partial correlation coefficient displayed here controls only for the independent variables that are already in the model. If the partial correlation coefficient is large in absolute value, the variable has unique information to contribute to the equation.

The column labeled *Tolerance* is the smallest tolerance for any independent variable in the model, if that variable enters next. SPSS will not enter a variable into the model if it results in a very small tolerance for any independent variable. The reason for this is that very small tolerances may cause computational problems.

By examining the statistics in Figure 23.13, you can determine the sequence in which variables enter into the model. Remember, each model in the sequence is the result of adding the variable that results in the largest change in R^2 to the previous model. Look at the statistics for *Model 1*. To get the largest change in R^2, you must enter the variable with the smallest observed significance level, which is the same as the variable with the largest *t* statistic or the variable with the largest absolute value of the partial correlation coefficient. The *lngdp* variable has the largest *t* value, and its observed significance level is less than 0.05 (the default criterion for adding a variable), so it enters next. Similarly, by looking at the statistics for *Model 2*, you see that the largest *t* value is for *lnradio*, so that's the variable added to *Model 2* to make *Model 3*. The statistics for *Model 3* show that there are only two variables still not in the regression equation: *lnbeds* and *urban*. You see that both of them have observed significance levels greater than 0.05, so neither is eligible to enter the next model. In fact, there is no next model since there are no more variables that meet the criteria for entering into a model.

Compare the five-independent-variables model shown in Figure 23.3 with *Model 3*. You see that there is a very small difference in the R^2 between the two models. If you look at the adjusted R^2, which takes into account the fact that adding variables to a model always increases R^2 or leaves it unchanged without necessarily improving the fit of the model in the population, you also see that the two additional variables are of little use in the regression model. The standard errors of the coefficients in the three-variable model are smaller than the standard errors of the same coefficients in the five-variable model. That's because the five-variable model includes poor predictors. They increase the variability of the regression coefficients without improving prediction.

Backward Elimination

In forward selection, you start with a model that has no independent variables and sequentially add variables to the model. In backward elimination, you start with a model that contains *all* of the independent variables, and at each step you eliminate the variable that causes the smallest change in R^2. By default, SPSS removes a variable from the model if the observed significance level for its coefficient is greater than 0.10.

In this example, backward elimination starts with the complete five-variable model. The sequence of models that are built is shown in Figure 23.14. *Model 1* contains all five independent variables. *Model 2* is *Model 1* with the ***urban*** variable removed. *Model 3* is *Model 2* with the ***lnbeds*** variable removed. The final model, *Model 3*, is exactly the same one that was produced by forward selection. It contains the ***lnradio***, ***lndocs***, and ***lngdp*** variables.

Figure 23.14 Variables entered/removed

Model	Variables Entered	Variables Removed	Method
1	URBAN, LNBEDS, LNRADIO, LNDOCS, LNGDP	.	Enter
2	.	URBAN	Backward (criterion: Probability of F-to-remove >= .100).
3	.	LNBEDS	Backward (criterion: Probability of F-to-remove >= .100).

The model summary statistics for each of the three models are shown in Figure 23.15. The R^2 is largest for *Model 1* since it contains all of the independent variables. At each step, R^2 decreases since an independent variable is removed. The coefficients for the three models are also shown in Figure 23.15. Look at *Model 1*. The largest observed significance level is for the ***urban*** variable. Since the observed significance level is larger than the default value of 0.1 for remaining in the model, ***urban*** is removed from the model. From the statistics for *Model 2*, you see that ***lnbeds*** now has the largest observed significance level (0.115). Since it's larger than 0.1, ***lnbeds*** is the next variable to be removed. This results in *Model 3*. All of the observed significance levels for the coefficients are less than 0.1 in *Model 3*, so no more variables are removed from the model. In this example, the forward selection and backward elimination techniques have resulted in the same model. That's not always the case.

Figure 23.15 Model summary and regression coefficients

Model	R	R Square	Adjusted R Square	Std. Error of the Estimate
1	.909	.827	.819	4.75
2	.909	.826	.820	4.73
3	.907	.823	.818	4.76

Model		Unstandardized Coefficients		Standardized Coefficients		
		B	Std. Error	Beta	t	Sig.
1	(Constant)	40.779	3.201		12.739	.000
	LNBEDS	1.174	.750	.097	1.565	.120
	LNDOCS	3.965	.568	.555	6.982	.000
	LNGDP	1.626	.619	.225	2.629	.010
	LNRADIO	1.541	.703	.130	2.191	.031
	URBAN	-6.56E-03	.033	-.015	-.201	.841
2	(Constant)	40.857	3.164		12.915	.000
	LNBEDS	1.184	.745	.098	1.589	.115
	LNDOCS	3.919	.517	.548	7.573	.000
	LNGDP	1.590	.589	.220	2.698	.008
	LNRADIO	1.510	.683	.128	2.211	.029
3	(Constant)	41.697	3.140		13.278	.000
	LNDOCS	4.123	.505	.577	8.168	.000
	LNGDP	1.871	.566	.259	3.306	.001
	LNRADIO	1.684	.679	.142	2.482	.015

In this example, when you use stepwise variable selection to build the model, you will again get exactly the same results as you did from forward selection and backward elimination. That's because when you added variables to the model, the coefficients for the variables already in the model stayed significant. Stepwise selection will differ from forward selection only if, at some step, a variable already in the model becomes "unimportant" when another variable is added.

? *What should I do if the methods result in different models?* You should examine the different models and choose among them based on how easy they are to interpret, how easily the values for the independent variables can be obtained, and how well the regression assumptions are met. If you have a large data file, you can split it into two parts (80% for estimating the coefficients and 20% for testing is a good split), estimate the coefficients from one part, and use those coefficients to predict values for the other part. You can see which model works best on the test part of the sample.

Summary

How do you build a regression model with more than one independent variable?

- A multiple linear regression model is used to predict values of a dependent variable from a set of independent variables.
- The coefficient for an independent variable in a multiple regression equation is called a partial regression coefficient. Its magnitude and observed significance level depends on the other independent variables in the model.
- The overall regression *F* test is used to test the null hypothesis that all of the population regression coefficients are 0.
- Forward entry, backward elimination, and stepwise variable selection help you select a regression model that contains only independent variables that meet certain criteria.

What's Next?

In this chapter, you learned how to interpret the results of a multiple regression model. You also developed a multiple regression model using different techniques for variable selection. Although the assumptions needed for testing hypotheses were outlined, you did not examine your data for violations of the requisite assumptions. In Chapter 24, you'll see how to use residuals and other diagnostic information to check for violations of the multiple regression assumptions.

How to Obtain a Multiple Linear Regression

This section provides information about obtaining a multiple linear regression analysis (a regression with more than one independent variable), using the SPSS Linear Regression procedure. For a basic overview of the Linear Regression procedure, see Chapter 20.

▶ To open the Linear Regression dialog box (see Figure 23.16), from the menus choose:

Analyze
  Regression ▶
    Linear...

Figure 23.16 Linear Regression dialog box

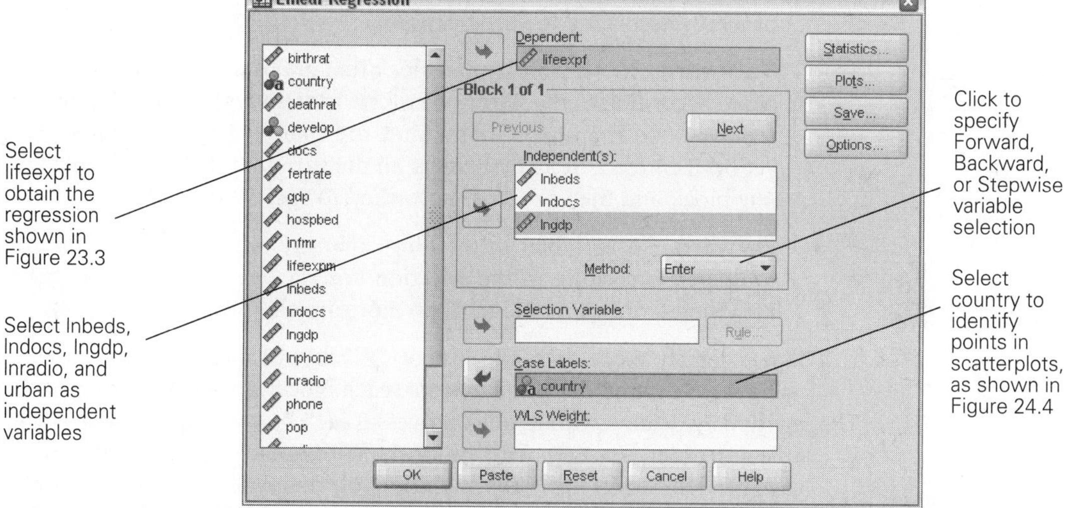

▶ Select the dependent variable and move it into the Dependent list.

▶ Move two or more independent variables into the Independent(s) list and click **OK**.

The list holds more variables than are visible; use the scroll bar to see all of them. To remove a variable from the Independent(s) list, click the same button that you used to add it to the list.

You can specify more than one list, or "block," of variables, using the **Next** and **Previous** buttons to display the different lists. You can specify up to nine blocks.

Method. SPSS offers a great deal of control over the way in which variables are entered into, and removed from, the regression equation. The Method alternatives available for a block of variables are:

Enter. All variables in the block are entered into the equation as a group.

Stepwise. Selection of variables within the block proceeds by steps. At each step, variables already in the equation are evaluated according to the selection criteria for removal; then variables not in the equation are evaluated for entry. This process repeats until no variable in the block is eligible for entry or removal.

Remove. All variables in the block that are already in the equation are removed as a group. (If no variables have yet been entered, SPSS first enters all the variables that are included in any block and then proceeds according to specifications.)

Backward. All variables in the block that are in the equation are evaluated according to the selection criteria for removal. Those eligible are removed one at a time until no more are eligible. (If no variables have yet been entered, SPSS first enters all the variables that are included in any block and then proceeds according to specifications.)

Forward. All variables in the block that are not in the equation are evaluated according to the selection criteria for entry. Those eligible are entered one at a time until no more are eligible.

If you specify more than one list (or "block") of independent variables using the Next and Previous buttons, each block of variables has its own method associated with it. SPSS processes each block of independent variables in order, starting with block 1. Within each block, SPSS selects variables for entry into the equation or removal from the equation according to the procedure specified by that block's Method alternative.

Options: Variable Selection Criteria

The stepwise, backward, and forward methods must choose variables one at a time for entry into the equation or removal from it. They do this according to statistical criteria that you can specify in the Linear Regression Options dialog box. In the Linear Regression dialog box, click Options to open the Linear Regression Options dialog box, as shown in Figure 23.17.

Figure 23.17 Linear Regression Options dialog box

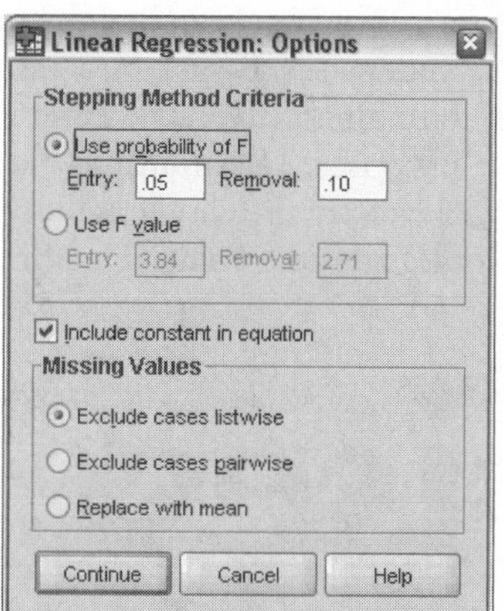

Stepping Method Criteria. In addition to the options discussed in Chapter 20, you can choose between two alternatives in the Stepping Method Criteria group. Both are based on the *F* statistic for the statistical significance of the change in R^2 when the variable in question enters or is removed from the equation. (The observed significance levels of the *F* statistic are the same as those for the *t* test for the coefficients. Similarly, if you square the *t* value, you get the *F* value described.) They are:

Use probability of F. The variable with the lowest probability of *F* is entered first provided that this probability is lower than the Entry value specified in this dialog box; or the variable with the highest probability of *F* is removed first provided that this probability is higher than the Removal value specified in this dialog box. You can change the Entry and Removal probabilities. Both values must be greater than 0 and less than or equal to 1, and the Entry value must be less than the Removal value.

Use F value. The variable with the highest *F* statistic is entered first provided that this value is greater than the Entry value specified in this dialog box; or the variable with the lowest *F* statistic is removed first provided that this value is less than the Removal value specified in this dialog box. You can change the Entry and Removal values. Both values must be greater than 0, and the Entry value must be greater than the Removal value.

The combination of blocks of independent variables with the variable selection methods can be confusing. Usually you can do what you want by moving all of your independent variables into a single block and choosing a suitable Method alternative. The enter and stepwise methods are sufficient for most purposes. Use multiple blocks of variables only when you need to control exactly the way in which your equation is built regardless of the statistical criteria for variable selection.

Exercises

Statistical Concepts

1. The following output is from a regression used to predict the percentage of defective items from the percentage of machines in operation, the percentage of employees present, and the volume produced:

Model	Variables Entered	Variables Removed	Method
1	PCTMACH	.	Enter
2	PCTSTAFF	.	Enter
3	VOLUME	.	Enter

Model	R	R Square	Adjusted R Square	Std. Error of the Estimate
3	.83867	.70338	.69767	4.01478

Model		Sum of Squares	df	Mean Square	F	Sig.
3	Regression	5962.48436	3	1987.49479	123.30556	.00000
	Residual	2514.47853	156	16.11845		
	Total	8476.96289	159			

Model		Unstandardized Coefficients		Standardized Coefficients	t	Sig.
		B	Std. Error	Beta		
3	(Constant)	-34.67142	9.50362		-3.64800	.00040
	PCTMACH	-.11889	.03542	-.15815	-3.35600	.00100
	PCTSTAFF	-.23843	.03142	-.37717	-7.58800	.00000
	VOLUME	.01893	.00183	.51860	10.35000	.00000

a. From the preceding output, write the multiple linear regression equation that predicts the percentage of defective items from the percentage of machines in operation, the percentage of employees present, and the volume produced.

b. What does the negative sign for two of the coefficients tell you?

c. Can you reject the null hypothesis that in the population all coefficients are 0? On what do you base your conclusion?

d. What percentage of the variability in the percentage of defective items is explained by the three independent variables?

e. From the coefficients, can you determine which variable is the single best predictor of the defective percentage?

f. If you built the model using forward or backward elimination, do you think the model would change? Why or why not?

2. For the regression in question 1, indicate what steps you would take to test whether the assumptions necessary for the linear regression are violated.

3. For the regression in question 1, what would you predict the defective rate to be when 90% of the machines are in operation, 90% of the staff is present, and 4000 items are produced?

4. Is the following regression output possible if a variable is added at each step and none are removed from the equation?

Step	Multiple R	R square
1	0.5821	0.3388
2	0.6164	0.3799
3	0.6025	0.3630
4	0.6399	0.4095

Data Analysis

Use the *gss.sav* data file to answer the following questions:

1. Run a linear regression model to predict a person's education (*educ*) from his or her father's education (*paeduc*) and the number of hours of television watched per day (*tvhours*). Save the Studentized residuals, the unstandardized predicted values, and the changes in the regression coefficients when a case is removed from the analysis.

a. Write the equation.

b. Can you reject the null hypothesis that all the population values for the regression coefficients are 0? On what do you base your conclusion?

c. Can you reject the null hypothesis that the population value for the proportion of variability in the dependent variable explained by the independent variables is 0? On what do you base your conclusion? What proportion of the sample variability in education is explained by the two independent variables?

d. Can you reject the null hypothesis that there is no linear relationship between a person's education and their father's education and the number of hours of television they watch a day? On what do you base your conclusion?

e. Can you reject the null hypothesis that the population value for the partial regression coefficient for father's education is 0? On what do you base your conclusion?

f. Can you reject the null hypothesis that the population value for the partial regression coefficient for television hours is 0? On what do you base your conclusion?

g. How many years of education would you predict for a person whose father has 14 years of education and who watches 3 hours of television a day?

h. Does the predicted value for education increase or decrease as television watching increases? As father's education increases?

i. How much does the predicted value for education change for a one-year increase in father's education? For a one-hour increase in hours of television watched?

2. Test the required regression assumptions for the regression in question 1.

a. Obtain a histogram and Q-Q plot of the Studentized residuals. Is the distribution of residuals approximately normal?

b. Plot the Studentized residuals against the predicted values and against the values of each independent variable. Do you see any patterns that concern you?

c. Plot the change in the coefficient for father's education when each case is removed from the computation of the coefficients. Identify the IDs of the cases that cause the largest changes in the coefficient. Go to the Data Editor and list their values for the dependent variable and both independent variables.

d. Repeat question 2c for the change in the coefficient for hours of television watched.

e. Give your assessment of how well the assumptions required for the regression model are met.

3. Use stepwise linear regression to obtain a model to predict years of education from the following independent variables: father's education (variable *paeduc*), spouse's education (*speduc*), hours of television watched per day (*tvhours*), age (*age*), income in dollars (*rincdol*), and hours worked (*hrs1*). Use the default criteria for variable entry and removal.

a. By including the variable *speduc*, you are restricting your analysis to what kind of people?

b. What is the first independent variable to enter the model? How is it selected?

c. What proportion of the variability in the dependent variable is explained by the independent variable? Can you reject the null hypothesis that the population value for the regression coefficient is 0?

d. Based on the model, what do you predict for years of education for a 50-year-old person whose father has 12 years of education, whose spouse has 14 years of education, who watches 3 hours of television a day, has an income of $30,000, worked 42 hours last week, and is male?

e. What variable enters the model at the second step? What is the change in R^2 when the variable enters? Write the regression equation and obtain the predicted years of education for the person described in question 3d. How has your prediction changed?

f. What variable enters at the third step? How much does R^2 change? Write the regression equation and obtain the predicted years of education for the person described in question 3d.

g. What is the last variable to enter the model? How much does R^2 change when it is entered? What is the predicted value for years of education for the person described in question 3d? How much has the predicted value for the person changed from question 3d to question 3g?

h. Why does variable selection stop?

i. What is the partial correlation between education and father's education, "controlling" for spouse's education and respondent's income in dollars?

j. Based on the size of the partial regression coefficients, is it reasonable to conclude that a person's income is least strongly related to education? Why or why not?

k. Consider the spouse's education variable. How has its coefficient changed at each step of model building? Why has this happened?

4. Look for violations of the regression assumptions for the model you built in question 3. Use the steps outlined in question 2 as a guide. Summarize your findings.

5. Use the Partial Correlations procedure to calculate the following partial correlation coefficients:

a. The partial correlation between *educ* and *paeduc*, controlling for *tvhours*.

b. The partial correlation between *educ* and *speduc*, controlling for *paeduc* and *rincome*.

c. The partial correlation between *educ* and *paeduc*, controlling for *maeduc* and *rincome*.

d. The partial correlation coefficient between *educ* and *speduc*, controlling for *tvhours*, *rincome*, *age*, and *hrs1*.

e. Interpret each of the coefficients and indicate what type of relationship the coefficients are measuring.

6. Develop a regression model to predict education for all adults, not just those who are married. That means you can't use any variables that are available only for married people (such as spouse's education). Check the regression assumptions and summarize your findings.

7. Develop a regression equation to predict hours worked last week for full-time employees only (variable *wrkstat* equals 1). Summarize your findings. Be sure to check for violations of the regression assumptions. If the assumptions appear to be violated, try taking the square root of the number of hours worked as the dependent variable. Rerun the model and check the diagnostics again. Do you see any improvement?

Use the *salary.sav* data file to answer the following questions:

8. Use the Regression procedure to obtain a multiple linear regression model that predicts the natural log of the beginning salary (variable *salbeg*) from age, gender, minority status, education level, and work experience (variables *age*, *sex*, *minority*, *edlevel*, and *work*).Write the multiple linear regression model. Remember, you're now predicting the natural log of the beginning salary, not the actual salary.

   a. Can you reject the null hypothesis that there is no linear relationship between the dependent variable and the independent variables?

   b. What proportion of the variability in transformed beginning salary is explained by the independent variables?

   c. For which variables can you reject the null hypothesis that the population values for the partial regression coefficients are 0?

   d. What does the negative sign for the *sex* and *minority* variables tell you?

   e. What can you tell from the size of the partial regression coefficients about the individual independent variables as predictors of beginning salary?

9. Rerun the equation without the age variable. Do the partial regression coefficients for the other variables change when age is removed from the model? Why? Obtain the correlation coefficients between all pairs of variables in question 8. Which variable is most strongly linearly related to the log of beginning salary?

10. For the variables described in question 8, use stepwise variable selection to develop a multiple linear regression model. Save the Studentized residuals, the predicted values, and the changes in the coefficients.

    a. Which variable enters the model first? Why? What proportion of the variability in the dependent variable is explained by this single variable?

b. How is the second variable to enter the model selected? Which variable enters? What proportion of the variability in the dependent variable is "explained" by the two independent variables?

c. What is the last variable to enter the model? Why does variable selection end? What proportion of the variability in the dependent variable is explained by all the independent variables in the model?

d. Look at the partial correlation coefficient between age and the log of the beginning salary. What does it tell you? Use the Partial Correlations procedure to calculate the same coefficient.

11. For the model in question 10, look for violations of the assumptions needed for multiple linear regression.

a. Make a histogram and Q-Q plot of the Studentized residuals. Is the distribution approximately normal? Do you see any outliers? If so, identify them in the Data Editor and look at their values for the dependent variable and the independent variables.

b. Make a plot of the Studentized residuals against the predicted values. What should you look for in this plot? Do you see any problems? Explain.

c. Plot the Studentized residuals against the values of each independent variable. Do these plots look all right?

d. Look at the impact of individual points on the estimates of each partial regression coefficient. Identify points that cause large changes in the coefficients.

Use the *country.sav* data file to answer the following questions:

12. Build a multiple linear regression model that predicts male life expectancy (variable *lifeexpm*) from percentage urban (variable *urban*), the natural log of doctors per 10,000 people (*lndocs*), the natural log of hospital beds per 10,000 people (*lnbeds*), the natural log of GDP (*lngdp*), and the natural log of radios per 100 people (*lnradio*).

a. Write the regression equation.

b. Can you reject the null hypothesis that there is no linear relationship between the dependent variable and the independent variables?

c. What proportion of the variability in male life expectancy is "explained" by the independent variables? Can you reject the null hypothesis that the population value for R^2 is 0?

d. For which variables can you reject the null hypothesis that their partial regression coefficients are 0?

e. Compare this model to the one developed in this chapter for predicting female life expectancy. Are the coefficients similar in value for the two models?

13. Using the Bivariate Correlations procedure, obtain the correlation matrix for the dependent and independent variables. Which variables appear to be most highly correlated with male life expectancy? For which variables can you reject the null hypothesis that the population value for the correlation coefficient is 0?

14. Using stepwise variable selection, build a model to predict male life expectancy from the independent variables listed in question 12. Save the Studentized residuals, the predicted values, and the changes in the coefficients.

a. What is the first variable to enter the regression model? How was it selected?

b. Why does the natural log of radios variable not enter the model?

c. Write the final regression model. Compare it to the model for female life expectancy. Are they basically similar?

d. What proportion of the variability in male life expectancy is "explained" by the natural log of doctors and the natural log of GDP?

e. Use the Partial Correlations procedure to estimate the partial correlation coefficient between the natural log of radios and male life expectancy, "controlling" for the natural log of hospital beds and the natural log of GDP. Where does this appear on the regression output? What does it mean?

15. Examine the regression assumptions for the model developed in question 14.

a. Obtain histograms and Q-Q plots of the Studentized residuals. Identify any outlying points. What countries do they represent? Are they the same countries that were unusual in the regression of female life expectancy?

b. Plot the Studentized residuals against the predicted values and against the values of each independent variable. Do you see any suspicious patterns?

c. Plot the changes in each of the regression coefficients against the sequence number. Which countries cause the largest changes in the coefficients?

16. Using the same variables as in question 12, use backward elimination to build the regression model. Does the model change?

Use the file *bodyfat.sav* to answer the following questions:

17. Build a multiple regression equation to predict percentage of body fat from both height and weight. Does this equation differ from the equation that predicts body fat from the variables individually? Why?

18. Estimate a regression equation to predict percentage of body fat from abdominal circumference.

a. Add the weight variable to the model. Does the coefficient for abdominal circumference change when weight is added? Why?

b. Calculate the residuals and check the regression assumptions.

19. Use one of the stepwise variable selection methods to predict percentage of body fat from the other variables in the data file. (Don't use *pctfat2*, since it's a different method for predicting percent body fat.)

a. Check the assumptions.

b. Summarize your results.

20. Repeat question 19 using a different method.

Multiple Regression Diagnostics

24

How can you check for violations of the multiple linear regression assumptions, and how can you identify cases that are influencing the regression results more than the other cases are?

- What can you learn by plotting residuals against the predicted values and the values of the independent variables?
- What is leverage and why is it important?
- Why is Cook's distance useful?
- What can you tell from a partial regression plot?

In Chapter 23, you built a multiple regression model to predict female life expectancy from per capita measures of the GDP and numbers of doctors and radios. In this chapter, you'll use residuals and other diagnostics to check for violations of the regression assumptions and to identify data points that are in some way unusual. All of the diagnostic techniques described in Chapter 22 are also useful for a multiple regression model. This chapter emphasizes techniques that are not described in Chapter 22.

▶ This chapter analyzes the residuals and other diagnostics for the multiple regression model described in Chapter 23. To duplicate the results shown, rerun the regression with only the three independent variables in the final model (the variables *lndocs*, *lngdp*, and *lnradio*). In the Linear Regression Save dialog box, save Studentized deleted residuals, predicted values, leverages, Cook's distance, and the standardized changes in the regression coefficients (SdfBetas), as shown in Figure 20.18 in Chapter 20. In the Linear Regression Plots dialog box, request all partial regression plots, as shown in Figure 20.17.

Examining Normality

You'll use Studentized deleted residuals to look for violations of the regression assumptions because they make it easier to spot unusual points. The Studentized deleted residual for a case is the Studentized residual when the case is excluded from the regression computations. If the regression assumptions are met, their distribution is a Student's *t* with $N-p-2$ degrees of freedom, where N is the number of cases and p is the number of independent variables in the model. Since your sample size is much larger than 30, the distribution of the Studentized deleted residuals should be approximately normal. (We'll refer to Studentized deleted residuals as simply residuals throughout the rest of this chapter.)

You can obtain stem-and-leaf plots using the Explore procedure, as discussed in Chapter 7. Select the variable sdr_1 in the Explore dialog box.

Figure 24.1 Plot of Studentized deleted residuals

```
Frequency    Stem &  Leaf

    1.00 Extremes    (=<-3.4)
    2.00       -2 .  02
    5.00       -1 .  57899
   12.00       -1 .  000012233444
   15.00       -0 .  555555666778889
   22.00       -0 .  0000000011111122334444
   28.00        0 .  0000000011111111122233334444
   18.00        0 .  555555556677888899
   11.00        1 .  00011122234
    5.00        1 .  55669
    2.00 Extremes    (>=2.7)

 Stem width:    1.00000
 Each leaf:        1 case(s)
```

Look first at the stem-and-leaf plot of the residuals in Figure 24.1. The shape of the distribution looks pretty good. It's symmetric and has a single peak. The three outlying values are identified in Figure 24.2, the boxplot of the residuals. (They're discussed in detail later.)

Figure 24.2 Boxplot of residuals

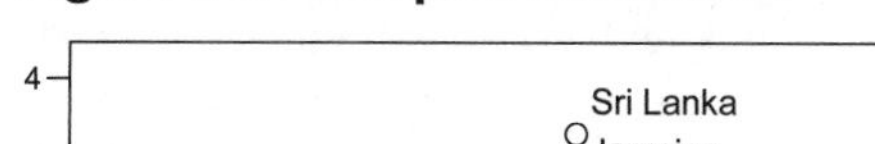

You can obtain boxplots using the Explore procedure, as discussed in Chapter 7. Select the variable sdr_1 in the Explore dialog box.

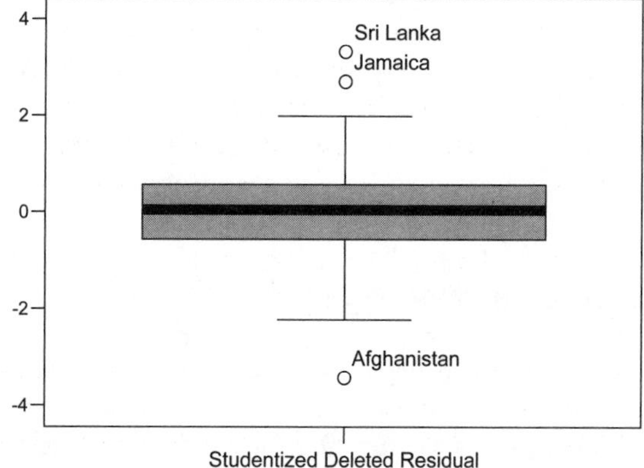

In the boxplot, you see that the distribution is fairly symmetric, since the median is in the middle of the plot. The middle half of the residuals are between –0.57 and +0.57, the values for the first and third quartiles. The corresponding quartiles for the normal distribution are –0.68 and +0.68, so the sample results match quite well. The whiskers extend to +2 and –2, which is what you expect for a sample from a normal population.

Figure 24.3 Q-Q plot of residuals

You can obtain normal probability plots using the Explore procedure, as discussed in Chapter 7. Select the variable sdr_1 in the Explore dialog box.

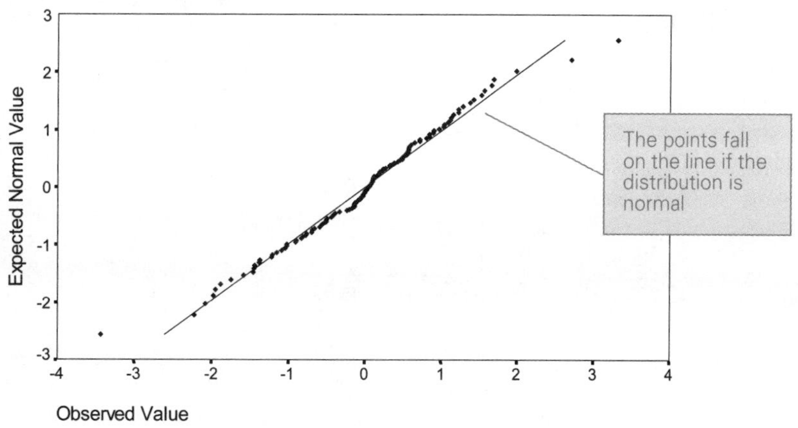

Figure 24.3 is the Q-Q plot of the residuals. If the residuals are from a normal population, they should fall close to the straight line. That's the case, except for the three outlying points. As you learned in Chapter 22, many different types of problems, such as unequal variances or violations of the linearity assumption, can cause the distribution of residuals to appear non-normal. That's why you should correct any other problems in your model before worrying about normality.

Scatterplots of Residuals

Figure 24.4 is a scatterplot of predicted and observed values of female life expectancy. Except for the cluster of points in the upper right corner, you see that the points are reasonably evenly distributed above and below the line. That's good, since clusters of points above or below the line indicate that you're consistently overpredicting or underpredicting for certain values of observed female life expectancy. That's an indication that a linear model might not be a good choice.

You can obtain scatterplots using the Graphs menu, as discussed in Chapter 9. Select the variables pre_1 and lifeexpf in the Simple Scatterplot dialog box.

Figure 24.4 Scatterplot of predicted and observed values

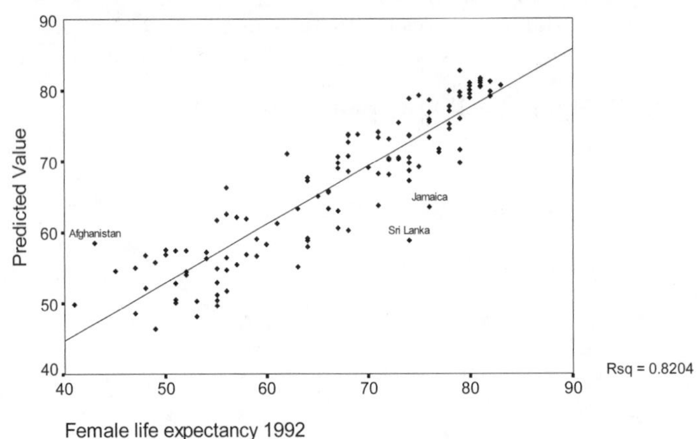

For each observed value of female life expectancy, you see a range of predicted values. If the assumption of equal variances is met, the range should be pretty much the same for all observed female life expectancies. For large observed values of female life expectancy, you see that the range of the predicted values is much narrower than elsewhere in the plot. That's probably because there is a natural limit on life expectancy that most of the highly developed nations have reached. There's just not much observed variability in life expectancy for the highly developed nations.

Their residuals are small and bunched together. That's an undesirable situation for the regression model, since it requires the variance of the residuals to be constant over the entire range of predicted values.

Note that the R^2 for the regression line of the predicted and observed life expectancies is the same as the multiple R^2 from the final stepwise model in Chapter 23. That's always the case. Multiple R is the correlation between the observed and predicted values of the dependent variable.

? *I distinctly remember that the multiple R^2 for the final model in Chapter 23 was 0.8225, not 0.8204; what gives?* When you estimated the stepwise regression model, only cases that had nonmissing values for all of the possible variables were included in the regression computations. That means that a country with missing values for any variable, regardless of whether it appears in the final model, is excluded from the computations. In this chapter, you estimated a model that includes only three independent variables. All cases with valid values for the three variables are included in the computations. It doesn't matter if they have missing values for other variables not in the model. This causes a slight change in the regression coefficients and in the multiple R^2 from Chapter 23. The multiple R^2 is really 0.8204.

When you have cases with missing values, you should rerun the regression after you select a model. That way, you include all cases that have values for the variables in the model. ■■■

In Figure 24.5, the residuals are plotted against the predicted values of female life expectancy. It's easier to see problems in this plot than in the previous one. You see that most of the residuals fall in a horizontal band around 0. Again, you see that the residuals for predicted values above 75 have somewhat less spread than the residuals for the smaller predicted values.

Sri Lanka, Jamaica, and Afghanistan stand out in Figure 24.5, since they have large absolute values for the residuals. The observed female life expectancy for Afghanistan (43 years) is low compared to the predicted value of 58.5 years. The residual of –3.4 has a probability of 0.0008, although if you multiply it by 121 to correct for picking out the largest residual from 121, the corrected probability of 0.10 is no longer particularly unusual. There are reasons why Afghanistan may be an outlier from the regression. It's been devastated by a long and fierce war, so it may be different in important ways from the other countries in the database. While it may appear to have some of the trappings of economic development, such as doctors, hospital beds, and radios, the war has probably had a large effect on the health of the population. Since the regression model doesn't include a variable for recent war, it's not so troublesome that Afghanistan's life expectancy is poorly predicted. It may make sense to eliminate Afghanistan from the model altogether and to restrict the model to countries that have not recently experienced unusual natural or man-made calamities.

Jamaica and Sri Lanka both have large positive residuals, meaning that their observed life expectancies are greater than those predicted by the model. The uncorrected probability of a residual in absolute value as large as that for Jamaica (2.7) is 0.008; for Sri Lanka (3.3), it is 0.001. If you correct for looking at many residuals, these two residuals are not unusually large.

Select the variables sdr_1 and pre_1 in the Simple Scatterplot dialog box.

Figure 24.5 Residuals versus predicted values

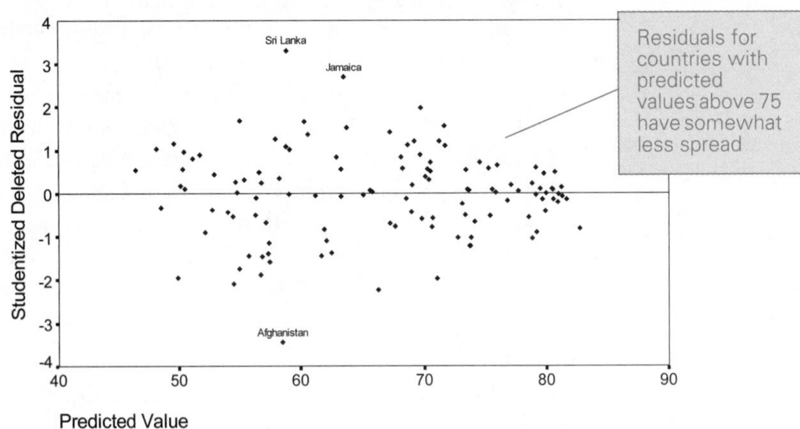

You can go to the Data Editor and examine the values of the independent variables for the three countries with the largest residuals in absolute value. You'll see that the individual values for GDP, doctors, and radios are not unusual. If they were, you would see them as outliers on stem-and-leaf plots or boxplots. What you really want to know, though, is whether these countries have unusual *combinations* of values. For example, does a country have a very high GDP and a very low number of radios? From the Data Editor, you can't easily tell if the observed combinations of values of the independent variable are unusual. However, there are diagnostic statistics in regression that make it easy for you to spot such cases.

? *I plotted the residuals against the observed values of the dependent variable, and they're not randomly distributed. What should I do?* Nothing. You don't expect to see a random distribution of points in this plot. The least-squares line for this plot will always have a slope of $1 - R^2$.

Leverage

You can use a statistic called the leverage to identify cases with unusual combinations of values of the independent variables. **Leverage** measures how far the values for a case are from the means of all of the independent variables. Leverage values computed by SPSS range in value from 0 to close to 1. Cases with high leverage values may have a large impact on the estimates of the regression coefficients. A rule of thumb is to look at cases with leverage values greater than $2p/N$ (0.0496 in this case—see the reference line in Figure 24.6), where p is the number of independent variables in the model and N is the number of cases. (For a small number of independent variables, this rule of thumb singles out too many points.

Figure 24.6 Leverage for points in the regression)

To obtain this scatterplot, select the variables lev_1 and sequence in the Simple Scatterplot dialog box.

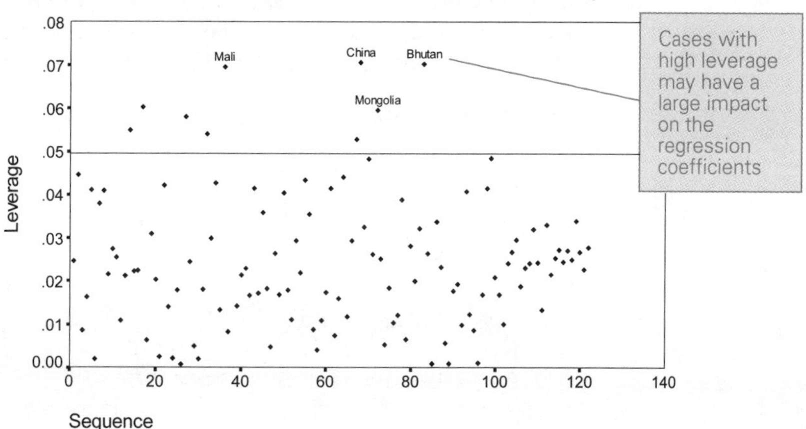

If you click on the points with the highest leverages in Figure 24.6, you'll see that they represent Mali, China, and Bhutan. Mali and Bhutan have the two smallest values for radios per 100 people. (Mali has 1.9 radios per 100 people, Bhutan has 1.6, and the United States has 200 radios per 100 people!) Mali and Bhutan are also "below average" for the other two independent variables. China has an unusual combination of values. It's above average on the number of doctors and below average on GDP and radios.

? *What should I do if I find points with large leverages?* If you find points that have large leverage values, look at the data values to make sure that the data file doesn't contain errors for these cases. That's one possible explanation for high leverage. If the values are correct, try to think of possible explanations for why the point is unusual. Often, you can learn a lot about a problem by scrutinizing cases with unusual values.

Problems with data collection may introduce unusual points. For example, it's possible that the statistics for the numbers of doctors, hospital beds, radios, etc., may not be equally accurate for all countries in the database. A good question to ask is: where did these numbers come from? Are they official government statistics, or were they obtained in a uniform manner from a reputable large-scale survey? That may explain some unusual findings.

Changes in the Coefficients

In Chapter 22, you identified points that change the value of the slope and the intercept. You calculated the slope with and without each case and saw how much the value changed. You can do the same thing in a multiple regression model. But now you have to look at the effect of removing a case on the values of each of the coefficients, including the constant.

To obtain this scatterplot, select the variables sdb_1 and sequence in the Simple Scatterplot dialog box.

There will be three sdb_1 variables, one for each of the independent variables in the model. To find out which corresponds to the doctor coefficient, check the notes in the Viewer.

Figure 24.7 Standardized changes in the doctor coefficient

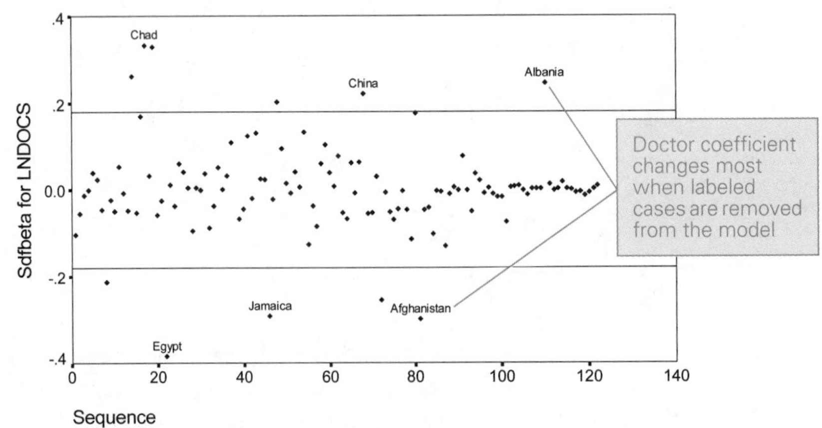

Look at Figure 24.7, which is a plot of the standardized changes in the doctor coefficient when each case is eliminated from the computations. Ideally, the points should fall in a horizontal band around 0. You see that there are several points that stick out from the rest. (A rule of thumb is to look at cases with absolute values greater than $2/\sqrt{N}$.) Egypt causes the largest standardized change in the doctor coefficient when it is eliminated from the analysis. Egypt has one doctor for every 616 people. The observed female life expectancy for Egypt, however, is not as large as you would predict based on all those doctors. If you remove Egypt from the model, the doctor coefficient changes from 4.19 to 4.38. (You have to compute DfBeta, the actual change in the coefficient, to know how much the coefficient changes.) When you have a large number of cases, it's unlikely that removal of a single case will change the actual value of a coefficient very much.

Cook's Distance

You can also compute Cook's distance (named after a statistician—nothing to do with kitchens or food!), which measures the change in all of the regression coefficients when a case is eliminated from the analysis. Cook's distance for a case depends on both the Studentized residual and the leverage values.

To obtain this scatterplot, select the variables coo_1 and sequence in the Simple Scatterplot dialog box.

Figure 24.8 Cook's distances

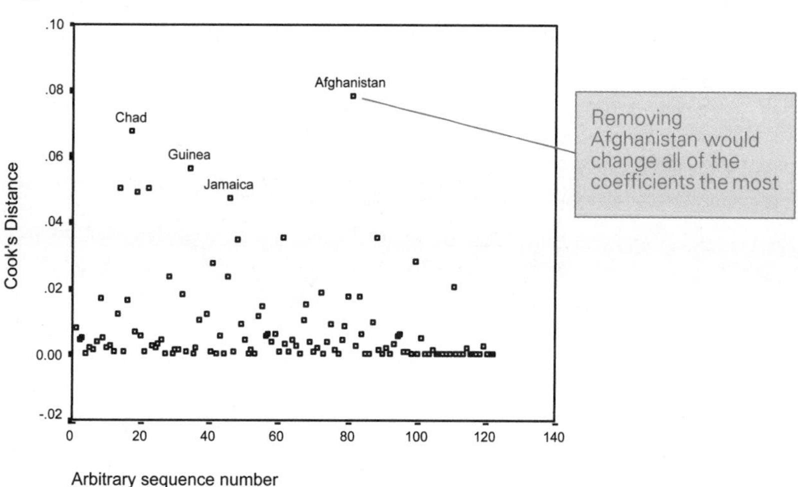

Figure 24.8 is a plot of Cook's distances for all of the cases in the regression. The largest Cook's distance is for Afghanistan, meaning that its removal would change all of the coefficients the most. That's not surprising, since Afghanistan has the largest residual. You've already seen that the life expectancy for Afghanistan is poorly predicted by the model. You can calculate probabilities for Cook's distances using the F distribution with $p+1$ and $N-p-1$ degrees of freedom, where p is the number of independent variables in the model and N is the number of cases. Cook's distances greater than 1 usually deserve scrutiny.

? *Why are you plotting the values of the diagnostic statistics against an arbitrary sequence number instead of getting stem-and-leaf plots of them?* If you plot the diagnostics using a scatterplot, you can readily identify the points. There's no easy way to do that in a histogram or a stem-and-leaf plot. However, you can identify outliers on boxplots or from lists of outliers in the Explore procedure. ■■■

Plots against Independent Variables

If the regression model is correct, when you plot the residuals against the values of the independent variables in the model, you should not see a pattern. If you do, the relationship between the dependent and the independent variable may not be linear.

Figure 24.9 Residual versus transformed doctors variable

To obtain this scatterplot, select the variables sdr_1 and lndocs in the Simple Scatterplot dialog box.

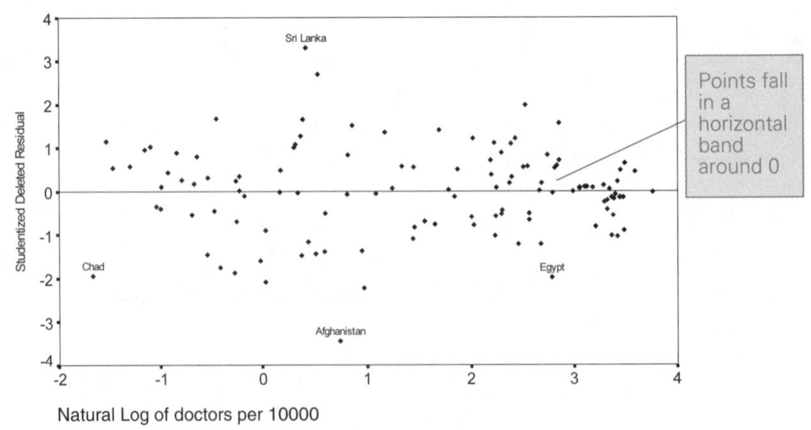

For example, Figure 24.9 is a scatterplot of the residual against the transformed doctors variable. You don't see a pattern—the points fall in a horizontal band around 0. If you didn't take logs of the doctors variable so that it would be linearly related to female life expectancy but instead used the original variable in the model, you would get the scatterplot shown in Figure 24.10. You see that the residuals are no longer randomly distributed in a horizontal band around 0. Instead, there is a definite pattern. That tells you there's something wrong.

Figure 24.10 Using untransformed doctor variable

To duplicate this plot, you need to repeat the stepwise regression described at the beginning of this chapter, substituting the untransformed variable docs for lndocs. Save the residuals again, and plot the resulting variable, sdr_2, in a scatterplot against the variable docs, as shown.

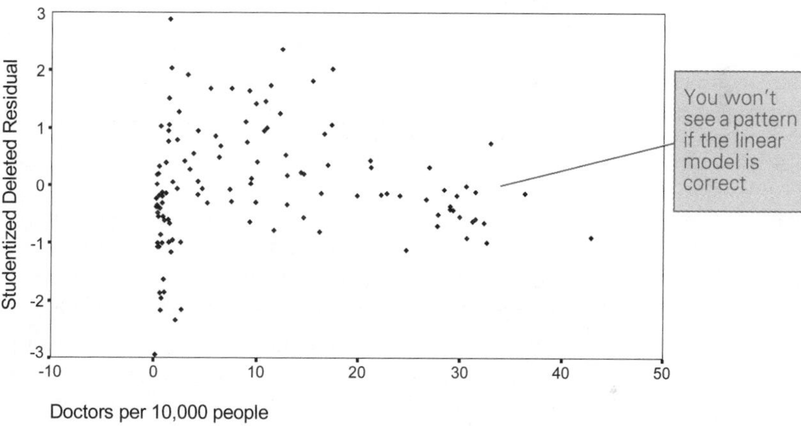

You can also plot the residuals against values of independent variables that are not in the model. If you see a relationship between the residuals and the values of the independent variable, you should consider including the variable in the model.

Figure 24.11 Residual versus log of phones variable

To obtain this scatterplot, select the variables sdr_1 and lnphone in the Simple Scatterplot dialog box.

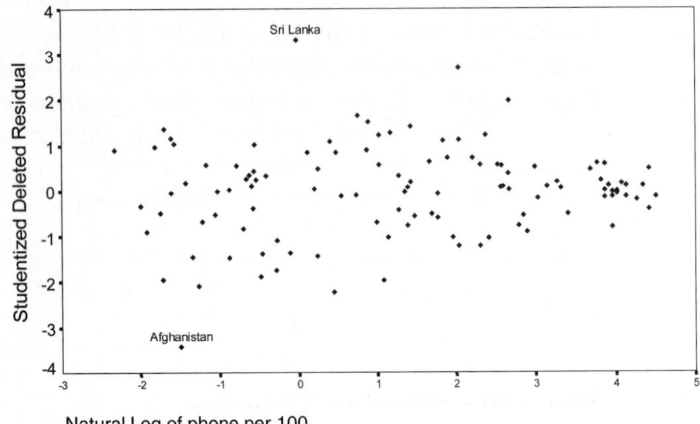

For example, Figure 24.11 is a plot of the residual against the natural log of the number of phones. You see that the points are scattered around a horizontal line through 0, indicating that it's probably unnecessary to include phones in the model. In contrast, look at Figure 24.12, which is a plot of the residuals against the death rate. Here you see that there is a relationship between the two variables. However, you probably don't want to include death rate in the model, since it's really just a somewhat different way of measuring life expectancy.

Figure 24.12 Residuals against death rate

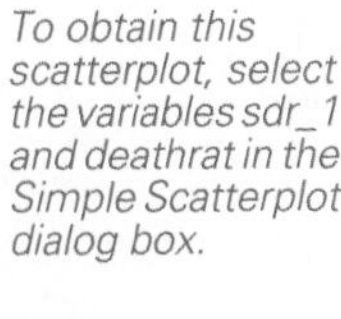
To obtain this scatterplot, select the variables sdr_1 and deathrat in the Simple Scatterplot dialog box.

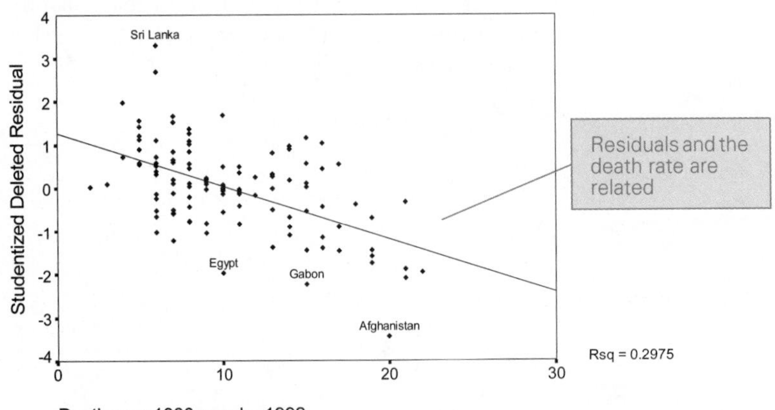

Partial Regression Plot

Another plot that helps you assess the adequacy of the regression model is the partial regression plot. Figure 24.13 is the partial regression plot for the doctors variable. It's a plot of two residuals. On the vertical axis, you plot the residuals from predicting female life expectancy from all of the independent variables except the doctors variable. On the horizontal axis, you plot the residuals from predicting the doctors variable from all of the other independent variables. By calculating the residuals, you remove the linear effects of the other independent variables from both the dependent variable and the independent variable.

If the assumption of linearity is met, as it appears to be in this example, the partial regression plot is linear. If it isn't, you may need to transform your independent variable or include additional terms, such as squares of the independent variable in the model.

Figure 24.13 Partial regression plot for lndocs

To obtain this scatterplot, select the variables lifeexpf and lndocs in the Simple Scatterplot dialog box.

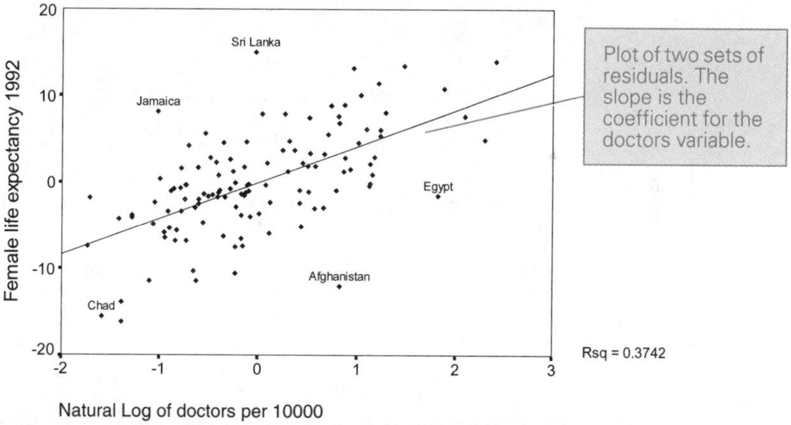

Plot of two sets of residuals. The slope is the coefficient for the doctors variable.

The slope of the regression line for the two residuals in Figure 24.13 is 4.19, the coefficient for the doctors variable in the multiple regression. The correlation coefficient between the two variables is the partial correlation coefficient—the correlation between the two variables when the other independent variables are held constant. From the square root of the R^2 value, you see that the partial correlation between female life expectancy and the doctors variable is 0.61.

The outliers on a partial regression plot are points that are influential in determining the coefficient of the independent variable. For example, you see that Afghanistan, Jamaica, and Sri Lanka are outliers on this plot as well.

Why Bother?

You've learned a lot of different ways you can examine the results of a regression analysis. It's important that you do so. Identify outliers and see if you can think of reasons why the model doesn't fit them. If possible, modify the model to include additional predictors that may improve the fit. If the independent variables aren't linearly related to the dependent variable, try to transform the variables. If you encounter points that are having a large influence on individual coefficients, worry about them. You want to have a regression model that doesn't depend heavily on one or two points. If you are careful in building the regression model, you'll have a useful summary of the relationship between a dependent variable and a set of independent variables.

Summary

How can you check for violations of the multiple linear regression assumptions, and how can you identify cases that are influencing the regression results more than the other cases are?

- By plotting residuals against the independent variables, you can see if there are additional variables that should be included in the model and if the variables that are in the model are linearly related to the dependent variable.
- Leverage is a measure of how much a case influences the regression.
- Cook's distance tells you how much the coefficients change when a case is removed from the model.
- A partial regression plot is a plot of two residuals. It's useful for assessing departures from the regression assumptions.

Statistical Concepts

1. What regression assumptions can you test with each of the following displays?

a. Histogram of residuals

b. Plot of residuals against predicted values

c. Plot of residuals against the independent variables

d. Partial regression plot

e. Plot of the residuals in the sequence the observations are taken

2. The following is a stem-and-leaf plot of Studentized residuals when salary is predicted from five independent variables:

```
Frequency     Stem &  Leaf

    1.00        -2 *  &
    4.00        -1 .  6&
   36.00        -1 *  000001122344
  102.00        -0 .  5555556666666777778888888889999999
  137.00        -0 *  000000000000000111111111112222222333334444444
   99.00         0 *  000000111111222222223333333344444
   48.00         0 .  5555667778888899
   27.00         1 *  0001122334
    6.00         1 .  56&
   10.00 Extremes     (2.0), (2.1), (2.2), (2.7), (3.0), (3.1), (3.4), (4.2)
    4.00 Extremes     (4.2), (4.9), (6.5), (9.1)

Stem width:    1.00000
Each leaf:         3 case(s)
```

Below is the same plot when the natural log of salary is the dependent variable:

```
Frequency     Stem &  Leaf

    1.00 Extremes     (-2.6)
    4.00        -2 *  3&
   12.00        -1 .  5667&
   44.00        -1 *  000011122233444
   89.00        -0 .  5555566666667777788888888999999
  101.00        -0 *  000001112222222233333333333444444
   94.00         0 *  00000001111111122222223334444444
   60.00         0 .  55566666677778888899
   42.00         1 *  00011222223344
   13.00         1 .  5677
    5.00         2 *  0&
    9.00 Extremes     (2.6), (2.7), (3.4), (3.5), (3.7), (3.8), (4.5), (4.7)

Stem width:    1.00000
Each leaf:         3 case(s)
```

Based on the distribution of residuals, which model do you prefer?

3. The following is the partial regression plot for years of education when the natural log of salary is predicted from age, work experience, education, gender, and minority status.

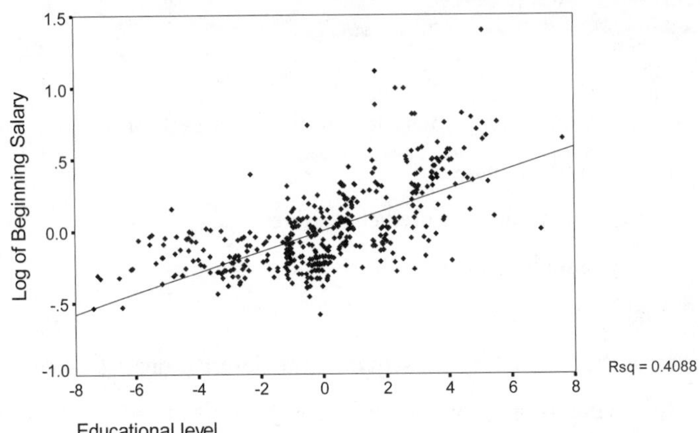

a. What is the partial correlation coefficient between log of salary and years of education?

b. Do you see any problems with the plot?

c. How can you try to remedy the problem?

Data Analysis

Use the *gss.sav* data file to answer the following questions:

1. Consider the regression model you developed in the exercises to Chapter 23 to predict a person's education from his or her spouse's education, father's education, income in dollars, and number of hours of television watched. Run the model again and produce all partial residual plots. Save the Studentized residuals, predicted values, leverages, Cook's distance, and standardized changes in the regression coefficients when cases are excluded from the model.

a. Check the normality assumption.

b. Plot Studentized residuals against predicted values, the values of the independent variables in the model, and mother's education. Do you see any disturbing patterns?

c. Plot the leverages against case ID numbers. Identify any points that have large leverage values.

d. Plot the standardized changes in the partial regression coefficients when cases are excluded from the analysis. Do you see any problems?

e. Plot Cook's distance. Do you see any points that have a large effect on the coefficients?

f. Look at each of the partial regression plots. Edit them to show the regression line and R^2. Do you see any problems with these plots?

2. Develop a multiple linear regression equation to predict husband's education from wife's education and other independent variables you have available. Check all of the regression assumptions. Look for influential points. Write a short paper summarizing your conclusions.

3. Build a multiple linear regression equation using any of the variables in the ***gss.sav*** file that interest you. Check all of the assumptions. Write a short summary of your model.

Use the ***salary.sav*** data file to answer the following questions:

4. Rerun the final equation you developed in the exercises to Chapter 23. Predict the natural log of beginning salary from gender, education level, work experience, and minority status (variables *sex*, *edlevel*, *work*, and *minority*). Save the leverages, Studentized residuals, standardized changes in the regression coefficients, and Cook's distance. Obtain all of the partial regression plots.

a. Plot the leverages against the ID variable. Are there any points that stand out from the rest?

b. Are there any unusual Cook's distances?

c. Look at each partial regression plot. Edit them to include the regression line and R^2. For the work experience variable, use the Partial Correlations procedure to show the relationship between R^2 for the partial regression plot and the partial regression coefficient.

d. Does the partial regression plot for education level look linear? What does it suggest?

5. Compute a new variable that is equal to the education level squared. Rerun the model in question 4, including education level and education level squared. Save the partial regression plots. Look at the partial regression plot for education level and education level squared. Does the plot appear more linear? How much has multiple R^2 increased with the inclusion of education level squared? Do you think this model is better than the model without education level squared? Explain.

6. What does the coefficient for the gender and minority status variables tell you? What reasons besides discrimination can explain the observed findings?

Use the *country.sav* data file to answer the following questions:

7. Build a multiple linear regression model to predict birthrate (variable *birthrat*) from the same variables that were used to predict female life expectancy in Chapter 23.

a. What percentage of the variability in birthrate can be "explained" by the other independent variables? Can you reject the null hypothesis that there is no linear relationship between the dependent variable and the independent variables?

b. Use either backward elimination or stepwise variable selection to build a model that contains fewer variables. Why are the partial regression coefficients for the variables negative? Are the same variables useful for predicting female life expectancy useful for predicting birthrate?

8. Check the regression assumptions for the model you developed in question 7b.

a. Make a stem-and-leaf plot of the Studentized residuals. Check for normality. Does the distribution appear more or less normal?

b. Plot the Studentized residuals against the predicted values and against the values of the independent variables. Do you see anything suspicious?

c. Obtain plots of leverage and Cook's distance against the sequence number. Identify the countries that have unusual values for the two statistics. Are they the same countries that were unusual when you predicted female life expectancy?

d. Plot the standardized changes in the regression coefficients against the sequence number. Do you see anything unusual?

e. Examine the partial regression plots for each of the independent variables. Edit them to include the regression line and R^2. Do you see departures from linearity in any of the plots? Use the Partial Correlations procedure to calculate the partial correlation coefficients that correspond to the values for R^2 displayed on the partial regression plots.

9. Develop a multiple linear regression model of your choice using the variables in the data file. Write a short paper explaining your model and showing the results of checking for departures from the regression assumptions.

Use the *schools.sav* data file to answer the following questions:

10. Build a multiple linear regression equation to predict 1993 ACT scores (variable *act93*). Be sure to check all of the assumptions and to identify points with a big effect on the partial regression coefficients. Write a short report outlining your final model, how you arrived at it, and what it means.

11. Repeat question 10 to predict 1993 graduation rates (variable *grad93*).

Use the file *bodyfat.sav* to answer the following question:

12. Accurate determination of percentage of body fat requires submerging a person under water and measuring water displacement. You have been retained as a consultant by a health club to suggest how percentage of body fat can be estimated from more easily obtainable measurements. Use the available data to prepare a report. Explain the use of multiple regression, the assumptions required for its use, and the impact of unusual observations. Use the methods outlined in this chapter to look for influential points in the dataset.

Obtaining Charts in SPSS A

Charts are extremely important in statistical analysis and are used throughout this book. Leaving discussions of how to interpret charts to the relevant chapters, this appendix focuses on how to obtain high-resolution charts using SPSS. This book describes the legacy dialog box interface to SPSS Graphics. The legacy dialog boxes are accessed by choosing Legacy Dialogs from the Graphs menu. SPSS 16.0 also offers a Chart Builder facility that makes the creation of charts easier by allowing the user to select a chart type and then move variables to their desired role in the chart. This is easier to use but more complex to describe in terse marginal annotations. To access it, choose Chart Builder from the Graphs menu.

Regardless of the chart type, the basic steps for creating and modifying a chart are similar. This appendix begins by demonstrating these steps, using bar charts as typical examples. Specific instructions on how to obtain some of the different charts discussed in this book are then provided.

Scatterplots are not covered in this appendix, since they are discussed in detail in Chapter 9. In addition, a number of the charts available in SPSS are not used in this book and are not discussed here. The online Help system describes in detail all of the charts available in SPSS.

Overview

Roughly speaking, the steps in working with charts are these:

▶ **Create a chart.** You can create charts using the Graphs menu. Many statistics procedures also create charts to accompany their statistical output.

- **Look at the chart.** Newly created charts appear in the Viewer.
- **Modify the chart.** To modify a chart, double-click the chart in the Viewer. This opens the chart in a Chart Editor window, where you can modify it.
- **Save the chart.** Charts are saved as part of an SPSS output document.
- **Print the chart.** Select the chart, then use the File menu (or the toolbar's Print icon) to print it. In the Print dialog box, be sure that the Print range is set to Selection (the default).

Creating Bar Charts

Regardless of the chart type, the basic steps for creating a chart are similar. What can be confusing is knowing exactly which options to select and understanding the relationship between chart structure and data structure. (This is easier to see if you use the Chart Builder.)

Creating a Chart Comparing Groups of Cases

Chapter 6 contains a clustered bar chart of Internet use for groups defined by age and education. To create this type of chart, follow these steps:

- Open the *gssnet.sav* data file.
- From the menus choose:

  Graphs
  Legacy Dialogs ▶
  Bar...

This opens the Bar Charts dialog box, as shown in Figure A.1.

Figure A.1 Bar Charts dialog box

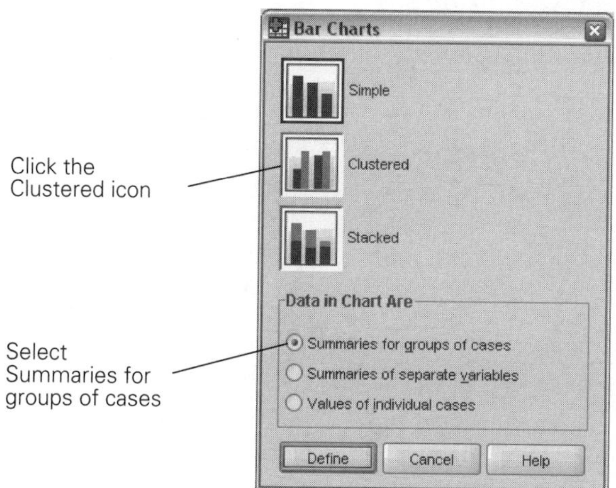

- Click the **Clustered** bar chart icon.
- Select **Summaries for groups of cases** if it is not selected already.
- Click **Define**.

This opens the Define Clustered Bar Summaries for Groups of Cases dialog box, as shown in Figure A.2.

Figure A.2 Define Clustered Bar Summaries for Groups of Cases dialog box

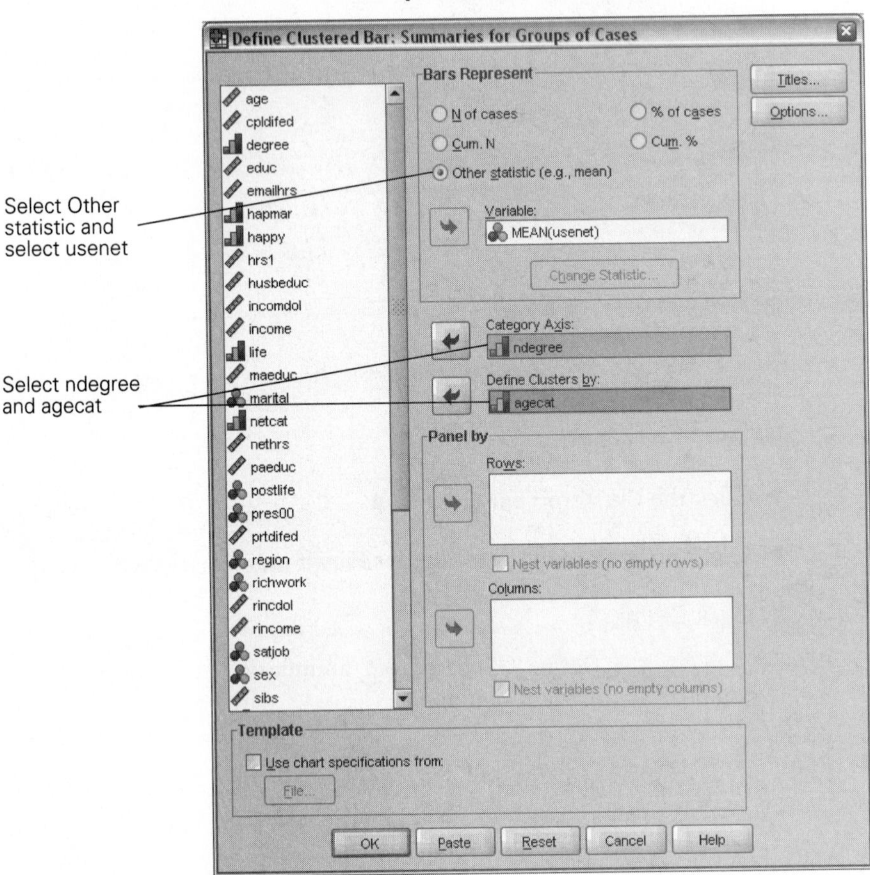

▶ In the Bars Represent group, select **Other statistic** and move *usenet* into the Variable box.

MEAN(*usenet*) appears in the Variable box. (The default statistic is the mean for the selected variable. You can change the statistic by clicking **Change Statistic** as described in "Changing the Summary Statistic" on p. 586.)

▶ Move *ndegree* into the Category Axis box.

▶ Move *agecat* into the Define Clusters By box.

▶ Click **OK**.

The resulting chart is shown in Figure A.3. Note that it shows the proportion of cases answering *yes*. To make a chart showing the percentage of cases answering *yes*, see “Changing the Summary Statistic” on p. 586.

Figure A.3 Bar chart of Internet usage by age and education

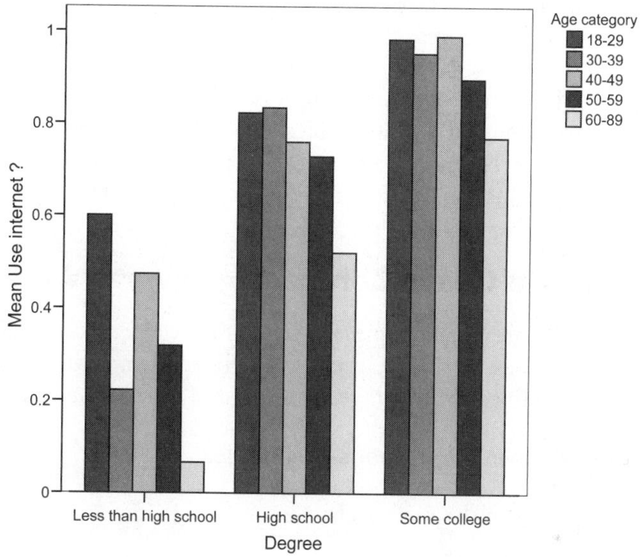

Data Summary Options

Most chart types have a dialog box similar to Figure A.1, where you specify the chart type and data structure for the chart you want. Choosing the right chart type icon is usually easy—just click the icon that looks most like the chart you want. The data summary option can be a little trickier; as you read the descriptions below, think about how your data are organized.

Summaries for groups of cases. Displays a summary statistic for different groups of cases. For example, in Figure A.3, a single variable—*usenet*—is summarized. Each bar within a cluster represents the mean for a different group of cases.

Summaries of separate variables. Displays summary statistics for different variables. In Figure A.6, four variables are summarized. Each bar represents the percentage of cases with values greater than 0.

Values of individual cases. Displays the actual values of a variable for different cases. No summary statistic (such as mean) is represented—rather, each bar represents the actual value of a case.

Click the Help button in any initial chart dialog box for more specific information and examples for each chart type. It may take a bit of trial and error, but don't worry too much. If you get the wrong chart the first time, just go back to the Graphs menu and try again.

Creating a Chart Comparing Several Variables

The previous example describes how to create a chart comparing groups of cases. In this example, you will create a chart in which several different variables are displayed in the same chart. You will generate Figure 4.7.

▶ Open the *gssnet.sav* data file.

▶ From the menus choose:

Graphs
  Legacy Dialogs ▶
    Bar...

This reopens the Bar Charts dialog box, as shown in Figure A.1.

▶ Click the Simple icon if it is not already selected.

▶ Select Summaries of separate variables.

▶ Click Define.

This opens the Define Simple Bar Summaries of Separate Variables dialog box, as shown in Figure A.4. (You'll notice that this dialog box is similar but not identical to the one shown in Figure A.2.)

Figure A.4 Define Simple Bar Summaries of Separate Variables dialog box

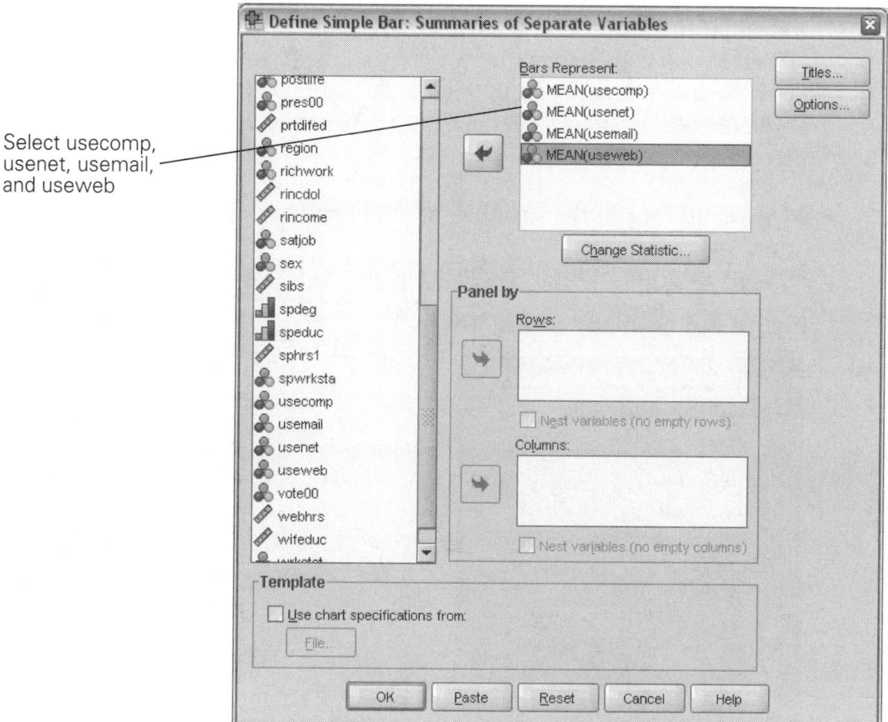

- ▶ Move *usecomp*, *usenet*, *usemail*, and *useweb* to Bars Represent.

- ▶ Click Options. Select Exclude cases variable by variable. This means that all cases that have valid values for a particular variable are included in the computation of the statistics for that variable.

MEAN(usecomp), MEAN(usenet), MEAN(usemail), and MEAN(useweb) appear in Bars Represent. If you click OK, you will obtain a bar chart that shows the means of each variable. Since the variables are coded *Yes*=1, *No*=0, this is the proportion of cases answering *yes*. The next section shows you how to display the percentage answering *yes* to each question.

Changing the Summary Statistic

Many charts include an option to select a summary statistic. The default statistic is the mean, but you can choose from many different statistics.

Before you can change the summary statistic, you must select the variables to be summarized. In Figure 4.7, the percentage of cases who answered *yes* is shown. To change the statistic from the mean to the percentage answering *yes*:

▶ Select all four of the variables by highlighting them.

▶ Click **Change Statistic** to open the dialog box shown in Figure A.5.

Figure A.5 Statistic dialog box

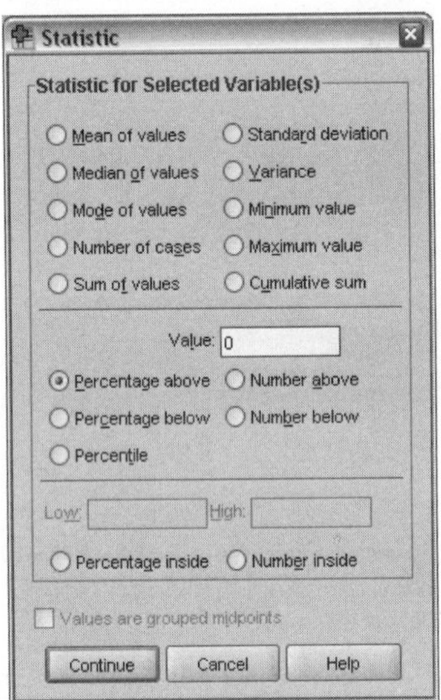

▶ Select **Percentage above.**

▶ Type **0** in the Value box to get the percentage of cases with values greater than 0.

▶ Click **Continue.**

▶ Click **OK** to create the chart shown in Figure A.6 and Figure 4.7.

▶ To show counts for each bar, in the Chart Editor, from the Elements menu choose **Show Data Labels.**

Figure A.6 Bar chart summarizing separate variables

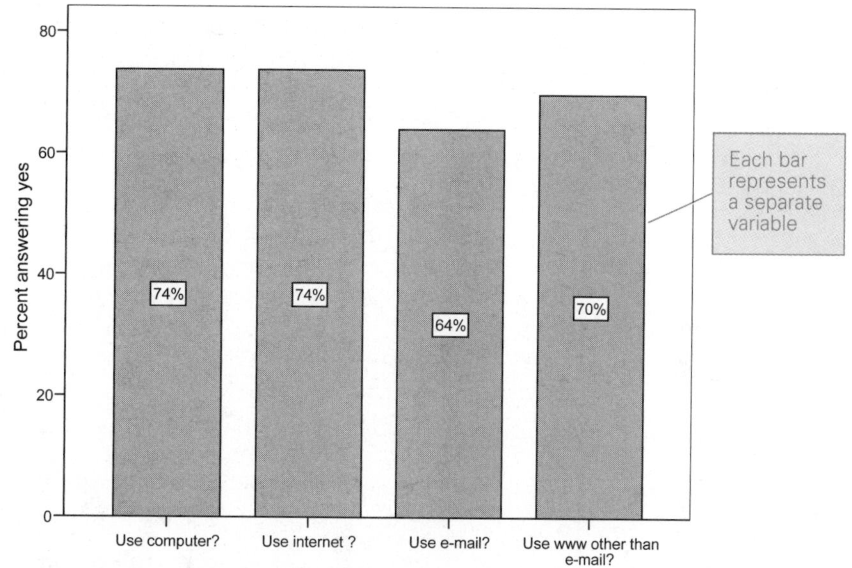

Options in Creating Charts

Most chart-definition dialog boxes contain a Titles button and an Options button.

- **Titles.** Opens the Titles dialog box, where you can enter a title, subtitle, and footnote for the chart. You can also change the titles and footnote in the Chart Editor after creating the chart.
- **Options.** Opens the Options dialog box, where you can control the way that missing values are processed in the creation of the chart.

Modifying Charts

When you create a chart, it appears with the rest of your output in the SPSS Viewer. To modify a chart, you must open it in the Chart Editor. You open a chart from the Viewer in the same way you open any other output object:

▶ To modify a chart that is displayed in the Viewer, double-click it in the display pane of the Viewer.

Figure A.7 Chart Editor

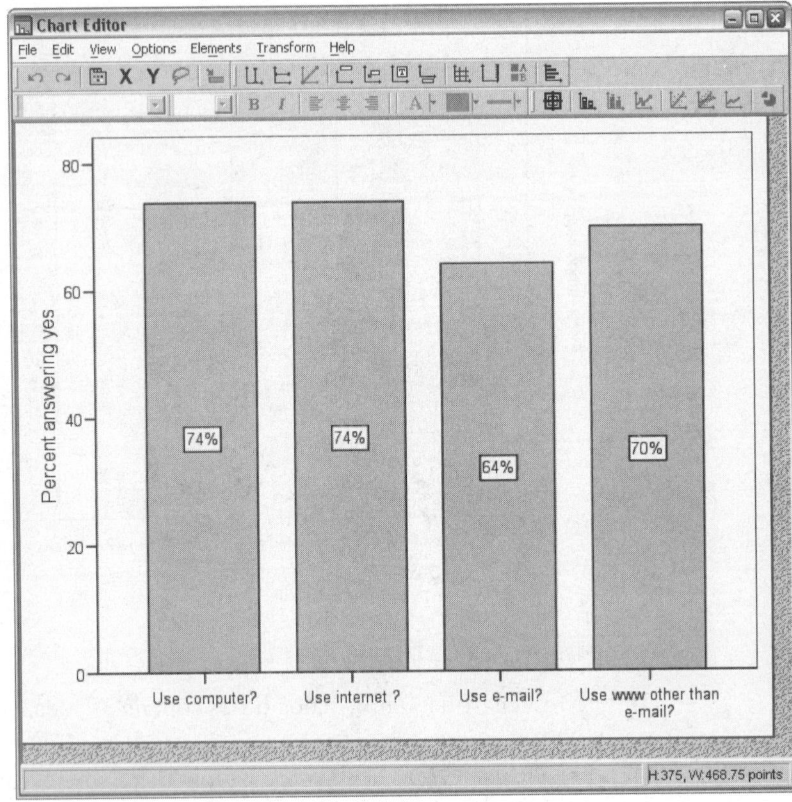

This moves the chart into a Chart Editor window, as shown in Figure A.7. A Chart Editor window has its own menu bar and toolbar, offering options for modifying charts.

Modifying Chart Options

Charts can be edited in many ways, depending on the type of chart. For example, in a scatterplot you can label points, while in a bar chart you can annotate bars with the number of cases each bar represents. You can add elements to a chart using the Options menu. You can control additional properties by making selections in the Properties dialog box and its associated tabs. For example, to collapse small slices of a pie:

- ▶ Double-click on the chart to open a Chart Editor window.

- Double-click on any slice of the pie.
- In the Properties dialog box, click the Categories tab. Select Collapse (sum) categories less than and type 10, as shown in Figure A.8.

Figure A.8 Properties dialog box, Categories tab

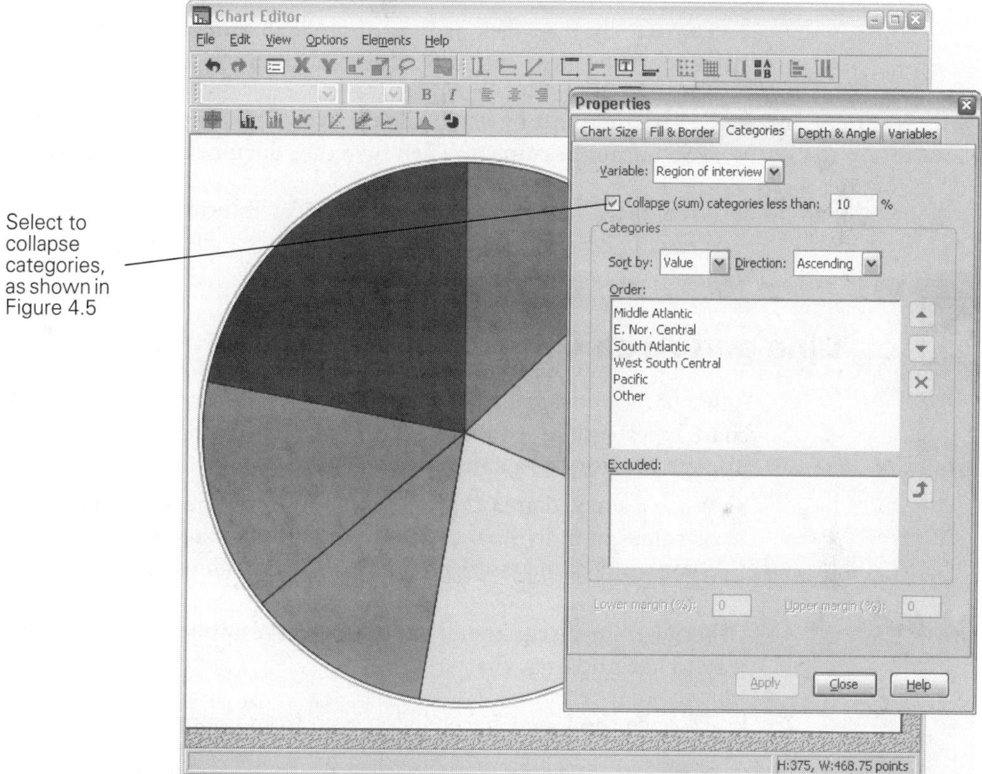

Hints on Editing Charts

Once you have moved your chart into the Chart Editor window, you must figure out how to edit it. That can be tricky. You can find editing options on the Options and Elements menus and in the Properties dialog box. The Properties dialog box can be opened directly from the Edit menu, or you can open it by double-clicking directly on a chart element, such as a bar. If you find the action that you want to perform on the Elements menu, choose it and the Properties dialog box will appear. If you want to select an action that is grayed out (disabled), you will have to double-click on a chart element to open the Properties dialog box. For more information, click on the Help menu in the Chart Editor.

Saving Chart Files

You save a chart to disk as part of an SPSS output document. You can also export the chart into any of several common formats.

▶ To save the chart by itself, you can copy and paste it into a new, empty SPSS output document. Then save that document.

▶ From the Chart Editor, you can select **Export Chart** from the File menu. The Export Chart dialog box offers a choice of file formats.

Line and Area Charts

Line charts and area charts are closely related to bar charts. All three chart types display counts, data values, or summary statistics for each discrete category of a categorical variable.

- When a stacked area chart is based on summaries of separate variables or values of individual cases, the variables or summaries should be on comparable scales so that it makes sense to cumulate them.

Otherwise, the discussion of bar charts above applies in a straightforward way to line and area charts.

▶ To open the Line Charts dialog box, from the menus choose:

Graphs
  Legacy Dialogs ▶
    Line...

▶ To obtain the multiple line chart in Figure 16.5, select the **Multiple** icon and **Summaries for groups of cases** in the Line Charts dialog box and click **Define**. Then select **Other statistic**, move ***hrs1*** into the Variable box, move ***degree*** into the Category Axis box, and move ***sex*** into the Define Lines By box.

Pie Charts

Pie charts are discussed in Chapter 4.

Pie charts display data in a common, easily understood form. The Pie Charts dialog box offers the same data alternatives discussed above, allowing you to compare categories, variables, or individual data values. A pie chart implies that the summary statistics can be regarded as parts of a whole, so counts, percentages, and sums are the most commonly used statistics.

The Chart Editor lets you "explode" pie segments for emphasis or collapse the smallest segments into one. (See "Modifying Chart Options" on p. 588.)

▶ To open the Pie Charts dialog box, from the menus choose:

Graphs
  Legacy Dialogs ▶
    Pie...

▶ To obtain the pie chart shown in Figure 4.4, select Summaries for groups of cases in the Pie Charts dialog box. Then select N of cases as the Slices Represent alternative and move *netcat* into Define Slices By.

▶ To collapse small categories to obtain the chart shown in Figure 4.5, open the chart in a Chart Editor window and double-click on any slice. Make the selections shown in Figure A.8.

▶ To "explode" a slice for visual emphasis, in the Chart Editor window from the Elements menu choose Explode Slice.

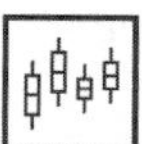

Boxplots

Boxplots are discussed in Chapter 7.

Boxplots are used to compare distributions. Each distribution is represented by a rectangular box whose length represents the variable's interquartile range, on a scale corresponding to the observed values of the variable. A line, or "whisker," extends from each end of the box to the variable's largest and smallest values, aside from those classed as outliers or extreme values. In SPSS, an **outlier** is a value that is more than one and one-half box lengths from the end of the box, while an **extreme value** is a value that is more than three box lengths from the end of the box. Outliers and extreme values are plotted individually and can be identified in the Chart Editor.

Boxplots are available from the Explore procedure (see Chapter 7) or from the Graphs menu.

▶ To open the Boxplot dialog box, from the menus choose:

Graphs
 Legacy Dialogs ▶
  Boxplot...

▶ To obtain the clustered boxplot in Figure 16.3, select the **Clustered** icon and **Summaries for groups of cases** in the Boxplot dialog box and click **Define**. Then move ***hrs1*** into the Variable box, move ***degree*** into the Category Axis box, and move ***sex*** into Define Clusters By.

Case Labels

Like scatterplots, you can identify individual cases in boxplots. Simply specify a variable for Label Cases By in the dialog box in which you define options for the boxplot. You will then be able to identify the individually plotted cases on a boxplot. (See "Changing the Markers to Represent Number of Cases" on p. 195 in Chapter 9.)

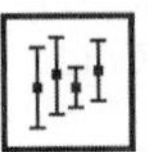

Error Bar Charts

See "Editing a Scatterplot" on p. 192.

Error bar charts are similar to boxplots but are usually used to compare confidence intervals or standard errors rather than the distributions. Each error bar is centered on the mean of a distribution and extends above and below to show a confidence interval or a specified number of standard errors or standard deviations.

▶ To open the Error Bar dialog box, from the menus choose:

Graphs
 Legacy Dialogs ▶
  Error Bar...

▶ To obtain the error bar chart in Figure 15.2, select the **Simple** icon and **Summaries for groups of cases** in the Error Bar dialog box and click **Define**. Then move ***hrs1*** into the Variable box and move ***degree*** into the Category Axis box. Be sure that **Confidence interval for mean** is selected in the Bars Represent group.

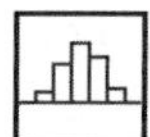

Histograms

Histograms are discussed in Chapter 4 and Chapter 7.

A histogram displays the distribution of a single variable. It resembles a simple bar chart, but each bar represents the number of cases falling into a range of values for the variable.

You can modify the smallest and largest values shown on a histogram, as well as the interval width. This is particularly useful when outlying points cause the majority of values to cluster in a few bars. For example, to produce Figure 4.10:

- From the menus choose:

  Graphs
  Legacy Dialogs ▶
  Histogram...

- Select the variable *nethrs*.
- Click OK.
- Double-click on the histogram to activate it.
- To change the scale in a histogram, from the Edit menu choose Select X Axis. This opens the Properties dialog box, as shown in Figure A.9. Click the Scale tab. Type 40 as the maximum value to be plotted.

Figure A.9 Properties dialog box, Scale tab

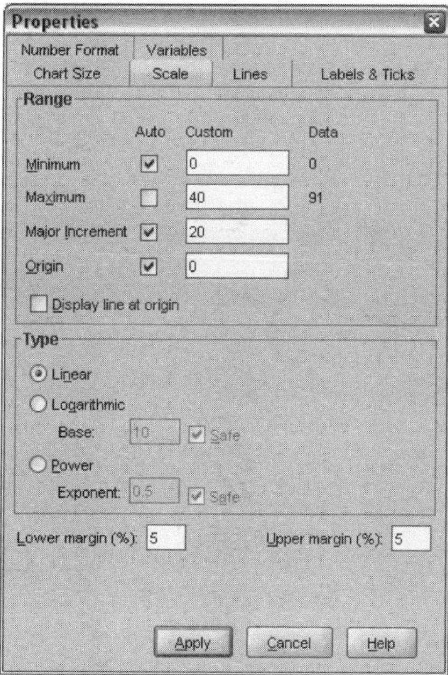

▶ To set the bin size, double-click the *x* axis on the plot and make the selections shown in Figure A.10.

Figure A.10 Properties dialog box, Binning tab

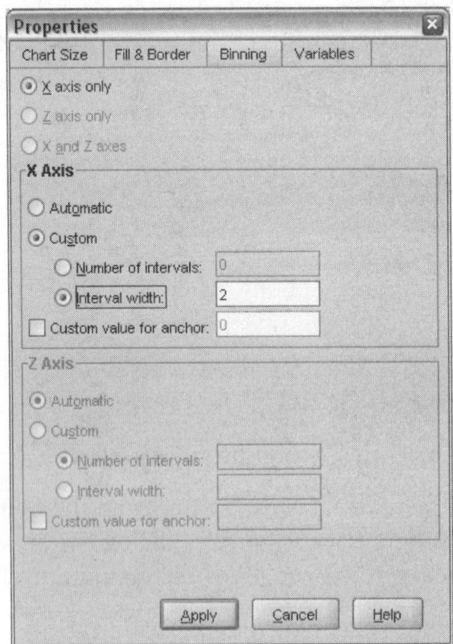

You can request histograms from the Frequencies procedure or the Explore procedure, or you can obtain them directly from the Graphs menu.

Normal Probability Plots

Normal probability plots are discussed in Chapter 7 and Chapter 22.

Normal probability plots are used to check whether distributions of variables follow the normal distribution. They are essentially scatterplots of a variable's distribution against a normal distribution. If the data are a sample from a normal distribution, points will cluster along a straight (diagonal) line. SPSS offers two types of normal probability plots:

Normal Q-Q. The Q-Q (quantile-quantile) plot shows the variable's observed values on the x axis and the corresponding predicted values from a standard normal distribution on the y axis.

Normal P-P. The P-P (proportion-proportion) plot shows the observed cumulative proportion of cases on the x axis and the predicted cumulative proportion of the normal distribution on the y axis.

Detrended normal plot. A detrended normal probability plot uses the difference between the case's observed value and its predicted value, rather than the predicted value itself, for the y axis. If the sample is from a normal distribution, the points are randomly scattered around a horizontal line through 0. SPSS always produces detrended plots to accompany normal probability plots.

The Explore procedure offers Q-Q plots. The Linear Regression procedure offers P-P plots. Both types are available directly from the Analyze menu when you choose Descriptive Statistics.

▶ To open the Q-Q Plots dialog box, from the menus choose:

Analyze
  Descriptive Statistics ▶
    Q-Q Plots...

See Appendix B for instructions on how to compute a new variable.

▶ To obtain the normal Q-Q plot in Figure 13.5, compute a variable *diff* equal to the difference between the pre-race and post-race measurements. In the Q-Q Plots dialog box, move *diff* into the Variables box.

▶ To obtain the normal Q-Q plots in Figure 22.3 and Figure 22.4, you must run the regression described in that chapter and save the Studentized residuals as a new variable as described in "Linear Regression Save: Creating New Variables" on p. 465 in Chapter 20. Then move *sre_1* (the Studentized residuals variable saved by the regression) into the Variables box.

Transforming and Selecting Data

SPSS includes a powerful set of facilities for transforming data values and selecting which cases should be analyzed. This appendix covers two general types of data manipulation: data transformation and case selection.

Data transformation procedures change the actual values of your variables or create new variables. For example, you can create a new variable that contains the natural log of an existing variable. **Case selection** procedures do not change data values but restrict the number of cases used in the analysis. For example, you can restrict your analysis to people who are married or who are holding full-time jobs.

SPSS also provides a number of advanced data manipulation utilities that are not used in this book and are not discussed here. These utilities are described, however, in the online Help system.

Data Transformations

Often you need to make modifications to your data before you can perform your analysis. For small changes, such as the urbanization of Bhutan in Chapter 9, it is easy to enter the corrected value into the Data Editor. But suppose you want to take the natural log of several variables, each with 1500 cases, as you do for the analysis in Chapter 23? SPSS provides data transformation facilities to handle such tasks easily and accurately.

Data transformations affect the values of existing variables or create new variables. Transformations affect only the working data file; the changes do not become permanent unless you save the working data file to your disk.

Transformations at a Glance

This appendix describes the following transformations, available using the SPSS Data Editor's Transform menu:

Compute. Compute calculates data values according to a precise expression. With this option, you can do anything from set a variable to 0 for all cases to calculate an elaborate expression involving the values of other variables. You can assign the computed values to a new variable, or you can assign them to an existing variable (replacing the current values). You can also request that the computation be carried out selectively based on a conditional expression.

Recode. Recode assigns discrete values to a variable, based solely on the present values of the variable being recoded. You can assign the recoded values to the variable being recoded, or you can assign them to a new variable. You can also request that the computation be carried out selectively based on a conditional expression.

Automatic Recode. Automatic recode assigns successive integer codes—1, 2, 3, and so on—to a new variable, based on the existing codes of another variable. This saves you the effort of specifying how the recoding should be carried out.

The following options are also available on the Transform menu but are not discussed in this book. These transformations are described in the online Help system.

Random Number Seed. Lets you reproduce the pseudo-random numbers generated by SPSS for sampling and certain functions in the transformation language.

Count. Creates a new variable that counts for each case the number of times certain specified values occur in other variables. You can count, for example, the number of times that values of 1 or 2 occur in a group of existing variables.

Rank Cases. Creates rank scores, which show each case's rank among all the cases in the file according to the values of a particular variable.

Create Time Series. Creates new time series, containing functions such as the differences between successive cases, in a time series data file.

Replace Missing Values. Supplies nonmissing values to replace missing values, according to any of several functions that might provide plausible values.

Run Pending Transformations. Forces SPSS to execute transformations that are pending as a result of the Transform & Merge Options setting. (See "Delaying Processing of Transformations" below.)

Saving Changes

Bear in mind when transforming your data that you are changing only the working data file.

- To make the changes permanent, save the working data file to your hard disk.
- To discard the changes, exit SPSS (or open a new data file) without saving the working data file.

Delaying Processing of Transformations

SPSS normally executes transformation commands as soon as you request them. However, since transformations can take several minutes to execute for a very large data file, there are times when you want to enter a dozen or more transformation commands one after another and then let the computer process them all at once.

- To prevent SPSS from processing transformations immediately, from the menus choose:

  Edit
   Options...

- In the tabbed SPSS Options dialog box, click the Data tab. This displays the SPSS Options Data tab, as shown in Figure B.1.

Figure B.1 SPSS Options Data tab

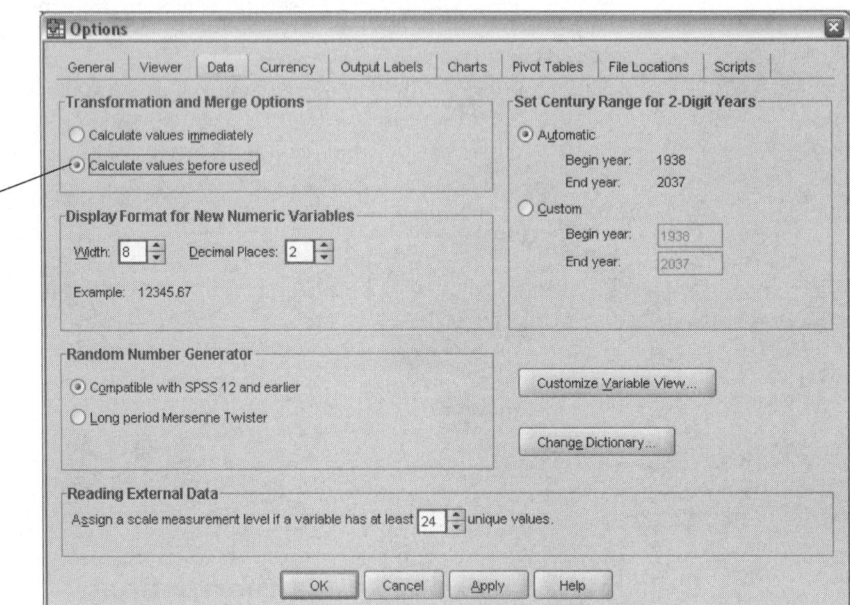

▶ Set Transformation and Merge Options to Calculate values before used and click OK.

With this setting, SPSS does not execute Compute and Recode transformations until it needs the data. In the meantime, the status bar displays the message Transformations pending and the results of transformations are not yet visible.

▶ To execute pending transformations, run a procedure that requires SPSS to use the data or choose Run Pending Transformations from the Transform menu.

When transformations are pending, the Data Editor will not allow you to make certain changes to your working data file.

Recoding Values

Recoding is done with a series of specifications, of the form, "If the old value is this, assign a new value of that." A case's existing value is checked against each of these specifications until one of them matches. Then the new value is assigned, and SPSS moves on to process the next case.

There are two Recode commands: Recode into Same Variables and Recode into Different Variables. The former changes the values of variables based solely on their existing values, while the latter creates new variables with values that depend only on the existing values of single variables.

- A case is never changed by more than one of a group of recode specifications.
- If a case doesn't match any of the recode specifications, its value remains unchanged (if recoding into same variable) or becomes system-missing (if recoding into a new variable).

Example: Recoding Age into Age Categories

This example recodes the variable *age* (age in integer years) into a new variable that contains age in one of three categories: 14 through 29, 30 through 49, and 50 or older. (If *age* is not an integer, you must modify the recode statement so that ages between 29 and 30, and between 49 and 50, are assigned to the proper groups.)

▶ Open the *salary.sav* data file.

▶ From the menus choose:

Transform
  Recode ▶
    Into Different Variables...

This opens the Recode into Different Variables dialog box, as shown in Figure B.2.

Figure B.2 Recode into Different Variables dialog box

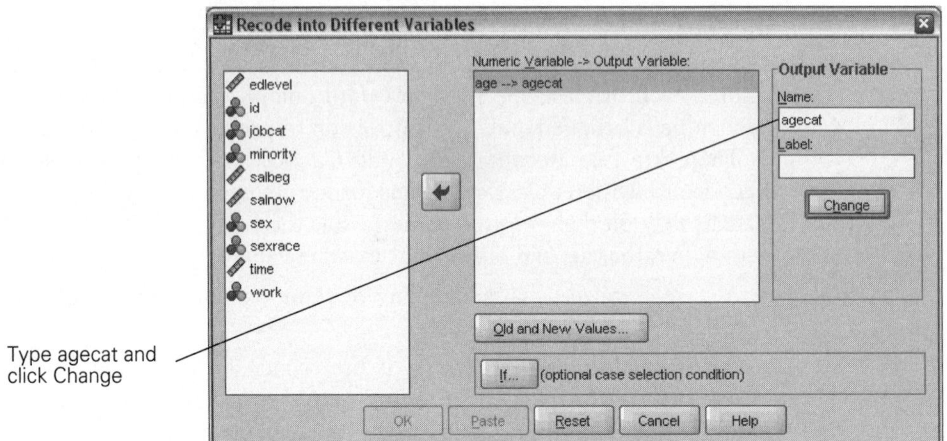

▶ Move *age* into the Input Variable -> Output Variable list. The name of the list changes to reflect that a numeric variable has been selected, as shown in Figure B.2.

▶ In the Output Variable Name box, type agecat for the output variable and click Change.

This adds agecat to the Numeric Variable -> Output list. A new variable *agecat* will be created, which contains the recoded values of *age*.

▶ Click Old and New Values.

This opens the Old and New Values dialog box, as shown in Figure B.3.

Figure B.3 Old and New Values dialog box

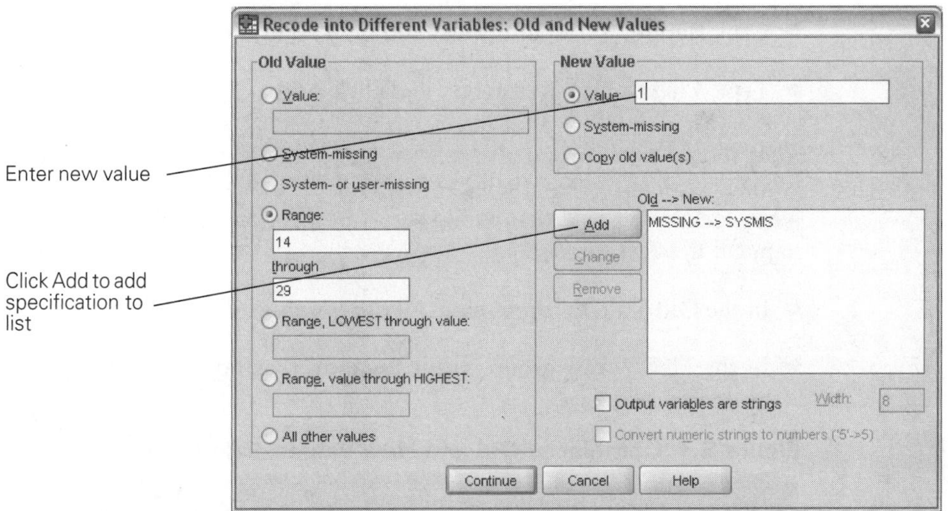

▶ In the Old Value group, select **System- or user-missing.**

▶ In the New Value group, select **System-missing.**

▶ Click **Add.**

The specification **MISSING -> SYSMIS** is added to the Old -> New list. It is always a good idea to take care of missing data first. Recode specifications treat user-missing values just like all other values, so it is easy to accidentally recode user-missing values (such as 99 for *age*) into a valid category otherwise.

▶ In the Old Value group, select the first **Range** alternative.

▶ Type 14 in the first range box and 29 in the second range box.

▶ Type 1 in the New Value box.

▶ Click **Add.**

The specification **14 thru 29 -> 1** is added to the Old -> New list. All ages between 14 and 29 will be coded 1 in the new *agecat* variable.

▶ Select **Range** again. Type 30 in the first range box and 49 in the second range box.

- Type 2 in the New Value box and click Add.
- Click Range: through highest and type 50 in the box.
- Type 3 in the New Value box and click Add.

That should take care of all age groups in this file. But what if someone is coded with an age less than 14? Since the file contains data about adults who work for a bank, that would surely be a coding mistake, but it could happen. It's best to be safe.

- In the Old Value group, select All other values.
- In the New Value group, select System-missing and click Add one more time.

Figure B.4 Completed Old and New Values dialog box

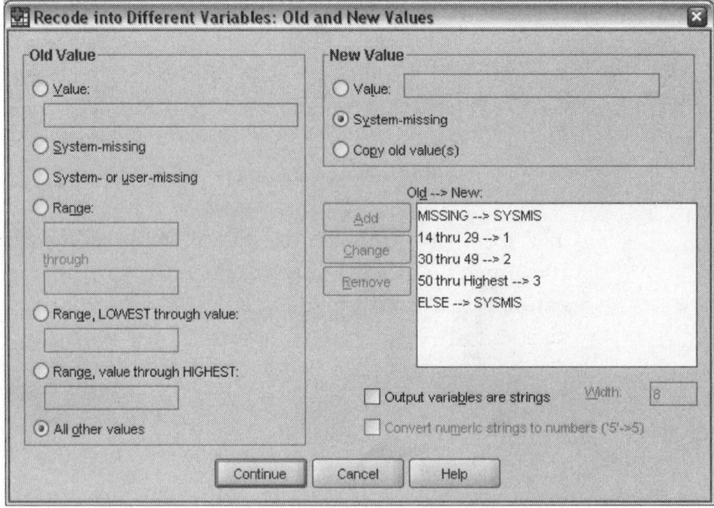

The Old and New Values dialog box should now look like Figure B.4. If it doesn't—if one of your specifications is incorrect—click the incorrect specification in the Old -> New list, make the needed correction, and click Change.

- Click Continue to return to the Recode into Different Variables dialog box. Then click OK to execute the transformation.

You have now changed the working data file; however, you don't want to make these changes a permanent part of the *salary.sav* data file.

▶ To avoid saving changes to the *salary.sav* data file, exit SPSS *without* saving changes or clear the Data Editor by selecting New from the File menu.

Computing Variables

The Compute Variable dialog box assigns the result of a single expression to a "target variable" for each case. The target variable can be a new variable or an existing variable (in which case the existing values will be overwritten). For example, you can compute standard scores for a variable, as described in the first example below. A great number of functions are available, so expressions can be quite complex.

Functions are grouped by category and category types are displayed in the Function group list. When you click on a category, the available functions for that category are listed in the Functions and Special Variables list. When you click on a function, its description is shown in the box under the keypad.

▶ To open the Compute Variable dialog box, as shown in Figure B.5, from the Data Editor menus choose:

Transform
  Compute...

Figure B.5 Compute Variable dialog box

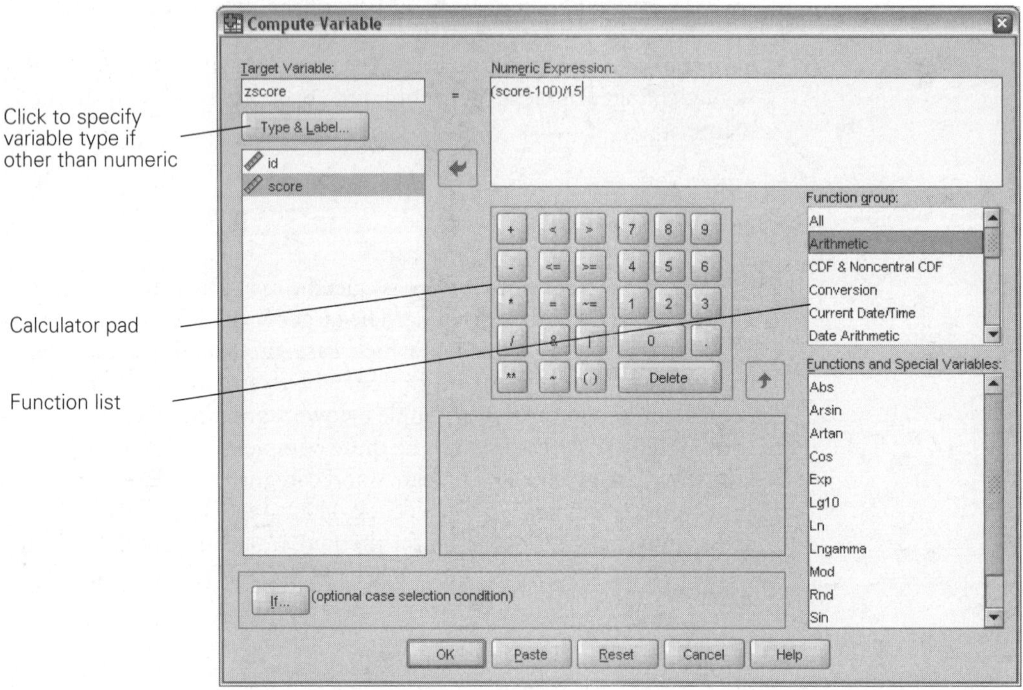

Unlike a spreadsheet, SPSS does not remember the formula used to compute data values or automatically update them. (In the example mentioned above, if you go back and change the values for the variable *score*, the *zscore* values will not be automatically recalculated to reflect the change.)

The Calculator Pad

The calculator pad allows you to paste operators and functions into your formula. You don't have to use the calculator pad; you can click anywhere in the Numeric Expression box and start typing. Often that's the simplest and quickest way to build an expression. The visual controls in the calculator pad are there to remind you of the possibilities and to reduce the likelihood that you won't remember how to spell one of the many functions available in SPSS.

To use the calculator pad, just click the buttons to paste symbols and operators at the insertion point. Use the mouse to move the insertion point.

See "Example: Cumulative Distribution Function" on p. 609 for an example of using functions in an expression.

To paste a function, select it in the scrolling list and click the [▲] button. You must then fill in the arguments, which are the values that the function operates on.

A few basic calculator pad operators are described in Table B.1. The Help system contains a more detailed description of the calculator pad, with definitions of all the functions.

Table B.1 Calculator pad operators

*	Multiply
/	Divide
**	Raise to power
+	Add
–	Subtract

Example: Computing Z Scores

Suppose you have a sample of IQ scores and you want to calculate standard scores (*z* scores) for the sample. Assuming that in the population IQ scores have a mean of 100 and a standard deviation of 15 (as was long assumed to be true), the formula is

$$zscore = (score - 100)/15$$

To compute standard scores for a variable according to this formula:

▶ Activate the Data Editor window.

▶ From the menus choose:

Transform
 Compute...

This opens the Compute Variable dialog box, as shown in Figure B.6.

Figure B.6 Compute Variable dialog box

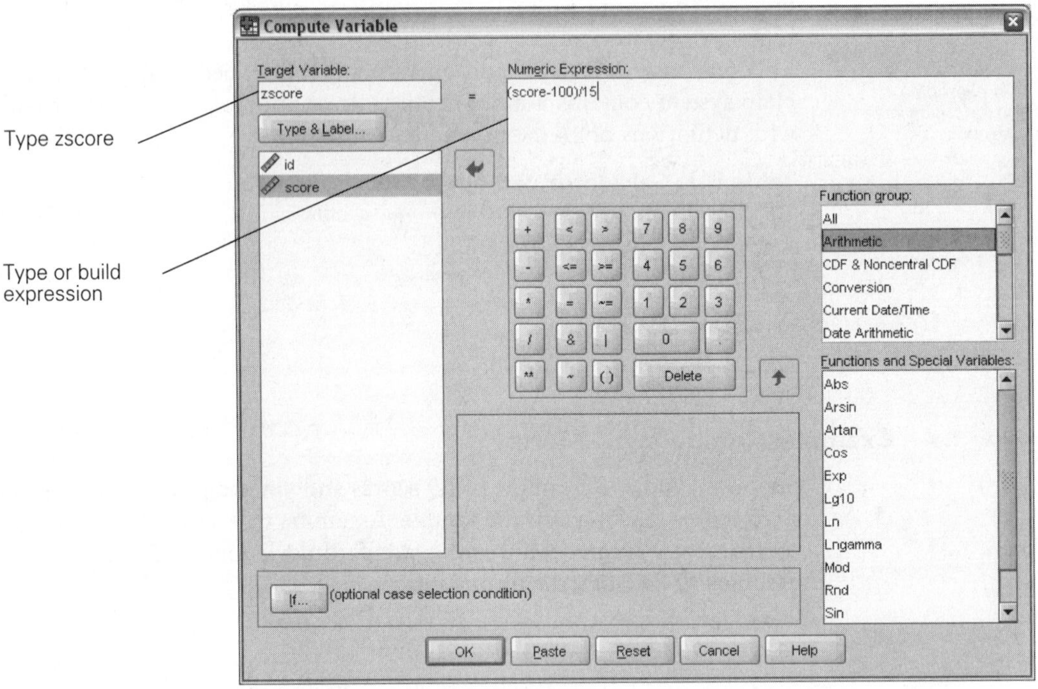

Type zscore

Type or build expression

▶ Click in the Target Variable box and type **zscore**.

▶ Click in the Numeric Expression box.

Since zscore is a numeric variable, you don't have to click the Type & Label button.

You can simply type the expression **(score–100)/15** directly in the Numeric Expression box or build it using the calculator pad, as follows:

▶ Select **score** in the variable list and click ▶.

The variable name ***score*** is pasted into the expression at the insertion point.

▶ Enter –100.

▶ Select the entire expression **score –100** and click the **( )** button.

The expression now reads **(score –100)**.

▶ Enter /15.

The expression now reads (score –100)/15.

▶ Click OK.

SPSS computes z scores for all cases in the working data file.

Example: Cumulative Distribution Function

You can calculate the proportion of the population with z scores greater in absolute value than each of the z scores in your sample, as discussed in question 10 in the exercises for Chapter 11. Assuming the variable ***zscore*** contains the z scores for your sample, the formula is

twotailp = 2 *(1 – cdfnorm(abs(zscore)))

▶ If you want to attempt this example, you can substitute any variable that contains z scores for the ***zscore*** variable named in the formula above. (You can use the Descriptives procedure to save z scores for any variable, as described in Chapter 5.)

▶ From the menus choose:

Transform
  Compute...

This opens the Compute Variable dialog box.

▶ Type twotailp in the Target Variable box.

Figure B.7 Compute Variable dialog box

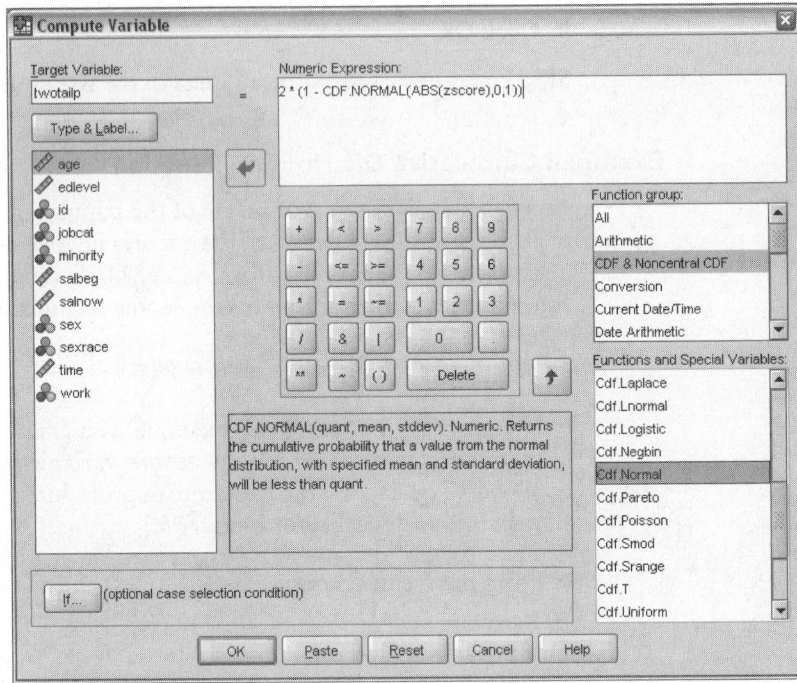

You can simply type the expression 2*(1-CDF.NORMAL(ABS(zscore))), as shown in Figure B.7, or build the expression as follows:

- Enter 2*(1–).
- With the cursor inside the right parenthesis, click on the function group CDF & Noncentral CDF.
- In the Functions and Special Variables list, click on CDF.NORMAL and click [▲].

The CDF.NORMAL function is pasted into the formula at the insertion point. The expression now reads 2*(1–CDF.NORMAL(?,?,?)). You must replace the question marks with the arguments for the CDF.NORMAL function. The description of the arguments is shown below the keypad.

▶ Select ABS from the Arithmetic function group and click [▲].

The expression now reads 2*(1–CDF.NORMAL(ABS(?),?,?)). Once again, the question mark is selected; you must now supply an argument for the ABS function.

▶ Select *zscore* in the variable list and click [▶].

The variable *zscore* is now pasted in as the argument for the ABS function. Type 0 and 1 in place of the two remaining question marks. The expression is now complete.

▶ Click OK.

SPSS computes the proportions for all cases in the working data file.

Automatic Recoding

Automatic Recode is particularly useful as a way of converting a string variable into a numeric variable.

SPSS's Recode facility is quite useful but requires you to enter detailed specifications. The Automatic Recode facility needs no specifications. It simply converts all the codes of a current variable into new codes—1, 2, 3, and so on—for a new variable.

Example: Creating Numeric Country Codes

In the *country.sav* data file, the string variable *country* contains the name of each country. Suppose you want to create numeric country codes. You can do this as follows:

▶ From the Data Editor menus choose:

Transform
  Automatic Recode...

This opens the Automatic Recode dialog box, as shown in Figure B.8.

Figure B.8 Automatic Recode dialog box

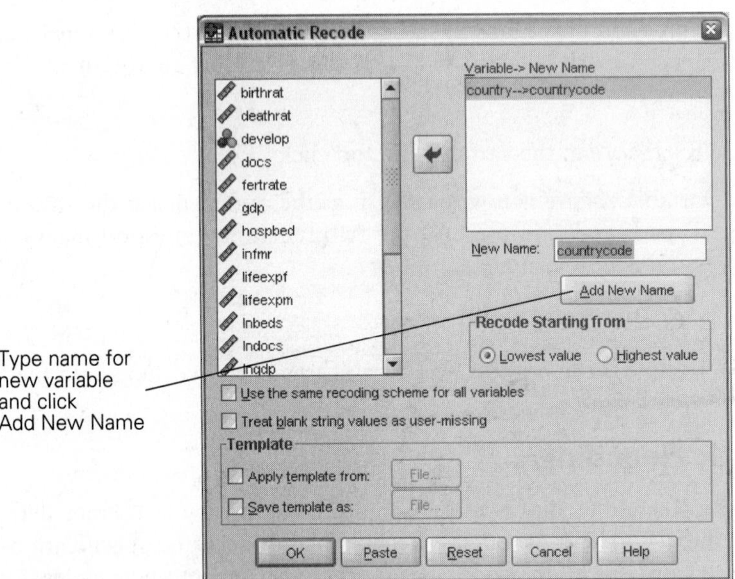

▶ Select *country* in the variable list and move it into the Variable -> New Name list.

▶ Type ctrycode in the New Name box and click New Name.

▶ Click OK.

SPSS creates the new variable *ctrycode*, which contains a unique numeric code for each country. The codes are assigned in sequence; the first country will have a code of 1, the second 2, and so on. If there were several cases for the same country, they would all be assigned the same code value.

Since the original variable *country* does not have value labels, the actual values of *country* (Afghanistan, Albania, Algeria, and so on) are used as value labels for the new variable *ctrycode*.

Conditional Transformations

If you want to transform the values of only some cases, depending on their data values, you need a **conditional transformation**, one that is carried out only if a logical condition is true. For example, you might want to transform *only* cases for people who are full-time workers.

The Compute Variable dialog box, both Recode dialog boxes, and the Count dialog box (not shown) allow you to specify such a logical condition.

▶ To specify a logical condition for a transformation, click If in the Compute Variable, Recode, or Count dialog box.

This opens a dialog box where you can specify a logical condition. For example, the Compute Variable If Cases dialog box is shown in Figure B.9.

Figure B.9 Compute Variable If Cases dialog box

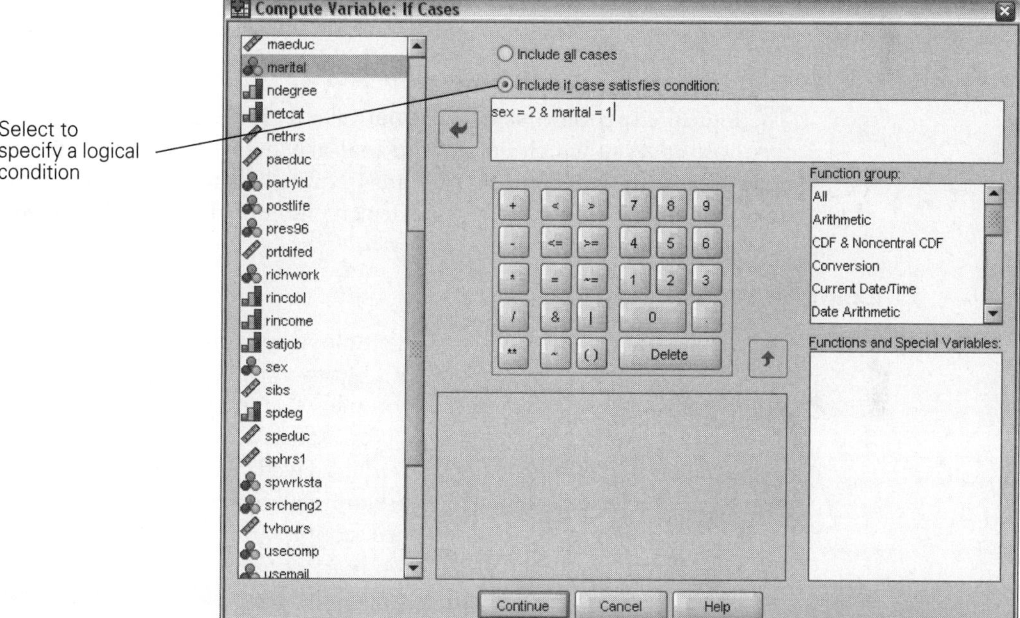

See "The Calculator Pad" on p. 606.

This dialog box contains the familiar calculator pad. Here you use it to build a logical condition, one that is either true or false for a case, depending on the case's data values. Table B.2 describes some operators that are particularly useful in building logical conditions.

Table B.2 Operators useful in logical expressions

<table>
<tr><td><</td><td>Less than</td></tr>
<tr><td>></td><td>Greater than</td></tr>
<tr><td><=</td><td>Less than or equal to</td></tr>
<tr><td>>=</td><td>Greater than or equal to</td></tr>
<tr><td>=</td><td>Equal to</td></tr>
<tr><td>~=</td><td>Not equal to</td></tr>
<tr><td>&</td><td>And</td></tr>
<tr><td>|</td><td>Or</td></tr>
<tr><td>~</td><td>Not</td></tr>
</table>

The logical expression sex = 2 & marital = 1, for example, is true only for those cases in which *both* conditions are met: the variable ***sex*** must equal 2 *and* the variable ***marital*** must equal 1. The logical expression sex = 2 | marital = 1, by contrast, is true if either of the conditions is met.

Example: Wife's Employment Status

The General Social Survey contains employment status questions for the respondent and for the respondent's spouse. The respondent could be either husband or wife, depending on who was interviewed. This means that for each household, the wife's work status could be coded in either the variable ***wrkstat*** (if the wife was interviewed) *or* in the spouse's work status variable ***spwrksta*** (if the husband was interviewed). To create a variable containing, for all married couples, the wife's employment status, you might proceed as follows:

▶ To open the Compute Variable dialog box (see Figure B.10), from the menus choose:

Transform
  Compute...

Figure B.10 Compute Variable dialog box

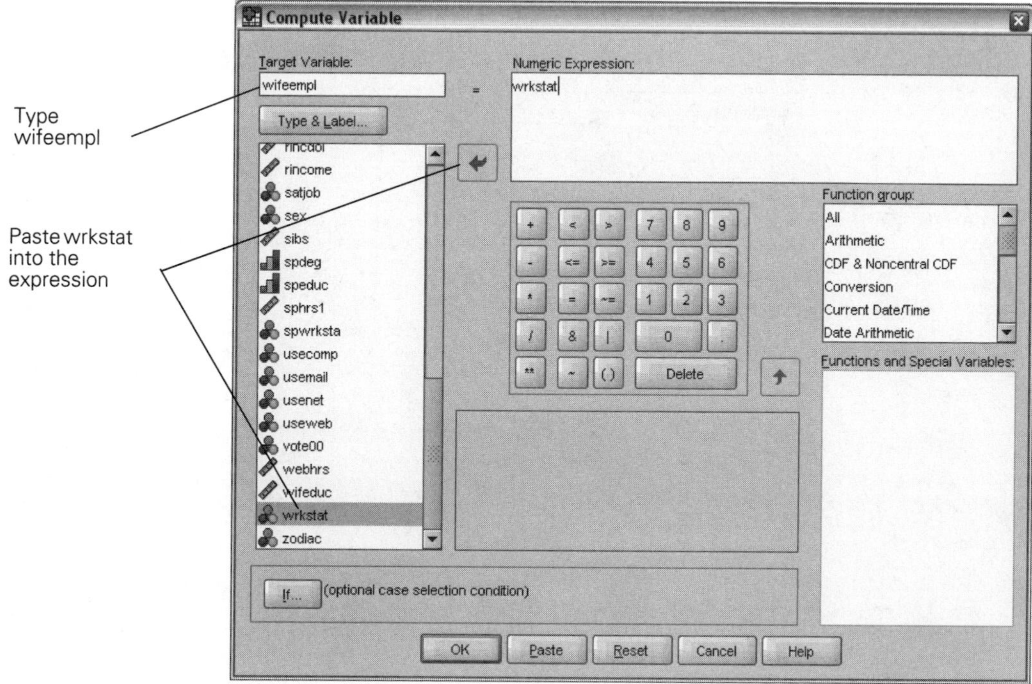

- In the Compute Variable dialog box, type **wifeempl** in the Target Variable box.
- Select *wrkstat* in the variable list and click [▶] to move it into the Numeric Expression box.

The new variable *wifeempl* will have the same value as the variable *wrkstat.* However, you must specify that this expression will only be evaluated for cases where the respondent is a married woman.

- Click **If**.

This opens the Compute Variable If Cases dialog box, as shown in Figure B.11.

Figure B.11 Compute Variable If Cases dialog box

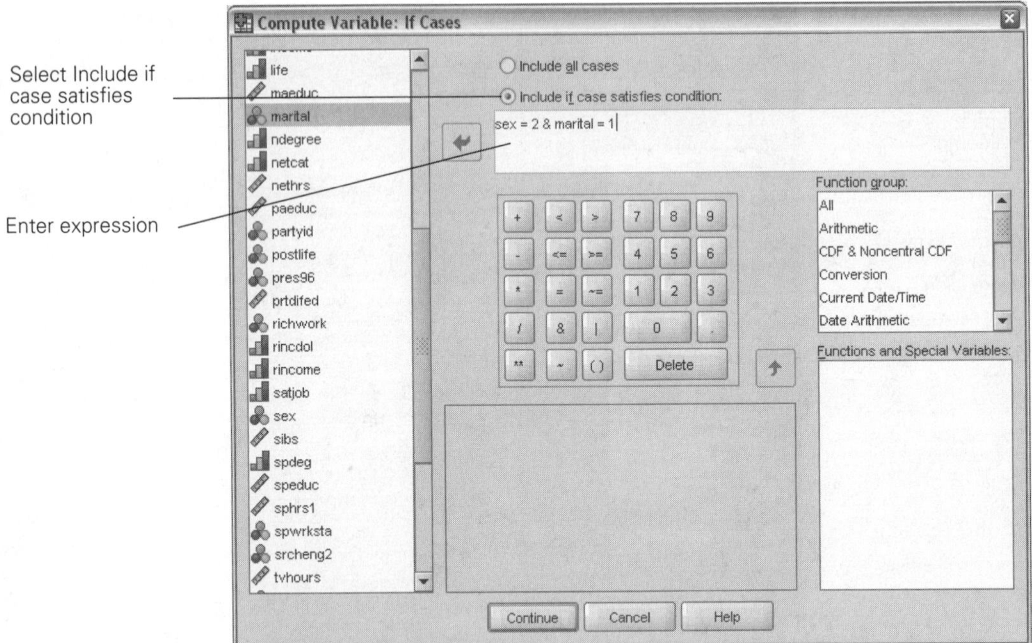

▶ Select Include if case satisfies condition.

▶ Using either the calculator pad or the keyboard, enter the condition sex = 2 & marital = 1.

This condition specifies that the new value should be computed only for cases for whom the value of the variable *sex* equals 2 (the code for female) *and* for whom the value of *marital* equals 1 (married). For cases that do not meet this condition, the new variable *wifeempl* will be equal to the system-missing value.

▶ Click Continue to return to the Compute Variable dialog box. Then click OK.

This creates a new variable *wifeempl*, which is equal to *wrkstat* for married women. For cases where the respondent is not married, or is a man, the value of *wifeempl* is not defined (system-missing).

At this point, you're halfway there. But what about respondents who are married men? In that case, the wife's employment status would be coded in the variable *spwrksta*, which contains the work status of the respondent's spouse.

▶ Open the Compute Variable dialog box again.

▶ Delete *wrkstat* from the Numeric Expression box.

▶ Select the variable *spwrksta* and paste it into the Numeric Expression box.

▶ Click If.

The logical expression still reads sex = 2 & marital = 1.

▶ Delete the 2 and type 1 in its place.

The expression now reads sex = 1 & marital = 1.

▶ Click Continue and then click OK.

This sets *wifeempl* equal to *spwrksta* for married men. To summarize, the first transformation creates a new variable *wifeempl*, which is equal to *wrkstat* for married women and not defined for others. The second conditional transformation sets *wifeempl* equal to *spwrksta* for married men. The end result is a variable equal to wife's employment status for all married couples. For unmarried respondents, neither transformation is executed and *wifeempl* is never changed. Since it's a new variable, it is assigned the system-missing value for the unmarried respondents.

Case Selection

Sometimes you want to analyze only part of your cases. For example, some of the analyses described in this book look only at full-time workers or only at college graduates.

The Select Cases dialog box allows you to restrict your analysis to a specific group of cases. There are a number of options for selecting cases:

- You can choose cases according to a logical condition based on their data values.
- You can select a random sample of the cases in your file.
- You can select a range of cases according to their order in the file.
- You can select those cases that are marked with a non-zero value for a "filter variable."

▶ From the menus choose:

Data
  Select Cases...

This opens the Select Cases dialog box, as shown in Figure B.12.

Figure B.12 Select Cases dialog box

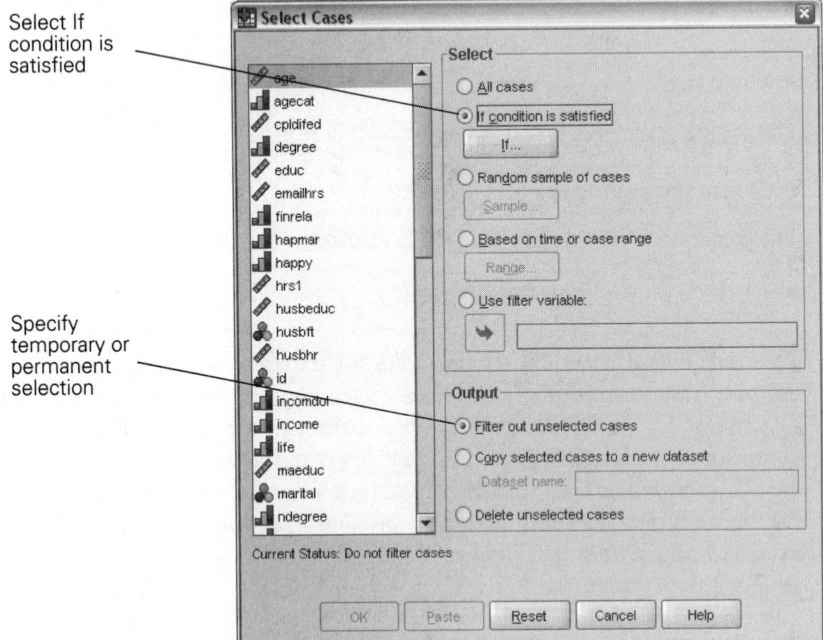

Temporary or Permanent Selection

The Select Cases dialog box offers a choice between filtering cases (selecting temporarily) or deleting cases (selecting permanently). The distinction between temporary and permanent case selection is important to understand.

- When you **filter** cases, or select a temporary subset, the unselected cases remain in the working data file. You can regain all the original cases at any time.
- When you **delete** cases, or select a permanent subset, SPSS deletes them forever from your working data file. If you save the working data file, replacing the copy on your disk, the deleted cases are gone forever from that file, too. This can be useful because it allows you to save a smaller data file.

If you haven't saved the working data file, you can often "undo" a permanent case selection by reopening the original data file. If you have saved the working data file there is no way to get back cases that have been deleted, unless you have a backup copy of the data file.

Example: Selecting College Graduates

Some of the analyses in this book that use the GSS data restrict the analysis to college graduates. In the *gssnet.sav* and *gssft.sav* data files, the variable *degree* indicates the highest degree earned by each respondent. Four-year college graduates are coded 3 (for bachelor's degree) or 4 (for advanced degree). To select people with bachelor's or advanced degrees:

▶ From the Data Editor menus choose:

Data
 Select Cases...

▶ Select If condition is satisfied in the Select Cases dialog box (see Figure B.12).

▶ Select Filtered in the Unselected Cases Are group.

This assures that the unselected cases will remain in the working data file if you want to use them in future analyses.

▶ Click If.

This opens the Select Cases If dialog box, as shown in Figure B.13. This dialog box, which strongly resembles the Compute Variable If Cases dialog box shown in Figure B.9, allows you to specify a conditional expression.

Figure B.13 Select Cases If dialog box

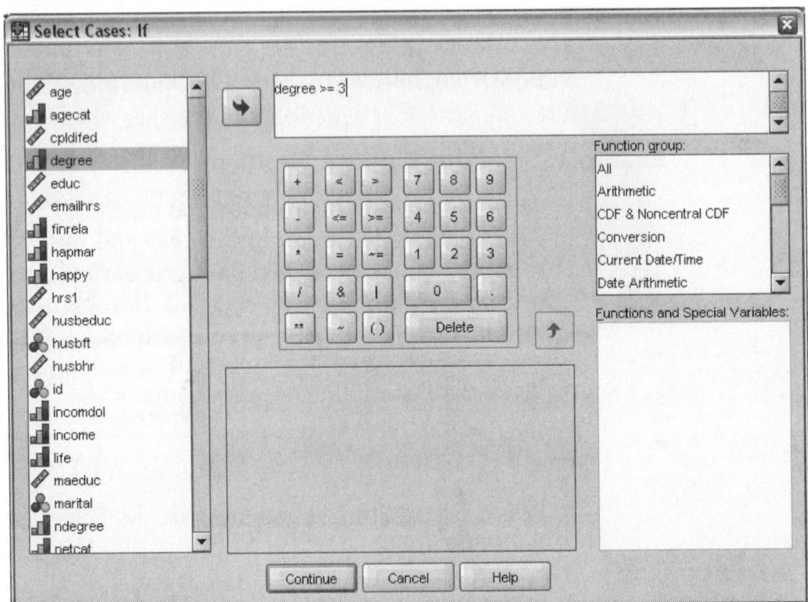

▶ Enter degree >= 3 in the Numeric Expression box.

This expression specifies that cases should be selected "if degree is greater than or equal to 3."

▶ Click Continue to return to the Select Cases dialog box.

▶ Click OK.

Cases for people who have college degrees are now selected. In the Data Editor, unselected cases are indicated by a slash mark over the row number.

▶ To turn off case selection, open the Select Cases dialog box again, select All cases, and click OK.

Other Selection Methods

Other options available in the Select Cases dialog box include:

Random sample. Sometimes you want a random subset of cases. You have no particular criterion for choosing which cases to process, but you don't want the whole data file.

Based on time or case range. Under some circumstances, it is desirable to select a range of cases according to the order of cases, as displayed in the Data Editor. This can be useful for time series data files.

Use filter variable. A filter variable is simply a variable that indicates whether or not a particular case should be selected. Cases for which the specified filter variable has a valid non-zero value are retained. Cases for which it is 0 or missing are dropped.

C The T Distribution

t value	0.0	0.25	0.50	0.75	1.00	1.25
df	Two-tailed Probability					
1	1.0000	.8440	.7048	.5903	.5000	.4296
2	1.0000	.8259	.6667	.5315	.4226	.3377
3	1.0000	.8187	.6514	.5077	.3910	.2999
4	1.0000	.8149	.6433	.4950	.3739	.2794
5	1.0000	.8125	.6383	.4870	.3632	.2666
6	1.0000	.8109	.6349	.4816	.3559	.2578
7	1.0000	.8098	.6324	.4777	.3506	.2515
8	1.0000	.8089	.6305	.4747	.3466	.2466
9	1.0000	.8082	.6291	.4724	.3434	.2428
10	1.0000	.8076	.6279	.4705	.3409	.2398
11	1.0000	.8072	.6269	.4690	.3388	.2372
12	1.0000	.8068	.6261	.4677	.3370	.2351
13	1.0000	.8065	.6254	.4666	.3356	.2333
14	1.0000	.8062	.6248	.4657	.3343	.2318
15	1.0000	.8060	.6243	.4649	.3332	.2305
16	1.0000	.8058	.6239	.4641	.3322	.2293
17	1.0000	.8056	.6235	.4635	.3313	.2282
18	1.0000	.8054	.6231	.4629	.3306	.2273
19	1.0000	.8053	.6228	.4624	.3299	.2265
20	1.0000	.8051	.6225	.4620	.3293	.2257
22	1.0000	.8049	.6220	.4612	.3282	.2244
24	1.0000	.8047	.6216	.4605	.3273	.2234
26	1.0000	.8046	.6213	.4600	.3265	.2224
28	1.0000	.8044	.6210	.4595	.3259	.2216
30	1.0000	.8043	.6207	.4591	.3253	.2210
35	1.0000	.8040	.6202	.4583	.3242	.2196
40	1.0000	.8039	.6198	.4576	.3233	.2186
45	1.0000	.8037	.6195	.4572	.3227	.2178
50	1.0000	.8036	.6193	.4568	.3221	.2171
∞	1.0000	.8026	.6171	.4533	.3173	.2113

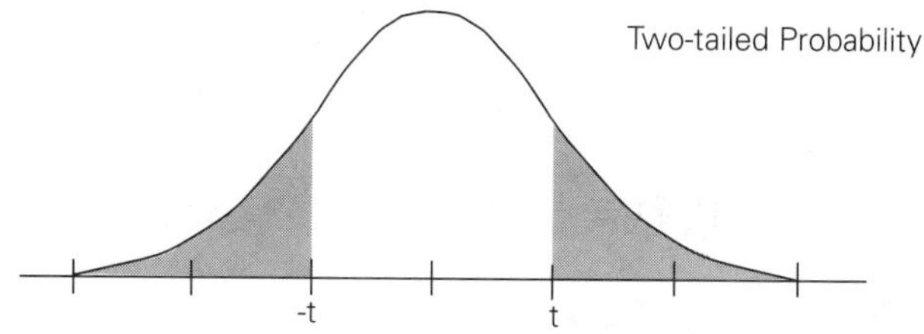

1.50	1.75	2.00	2.25	2.50	2.75	3.00	*t* value
Two-tailed Probability*							**df**
.3743	.3305	.2952	.2662	.2422	.2220	.2048	1
.2724	.2222	.1835	.1534	.1296	.1107	.0955	2
.2306	.1784	.1393	.1099	.0877	.0707	.0577	3
.2080	.1550	.1161	.0876	.0668	.0514	.0399	4
.1939	.1405	.1019	.0743	.0545	.0403	.0301	5
.1843	.1307	.0924	.0654	.0465	.0333	.0240	6
.1773	.1236	.0856	.0592	.0410	.0285	.0199	7
.1720	.1182	.0805	.0546	.0369	.0251	.0171	8
.1679	.1140	.0766	.0510	.0339	.0225	.0150	9
.1645	.1107	.0734	.0482	.0314	.0205	.0133	10
.1618	.1079	.0708	.0459	.0295	.0189	.0121	11
.1595	.1056	.0687	.0440	.0279	.0176	.0111	12
.1575	.1037	.0668	.0424	.0266	.0165	.0102	13
.1558	.1020	.0653	.0411	.0255	.0156	.0096	14
.1544	.1005	.0639	.0399	.0245	.0149	.0090	15
.1531	.0993	.0628	.0389	.0237	.0142	.0085	16
.1520	.0981	.0617	.0380	.0229	.0137	.0081	17
.1510	.0971	.0608	.0372	.0223	.0132	.0077	18
.1500	.0963	.0600	.0365	.0217	.0127	.0074	19
.1492	.0954	.0593	.0359	.0212	.0123	.0071	20
.1478	.0941	.0580	.0348	.0204	.0117	.0066	22
.1467	.0929	.0569	.0339	.0197	.0111	.0062	24
.1457	.0919	.0560	.0331	.0191	.0107	.0059	26
.1448	.0911	.0553	.0325	.0186	.0103	.0056	28
.1441	.0903	.0546	.0319	.0181	.0100	.0054	30
.1426	.0889	.0533	.0308	.0173	.0094	.0049	35
.1415	.0878	.0523	.0300	.0166	.0089	.0046	40
.1406	.0869	.0516	.0294	.0161	.0086	.0044	45
.1399	.0863	.0509	.0289	.0157	.0083	.0042	50
.1336	.0801	.0455	.0244	.0124	.0060	.0027	∞

*For one-tailed probability, divide by 2

Areas under the Normal Curve

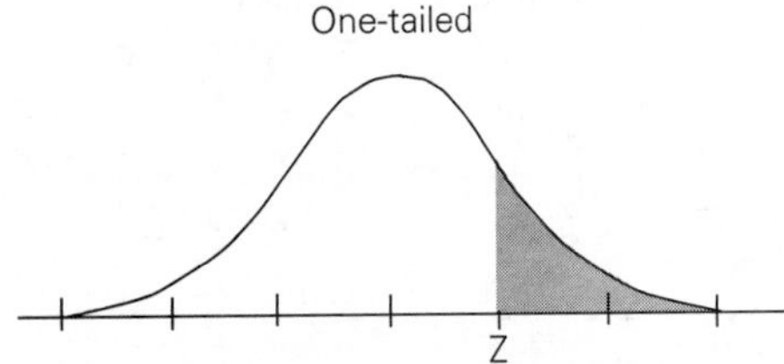

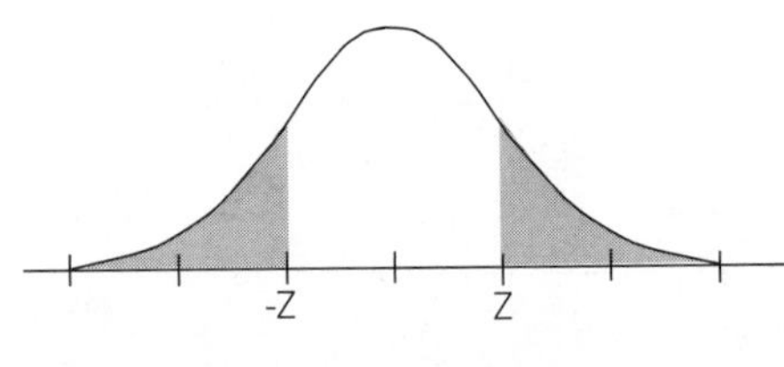

Z Scores	Probability	
	One-tailed	Two-tailed
.0	.50000	1.00000
.1	.46017	.92034
.2	.42074	.84148
.3	.38209	.76418
.4	.34458	.68916
.5	.30854	.61708
.6	.27425	.54851
.7	.24196	.48393
.8	.21186	.42371
.9	.18406	.36812
1.0	.15866	.31731
1.1	.13567	.27133
1.2	.11507	.23014
1.3	.09680	.19360
1.4	.08076	.16151
1.5	.06681	.13361
1.6	.05480	.10960
1.7	.04457	.08913
1.8	.03593	.07186
1.9	.02872	.05743

Z Scores	Probability	
	One-tailed	Two-tailed
1.96	.02500	.05000
2.0	.02275	.04550
2.1	.01786	.03573
2.2	.01390	.02781
2.3	.01072	.02145
2.4	.00820	.01640
2.5	.00621	.01242
2.6	.00466	.00932
2.7	.00347	.00693
2.8	.00256	.00511
2.9	.00187	.00373
3.0	.00135	.00270
3.1	.00097	.00194
3.2	.00069	.00137
3.3	.00048	.00097
3.4	.00034	.00067
3.5	.00023	.00047
3.6	.00016	.00032
3.7	.00011	.00022
3.8	.00007	.00014

Z Scores	Probability	
	One-tailed	Two-tailed
3.9	.00005	.00010
4.0	.00003	.00006
4.1	.00002	.00004
4.2	.00001	.00003
4.3	.00001	.00002
4.4	.00001	.00001

Descriptions of Data Files

bodyfat.sav Body measurements from 252 men. Percentage of body fat was measured using two different methods that required underwater weighing. See Johnson (1996).

buying.sav Information about buying behavior for 100 married couples. One of the goals of the study was to examine husband and wife agreement in product purchases. See Davis and Ragsdale (1983).

country.sav Demographic information for 122 countries. Most of the data are from 1992. See *The World Almanac and Book of Facts* (1994).

cntry15.sav A random sample of 15 countries from *country.sav*.

crimjust.sav Data from a telephone survey of Vermont residents regarding their views about the criminal justice system in Vermont. See Doble and Greene (1999).

electric.sav A sample of 240 men from the Western Electric Study. Half of the men experienced a cardiac event and half did not. The Western Electric Study was a prospective study of factors related to the incidence of coronary heart disease. The procedures according to which participants were selected, examined, and followed are described in Paul et al. (1963).

endorph.sav Beta endorphin levels before and after a half-marathon run for 11 men. See Dale et al. (1987).

gss.sav Contains 1419 randomly selected cases from the 2000 General Social Survey. Some variables have been recoded or otherwise modified. Cases were omitted because of case-number limitations for the student version of SPSS. See Davis and Smith (2000).

gssft.sav	A sample of people from the complete 2000 General Social Survey who indicated that they worked full time (*wrkstat* = 1).
gssnet.sav	A sample of people from the 2004 General Social Survey who were asked the questions on Internet use.
library.sav	Data from a telephone survey about the impact of the Internet on public library use. See Rodger et al. (2000).
lib1500.sav	A sample of 1500 cases from *library.sav.*
manners.sav	Data from an ABC News telephone survey on manners in the United States. See ABC News (1999).
marathon.sav	Completion time, age, and gender for participants in the 2001 Chicago marathon.
mar1500.sav	A sample of 1500 cases from *marathon.sav.*
renal.sav	Contains medical data on 84 patients, all undergoing cardiac surgery. Forty-two patients experienced acute renal failure; the other 42 did not. The study is described in Corwin et al. (1989).
salary.sav	Information about 474 employees hired by a Midwestern bank between 1969 and 1971. The bank was subsequently involved in EEO litigation. For additional information, see Roberts (1979, 1980).
schools.sav	Demographic and performance data for 64 Chicago high schools for 1993 and 1994. See the *Chicago Sun-Times* (1993, 1994).
simul.sav	Computer-generated data.
siqss.sav	Time-diary data from a study of the impact of the Internet on social interactions. See Nie et al. (2003).

Answers to Selected Exercises

F

The following answers are provided to the even-numbered problems in the statistical concepts portion of the exercises that appear at the end of each chapter. The output is produced using SPSS for Windows. If you are using another operating system, the appearance may be different, but the statistical results should be the same.

Chapter 4: Counting Responses

Statistical Concepts

2. a, d.

4. a.

		Frequency	Percent	Valid Percent	Cumulative Percent
Valid	Very safe	351	28.0	29.2	29.2
	Somewhat safe	414	33.0	34.4	63.5
	Not very safe	251	20.0	20.8	84.4
	Not at all safe	188	15.0	15.6	100.0
	Total	1204	96.0	100.0	
Missing	Not sure	50	4.0		
Total		1254	100.0		

b. Probably not, since "not sure" represents a level of certainty. People who don't have strong opinions on the topic are likely to select this response. However, it is also possible that "not sure" is selected by people who are totally ignorant or uninterested in the safety of savings and loans, and "not sure" is a substitute for "leave me alone." In this case, "not sure" would be considered a missing value.

c.

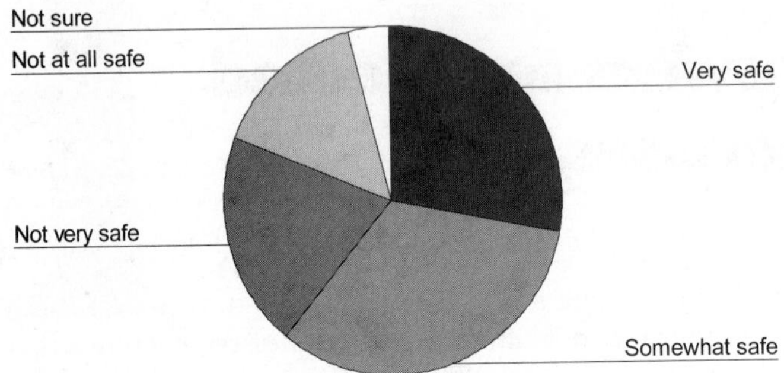

Chapter 5: Computing Descriptive Statistics

Statistical Concepts

2. **a.** The University of Texas is the mode.

b. The median is not used for nominal variables.

c. The mean is not used for nominal variables.

d. The variance is not computed for nominal variables.

4. The median salary is not affected by extreme values, so it would better represent the pay of the majority of employees than would the mean. Since there are several large salaries, the mean would be larger than the median. The bargaining agent for employees would prefer the median salary, while the employer would prefer the mean.

6. The mean is 50; the range is 50; the number of cases is 100; the minimum is 20; and the maximum is 70.

8. **a.** Histogram.

b. Bar chart.

c. Bar chart.

d. Histogram.

e. Bar chart.

f. Histogram.

Chapter 6: Comparing Groups

Statistical Concepts

2. a.

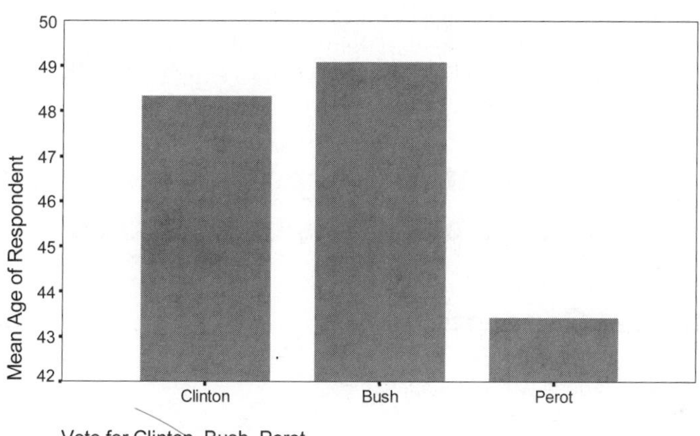

b.

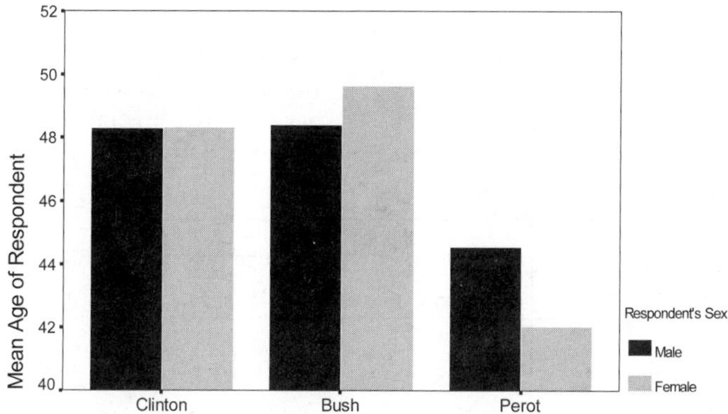

Chapter 7: Looking at Distributions

Statistical Concepts

2. a. 8.

b. 10.

c. 22.

d. 2.

e. Warehouse 1 is preferable to the others because the time required to ship products is relatively short and the variability is small. There are no long shipping times. This means you can tell customers when to expect items.

Chapter 8: Counting Responses for Combinations of Variables

Statistical Concepts

2. The independent variables are:

a. Race.

b. Gender.

c. Astrological sign.

d. Mother's degree.

e. Either.

4. a.

Count

		SATISFY			
		not satisfied	satisfied	very satisfied	Total
OWNER	own	5	1	2	8
	rents		2	10	12
Total		5	3	12	20

b.

			SATISFY			
			not satisfied	satisfied	very satisfied	Total
OWNER	own	Count	5	1	2	8
		% within OWNER	62.5%	12.5%	25.0%	100.0%
		% within SATISFY	100.0%	33.3%	16.7%	40.0%
	rents	Count		2	10	12
		% within OWNER		16.7%	83.3%	100.0%
		% within SATISFY		66.7%	83.3%	60.0%
Total		Count	5	3	12	20
		% within OWNER	25.0%	15.0%	60.0%	100.0%
		% within SATISFY	100.0%	100.0%	100.0%	100.0%

c. 40%.

d. 60%.

e. 25% and 83.3%.

6. You should use row percentages to describe these data. Knowing that 63.2% of the very satisfied people are married doesn't tell you anything unless you know what percentage of the entire sample are married. For example, if you observed these results in a sample where 90% of the people are married, you would conclude that married people are *less likely* to be very satisfied. Conversely, if only 10% of the sample were married and 63.2% of the very satisfied people were married, you would conclude that married people are *more likely* to be very satisfied.

8. Only **b.**

Chapter 9: Plotting Data

Statistical Concepts

2.

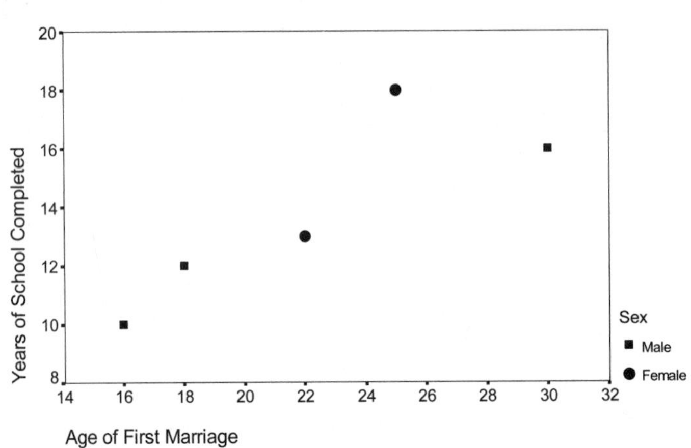

Chapter 10: Evaluating Results from Samples

Statistical Concepts

2. a. You must find out how often you would expect to see 19 schools improve and 37 schools worsen if scores have remained the same. If scores have remained the same, the probability that a school improves is the same as the probability that a school worsens. Another way of saying this is that the probability of improvement is 0.50 and the probability of worsening is 0.50.

b. No. The binomial test tells you that your observed results are unlikely if scores have not changed. Sample results as unusual as the ones observed would occur only 2 times in 100 if the scores have not changed. Since almost twice as many schools worsened as improved, the observed results don't support the claim that scores have improved.

c. If scores have not changed between the two time periods, you would expect to see a split as great as 19 to 37 only about 2 times out of 100. So there is a chance, although not very large, that scores haven't changed.

d. 0.0231.

Chapter 11: The Normal Distribution

Statistical Concepts

2. You would not expect the rate to be exactly 50%. Since the occupancy rate for each hotel is not exactly 50, the average of 5 rates need not be 50.

4. If you assume that vacancy rates for a single Saturday night at a hotel are normally distributed with a mean of 50% and a standard deviation of 15%, then the average for five nights should also be normally distributed with a mean of 50% and a standard deviation of 6.71%. From the histogram, you see that a value of 75% is very unlikely if the true mean is 50%.

- **a.** The hypothesis is that the average Saturday night vacancy rate is 50%.
- **b.** The means should be normally distributed with a mean of 50% and a standard deviation of 2.37%:

$$15/(\sqrt{40}) = 2.37$$

- **c.** No. Means based on 40 nights should be quite close to 50%. A value of 60% is more than 4 standard deviations above the mean. It's an unlikely result if the true vacancy rate is 50%.

6. For any range of nostril widths, you can figure out exactly what percentage of the population has nostril widths within that range. For example, 95% of the population has nostril widths between 0.51 inches and 1.29 inches (plus or minus 1.96 standard deviations).

8. **a.** The standard deviation tells you how much the individual lengths of stay vary. The standard error of the mean tells you how much means calculated from a sample of a particular size vary. For samples from the same population, means always vary less than individual observations.

b. The standard error of the mean would be $3/(\sqrt{50})$ (approximately 0.42) for samples of 50; it would be $3/(\sqrt{10})$ (approximately 0.95) for samples of 10.

10. a. Show *score*, *zscore*, and *twotailp* (see table below for values).

score	zscore	cumprob	twotailp	onetailp
55.00	−3.00	.0013	.0027	.0013
70.00	−2.00	.0228	.0455	.0228
85.00	−1.00	.1587	.3173	.1587
100.00	.00	.5000	1.0000	.5000
115.00	1.00	.8413	.3173	.1587
130.00	2.00	.9772	.0455	.0228
145.00	3.00	.9987	.0027	.0013

b. The probability of a score less than 85 or greater than 115 is 0.3173; greater than 145 is 0.0013; less than 100 is 0.50; less than 55 or greater than 145 is 0.0027; and between 55 and 145 is 0.9973.

Chapter 12: Testing a Hypothesis about a Single Mean

Statistical Concepts

2. a. You must figure out how often for an individual pizza you would expect to see a difference of at least 12 minutes from the promised delivery time. The standard score is 1.2 ((42 – 30)/10). This is not an unlikely standard score, so you don't have enough evidence to doubt the chain's claim.

b. You must figure out how often for a sample of 20 pizzas you would expect to see an average delivery time at least 12 minutes from the promised delivery time. The standard error of the mean for 20 pizzas is 2.24 $(10/(\sqrt{20}))$. The standard score is 5.4 ((42 – 30) / 2.24). Since the standard score is large in absolute value, you have reason to suspect that the chain's claim is false.

c. The 95% confidence interval for the true delivery time is $42 \pm 1.96(2.24)$. The lower limit of the confidence interval is 37.61 minutes, and the upper limit of the confidence interval is 46.39 minutes. Note that the chain's claim of 30 minutes is not in the 95% confidence interval. It is not a plausible value for the population mean.

d. To obtain the 95% confidence interval for the true difference between the real average delivery time and the chain's claim, subtract 30 minutes (the claim) from the lower and upper limits of the confidence interval calculated in question 2c. The lower limit for the difference is 7.61 minutes, and the upper limit is 16.39 minutes. Note that the value of 0 is not included in the interval.

Chapter 13: Testing Hypotheses about Two Related Means

Statistical Concepts

2. The advantage of using identical twins in such studies is that they have the same genetic heritage. Thus, any differences you see between members of a pair may be more readily attributable to differences in parental influence. However, you should also look at pairs of twins who are not raised apart to see how much they differ in intellectual development. This will give you some idea of how much variability there is in intellectual development when both influences are supposedly the same.

Chapter 14: Testing a Hypothesis about Two Independent Means

Statistical Concepts

2. **a.** There is a statistically significant difference between the average age of purchasers and nonpurchasers, based on the two-independent-samples *t* test ($p < 0.0005$).

b. No.

c. The plausible range is from –12.883 to –4.223.

d. It would become smaller.

4. Whether an observed difference is "statistically significant" depends on the magnitude of the difference, the variability of the observations, and the sample size. It's possible that a difference that is not statistically significant when the sample size is small would be statistically significant when the sample size increases.

Chapter 15: One-Way Analysis of Variance

Statistical Concepts

2. **a.** No.

b. There is insufficient evidence to reject the null hypothesis.

c. Pretty similar.

4. **a.** The null hypothesis is that average standardized test scores are the same for the four teaching methods.

b. One-way analysis of variance.

c. Each group must be a random sample from a normal population. The variances for all groups must be the same.

Chapter 16: Two-Way Analysis of Variance

Statistical Concepts

2. The problem is that the observations are not independent because the same person is measured four times.

4.

Dependent Variable: HOURS PER DAY WATCHING TV

Source	Type III Sum of Squares	df	Mean Square	F	Sig.
Corrected Model	35.631[1]	2	17.815	3.591	.028
Intercept	6411.570	1	6411.570	1292.331	.000
ANOMIA5	13.179	1	13.179	2.656	.103
SEX	21.174	1	21.174	4.268	.039
Error	4718.143	951	4.961		
Total	12704.000	954			
Corrected Total	4753.774	953			

1. R Squared = .007 (Adjusted R Squared = .005)

Chapter 17: Comparing Observed and Expected Counts

Statistical Concepts

2. a. You would reject the null hypothesis that they are independent.
   b. 4.95 times out of 100.
   c. No.

4. Variable pairs **a** and **d** are probably independent; the others are probably dependent.

Chapter 18: Nonparametric Tests

Statistical Concepts

2. a. Taking the same test twice does not change the average score.
   b. Taking the same test twice does not change the average score.
   c. People who approve of a candidate before a debate are equally likely to change their approval rating as people who don't approve of a candidate.
   d. Four cereals have the same number of raisins in a cupful of cereal.

Chapter 19: Measuring Association

Statistical Concepts

2. This would simply reverse the sign of gamma.

Chapter 20: Linear Regression and Correlation

Statistical Concepts

2. You cannot use the correlation coefficient when one of the variables is nominal.

4. a. Yes. Positive.
   b. Defect rate is dependent. Volume is independent.

c. Defect rate = –97.07 + 0.027 x Volume.

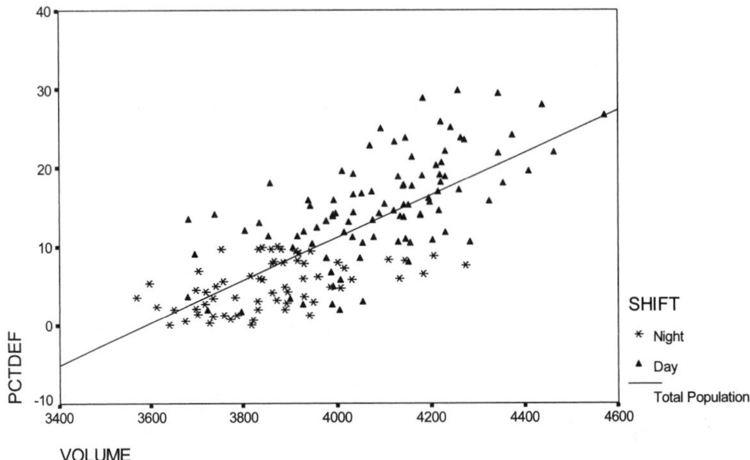

d. 54.8%.

e. 10.9%.

f. 145.9%. However, you should not predict values outside of the range of the observed values. Also, you cannot have more than 100% of items defective!

g. No.

h. 4.92.

i. –0.9%.

6. You could use linear regression for **b** and **d**.

Chapter 21: Testing Regression Hypotheses

Statistical Concepts

2. a. No.

b. The observed significance level indicates that 3 times in a 100 you would see a sample slope as large at 0.12 in absolute value if the population value is 0.

c. No. No.

d. No. The tests are equivalent.

4. Without the correlation coefficient, you cannot tell.

6.

Model		Unstandardized Coefficients		Standardized Coefficients	t	Sig.
		B	Std. Error	Beta		
1	(Constant)	4.300	2.728		1.576	.213
	X	1.700	.823	.766	2.067	.131

You would not reject the null hypothesis, since the observed significance level is larger than 0.05.

Chapter 22: Analyzing Residuals

Statistical Concepts

2. **a.** Whether the distribution is normal.

b. Whether the observations are independent.

c. Whether the relationship is linear.

d. Whether the linear model fits.

Chapter 23: Building Multiple Regression Models

Statistical Concepts

2. You would obtain histograms and plots of the residuals against the predicted values and the values of the independent variables.

4. No, because R^2 must increase at each step.

Chapter 24: Multiple Regression Diagnostics

Statistical Concepts

2. The distribution of residuals from the equation that uses the natural log of salary appears to be more normal.

Bibliography

ABC News. 1999. *ABC News Manners Express Poll*, May 1999 (computer file). ICPSR version. Horsham, Pa.: Chilton Research Services (producer). Ann Arbor, Mich.: Inter-university Consortium for Political and Social Research (distributor).

Cedercreutz, C. 1978. Hypnotic treatment of 100 cases of migraine. In: *Hypnosis at Its Bicentennial*, F. H. Frankel and H. S. Zamansky, eds. New York: Plenum.

Chicago Sun Times. 1993. Demographic and performance data for 64 Chicago high schools, November 16.

_____. 1994. Demographic and performance data for 64 Chicago high schools, October 27.

Corwin, H. L., S. M. Sprague, G. A. DeLaria, and M. J. Norušis. 1989. Acute renal failure associated with cardiac operations. *Journal of Thoracic and Cardiovascular Surgery*, 98:6, 1107–1112.

Dale, G., J. A. Fleetwood, A. Weddell, and R. D. Ellis. 1987. β-endorphin: A factor in "fun run" collapse? *British Medical Journal*, 294: 1004.

Davis, H., and E. Ragsdale. 1983. Unpublished working paper. Graduate School of Business, University of Chicago.

Davis, J. A., and T. W. Smith. 1972–2004. General Social Surveys. Chicago: National Opinion Research Center.

Doble, J., and J. Greene. 1999. *Attitudes toward crime and punishment in Vermont: Public opinion about an experiment with restorative justice*, 1999 (computer file). Englewood Cliffs, N.J.: Doble Research Associates, Inc. (producer), 2000. Ann Arbor, Mich.: Inter-university Consortium for Political and Social Research (distributor), 2001.

Johnson, R.W. 1996. Fitting percentage of body fat to simple body measurements. *Journal of Statistics Education*, 4:1.

Markoff, J. 2002. "How Lonely Is the Life That Is Lived Online?" *New York Times*, January 21, 3(C).

Nie, N. H., D. S. Hillygus, and L. Erbring. 2003. Internet use, interpersonal relations and sociability: A time diary study. In: *The Internet in Everyday Life,* B. Wellman and C. Haythornthwaite, eds. Oxford: Blackwell Publishers.

Norušis, M. J. 2008. *SPSS 16.0 Advanced Statistical Procedures Companion.* Upper Saddle River, N.J.: Prentice Hall Inc.

_____. 2008. *SPSS 16.0 Statistical Procedures Companion.* Upper Saddle River, N.J.: Prentice Hall Inc.

Paul, O., et al. 1963. A longitudinal study of coronary heart disease. *Circulation*, 28: 20–31.

Roberts, H. V. 1979. An analysis of employee compensation. *Report 7946*. Center for Mathematical Studies in Business and Economics, University of Chicago.

_____. 1980. Statistical bases in the measurement of employment discrimination. In: *Comparable Worth: Issues and Alternatives*, E. R. Livernash, ed. Washington, D.C.: Equal Employment Advisory Council.

Robey, B., S. O. Rutstein, and L. Morris. 1993. The fertility decline in developing countries. *Scientific American*, 269:6, 60–67.

Rodger, E. J., G. D'Elia, C. Jorgensen, and J. Woelfel. 2000. *The impact of the Internet on public library use: An analysis of the current consumer market for library and Internet services*. Final report submitted to the Institute of Museum and Library Services.

Santelmann, N. 1991. The FYI CEO cholesterol level contest. *Forbes*, 147: 39–40.

Stevens, S. S. 1946. On the theory of scales of measurement. *Science*, 103: 677–680.

Velleman, P., and L. Wilkinson. 1993. Nominal, ordinal, interval, and ratio typologies are misleading. *The American Statistician*, 47:1, 65–72.

Walker, K., and M. Woods. 1976. *Time use: A measure of household production of family goods and services*. Washington, D.C.: American Home Economics Association.

The World Almanac and Book of Facts. 1994. R. Famighetti, ed. Mahwah, N.J.: Funk & Wagnalls.

Index